INTERNATIONAL TECHNOLOGICAL UNIVERSITY
This Book is Donated by:
PROF. WAI-KAI CHEN

Date:

PHYSICS OF
MAGNETIC MATERIALS

Proceedings of the 4th International Conference on

PHYSICS OF MAGNETIC MATERIALS

Szczyrk-Biła (Poland), September 4–10, 1988

Editors:
W. Gorzkowski
H. K. Lachowicz,
H. Szymczak

World Scientific
Singapore • New Jersey • London • Hong Kong

Published by

World Scientific Publishing Co. Pte. Ltd.
P O Box 128, Farrer Road, Singapore 9128

USA office: World Scientific Publishing Co., Inc.
687 Hartwell Street, Teaneck, NJ 07666, USA

UK office: World Scientific Publishing Co. Pte. Ltd.
73 Lynton Mead, Totteridge, London N20 8DH, England

Library of Congress Cataloging-in-Publication Data

International Conference on Physics of Magnetic Materials (4th: 1988
: Biła, Szczyrk, Poland)
    Proceedings of the 4th International Conference on Physics of Magnetic
Materials, Szczyrk-Biła, Poland, September 4–10, 1988/editors, W.
Gorzkowski, H.K. Lachowicz, H. Szymczak.
      p.   cm.
    ISBN 9971506963
    1. Magnetic materials--Congresses. I. Gorzkowski, W. II. Lachowicz,
Henryk K. III. Szymczak, H. IV. Title.
QC764.5.I58 1988
538'.4--dc 19   89-5410
                                                                                 CIP

Copyright © 1989 by World Scientific Publishing Co. Pte. Ltd.

All rights reserved. This book, or parts thereof, may not be reproduced in any form or by any means, electronic or mechanical, including photocopying, recording or any information storage and retrieval system now known or to be invented, without written permission from the Publisher.

Printed in Singapore by JBW Printers & Binders Pte. Ltd.

Motto:

"He who learns from one occupied in
learning, drinks from a running stream.
He who learns from one who learnt all
he has to teach, drinks the green mantle
of a stagnant pool."

Scott,
the first Principal of Owens College,
later the University of Manchester.

FOREWORD

We believe that the present Proceedings will give a good chance to those who are teaching, for further refreshing the stream of knowledge in the field of magnetism.

The fourth International Conference on Physics of Magnetic Materials, 4ICPMM, was held at the "Klimczok" rest-house, Szczyrk, Poland, from September 4 through September 10, 1988, the very same place where the previous meeting of the series was held in 1986. The Conference was, as usual, organized by the Institute of Physics of the Polish Academy of Sciences, in cooperation with the Research Laboratory of the "Polfer" Plant of Magnetic Materials. This time, the meeting was held under the auspices of the European Physical Society.

The decision to organize the 4ICPMM at the place where the previous Conference was held went against the original intention of the organizers to hold every next meeting in a new place. However, those non-Polish participants who have already learnt at the previous meeting how to pronounce the strange-sounding name of Szczyrk profited from this decision.

The Conference addressed a wide range of topics in the field of physics of magnetic materials, as well as in the field of application of the most modern magnets, but also broadened its traditional coverage to include sessions of the high T_C superconductors, mainly devoted to the magnetic properties of these new materials. It is worth noting that at one of the plenary sessions on the high T_C superconductors, Alex Braginski (Pittsburgh) gave us a superb talk, showing not only the state of art of these materials but also prospects of their application evaluated on the basis of some economic factors, too. It was a really good lesson for us listeners, mostly academic reaserchers, who usually do not pay great attention to such factors.

There were 253 registered participants from 19 countries, the majority of them representing academic institutions.

The Conference format consisted of plenary sessions held twice or three times a day and poster sessions. At the plenary sessions, 36 invited talks have been given, whereas at poster sessions 206 contributed papers have been presented, 70 of them being accepted for publication in two issues of the Acta Physica Polonica, due to appear in the middle of 1989 (additionally, 10 contributed papers will be published as a regular paper in the same journal). The full list of contributed papers is given in this volume.

Of the invited papers, the majority (13 talks) were devoted to magnetic thin films and multi-layered structures. This fact is not surprising since these topics seem to be the hottest ones in the field of magnetic materials research. The strong activity presently observed in this area has its motivation in both basic scientific interest and in very promising prospects of application. This is also confirmed by a number of contributed papers devoted to the topics mentioned above and presented at the Conference (1/3 of the total number).

This Proceedings contains only 28 of the 36 invited papers given at the Conference. Manuscripts of eight of them did not, unfortunately, reach us in time. However, in order to present as complete a view of the Conference as possible, we feel, we should at least present the titles of these talks. They are: (1) "Thermomagnetic phenomena in rare earth magnets"(by K.P.Belov, Moscow), (2) "Giant magnetostriction in iron-rare earth alloys" (by A.E.Clark, White Oak, the paper was presented by H.T.Savage), (3) "Surface anisotropy in multilayer magnetic films" (by E.M.Gyorgy, Stockholm), (4) "Specific heat and transport in high T_C superconductors" (by S.von Molnar, Yorktown Heights), (5) "New superconductors: promises and complexities" (by K.V.Rao, Stockholm) (6) "Exchange coupling through a tunnelling barrier" (by J.C.Slonczewski, Yorktown Heights), (7) "Overview of superconducting oxide thin films" (by K.Tsushima, Stockholm), (8) "Soft ME modes in magnetic films" (by E.A.Turov, Sverdlovsk).

On the other hand, two invited papers which have been not presented at the meeting (at the very last moment it turned out that the authors are not able to come) are included in this volume. They are: (1) "The phenomenological theory of relaxation processes in magnetics" (by V.G.Bar'yakhtar), (2) "Stress dependence of magnetostriction in metallic glasses, a consequence of the coexistance of two amorphous phases ?" (by A.Hernando).

To make the Conference more attractive, a poster competition has been introduced at this meeting. The International Poster Evaluating Committee has awarded the best poster (considering both the merits and originality of presentation) of the posters presented each day of the meeting. Six awards have been granted. Success of the work of this Committee would have been impossible without active participation of its members, particularly its chairman, Karel Zaveta, who unstintingly gave many hours of their time during the Conference. We would like to express our thanks to all the members of this Committee for their superb, effective, and time-consuming work.

Our impression is that the Conference has proceeded in a very friendly and informal atmosphere, and was useful to the participants. We hope the majority of them share this opinion.

We are very much indebted to all the members of the International Advisory Committee for their valuable help in making this Conference a success.

We also want to thank all the members of the Publication Committee who helped to review the contributed papers.

The perfect work and personal interest of our colleagues from the Organizing Committee have made the Conference go smoothly, and are gratfully acknowledged, particularly that of Anna Pajączkowska, the vice-chairlady of the Committee, and of Marek Gutowski, the Conference Secretary, whose patience, calm authority and long experience proved a key element in the smooth running of the meeting.

We hope that there will be an opportunity for all the participants of the 4ICPMM to meet each other again at the fifth meeting, planned for 1990, possibly at the same place as the last one.

Henryk Szymczak
Henryk K.Lachowicz

co-chairmen of the 4ICPMM

4th INTERNATIONAL CONFERENCE on PHYSICS of MAGNETIC MATERIALS
held under the auspices of EPS - European Physical Society
Szczyrk-Biła, Poland, September 4-10, 1988

Previous Conferences:

  1st ICPMM - Jaszowiec, Poland
  2nd ICPMM - Jadwisin, Poland
  3rd ICPMM - Szczyrk-Biła, Poland

ORGANIZING COMMITTEE:

H.K. Lachowicz - Chairman
A. Pajączkowska - Vice-chairman
T. Postupolski - Vice-chairman
M.W. Gutowski - Secretary
K. Piotrowski - Finances
W. Gorzkowski - Publications
A. Reich - Publications

M. Wójcik - Publications
L. Borg
W. Cieśliński
K. Łukawski
A. Szewczyk
L. Załuski

INTERNATIONAL ADVISORY COMMITTEE:

H. Szymczak - Poland
    Chairman
P. Allia - Italy
V. Baryakhtar - USSR
K.P. Belov - USSR
A.S. Borovik-Romanov - USSR
J.I. Budnick - USA
E. Burzo - Romania
V.V. Eremenko - USSR
A.J. Freeman - USA
P.J. Grundy - UK
H. Kirchmayr - Austria
J. Klamut - Poland

R. Krishnan - France
H. Kronmüller - FRG
S. Krupička - Czechoslovakia
R. Lemaire - France
T. Postupolski - Poland
J. Schneider - GDR
J.C. Slonczewski - USA
M. Takahashi - Japan
T. Tarnóczi - Hungary
K. Tsushima - Japan
E.P. Wohlfarth - UK
W. Zinn - FRG

PUBLICATION COMMITTEE:

L. Kowalewski - Poznań
K. Krop - Kraków
H.K. Lachowicz - Warszawa
J. Pietrzak - Poznań
T. Postupolski - Warszawa

A. Sukiennicki - Warszawa
W. Suski - Wrocław
H. Szymczak - Warszawa
B. Wysłocki - Częstochowa

CONTENTS

Foreword	v
Committees	viii
The Evergreen Magnetism: Some Remarks Concerning its Past, Presence and Future S. Krupička	1
The Phenomenological Theory of Relaxation Processes in Magnetics V. G. Bar'yakhtar	21
Chaotic States in Systems Described by the Landau-Lifshitz Equation A. Sukiennicki & J. J. Żebrowski	43
Nonlinear Responses in Magnetic Materials P. E. Wigen & R. D. McMichael	58
Physics of Materials for Vertical Bloch Line Data Storage Devices W. Clegg	78
Multilayers: Past, Present and Future I. B. Puchalska & H. Niedoba	98
Coherent Spin Waves in Magnetic Films A. G. Gurevich	109
Coupled Ferromagnetic Resonances of Iron Films M. Pomerantz, J. C. Slonczewski & E. Spiller	134
Enhanced Magneto-Optical Effects Observed in Compositionally Modulated Films T. Morishita	151
Effect of Local Anisotropy Fluctuation on the Permeability of Sputtered Fe-Si-Al Alloy Films M. Takahashi	165

Multilayered Magnetic Films — Properties and Applications 195
   P. J. Grundy & M. Ohkoshi

Magneto-Optical Materials for Data Storage Applications 213
   R. Carey, D. M. Newman & B. W. J. Thomas

Theory of Magnetostriction in Amorphous Ferromagnets 228
   M. Fähnle, J. Furthmüller, R. Pawellek & G. Herzer

Stress Dependence of Magnetostriction in Metallic Glasses,
a Consequence of the Coexistence of Two Amorphous Phases? 248
   A. Hernando

Magnetic and NMR Spin Echo Studies in Some Mn Containing
Metallic Glasses 268
   R. Krishnan, H. Lassri, P. Rougier, K. Le Dang & P. Veillet

New Results in the Application of Soft Magnetic Metallic Glasses 280
   G. Konczos & T. Tarnóczi

Application of Magnetically Soft Amorphous Alloys in Electronic
and Electrical Devices 298
   Ying-Shan Yang

Coercivity and Magnetization Processes in NdFeB-Magnets 313
   J. Schneider, A. Handstein, D. Eckert & K.-H. Müller

Exchange and Crystal-Field Interactions in Ho_2Co_{17} and $Ho_2Fe_{14}B$:
Two Examples of R-T Intermetallics 336
   J. J. Franse & R. J. Radwański

Spin Glass and Invar Properties of Iron-Reach Amorphous Alloys 354
   K. Fukamichi, T. Goto, H. Komatsu & H. Wakabayashi

Non-Exponential Relaxation in Spin Glasses and Other Disordered
Systems 382
   I. A. Campbell, J. M. Flesselles, R. Botet & R. Jullien

Spin Glass Approach to Neural Network Model 390
   S. Kobe

Acoustic Emission, Domain Walls and Hysteresis in Various Ferro-
and Ferrimagnets 400
   M. Guyot, T. Merceron & V. Cagan

Instability of Itinerant Antiferromagnetism in Manganese
Laves Phases 427
   R. Ballou, J. Déportes, R. Lemaire, B. Ouladdiaf & P. Rouault

Magnetoelastic Properties of Zircon Structure Compounds Containing
Rare Earth Ions with Orbital Degenerate Electronic State 444
   V. I. Sokolov & Z. A. Kazei

Magnetic Order in Organic Compounds 478
   J. Pietrzak

Magnetic Properties of Cerium Monopnictides 495
   L. Kowalewski

Prospects for Applications of Oxide Superconductors (extended abstract) 521
   A. I. Braginski

Magnetic Order and Superconductivity in $RBa_2Cu_3O_2$ 527
   I. Felner, Y. Wolfus, E. R. Bauminger & I. Nowik

Energy Gap in High-T_c Superconductors Studied by Means of Electron
Tunneling Spectroscopy 559
   J. Raułuszkiewicz

Heat Capacity and Transport Properties of Copper Oxide Superconductors 573
   S. von Molnár, J. M. D. Coey & P. Strobel

List of Contributed Papers 603

List of Participants 629

THE EVERGREEN MAGNETISM; SOME REMARKS CONCERNING ITS PAST, PRESENCE AND FUTURE

Svatopluk Krupička

Institute of Physics, Czechoslovak Acad. Sci.
Na Slovance 2, CS-180 40 Praha 8, CZECHOSLOVAKIA

When Professor Rathenau opened the ICM conference in 1976 in Amsterdam [1], he called his talk "Quo vadis magnetism." He presented many interesting data mainly of statistical character showing the increasing number of papers and participants, the increase of collectivity in the work and that of international cooperation, the percentage of papers from industry etc. From such an analysis some useful conclusions could be drawn exhibiting the overall vitality of magnetism and its subfields and judging how further development can be like. In this talk I would like to try to get to similar conclusions but using another way. I shall call your attention rather to inner aspects, to the unique position magnetism has had during all the history of physics, to the enormous multitude of effects, materials, ideas and problems that have emerged from the research in magnetism and still are of interest at present. I will briefly mention the interplay of magnetism with other branches of physics, the broadness and importance of applications of magnetism both in science and technical development and, also, I would like to indicate some promising subfields for further research.

1. THE OLD HISTORY

Let me make first a short excursion to the history. It is well known that the story on magnetism begins with magnetite (or lodestone) and its marvellous property, i.e. the ability to attract pieces of iron. This was followed by invention of compass that in fact was the first practical application of a physical effect of other than mechanical origin. Both the lodestone and compass were discovered by ancient cultural nations - Chinese nad Greeks - and the discussions about the priority seem to be irrelevant due to lack of interaction between both cultures. The Chinese literal sources indicate the construction of a primitive compass in the form of a selfbalanced spoon made of lodestone and lying on a smooth copper plate in 3000-2500 B.C. [2]. In Greek writings the lodestone was mentioned to be known by the year 800 B.C. [3]. Due to its strange behaviour, the lodestone was believed to posess the soul. This first "theory" of magnetism expressed by Thales of Miletus, Anaxagoras and other Greek animists has survived surprisingly until renaissance together with the poetic tales on the mutual attraction of iron and lodestone desiring to embrace each other.

An important period for the history of magnetism was 13th-17th century marked by the names of Peregrinus, Gilbert and Descartes. The first two were real pioneers of the experimental work in physics. It is not necessary to stress the importance of this then really novel approach to the exploration and understanding of the nature for all the physics and science - not only for magnetism. On the other hand, their work concerns magnetism and the contribution they made to it was also important. Peregrinus [4] made experiments on a spherical lodestone he called "terella" which might be regarded as a model of the earth. Using an oblong piece of iron he found the course of the magnetic lines of force at its surface and introduced the term poles of the magnet for two points where these lines crossed. Gilbert, living about 300 years later than Peregrinus, not only repeated and completed his experimental work but collected and critically commented on all the knowledge concerning magnetism in the famous book "De Magnete" published in 1600 [5]). Here the lines of force (magnetic induction lines) and poles of a magnet were defined and among other original observations also the loss of magnetism by heating to sufficiently high temperatures anticipated the existence of the Curie temperature. Rather surprisingly, even though he tried to purify the magnetism from all superstitions when stressing the necessity of experimental proof he still shared the animists' believe in the soul in the lodestone (and in the earth as well) as the origin of magnetic effects. This was definitely abandoned in the first theory of magnetism created by Descartes (1596-1650) and published in his "Principia". This theory based on the existence of the s.c. "threaded parts" of two kinds: they enter the earth at its two poles, pass it through and get off at the oposite pole returning back through air and so it is also with other magnets and magnetized bodies. Although such way of reasoning now must seem to us as purely speculative without taking care of and discussing various questions that could be raised (or even were in Gilbert's work) the Descartes' theory influenced all thinking concerning magnetism for about further 100 years.

After Gilbert and Descartes there were some other works, some of them quantifying the older observations but nothing fundamental happened. In the 2nd half of the 18th century the character of the science begun to change, the theory and experiment became cooperative. Magnetism, and also electricity, started their new developments, more or less independently, even though the fluid-hypothesis and especially the torsion balance measurements by Coulomb who discovered the inverse square law of force indicated some kind of affinity between them.

The mutual relation between electricity and magnetism was firmly established by Oersted's famous experiment in 1820 after many years of his abortive pains. That what followed had not had any analogy before: the best physicists immediately started their own experiments and theoretical work and discovered many new important effects and laws. The way to create a new physical discipline - electromagnetism - was open, connected with the names as Arago, Biot, Savart, Ampère, Laplace, Poisson, Fourier, Fresnel, Davy, Faraday, Weber and others. Its final triumph were the Maxwell equations (1873) and Hertz's discovering and exploring the electromagnetic waves (1887) predicted by Maxwell's theory.

2. FROM ELECTROMAGNETISM TO MODERN MAGNETISM

All this as well as the discovery of the electron towards the end of the century created a new firm basis for straightforward progress not only in magnetism but for all the physics in the 20th century including many branches of applied science. As far as the magnetism concerns I would like to make the following remark: Even though it lost something of its independence being interrelated with the electricity in a unified electromagnetic theory, its role in discovering the laws of electromagnetism was essential. Moreover the material-dependent relations in the Maxwell equations are separate for both fields:

$$\vec{D} = \varepsilon \vec{E}, \quad \vec{B} = \mu \vec{H}, \quad \vec{j} = \sigma \vec{E}.$$

This stirred new studies of the magnetic properties of various substances and materials on a quantitative basis and actually opened new era of magnetism. Let us remind here the work of P. Curie and P. Weiss! Last but not least, as the Maxwell equations include the magnetostatic case, they have been used - and still they are - in all cases where magnetization vector may be regarded as a continuous quantity: micromagnetism, domains, domain walls, magnetization processes, and, together with the Landau-Lifshitz equation of motion [6]), when solving dynamical problems such as resonance modes, propagation of the electromagnetic waves in gyrotropic media etc.

There is not enough space to continue with the historical survay of magnetism with all details up to now. We must realize that there is no unique stream in the nowadays' magnetism as many special branches gradualy split off from the main tree represented by magnetic materials. This is illustrated in Fig. 1, that partly follows the picture by Enz [7]). The most important discoveries and developments are listed in the Table.

TABLE

~1890 - ~1920 Foundations of new magnetism

    Curie law (1895); Zeeman effect (1896); Langevin's theory of dia and paramagnetism (1905); Weiss' molecular field; hypothesis of domains (1907); Barkhausen effect (1919)

1913 - ~1926 "Old" quantum theory: magnetism and atomic spectra

    Spatial quantization; Bohr magneton (1920); Stern and Gerlach exper. (1922); electron spin (Compton 1921, Goudsmit and Uhlenbeck 1925); Pauli exclusion principle (1925); Hund's rules (1927)

1925 Ising: first statistical model (1-D)

1926 - ~1932 "New" quantum mechanics
Pauli matrices (1927); Dirac's theory of electron (1928); Exchange and ferromagnetism (Heisenberg, Dirac, VanVleck, Frenkel, Slater 1926 - 1930); Spin waves (Bloch, Slater 1930); Crystal field theory (Bethe 1929, VanVleck 1932)

1931 - ~1935 First theories and observations of domains
(Heisenberg, Bloch, Bitter, Elmore, Landau-Lifshitz, ...)

1932 Antiferromagnetism (Néel)

1935 Landau-Lifshitz equation of motion

1938 Stoner theory of itinerant magnetism

1945 - 1946 Magnetic resonance experiments
paramagnetic (Zavoisky 1945), ferromagnetic (Griffiths 1945), nuclear (1946)

1946 - 1956 Ferrites and similar new oxidic magnetics
ferrospinels (1946); hexaferrites (1952); garnets (1956); Néel theory of ferimagnetism (1948)

~1950 ⟶ Mag. structures by neutron diffraction

1958 Mössbauer effect

1960 - 1964 Amorphous magnetics
Gubanov 1960, Duwez 1964

1960 - 1967 ⟶ Bubbles and bubble memories
Enz, Kaczér 1960, Bobeck 1967, ...

1979   Magnetic fluctuation theory
        Prange, Korenman, Shirane, Moryia

1980 ⟶ Surface magnetism
        monolayers, concentration modulated films,
        multilayers, artificial superlattices
        (Gradmann, Freeman, Zinn, ...)

1986 ⟶ High T_c superconductors

    As the work in magnetism especially the last few decades have been so extended only a selection could be done which naturally is subject of the personal taste and view. At least one important thing, however, may be judged from it. All the time the magnetism has been developed in close contact with the main progress in physics as its integral and vivid part. Many fundamental concepts in quantum, statistical and solid state physics were anticipated or elaborated in close connection with problems of magnetism and their consequences tested on magnetic systems. Perhaps most important and firmest relations developed between magnetism and statistical physics. For example, the study of the transition from ferromagnetism to paramagnetism gave a basic knowledge and an example of a cooperative effect which could be later generalized in the theory of 2nd order phase transitions that reached its height in the renormalization group method. Also various types of magnetic systems characterized by an effective spin hamiltonian (Heisenberg, Ising, etc.) have been - and still are - used as model systems for many statistical calculations. In addition let us mention that the Weiss' molecular field theory is a special case of the mean-field approximation frequently used in statistical problems. We find also that some quantum statistical methods originally

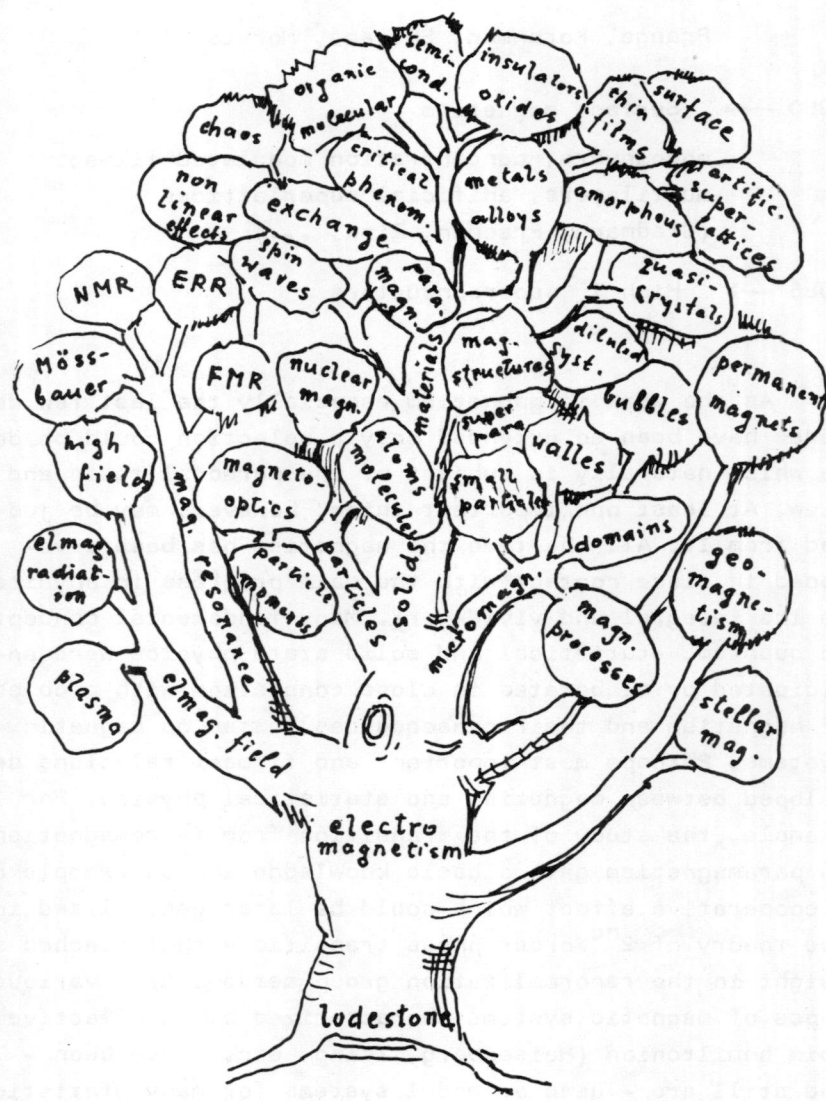

Fig. 1. Various branches and subfields of magnetism as developed during centuries.

elaborated for solving magnetic problems have been transformed to become suitable for the field theory and particle physics [8]).

But also in other branches of physics the magnetism played an important role. Let us mention: the importance of magnetism in discovering the electronic structure of atoms (Zeeman-effect, the existence of the spin-angular momentum, the spin-orbit coupling), the spin-dependent interactions (exchange etc.) in atoms, molecules and solids, EPR studies of paramagnetic impurities and crystal field effects, pioneering work on the correlation effects in electronic structure of solids due to the spin-dependent interactions, a deep insight into the character of the d- and f-band states in the transition metals, rare earths and actinides (localized resonance, Kondo effect, heavy fermions, mixed valency compounds), the contribution to the physics of non-linear processes (the equations for magnetization and spin dynamics are essentially non-linear) etc.

And vice versa, we can see that many of new theoretical methods and experimental technics introduced during last decades into physics were immediately used and eventually further developed when solving magnetic problems. Let us mention e.g. the application of Green's functions in magnetism, the "magnetic" variant of the local electronic charge density calculations - i.e. the local spin density method for determining the ground-state distribution of magnetic moments and calculations of the macroscopic measurable quantities of transition metals and some compounds from first principles, or the attempts to use the theory of solitons for Bloch-wall problems or in describing the non-linear excitations at higher temperatures. From experimental technics the neutron diffraction for determination of magnetic structures, the use of MBE technic for preparing well defined thin and composed films, elaboration of various spin-resolved methods for studying magnetic surfaces. On the other hand, many in principle magnetic effects and methods surpassed the limits of physics itself and were recognized as very useful tools in other sciences (chemistry, biology): EPR and NMR, Mössbauer spectroscopy, sensitive magnetometry.

3. THE APPLICATIONS

Now, before going over to the topics that may be guessed to play an important role in the future, let me say few words about the applications of magnetism - how they have developed during decades of years to the present imposing extension and importance. The oldest application - the permanent magnets - is still very important and during years a big variety of possibilities has been found how to use them, e.g. in electrical and radio engineering, in both industry and practical life as mechanical tools, in various devices etc. The quality of permanent-magnet materials as defined by $(BH)_{max}$ values has increased 200x during last 100 years very closely following an exponential curve (Fig. 2). Another old field are soft magnetic materials and primarily those for transformer cores (electric steel sheets, amorphous ribbons). These materials in big volumes - hundreds of thousands of tons per year - especially for purposes of electric power production and distribution. From the important parameters the decrease of 60-Hz losses of laminated transformer core materials with time is shown in Fig. 3.

In both cases the trends indicate the possibility of an optimistic extrapolation for the next future. Of course, theoretical limits do exist and were analysed as well as the optimizing procedures taking into account mutual relations among different material parameters [9] [10]). E.g., the maximum value $(BH)_{max} \approx 50$ MGOe measured on a poly-

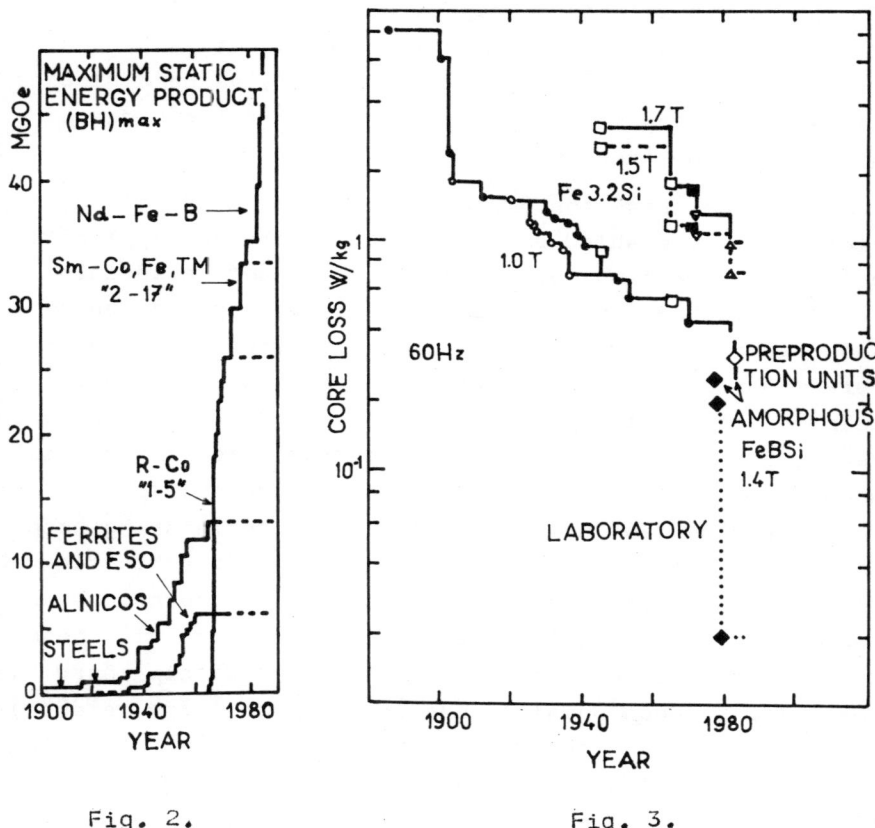

Fig. 2.   Fig. 3.

Fig. 2. Development of permanent magnets; $(BH)_{max}$ is plotted (after Strnat - see 9)

Fig. 3. Decrease of magnetic losses in transformer-core materials (after Luborsky - see 9)

crystalline $Fe_{14}R_2B$ magnet represents about 76 % of the maximum theoretical value $(4\pi M_s)^2/4$ for these materials. Therefore, in a longer perspective materials with higher M_s are to be looked for, if this trend ought to be continued. On the other hand, the laboratory value of losses in FeBSi ribbons in Fig. 3 are still about one order of magnitude above the theoretical limit.

In some cases the discovery of a new class of magnetic materials enabled and stimulated quite new applications. Let us mention, e.g., ferrites: as soft materials they enabled new radiofrequency applications, as square loop material they revolutionized the computer by ferrite core memories and, as materials for microwaves they stimulated construction of new devices. Another example, epitaxial garnet films and bubble memories, etc. In this way magnetism has proved to be also an ineplaceable part of the development of new technologies and engineering both in the past and at present showing an uncomparable flexibility in matching the needs posed by the technical and social progress. But in order to be able to fulfil these tasks, magnetism had to get a rather interdisciplinary character between physics, chemistry, metallurgy, electronics and perhaps others. After Enz [7] this fact makes important implications concerning the optimal organization of research in magnetism: a direct cooperation between physicists, chemists, electronic engineers etc in one laboratory is desirable, usually very fruitful and brings success.

Before closing the remarks on the present status in the application field it is necessary to comment on one important topic: Even though the soft magnetic materials for electric power purposes represent a biggest production and no appriciable change during next decade is expected the most dynamical area has become the information storage and manipulation. Here not only new materials are being developed but also various new principles are invented or newly tailored to be able to compete each other. So the magnetic discs have defended their position against bubble memories and still have reserves for further improvement, in particular when the perpendicular magnetic recording and magnetic thin films will be used. At the same time the magnetcoptic discs, erasable like the normal magnetic discs, have appeared and could-in principle-represent a further step in development. In parallel, the particulate recording media as γ-Fe_2O_3 undergo gradual innovations (hexagonal ferrites, elongated particles of iron). Because both technical and economical aspects play role in this competition it is difficult to fore cast the future. This is the reason for which the research is done on a very broad front when several variants are studied in parallel leaving the final choice to the future.

4. SOME PERSPECTIVE TOPICS

In the last part of my talk I would like to come back to the basic problems of the present magnetism and to try to select some topics that seem to represent new challenge for further research. Our evergreen tree of magnetism has grown with magnetic materials as its trunk based on lodestone at its roots. Hence the problems of the type whether egg or hen was the first need not appear. We may suppose that even in the future the magnetic materials or, let us say, the study of magnetic properties of substances will continue to fulfil their triplet role: to keep all branches firmly together, to stimulate further deepening of the knowledge and, to ensure new applications which is necessary when research in magnetism shall be rich in both the respect and money.

Even though the possibility of combining elements of the periodic table to get various magnetic substances by traditional preparation methods seems to be almost inexhaustible, strong tendency has recently appeared to broaden the spectrum of materials by using unconventional technologies as rapid quenching, ion implantation, molecular beam epitaxy etc. This enabled preparation of materials as nonequilibrium phases (amorphous, quasicrystaline), very thin layers with well defined thickness (including monolayers), sandwiches and multilayers (artificial superlattices) or compositionaly modulated films. This certainly is a very important development of the last few years and seems also to be a very promising field for the future. By preparation of monolayers e.g. we get practically ideal 2-D structures that are now often investigated for the fundamental reasons. In other cases we can study range of exchange coupling between two layers separated by a proper nonmagnetic film of well defined thickness or resonance modes of layered structures and so on.

It is very important that besides advanced technologies of preparation there are both theoretical and experimental methods available [11] enabling to study the real situation in both ways in parallel and to confront the calculated results with experiment. One of the interesting theoretical calculations using local spin density method shows that in some 3d transition metals the surface atoms have enhanced magnetic moments. This effect is particularly strong in Cr (2.49 μ_B compared with 0.59 μ_B in bulk) which is connected with a surface FM transition [12]; the result is in agreement with angle-resolved photoemission experiment [13]. On the other hand, the moment of Fe in single monolayer was experimentally (spin polarized Auger spectroscopy) found the same as for thicker layers or bulk, without any enhancement [14]. In experiments on artificial superlattices (layered structures) very often

noble metals Ag, Au, or Cu are used in combination with $3d^n$ metals. Very interesting should be to use Pd or Pt as nonmagnetic metal as they actually are strongly exchange-
-enhanced paramagnets. It is known that in these metals magnetic impurities may cause strong polarization in their vicinity which manifests itself in appearance of the s.c. giant moments at the impurity. In the case of superlattices Fe/Pd and Co/Pd the existence of strong interface anisotropy perpendicular to the layer surface [15] was observed which could be of interest for application e.g. in recording (see also [16]). Besides $3d^n$ elements also superlattices with rare earths were prepared and studied.

Another reserve of new magnetic materials is represented by organic substances on molecular basis. There are two classes: in one the carriers of magnetism are transition metal atoms built in the molecules; the others are purely organic and the origin of magnetism may be sought in coupling of free radicals or biradicals. An example of the latter are some polydiacetylens [17] obtained by polymerization of diacetylens. Ferro, ferri and antiferromagnets with eventual weak ferromagnetism have been obtained [18] - [20]. A promising feature of the organic magnetically ordered materials is almost inexhanstible variety of compounds and relatively easy synthetization. It has been shown that a prerequisite of such materials is a high-spin electronic ground state of the respective molecule which requires that it has singly occupied molecular orbitals. From physical point of view the organic ferro- and antiferromagnets give us new examples of low-dimensional magnetics sometimes with rather unusual exchange interactions.

There is perhaps no hotter problem in solids at present then the high temperature superconductors. But what are these marvellous materials - La_2CuO_4: Ba, Sr; $YBa_2Cu_3O_{7-x}$; $CaBi_2Sr_2Cu_2O_8$; $CaTl_2Ba_2Cu_2O_8$ and some related compounds? All these are oxides with crystal structures related to perovskites with one common feature which is the presence of planes of approximately square lattice of copper and oxygen ions. And because the Cu ions, expected to be present mostly as Cu^{2+} are found to be subject of very strong antiferromagnetic mutual coupling, we have to do with potentially magnetic systems. Even though the mechanism of the HTS has not been yet explained lot of papers have been published that discuss the role of magnetic interactions in this mechanism [21] [22]. To sketch briefly the situation: In Fig. 4 the environment of an

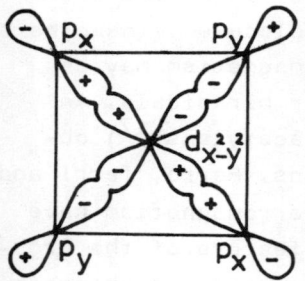 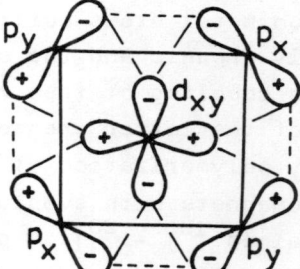

Fig. 4. Cu and O orbitals lying in the Cu-O plane. The hybridization may lead to forming
a) $\sigma_{x^2-y^2}$ orbital, b) the "ring" π-oxygen orbital [23]

Cu^{2+} ion in the plane is shown. The oxygen p states are split into lower p_σ (hybridization with $d_{x^2-y^2}$) and higher p_π states forming bands that eventually overlap. With the oxygen stoichiometry an antiferromagnetic state

with very strong n.n. antiferromagnetic interactions
($J \approx 10^3$ cm^{-1}) appears (T_N = 410 K in YBa$_2$Cu$_3$O$_6$). When increasing the oxygen content p-holes appear (i.e. O$^-$ states) which gradually destroy the 3-D ordered magnetic state and superconductivity appears. The latter is believed to be carried by pairs of p$_\pi$ holes. But even in the superconducting state very large 2-D magnetic fluctuations seem to exist. In such a situation the exchange or superexchange interactions including both Cu^{2+} and O$^-$ states (the latter are magnetic) may play an active role in forming the above mentioned pairs. But whatever the exact mechanism is, the study of magnetic effects and magnetic interactions should be fully in the scope of the present struggle for understanding of the HTS.

The last topic I would like to mention here is even more distant from that we are used to include into magnetism. I mean the nuclear magnetism. We usually think of ourselves as being familiar with nuclear spins, much less we are with nuclear magnetic moments and we know also something about the hyperfine interaction. But what is the nuclear spin, how is the nuclear moment formed and why is its value often so different from that one expected by a simple theory? We just ignore such questions and leave them to be answered by nuclear physicists. Well, it's true, we are living (or we think that we are) in the time of specialization. But still that is a problem of magnetism. Moreover a very important problem because the nuclear magnetic moments may help to solve problems of nuclear structure in a similar way as did the atomic moments with respect to the electronic structure of atoms. In fact, there is a close similarity between both but the nuclear structure is more complicated. All three fundamental interactions - i.e. strong, weak and electromagnetic - are taking part in it, the magnetic moments arise not only by summing up moments of individual nuclides but

there also are currents connected with exchange of mezons
and other complications which, on the other hand, make
the problem even more important. Beside this, the existence of quarks and gluons as basic constituents of proton,
neutron and other hadrons may not only change our conception of magnetic moments of these particles but also complicate the problem of nuclear magnetic moments and nuclear structure if they take part in the internucleon interaction inside nuclei. For these reasons some experiments have been recently prepared using Λ-hyperons in order to see whether its magnetic moment will change when
the Λ-particle will penetrate into a nucleus [23]. Anyway,
I believe that the collaboration of people experienced in
magnetism in solving such and other problems of nuclear
magnetism would be useful and also appreciated. It is
a real challenge.

5. CONCLUSION

I hope that the examples and remarks concerning the
history as well as the present status of magnetism and
showing some possibilities of its further development
could strenghten - if it was necessary - your conviction
of lasting health and vitality of magnetism, its enormous
inner richness and its applicability. In fact, you can
hardly find any other area of physics that includes such
a broad scale of materials and effects and where the relevant parameters may change by so many orders of magnitude.
And what seems to me even more fascinating: all this immense variety of properties is brought about by only very
few types of fundamental interactions and their interplay.
For those who do not content themselves with the knowledge
of the fundamental laws of nature but who yearn for pursueing the ways in which these laws are working in creating its richness the magnetism brings perhaps greatest
satisfaction.

6. REFERENCES

1. Rathenau, G.W., Proceedings of the ICM'76, Amsterdam 1976, Part I, North-Holland Publ.Comp., p.1.
2. Yu-Quing Yang, Physics of Magnetic Materials, ed. W. Gorzkowski, K.H. Lachowicz and H. Szymczak, World Scientific, 1986, p. 4.
3. see e.g. Mattis,D.C.: The Theory of Magnetism I, Springer-Verlag Berlin, Heidelberg, New York 1981, Chaper 1; this chaper was very stimulating for me when preparing the lecture.
4. Peregrinus,P., "Epistola Petri Peregrini de Manicourt and Sygerum de Foucaucourt Militem de Magnete" 1289 (after 3).
5. Gilbert,W., De Magnete, trans., rev. ed. Baric Books, New York 1958.
6. Landau,L. and Lifshitz,E., Phys.Z.Sowjetunion $\underline{8}$, 153, (1935).
7. Enz,U. in Ferromagnetic Materials, Vol. 3, E.P. Wohlfarth ed., North Holland Publ. Comp., Amsterdam-New York-Oxford 1982.
8. Kogut,J., Revs. Mod. Phys. $\underline{51}$, 659 (1979); see 3 p. 37-38.
9. Livingston, J.D., Proceedings of the International Symposium on Physics of Magnetic Materials, Sendai 1987, p. 3.
10. Kronmüller, H., ibid. p. 17.
11. Wimmer, E., Krakauer, H., Weinert, M. and Freeman, A.J., Phys. Rev. $\underline{B24}$, 864 (1981).
12. Fu, C.L. and Freeman, A.J., Phys. Rev. $\underline{B33}$, 1755 (1986).
13. Klebanoff, L.E., Robey, S.W., Liu, G. and Shirley, D.A., Phys. Rev. $\underline{B30}$, 1048 (1984).
14. Taborelli, M., Paul, O. and Landolt, M., ICM 88, Paris, paper 1 B-4.

15 Garcia, P.F., Proceedings of the International Symposium on Physics of Magnetic Materials, Sendai 1987, p. 240.
16 Broeder, den, F.J.A., Kuiper, D. and Donkersloot, H.C., ICM 88, Paris, paper 1 B-6.
17 Korshak, Ju.V., Ovshinnikov, A.A., Shapiro, A.M., Medviedeva, T.V. and Spektor, V.H., Pisma v ZhETF 43, 309 (1986).
18 Iwamura, H., ICM 88, Paris, paper 2 I-2.
19 Journaux, Y., Van Koningsbrugen, P., Lloret, F., Nakatani, K., Pei, Y., Kahn, O. and Renard, J.P., ibid, paper 1 P F-1.
20 Epstein, A.J., Chittipeddi, S. and Miller, J.S., ibid, paper 2 I-1.
21 Shirane, G., Endoh, Y., Birgeneau, R.J., Kastner, M.A., Hidaka, Y., Oda, M., Suzuki, M. and Murakami, T., Phys. Rev. Letters 59, 1613 (1987).
22 Shirane, G., ICM 88, Paris, paper 5 B-1.
23 Chakraverty, B.K., Feinberg, D., Hang, Z. and Avignon, M., Solid St. Commun. 64, 1147 (1987).
24 Baudō, H., Motoba, T., Sotona, M. and Žofka, J., to be published; see also Ejiri, H., et al., Phys. Rev. C36, 1435 (1987).

THE PHENOMENOLOGICAL THEORY OF RELAXATION PROCESSES IN MAGNETICS

V.G. Bar'yakhtar

Institute of Metal Physics, Academy of Sciences
Ukrainian SSR, Vernadskii st. 36, 252180 Kiev 142,
USSR

ABSTRACT

A method is proposed for determining the relaxation terms on the bases of the symmetry of the crystal lattice. The hierarchy of the relaxation constants is noted. Expressions are obtained for the dissipative functions of both the exchange and relativistic origins

1. INTRODUCTION

An equation of magnetic - moment mition has been established by Landau and Lifshits[1] in 1935. The simplest variant of a dissipative term describing the relaxation processes owing to relativistic interactions has been proposed also by them. The microscopic character of the relaxation processes in magnetics has been estabkished by Akhiezer[2] in 1946.

In author's work[3] a type of the dissipative term in the equation of the magnetic-miment motion describing the exchange-character relaxation has been determined, and in works[5] the general method for constructing the relaxation terms of relativistic character in present-symmetry crystal is given, and the type of this term for both "the uniaxial" and cubical crystals is found. The specific type of the relaxation term, determined in[5] for uniaxial crystal differs essentially on one propoded by Landau and Lifshits in due time, as well as by Gilbert.

An approach, developed in[5], allows to define a type of the relaxation terms in the Landau-Lifshits motion equation as for usual undegegenerate vacuum systems as for spontaneously brouken - symmetry system, in which vacuum degeneration as described by a continuons parameter (parameters).

This review is devoted to a statement of main results about constructing the relaxation terms in the dynamic equations of magnetic crystals and using them for a description of the spinwaves attenuation, the line widths of homogeneous resonance, the magnetic solitons, domain boundaries and the Bloch lines.

2. DESCRIPTION OF THE RELAXATION IN FERROMAGNETIC

2.1. The Motion Equation.
Dissipative Function.

For constructing the dissipative terms in the dynamic equations of magnetics we shall issue from the Onsager kinetic equations. As known, for writting the Onsager kinetic equations, first, an entropy production S is to be found and then by using this the generalized forces and fluxes. In rht case of ferromagnetic an enternal energy variation δW is connected with magnetization variations $\delta \vec{M}$ by the formula [6]

$$\delta W = -\int \vec{H}_n(\vec{x},t)\, \delta \vec{M}(\vec{x},t)\, d^3x \qquad (2.1)$$

where, $\vec{H}_n(\vec{x},t)$ is the effective magnetic field. Hence

$$\frac{dW}{dt} = -\int \vec{H}_n(\vec{x},t) \frac{\partial \vec{M}(\vec{x},t)}{\partial t} d^3x \qquad (2.2)$$

The dynamic part of equation of motion for the magnetization provides, obviously, for conservation of the internal energy (dW/dt =0). The difference dW/dt from zero is associated with the dissipative terms in the equation of motion, Therefore

$$\frac{dW}{dt} = -T\frac{dS}{dt} \qquad (2.3)$$

Comparing (2.2), (2.3), we find

$$\frac{dS}{dt} = \frac{1}{T}\int \vec{H}_m(\vec{x},t)\frac{\partial \vec{M}(\vec{x},t)}{\partial t}d^3x \qquad (2.4)$$

Considering this formula, we see, that the effective magnetic field $\vec{H}_m$ is to be taken as the generakezed forces, and $\partial \vec{M}/\partial t$ as the generalized fluxes.

Taking account of both the time and space dispersions, the Onsager equations will be written in the form

$$\frac{\partial M_i(\vec{x},t)}{\partial t} = \int_{-\infty}^{t} dt' \int d^3x' \Gamma_{i\kappa}(\vec{x}-\vec{x}', t-t') H_{m,\kappa}(\vec{x}',t) \qquad (2.5)$$

Integration of time limits in this formula corresponds to causality principle. If the typical frequencies are low as compared with the relaxation ones, the time dispersion may be neglected, and Eq. (2.5) takes the form

$$\frac{\partial M_i(\vec{x},t)}{\partial t} = \int d^3x' \Gamma_{i\kappa}(\vec{x}-\vec{x}') H_{m,\kappa}(\vec{x}',t) \qquad (2.6)$$

Assuming, that a variation of the magnetization in a space is smooth sufficiently, Eg. (2.6) will be rewritten in the form

$$\frac{\partial M_i}{\partial t} = \Gamma_{i\kappa} H_{m,\kappa} - \Gamma_{i\kappa,sp}\frac{\partial^2 H_{m,\kappa}}{\partial x_s \partial x_p} \qquad (2.7)$$

In this equation

$$\Gamma_{i\kappa} = \int d^3x\, \Gamma_{i\kappa}(\vec{x}); \quad \Gamma_{i\kappa,sp} = \int d^3x\, x_s x_p \Gamma_{i\kappa}(\vec{x}) \qquad (2.7')$$

The tensors $\Gamma_{i\kappa}$ and $\Gamma_{i\kappa,sp}$ in Eq. (2.7) are the zeroth and the second terms of power expansion of space gradients as well as the zeroth terms of time-derivative expansion of the coefficients being the members of Eq. (2.5)[*]. Indexes of the tensors $\Gamma_{i\kappa}$ and $\Gamma_{i\kappa,sp}$ designate coordinate axes (i = 1,2,3; 1 = x ; 2 = y; 3 = z).

---

[*] The first term of space gradients expansion is reduced to zero for the crystals with a center of symmetry.

The kinetic coefficients Γ satisfy the Onsager symmetry relations

$$\Gamma_{i\kappa}(\vec{M}) = \Gamma_{\kappa i}(-\vec{M}); \quad \Gamma_{i\kappa,s\rho}(\vec{M}) = \Gamma_{\kappa i,s\rho}(-\vec{M}) \tag{2.8}$$

Introducing an operator

$$\mathcal{L}_{i\kappa}(\vec{M}) = \Gamma_{i\kappa}(\vec{M}) - \Gamma_{i\kappa,s\rho}(\vec{M}) \frac{\partial^2}{\partial x_s \partial x_\rho} \tag{2.9}$$

we shall rewrite Eq.(2.7) in the form

$$\frac{\partial M_i(\vec{x},t)}{\partial t} = \mathcal{L}_{i\kappa}(\vec{M}) H_{M,\kappa}(\vec{x},t) \equiv \mathcal{D}_i + R_i \tag{2.10}$$

where

$$\mathcal{D}_i = \frac{1}{2}(\mathcal{L}_{i\kappa} - \mathcal{L}_{\kappa i}) H_{M,\kappa} \equiv \mathcal{Y}_{i\kappa} H_{M,\kappa}$$

$$R_i = \frac{1}{2}(\mathcal{L}_{i\kappa} + \mathcal{L}_{\kappa i}) H_{M,\kappa} \equiv \Lambda_{i\kappa} H_{M,\kappa}$$

The operator $\mathcal{Y}_{i\kappa}$ being asummetrical on its indexes, determines a nondissipative part of the equastion of mition for the magnetization, and the operator $\Lambda_{i\kappa}$, being symmetrical on its indexes, determines a dissipative term in the equation of motion as well as a dissipative function of the magnetic. According to Eqs.(2.4), (2.7), (2.10) we have

$$Q = \frac{1}{2} T \frac{dS}{dt} = \frac{1}{2} \int H_{M,i} \Lambda_{i\kappa} H_{M,\kappa} d^3x =$$
$$= \frac{1}{2} \int d^3x \left\{ H_{M,i} \lambda_{i\kappa} H_{M,\kappa} + \lambda_{i\kappa,s\rho} \frac{\partial H_{M,i}}{\partial x_s} \frac{\partial H_{M,\kappa}}{\partial x_\rho} \right\} \tag{2.11}$$

where

$$\lambda_{i\kappa} = \frac{1}{2}(\Gamma_{i\kappa} + \Gamma_{\kappa i}); \quad \lambda_{i\kappa,s\rho} = \frac{1}{2}(\Gamma_{i\kappa,s\rho} + \Gamma_{\kappa i,s\rho})$$

Eq.(2.7) together with given symmetry relations (2.8) is, strictly speaking, valid in a low-frequency (relaxation) approximation for the linear deviations $\partial \vec{M}/\partial t$ and $\vec{H}_M$ from their equilibrium values. We remind, that for a thermodynamic equilibrium values. We remind, that for a thermodynamic equilibrium state $\vec{M} = \vec{M}_o$, $\vec{H}_M = 0$ (minimum internal energy condition), $(\partial \vec{M}/\partial t) = 0$ (stationarity condition), Constructing the dissipative terms in the equations of dynamics of magnetic media, we shall assume, that the relations (2.9) and Eq.(2.7) (or a type of Eq.(2.7) and a type of relation (2.9)) were valid for small deviations from an equilibrium state in the high-frequency limit. In another words, we shall assume, that a type of the relaxation terms being determined in a low-frequency linear approximation, is also valid for a description of nonlinear synamics of the magnetic medium.

In a number of cases the field $\vec{H}_M$ is to be expressed through $\partial \vec{M}/\partial t$. For this purpose we shall introduce the Green function $G_{ik}(\vec{x})$ according to equation

$$\mathcal{L}_{ik} G_{ks}(\vec{x}) = \delta_{is} \delta(\vec{x}) \tag{2.12}$$

By using the Green function we find

$$H_{M,K}(\vec{x}, t) = \int d^3 x' G_{K,n}(\vec{x} - \vec{x}') \frac{\partial \vec{M}_n(\vec{x}', t)}{\partial t} \tag{2.13}$$

Let us discuss Eq.(2.12) for $G_{Kn}(\vec{x})$ and the method for its solution. As the magnetization depends on coordinates $\vec{M} = \vec{M}(\vec{x}, t)$, Eq.(2.12) is the equation with variable coefficients. Owing to a smooth change of $\vec{M}$ with the coordinates mentioned above the coefficients Γ_{ik} and $\Gamma_{ik,sp}$ are the snooth functions of the coordinates. For this reason to solve Eq.(2.12) the well-developed methods for soluting the Schrödinger equation in quasiclassical approximation can be used.

In order to determine the explicit expression for the Green function G_{ik} in zero approximation, when the dependence between the magnetization and coordinates may be neglected, we shall proceed to the Fourier components by the coordinates

$$G_{in}(\vec{x}) = \frac{1}{(2\pi)^3} \int G_{in}(\vec{\kappa}) e^{i \vec{\kappa} \vec{x}} d^3 \kappa$$

$$\delta(\vec{x}) = \frac{1}{(2\pi)^3} \int e^{i\vec{\kappa}\vec{x}} d^3\kappa$$

Then Eq.(2.12) takes the form

$$[\gamma_{in} + \gamma_{in,sp} \kappa_s \kappa_p] G_{nm} = \delta_{im} \qquad (2.14)$$

Wherefrom

$$G_{nm} = [\gamma + \gamma_{,sp} \kappa_s \kappa_p]^{-1}_{nm} \qquad (2.15)$$

Using Eq.(2.13),   can be expressed through $(\partial \vec{M}/\partial t)$

$$Q = \frac{1}{2} \int d^3x \int d^3x' \frac{\partial M_i(\vec{x},t)}{\partial t} G_{ik}(\vec{x}-\vec{x}') \frac{\partial M_k(\vec{x}',t)}{\partial t} \qquad (2.16)$$

Taking into account the fact, that $M_i(\vec{x},t)$ is smoothly changed with the coordinates, it allows that the integral expression for a density of the dissipative function goes over into a local expression for Q. Expressing as a power series in $\vec{x}'-\vec{x}$ and limiting by the first three terms, we obtain

$$Q = \frac{1}{2} \int d^3x \left\{ \frac{\partial M_i}{\partial t} Z_{ik} \frac{\partial M_k}{\partial t} + \frac{\partial^2 M_i}{\partial t \partial x_s} Z_{ik,sp} \frac{\partial^2 M_k}{\partial t \partial x_p} \right\} \qquad (2.17)$$

where

$$Z_{ik} = \int d^3x\, G_{ik}(\vec{x}); \quad Z_{ik,sp} = \int d^3x\, x_s x_p G_{ik}(\vec{x}) \qquad (2.18)$$

Of cource, only the parts of the tensors Z_{ik} and $Z_{ik,sp}$ being symmetrical jn the first two indexes provide a contribution into Eq.(2.17) for Q. The tensors Z as well as ones Γ satisfy a symmetry principle of the kinetic coefficients.

$$Z_{ik}(\vec{M}) = Z_{ki}(-\vec{M}); \quad Z_{ik,sp}(\vec{M}) = Z_{ki,sp}(-\vec{M}) \qquad (2.19)$$

The tensors Γ and Z can be shown to be connected by the relations

$$Z_{ik} = \Gamma^{-1}_{ik}; \quad Z_{ik,sp} = \Gamma^{-1}_{in} \Gamma_{nm,sp} \Gamma^{-1}_{mk} \qquad (2.20)$$

Obtaining these relations we used essentially the fact, that the derivatives from the coordinate magnetization are small.

Using the formulae (2.10), (2.11), we obtain

$$R_i = \Lambda_{ik} H_{M,K} = \delta Q/\delta H_{M,i} \tag{2.21}$$

The antisymmetric tensor $\mathcal{Y}_{ik}$ by the first two indexes determines the dynamics of the ferromagnetic. As the components with $\Gamma_{ik,sp}$ describes the corrections to the main term of expansion in Eq.(2.7), then with describing the dynamics according to a traditional approach they will be not condidered, and we assume latter that $\mathcal{Y}_{ik} = \frac{1}{2}(\Gamma_{ik} - \Gamma_{ki})$.

Introduceng the tensor g_{ik} by using the magnetization vector $\vec{M}$, the tensor $\mathcal{Y}_{ik}$ will be presented in the form

$$\mathcal{Y}_{ik} = \varepsilon_{ikl} g_{ln} M_n \tag{2.22}$$

This formula allows to write the dynamic term $\mathcal{D}$ in the right part of Eg.(2.7) in the form

$$\mathcal{D}_i = \mathcal{Y}_{ik} H_{M,K} = [\hat{g}\vec{M}, \vec{H}_M]_i \tag{2.23}$$

Thus Eq.(2.7) takes the form

$$\frac{\partial \vec{M}}{\partial t} = \vec{\mathcal{D}} + \vec{R} = [\hat{g}\vec{M}, \vec{H}_M] + \hat{\Lambda}\vec{H}_M$$
$$(\hat{\Lambda}\vec{H}_M)_i = (\Lambda_{ik} - \Lambda_{ik,sp}\frac{\partial^2}{\partial x_s \partial x_p})H_{M,K} \tag{2.24}$$

2.2. The Exchange Relaxation

Owing to the Onsager relations the tensors Λ and g are the even functions of magnetization,

We shall consider, firs of all, the properties of Λ and g in a main exchange approximation without external magnetic field. In this approximation a medium is to be considered as isotropic one relative to arbitrary rotations of the magnetization. For this reason[6]

$$W_{ex} = \frac{1}{2}\int d^3x \left\{ f(M^2) + \alpha_{ik}\frac{\partial \vec{M}}{\partial x_i}\frac{\partial \vec{M}}{\partial x_k} \right\} \tag{2.25}$$

$$\vec{H}_M = -f_1(M^2)\vec{M} + \alpha_{ik}\frac{\partial^2 \vec{M}}{\partial x_i \partial x_k}$$

where, α_{ik} is the tensor of the inhomogeneous exchange interaction. The components of the tensors λ and g in the exchange approximation are the functions M^2.

The dependence on the first pair of the indexes of the tensors λ and g is to be constructed in the same manner as for an isotropic space, fpr these indexes belong to $\vec{M}$ and $\vec{H}_M$. The dependence on the second pair of the indexes for the tensor $\lambda_{ik,sp}$ is to correspond to a symmetry of the lattice, for this pair of indexes is related with coordinate derivatives.

We note finally, that in exchange approximation the equation of magnetic - moment motion is to be the form of low of conservation of the total moment[7]

$$\frac{\partial M_i}{\partial t} + \frac{\partial \Pi_{ik}}{\partial x_k} = 0 \tag{2.26}$$

All these considerations result in the fact, that the tensors λ and g take the form

$$g_{ik} = g\delta_{ik}; \quad \lambda_{ik} = 0; \quad \lambda_{ik,sp} = \lambda^e_{sp}\delta_{ik} \tag{2.27}$$

Therefore Eq.(2.24) in the exchange approximation take the form

$$\frac{\partial \vec{M}}{\partial t} = g[\vec{M},\vec{H}_M] - \lambda^e_{ik}\frac{\partial^2 \vec{H}_M}{\partial x_i \partial x_k} \tag{2.28}$$

Taking into account Eq.(2.25), we obtain [7], [4]

$$\Pi^{dyn}_{ik} = g[\vec{M}, \alpha_{ks}\frac{\partial \vec{M}}{\partial x_s}]_i$$
$$\Pi^{diss}_{ik} = \lambda^e_{ks}\frac{\partial H_i}{\partial x_s} = \lambda^e_{ks}\frac{\partial}{\partial x_s}\left[-f_1\vec{M} + \alpha_{nm}\frac{\partial^2 M_i}{\partial x_n \partial x_m}\right] \tag{2.29}$$

Let us analyse the conservation low of Eq.(2.28). We see, that, in nondissipative approximation ($\lambda^e = 0$) this equation together with a total magnetic moment

$$\vec{\mathcal{M}}(t) = \vec{\mathcal{M}}_o = \int d^3x\, \vec{M}(\vec{x}, t) \qquad (2.30)$$

has local integral motion

$$M^2(\vec{x}, t) = M_o^2(\vec{x}) \qquad (2.31)$$

The motion integral (2.31) corresponds to lacking dynamic degree of freedom associated with a change of the value of the magnetization.

More general equation (2.25) at = 0 results in dynamic relation

$$\left(\frac{\partial \vec{M}}{\partial t},\, \hat{g}\vec{M}\right) = 0 \qquad (2.32)$$

which is a generalized motion integral (2.31).

The exchange relaxation term in Eq. (2.28) provides for a diffusion of the value of the magnetic moment by the crystal. It may be as an usual atom diffusion (in this case $\lambda^e \sim D$, D is the coefficient of the atom diffusion), as "spin diffusion", when a local spin disorientation vanishes. In this case shortwave excitations dissipate quite quickly, and long-wave excitations survive. These are corresponded by quasiuniform distribution of the space magnetization.

Taking occount of the relaxation terms in the exchange approximation already, the whole dynamic degrees of freedom are regenerated. Owing to the local conservation low (2.31) an oscillation frequency of the value of the magnetization is dissipative.

In the exchange approximation, as known, the main state of the ferromagnetic is degenerated with two continuous parameters of degeneration: with azimuthal and polar angle of the magnetization vector. This degeneration is corresponded by a type of the relaxation term in Eq. (2.28).

In conclusion the expression for the dissipative function in exchange approximation is given. According to formula (2.11) we have

$$Q^e = \frac{1}{2}\int d^3x\, \lambda^e_{mn}\, (\partial \vec{H}_m/\partial x_m)(\partial \vec{H}_m/\partial x_n) \qquad (2.33)$$

In going from $\vec{H}_M(\vec{x},t)$ to the Fourier component

$$\vec{H}_M(\vec{x},t) = \frac{1}{V^{1/2}} \sum_{\vec{\kappa}} \vec{H}_M(\vec{\kappa},t) e^{i\vec{\kappa}\vec{x}} \tag{2.34}$$

we shall rewrite the formula for Q^e in the form

$$Q^e = \frac{1}{2} \sum_{\vec{\kappa}} |\vec{H}_M(\vec{\kappa},t)|^2 \lambda^e(\vec{\kappa}) \tag{2.35}$$

where

$$\lambda^e(\vec{\kappa}) = \lambda^e_{nm} \kappa_n \kappa_m; \quad \vec{H}_M(\vec{\kappa},t) = \vec{H}_M(-\vec{\kappa},t)$$

Besides Eqs. (2.33), (2.35), it is convenient to have the type (2.27) of expression for the dissipative function in the exchange approximation. In is easy done in a linear approximation with small deviations from a homogeneous state. Assuning, that $\vec{M} = \vec{M}_0 + \vec{m}$, Eq.(2.28) will be rewritten in the form

$$\frac{\partial \vec{m}}{\partial t} = \mathcal{L} \vec{H}_M \tag{2.36}$$

where,

$$\mathcal{L}_{in} = -g \varepsilon_{ins} M_n + \delta_{in} \lambda^e(\vec{\kappa})$$

Hence,

$$\vec{H}_M = \mathcal{L}^{-1} \frac{\partial \vec{m}}{\partial t} \tag{2.37}$$

Substituting this expression instead of $\vec{H}_M$ in Eq. (2.35), we obtain

$$Q^e = \frac{1}{2} \sum_{\vec{\kappa}} \frac{\lambda^e(\kappa)}{(gM_0)^2 + [\lambda^e(\kappa)]^2} \left|\frac{\partial \vec{m}(\vec{\kappa},t)}{\partial t}\right|^2$$

Taking into account, that $\lambda^e(\kappa) \ll gM_0$, we have

$$Q^e = \frac{1}{2(gM_0)^2} \sum_{\vec{\kappa}} \lambda^e(\vec{\kappa}) \left|\frac{\partial \vec{m}(\vec{\kappa},t)}{\partial t}\right|^2 \tag{2.38}$$

Comparing this formula with one (2.17), in going to the Fourier components if the magnetization in advance

$$\vec{m}(\vec{x},t) = \frac{1}{V^{1/2}} \sum_{\vec{\kappa}} e^{i\vec{\kappa}\vec{x}} \vec{m}(\vec{\kappa},t)$$

we find, that in the exchange approximation

$$Z_{in} = 0; \quad Z_{in,sp} = \delta_{in} \lambda^e(\vec{k})(gM_0)^{-2} \qquad (2.39)$$

2.3. Relativistic Relaxation

Taking into account the relativistic interactions, it results in as dectructing the motion integrals $\mathcal{M}_i$ as removing the main-state degeneration. If the latter appears to be degenerated, this degeneration is described by the discrete degeneration pframeter but no continuous one.

As known, the symmetry of the crystal determines a type and hierarchy of the relastivistic interactions in an internal energy of the magnetics. We shall remind this fact by using, as an wxample, the energy density of the magnetic anisotropy W_A. One will be presented in the form of ghe expansion

$$W_A(\vec{M}) = \frac{1}{2} K_{ik}^{(2)} M_i M_k + \frac{1}{4} K_{ik,sp}^{(4)} M_i M_k M_s M_p + \ldots \qquad (2.40)$$

The coefficients of this expansion are decreased with increasing number of the expansion term being proportional to $(v/c_0)^{27)}$

$$K^{(2n+2)}/K^{(2n)} \approx (v/c_0)^2 \qquad (2.41)$$

where v is the velocity of atom's electrons and Co is the light velocity.

The number of the independent components of the tensors is determined by a group of the symmetry of the fact, that the tensors K are symmetrical on their indexes.

Let us exspand analogously the kinetic coeffecients in a power series of the magnetization

$$\lambda_{ik}(\vec{M}) = \lambda_{ik}^{(0)} + \lambda_{ik,\ell m}^{(2)} M_\ell M_m \qquad (2.42)$$

Osing to (2.8) the terms of this expansion contain only the components with even powers M_i. The coefficients of these expansions are decreased with increasing number of the component of M by the pframeter $(v/c_0)^2$. In order to explain this significant circumstance, we shall remind how corrections to the expression for spin-wave spectrum, magnetic anisotropy energy spin-wave attenuation are obtained from the microscogic Hamiltoniam describing the relativistic interactions. The modern theoretical calculations of the values are carried out by the diagram methods. An order of the deagram is determined by the number of vertexes of this diagram. The higher its order is, the higher power of low parameter $(v/c_o)^2$, which is, strictly speaking, the relation between weak relativistic interactions and main exchange one. Begining from the second order the same diagrams determine the corrections to the energy of spin-waves, their attenuation and the energy of magnetic enisotropy. Therefore, the following terms of the expansions for W_A and λ are of the same ifinitesimal order, i.e., $\lambda^{(2n+2)}/\lambda^{(2n)} \approx (v/c_o)^2$. That circumstance may be charfcteristic of the diagram, that owing to symmetry considerations a nymber of vortexes can be vanished. As an example, the vortex can be, which describes the exchange scattering of two spin waves. Owing to the exchange interaction symmetry relative to arbitrary homogeneous turns in a spen space and according to the low of total-spin conservation, connected with this symmetry, this vortex is vanished, when spin wave impulses tend to zero.

Substituting Eq.(2.42) into formula (2.11), we obtain

$$Q^r = \frac{1}{2}\int d^3x \left\{ \lambda^{(o)}_{ik} H_{m,i} H_{m,k} + \lambda_{ik,mn} H_{m,i} H_{m,k} M_n M_m^+ \ldots \right\} \quad (2.43)$$

The dissipatine function Q^r (as Q^e) is to be invariant relative to symmetry transformations of the crystal in the paramagnetic phase. In anther words, a type of the tensors λ and the nuber of their independent components is determined by this symmetry, as at $M = 0$ they are derivaties $\lambda(\vec{M})$ by $\vec{M}$.

Finally we note, that the terms in Q^r containing the derivatives $\partial \vec{H}_m/\partial x_k$ are small by two pframeters: by a small parameter of the relativistic interactions and small gradient of the magnetization. Therefore, they are not taken into account in Eq.(2.43), and they will not by taken into account later. This situation is analogous that one, that in the energy of the magnetic anisotropy the components including the magnetic anisotropy the components inckuding the magnetization gradients are not taken into account.

The condition given allows to proporse the following phenomenological method for constructing the dissipative function Q^r. From the vectors $\vec{H}_M$ and $\vec{M}$ the combinations being qudratic by $H_{M,i}$ and with even powers M_K are constructed, which are invariant relative to a group of the symmetry of the crystal in the pfrfmagnetic pgase. Later a sum of such invariants with arbitraty coefficients is taken. These coefficients will be the relaxation constants. The higher the power M_i in the invariant, the smaller coefficient is to be considered at a corresponding component. In order to provide for the relaxation of all the components of the magnetization to a ground state, the number of the components in dissipative function is to be corresponding.

Hierarchy of the relaxation constants is somewhat differed from one of constants of the magnetic anisotropy, as by using such a method for constructing the dissipative function the laws of conservation of the components (or component) for the magnetic moment are not taken into account, which are valid in an exchange approximation or in an uniaxial crystal one. In order to restore the exact laws for conserving the magnetic moment, some number of the relaxation constants have to be zero. This requirement shows indirectly a symmetry of the internal energy in dissipative function properties.

The construction of the invarianrs from the vector components $\vec{H}_M$ and $\vec{M}$ is relieved by the fact, that $\vec{H}_M$ is transformed in the same manner as $\vec{M}$ under action of the symmetry elements of the crystal. It is due to determining $\vec{H}_M = -\delta W/\delta \vec{M}$ and invariance of the internal energy W.

2.3.1. <u>Approximation of the Uniaxial Crystal</u>. Let us consider, as an sxample, the hexagonal, rhombohedral and tetragonal crystals. Anisotropy in a basic plane of these crystals is described by the sixthor fourthorder terms by the magnetization compontnts. The corresponding components of the anisotropy qre much lower than those of the constant describing the anisotropy relative to a main axis of the symmetry, i.e., the axis Z[7].

If neglecting the anisotropy in the basic plane, we obtain the "uniaxial" crystal model with the symmetry axis C_∞. In this model there is one motion integral $\mathcal{M}_z$ and the main state, generally speaking, has one continuous degeneration parameter, i.e., the azimuthal angle of the vector $\vec{M}$.

In the previous section a part of the dissipative function being due to exchange interaction has been determined. Its density is equal

$$q^e = \frac{1}{2} \lambda^e_{ik} (\partial \vec{H}_M / \partial x_i)(\partial \vec{H}_M / \partial x_k)$$

For uniasial crystals the tensor λ^e is diagonal and has two independent components

$$\lambda^e = diag\left(\lambda^e_{11}, \lambda^e_{11}; \lambda^e_{33}\right) \tag{2.44}$$

Let us determine a type of the dissipative function Q^r. It is easy seen, that by using the values $\vec{H}_M$ for the crystal, having the symmetry C_∞, two invariants can be constructed

$$\vec{H}^2_{M\perp} ; H^2_{Mz}$$

where, $\vec{H}_{M\perp} = \vec{H}_M - \vec{e}_z H_{Mz}$ and the unit vector $\vec{e}_z$ is along the symmetry axis. Formally

$$q^r = \frac{1}{2}\left[\lambda_{11} \vec{H}^2_{M\perp} + \lambda_{33} H^2_{Mz}\right] \tag{2.45}$$

The constant λ_{11} describes the magnetization components relaxation in the basic plane, i.e., the rwlyxation $\vec{M}_\perp$. The constant λ_{33} describes the relaxation of the component M_z, including a homogeneous dixtribution of M_z in the space. This fact is at contradiction with the conservation low of the components $\mathcal{M}_z$ of the total magnetic moment in a body. Therefore the relaxation constant λ_{33} is to be zero, and we have for total density of the dissipative function

$$q_{tot} = q^e + q^r = \frac{1}{2}\lambda^e_{ik}\frac{\partial \vec{H}_M}{\partial x_i}\frac{\partial \vec{H}_M}{\partial x_k} + \frac{1}{2}\lambda^r_{11}\vec{H}^2_{M\perp} \tag{2.46}$$

Hence,

$$\vec{R}_\perp = \left(\lambda^r_{11} - \lambda^e_{ik}\frac{\partial^2}{\partial x_i \partial x_k}\right)\vec{H}_{M\perp} \tag{2.47}$$

$$R_z = -\lambda^e_{i\kappa} \frac{\partial^2 H_{mz}}{\partial x_i \partial x_\kappa}$$

We see, that the relaxation term R_z has the form of dissipative-tensor divergence of the flux

$$\Pi_{z\kappa} = \lambda_{\kappa i}(\partial H_{mz}/\partial x_i)$$

With constructing q^r we are restricted ouself at present by the zeroth-order terms by magnetization. The terms of higher order smallness in q^r are not taken into account, as $\lambda_{\prime\prime}$ describes the relaxation $\vec{M}_\perp$, but the homogeneous component M_z in uniaxial crystal model does not relax.

We shall be convinced below, that the use of the higher-order terms by $\vec{M}$ is only needed, when a group of the crystal symmetry G_κ is lower than that of the group C_∞.

Let us write out now the effective magnetic field of uniaxial crystal and the equation of magnetic-moment motion. For this purpose the effective field of the magnetic anisotropy is to be added to the effective magnetic field $\vec{H}^{ex}_m$ determined by formula (2.25). Taking the magnetic anisotropy energy W_A in the simplest form

$$W_A = \tfrac{1}{2} K \vec{M}^2_\perp \tag{2.48}$$

We determine

$$\vec{H}_A = -K\vec{M}_\perp \tag{2.49}$$

and the total effective field is equal

$$\vec{H}_M = \vec{H}^{ex}_M + \vec{H}_A = -f_t M + d_{i\kappa} \frac{\partial^2 \vec{M}}{\partial x_i \partial x_\kappa} - K\vec{M} + \vec{H} - 4\pi \hat{N}\vec{M} \tag{2.50}$$

where, $\vec{H}$ is the external magnetic field, $\hat{N}$ is the tensor of the demagnetizing constants.

If to be within the framework of uniaxial crystal approximation, the equations of the motion for the magnetization components have the form

$$\frac{\partial \vec{M}_\perp}{\partial t} = g[\vec{M}, \vec{H}_M]_\perp + \lambda_{11} \vec{H}_{M\perp} - \lambda^e_{ik} \frac{\partial^2 \vec{H}_{M\perp}}{\partial x_i \partial x_k} \qquad (2.51)$$

$$\frac{\partial M_z}{\partial t} + \frac{\partial}{\partial x_k}\left(\Pi^{dyn}_{zk} + \Pi^{diss}_{zk}\right) = 0 \qquad (2.52)$$

Let us give the expression for the dissipative function

$$Q = \frac{1}{2}\int d^3x \left\{ \lambda^e_{ik} \frac{\partial \vec{H}_M}{\partial x_i} \frac{\partial \vec{H}_M}{\partial x_k} + \lambda_{11} \vec{H}^2_{M\perp} \right\} \qquad (2.53)$$

The index $\perp$ for the vectors means

$$\vec{A}_\perp = \vec{A} - \vec{e}_z(\vec{e}_z \vec{A}) = A_x \vec{e}_x + A_y \vec{e}_y$$

We note once more, that, as seen from Eqs. (2.51), (2.52), the relaxation of the components $\vec{M}_\perp$ is determined by both the relativistic and exchange constans, but M_z is only done by the exchange ones. The form of Eq. (2.52) provides for the conservation low of the total-magnetization component along the axis Z.

2.3.2. <u>The cubic-symmetry crystals</u>. Let us consider a cubic crystal now. As well known, the energy of the magnetic anisotropy in it has the form

$$W_A = \frac{1}{4}K_4\left(M^4_x + M^4_y + M^4_z\right) \qquad (2.54)$$

Such a kind of the magnetic anisotropy energy destroyed the motion integral $\vec{\mathcal{M}}$. Therefore the relativistic part of the dissipative function can be given in the form

$$Q^r = \frac{1}{2}\int d^3x\, \lambda_{11} \vec{H}^2_M \qquad (2.55)$$

For the total dissipative function we obtain

$$Q_{tot} = Q^e + Q^r = \frac{1}{2}\int d^3x \left\{ \lambda^e(\partial \vec{H}_M/\partial x_i)^2 + \lambda_{11} \vec{H}^2_M \right\} \qquad (2.56)$$

Whence

$$\vec{R} = \lambda_{11} \vec{H}_M - \lambda^e \Delta \vec{H}_M \qquad (2.57)$$

(we took into account, that in the cubic crystal $\lambda^e_{i\kappa} = \lambda^e \delta_{i\kappa}$). A contribution of the magnetic anisotropy into the field $\vec{H}_M$ is equal

$$\vec{H}_A = K_4 [\vec{e}_x M_x^3 + \vec{e}_y M_y^3 + \vec{e}_z M_z^3] \tag{2.58}$$

where e_x, e_y, e_z are the unit vectors along the axes x, y, z (the fourth-order axes), respectively. A set of eqyations for the magnetic-moment motion for the cubic crystal takes the form

$$\frac{\partial M}{\partial t} = \lambda_{\shortparallel} (\vec{m} \vec{H}_m) - \lambda^e (\vec{m} \Delta \vec{H}_m)$$

$$\frac{\partial \vec{m}}{\partial t} = g[\vec{m}, \vec{H}_m] - \frac{\lambda_{\shortparallel}}{M}[\vec{m}(\vec{m}, \vec{H}_m)] + \frac{\lambda^e}{M}[\vec{m}(\Delta \vec{H}_m, \vec{m})] \tag{2.59}$$

where, $\vec{m} = \vec{M}/M$ and M is the value of the magnetic moment.

If to assume in this set of equations, that $\lambda^e = 0$, that the second equation is coincided with the classical one which were proposed by Landau and Lifshits. However, a set of equations (2.67) remains a set of coupled ones, as $(\vec{m} \vec{H}_m) \neq 0$.

3. SPIN-WAVE DAMPING

In this chapter in the framework of the phenomenological theory of one dissipative function we shall consider an attenuation of the spin waves, and the linewidth for the ferromagnetic resonances (FMR), as well as we shall carry out a comparison between the phenomenological calculations and microscopic results on spin-wave attenuation.

3.1. Exhange Attenuation

First of all, we shall consider an exhange attenuation of the spin waves in ferromagnetics. For this we assume

$$\vec{M}(\vec{x}, t) = \vec{M}_o + [\vec{m}(\vec{k}) e^{i \vec{k} \vec{x} - i \omega(\vec{k}) t} + k c] \tag{3.1}$$

where, $\vec{M}_o$ is the equilibrium value of the magnetic moment, $\vec{m}(\vec{k})$ is the amplitude of the spin waves with the wave vector $\vec{k}$, ω_κ is the frequency of these waves. Substituting this expression for $\vec{M}(\vec{x}, t)$ in the dissipative function (2.38), we obtain

$$Q^e = \frac{1}{2} \frac{\lambda(\kappa) \omega_\kappa^2}{(gM_o)^2} |\vec{m}(\vec{\kappa})|^2 \qquad (3.2)$$

In the same manner by using Eq. (2.25) for the exchange energy of FM and taking into account the nondissipative approximation $M^2 = M_o^2$, we obtain

$$W_e = \frac{1}{2} \frac{\omega_\kappa}{gM_o} |\vec{m}(\vec{\kappa})|^2 \qquad (3.3)$$

As known, the damping rate $\gamma(\kappa)$ of waves is equal to the relation between the dissipative function Q and doub-led wave energy. Therefore for $\gamma_e(\kappa)$ of spin wawes we have

$$\gamma_e(\kappa) = \frac{Q_e}{2W} = \frac{\lambda^e(\kappa)\omega_\kappa}{2gM_o} \qquad (3.4)$$

Let us remind, that in the exhange approximation

$$\omega_\kappa = gM_o \alpha \kappa^2; \quad \lambda^e(\kappa) = \lambda^e \kappa^2$$

We shall rewrite formula (3.4) in the folloeing form

$$\gamma_e(\kappa) = \frac{1}{2} \lambda^e(\vec{n}_\kappa) \alpha(\vec{n}_\kappa) \kappa^4 \qquad (3.5)$$

We shall give the expression for $\gamma_e(\kappa)$ which is obtained by the Green function method on the basis of the Heisenberg Hamiltonian [8], [9] for the cubic crystals

$$\gamma_e(\kappa) = \frac{3}{(2\pi)^3 S^2} \frac{SJ_o}{\hbar} \left(\frac{T}{SJ_o}\right)^2 (a\kappa)^4 \ln^2 \frac{T}{\hbar \omega_\kappa} \qquad (3.6)$$

where, $\hbar$ is the Plank constant, a is the lattice constant, T is the temperature, J_o is the exchange integral betveen the closest neighbors, S is the atomic spin. Comparing Egs. (3.5) and (3.7) we see, that with logarithmic accuracy the dependence on the value of the wave vector, provided by both the phenomenological and microscopic theories, is the same. Finally, comparing Eqs. (3.5) and (3.7), it follows, that

$$\lambda \alpha = \frac{3}{(2\pi)^3 S^2} \frac{SJ_o}{\hbar} \left(\frac{T}{SJ_o}\right)^2 a^4 \qquad (3.7)$$

In order of the value the exchange constant α is $\alpha \sim (T_0/hgM_0)a^2$
and
$$\lambda \simeq 10^{-2}(gM_0)(T/T_0)^2 a^2 \qquad (3.8)$$

3.2. The Linewidth of FMR

Let us consider now the linewidth of FMR. Having in mind iron-yttrium garnet, we assume, that the crystal is the cubic symmetry. The exchange dissipative function makes no contribution into the linewidth of FMR, as it is corresponded by uniform oscillations of the mfgnetization. Therefore f_r will to be considered only. We rewrite Eq. (2.60) in the form

$$Q = Q_r = \frac{1}{2} V \frac{\partial m_i}{\partial t} \tilde{\lambda}_{ik} \frac{\partial m_k}{\partial t} \qquad (3.9)$$

where, $m_i(t) = m_i(\kappa,t)|_{\kappa=0}$

$$\tilde{\lambda}_{ik} = diag\left(\frac{\lambda_{22}\lambda_{33}}{\mathcal{D}}, \frac{\lambda_{11}\lambda_{33}}{\mathcal{D}}, \frac{1}{\lambda_{33}}\right)$$

$$\mathcal{D} = \lambda_{33}[\lambda_{11}\lambda_{22} + (gM_0)^2] \simeq \lambda_{33}(gM_0)^2$$

The magnetization in the case of FMR may be presented in the form

$$\vec{M}(t) = \vec{M}_0 + (\vec{m}\,e^{-i\omega_0 t} + \vec{m}^* e^{i\omega_0 t}) \qquad (3.10)$$

Here, $\vec{M}_0$ is the magnetization in the ground state, directed along an easy magnetization axis [001], V is the sample, ω_0 is the frequency of FMR, We shall consider that the sample is the elleptic form, which one of axes is directed along the axis [001]. In this case

$$\omega_0^2 = \omega_1 \omega_2 \qquad (3.11)$$

where

$$\omega_1 = g[H_0 + H_{A1} - 4\pi N_3 M_0]; \quad \omega_2 = g[H_0 + H_{A2} - 4\pi N_3 M_0] \qquad (3.12)$$

$$H_{A1} = H_A + 4\pi N_1 M_0; \quad H_{A2} = H_A + 4\pi N_2 M_0$$

H_o is the external magnetic field directed along the axis $[001]$; $H_A = |K_4|M_o^2$, N_1, N_2, N_3 are the demagnetizing factors. A polariuation of the variable component of the magnetization is determined by the formula

$$(m_x/m_y) = -i(\omega_2/\omega_1) \qquad (3.13)$$

Substituting Eq. (3.10) in (3.9), we find

$$Q = \frac{1}{2} V \frac{\omega_o^2}{\omega_2 (gM_o)^2} [\lambda_{11}\omega_1 + \lambda_{22}\omega_2]|m_x|^2 \qquad (3.14)$$

The expression for the FMR energy may be presented in the form

$$W = V\left\{\frac{1}{2} f(M^2) - \vec{H}_o \vec{M} + 2\pi(N_1 M_x^2 + N_2 M_x^2 + N_3 M_z^2) + \frac{1}{4} K_4 (M_x^4 + M_y^4 + M_z^4)\right\} \qquad (3.15)$$

Using Eq. (3.10) and taking into account, that $M_o(df/dM^2)_o = H_o - 4\pi N_3 M_o$, we obtain

$$W - W_o = 2V \frac{\omega_o^2}{gM_o\omega_2} |m_x|^2 \qquad (3.16)$$

In this formula W_o is the energy of the basic state FM.

Fotmulae (3.14), (3.16) allow to obtain the expression for time attenuation of the amplitude oscillations at FMR

$$\gamma_o = \frac{Q}{2(W-W_o)} = \frac{1}{8gM_o} [\lambda_{11}\omega_1 + \lambda_{22}\omega_2] \qquad (3.17)$$

Formulae (3.11)-(3.17) are related, in fact, not, only to the cubic crystals but also to the $\mathcal{D}_2$ -symmetry ones. In is associated with the fact, that the symmetry of the body (threeaxial ellipsoid of rotation) ia coincided with the $\mathcal{D}_2$ symmetry. A transition to biaxial, the axes of symmetry of which are coincided with the axes of ellipsoid, is accomplished by changing the anisotropy energy by expression

$$W_A = \frac{1}{2}(K_{11}M_x^2 + K_{22}M_y^2) \qquad (3.18)$$

in Eq. (3.15), the fields H_{A1}, H_{A2} also are done by expression

$$H_{A1} = (K_{11} + 4\pi N_1)M_0 \; ; \quad H_{A2} = (K_{22} + 4\pi N_2)M_0 \qquad (3.19)$$

Let us return, however, to iron-yttrium garnet. For it owing to the cubic symmetry $\lambda_{11} = \lambda_{22}$ and

$$\gamma_0 = \frac{\lambda_{11}}{8(gM_0)}(\omega_1 + \omega_2) = \frac{\lambda_{11}}{4gM_0}\left[H_0 - 4\pi N_3 M_0 + \frac{1}{2}(H_{A1} + H_{A2})\right] \qquad (3.20)$$

the frequency attenuation γ_0 is corresponded by the magnetic-field linewidth. This linewidth is experimentally measured. It is connected with γ_0 by the relation

$$\gamma_0 = \Delta\omega = (d\omega_0/dH)\Delta H$$

Using formulae (3.11), (3.12) and (3.20), we obtain

$$\Delta H = \frac{\lambda_{11}\omega_0}{4g^2 M_0} \qquad (3.21)$$

REFERENCES

1. Landau L.D. Lifshits E.M.   Sov. Phys. $\underline{8}$, 153 (1935)
2. Bloch F.Z.   Physik, $\underline{61}$, 206 (1930)
3. Akhiezer A.I.   T. Phys. USSR, $\underline{10}$, 217 (1946)
4. Baryakhtar V.G.   ZhETE $\underline{87}$ 1501 (1984)
5. Baryakhtar V.G.   ZhETE $\underline{94}$ 196 (1988)
6. Akhiezer A.I., Baryakhtar V.G., Peletminskii S.V. "Spin waves", North Holland (1968)
7. Lifshits E.M., Pitaevskii L.P.   Statisticheskaya Physica, $\underline{2}$, M., Nauka, (1978)
8. Dyson F.   Phys.Rev. $\underline{102}$, 1230 (1956)
9. Kaszczeev V.N., Krivoglaz M.A.   FTT $\underline{3}$, 1541 (1961)

CHAOTIC STATES
IN SYSTEMS DESCRIBED BY THE LANDAU-LIFSHITZ EQUATION[+]

A.Sukiennicki and J.J.Żebrowski

Institute of Physics, Warsaw Technical University,
ul.Koszykowa 75, 00-662 Warszawa,
POLAND

ABSTRACT

A brief survey is given of deterministic chaos in magnetic systems described by the Landau-Lifshitz-Gilbert equation of motion.

1. INTRODUCTION

It has been known for some time that the Landau-Lifshitz equation without damping and without a driving force is completely integrable [1]. On the other hand, it has also been shown that such completely integrable equations as the nonlinear Schroedinger equation [2] and the sine-Gordon equation [3] perturbed by damping and a

---

[+]Supported by the University of Łódź under project CPBP 01.08.B1.1

deterministic driving force may exhibit chaotic behavior.

In this paper it is shown that a similar nonlinear behavior is obtained for the Landau-Lifshitz equation with damping in the Gilbert form and with a driving force included. Some examples of different physical systems described by such an equation are discussed. The main interest is focused on phenomena which occur in magnetic domain wall dynamics. Section I describes chaotic and periodic phenomena in Bloch wall containing horizontal Bloch lines. Section II discusses preliminary results on the chaotic behavior of a Bloch wall containing vertical Bloch lines. The above sections treat deterministic chaos in a Bloch wall extending in one of two directions in a plane parallel to the plane of the wall in materials with a large uniaxial anisotropy. Section III gives a short resumé of results obtained by other groups on the Bloch wall treated as a one dimensional chain lying in a direction perpendicular to the plane of the wall.

2. SECTION I

The Landau-Lifshitz equation of motion with damping in Gilbert form is:

$$\dot{\vec{M}} = - \vec{M} \times \frac{\delta\sigma}{\delta\vec{M}} - \frac{\alpha\vec{M} \times \dot{\vec{M}}}{M}$$

where α is the Gilbert phenomenological damping constant, the dot over the symbol signifies differentiation over the time and the symbol δ - the functional derivative. σ is the volume energy density:

$$\sigma = - K\sin^2\theta - A[(\nabla\theta)^2 + \sin^2\theta(\nabla\varphi)^2]$$
$$- 2\pi M^2\sin^2\varphi\sin^2\theta + MH_z\cos\theta$$

with K the anisotropy energy, A the exchange constant, $4\pi M$ the saturation magnetization and H_z the externally applied drive field. $\theta(z,t)$ and $\varphi(z,t)$ are the polar and azimuthal Euler angle, respectively, describing the position of the vector of magnetization in space. Using these angles, the Landau-Lifshitz equation may be written in the from:

$$\dot{\theta}\sin(\theta) = M^{-1}\frac{\delta\sigma}{\delta\varphi} + \alpha\dot{\varphi}\sin^2(\theta)$$

$$-\dot{\varphi}\sin(\theta) = M^{-1}\frac{\delta\sigma}{\delta\theta} + \alpha\dot{\theta}$$

To describe a wall separating two magnetic domains in a large anisotropy material the following assumptions are usually made [4]. The magnetization in domains is assumed to lie in either the positive or the negative z-direction. The Euler angles θ and φ are assumed to be independent of each other. The distribution of the angle θ in the direction y perpendicular to the wall is assumed to be equal to that in the static Bloch wall:

$$\tan\frac{\theta}{2} = \exp\frac{(y-q)}{\Delta}$$

and uniform along the surface of the wall. q is then the position of the center of the wall and $\Delta = \sqrt{(A/K)}$ is the Bloch wall width parameter. The angle φ is taken to be a constant over the thickness of the wall. Next, an integration over y is carried out so that the equations of motion of the Bloch wall become:

$$\dot{q} = \gamma\Delta[2\pi M\sin(2\varphi) - \frac{2A}{M}\frac{\partial^2\varphi}{\partial z^2}] + \alpha\dot{\varphi}$$

$$\dot{\varphi} = \gamma [H_z + \frac{2A}{\Delta M} \frac{\partial^2 q}{\partial z^2}] - \alpha \frac{\dot{q}}{\Delta}$$

where $q(z,t)$ is the position of the Bloch wall surface while $\varphi(z,t)$ is the azimuthal angle of the magnetization measured with respect to the surface of the wall. z is the coordinate along the direction of the applied field H_z. Note that the above derived equations of motion for the Bloch wall are the same as those derived by Slonczewski[5] for the twisted domain wall of bubble garnet films but the surface stray field is neglected here.

The equations of motion of the Bloch wall were solved numerically using a vector version of the DuFort-Frankel explicite finite difference scheme[3]. Force-free boundary conditions were assumed at the end points of the region of integration along the z-axis ("film surfaces"). The time step was 0.05 ns and 53 evenly distributed spatial grid points were used throughout this paper although as many as 200 have been also used as needed. The results were found to be numerically stable up to at least 3500 ns. For initial conditions $\varphi(z,0)=0$ and $q(z,0)=0$ were used (flat initial conditions).

For all drive fields smaller than the Walker critical field H_w the solutions of the equations of motion of the Bloch wall always reproduced the stationary motion of the Walker model[6]. The solutions then always converged onto a fixed point attractor[7].

When the drive field was larger than the Walker field the running oscillatory motion of the wall was obtained[6]. In a properly chosen phase space the running oscillatory state of the wall is represented by a fixed point. However, for this range of the drive field the wall may be also shown to exhibit a large sensitivity to initial conditions. To attain this we used some perturbation tests. If a perturbation above a certain very small threshold value is used a chaotic state of the wall was obtained. As perturbation procedure we simulated the effect on the motion of the wall of natural small fluctuations of material parameters (these are bound to occur in a real physical situation) by requiring that during the calculation at t=20 ns at a single grid point the drive field value be decreased by 8 % for the very short time of only 0.1 ns. The result of such a numerical procedure for H_z=12 Oe > H_w=10.92 Oe is given in fig.1.

In part a of fig.1 the structure of the wall $\varphi(z,t)$ is shown depicted every 10 ns. At t = 20 ns a collision occurs with a point defect situated as marked by the central tickmark on the horizontal axis. We stress that, during the collision, for two time steps (i.e. 0.1 ns) an 8% decrease of the drive field H_z occurs only at the point defect. Such a perturbation excites solitary wave like states in the form of horizontal Bloch lines in the wall structure. It can be seen that by t = 70 ns four fully formed Bloch lines are present in the wall structure. The Bloch lines connect minimum energy orientations of the magnetic moment; these orientations are π-distant from each other. The centers of the Bloch lines at t = 70 ns are marked by arrows which denote the proper direction of motion of each Bloch line. As the wall moves, the Bloch lines are driven towards each other and finally collide to pass through each other. It can be seen that not all Bloch lines survive such a collision. Thus the number of Bloch lines in the wall structure changes during wall motion: at a given moment in time there may be three, four and occasionally even only one Bloch line in the wall structure.

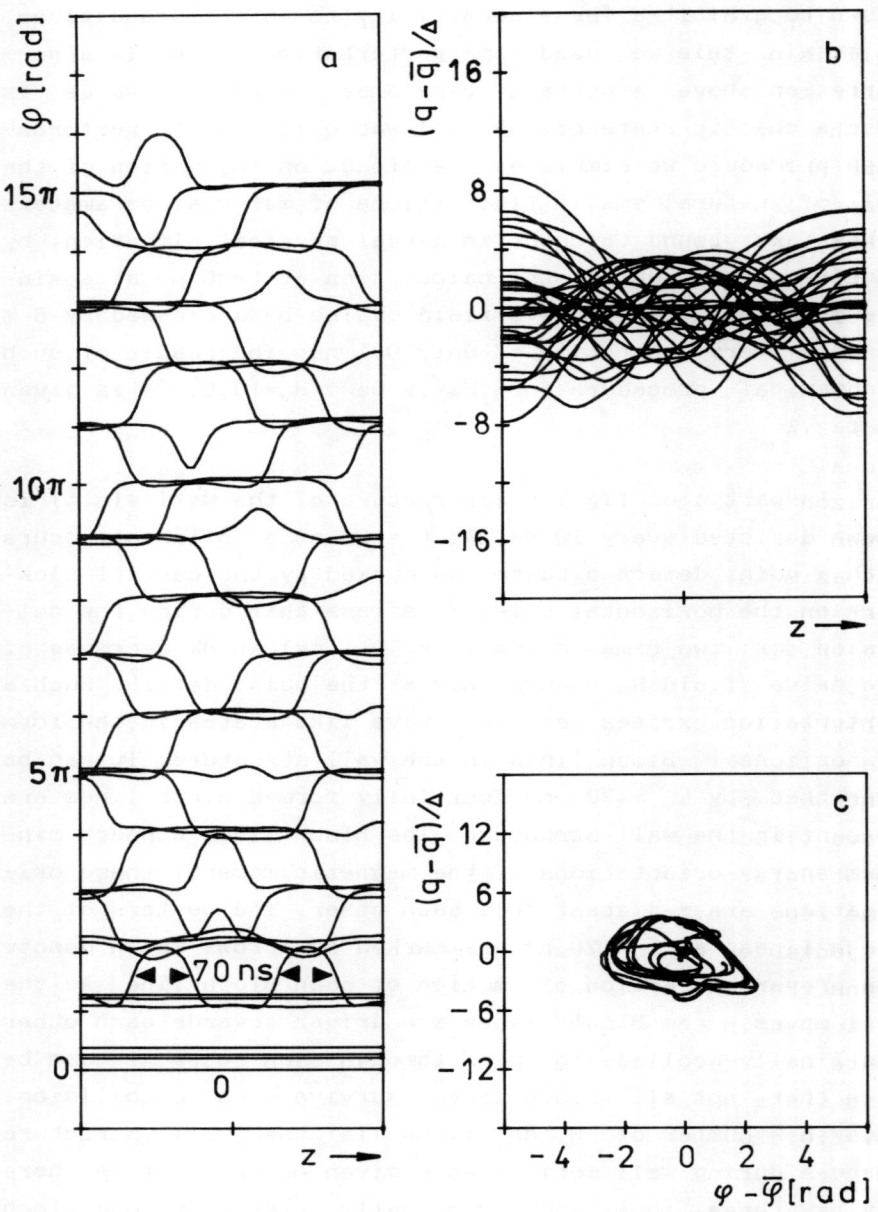

Fig.1 Chaotic state of the Bloch wall in the bulk case

In part b of fig.1 the shape of the surface of the wall is depicted: at each point the spatial average $\bar{q}(t)$ is subtracted from the local value q(z,t) so that the motion of the wall as a whole is omitted in fig.1b. This is done here primarily to show stable nodal points in the wall shape (if any). The position of the wall q(z,t) and the spatial average $\bar{q}(t)$ are both divided by the wall width Δ so that the quantity plotted is unitless. It can be seen in fig.1b that the shape of the wall is very complicated.

The phase portrait is depicted in fig.1c: the image shown is a two dimensional projection of the four dimensional trajectory of the center point in the wall. On the horizontal axis of the phase portrait the difference between $\varphi - \bar{\varphi}$ is plotted, where φ is the spatial average of the azimuthal angle. By plotting this difference instead of just the value of the angle it is possible to get rid of the trivial component $\gamma H_z t$ which is the same at all points of the wall. The center point of the wall was chosen here arbitrarily and the results obtained for other points in the wall are similar. The trajectory shown in fig.1c represents a strange attractor: e.g. if instead of by 8 % the drive field is decreased during the collision with the point defect by 9 % the attractor obtained occupies the same region of the phase space as in fig.1c but neighboring trajectories of the two attractors diverge[8]. This result means that the state of the wall is chaotic. Note that if a perturbation of less than 5 % was used the fixed point state was reproduced after a certain time which shows that this state is also a weak attractor above the Walker critical field.

The system appears to be relatively insensitive to its size (i.e. the distance between the end points) down to about 3.25 μm. For all the sizes larger than this the be-

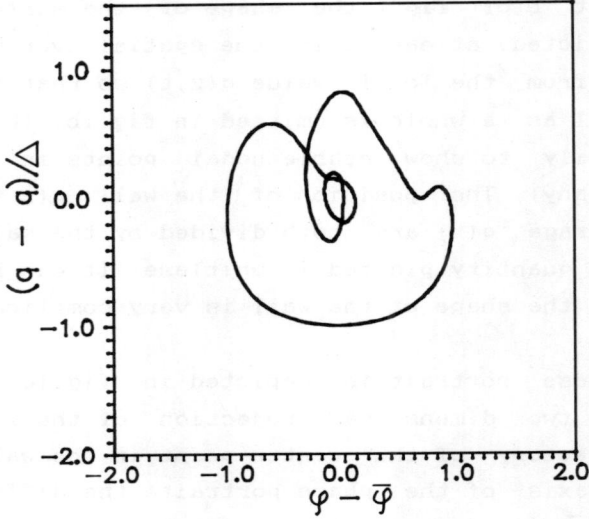

Fig.2 Phase portrait for the period-1 attractor.

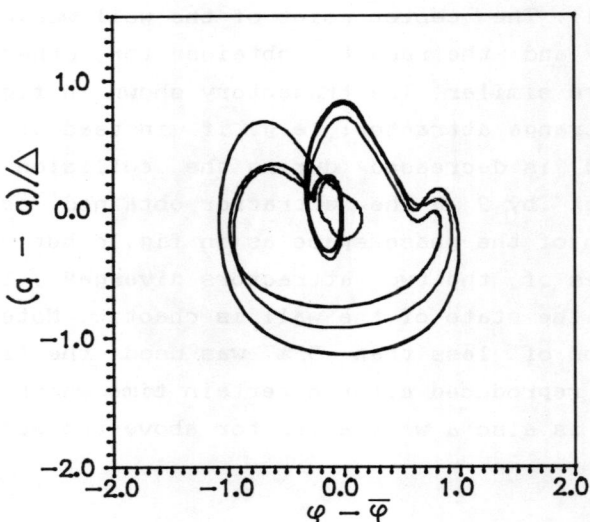

Fig.3 Phase portrait for the period-3 attractor.

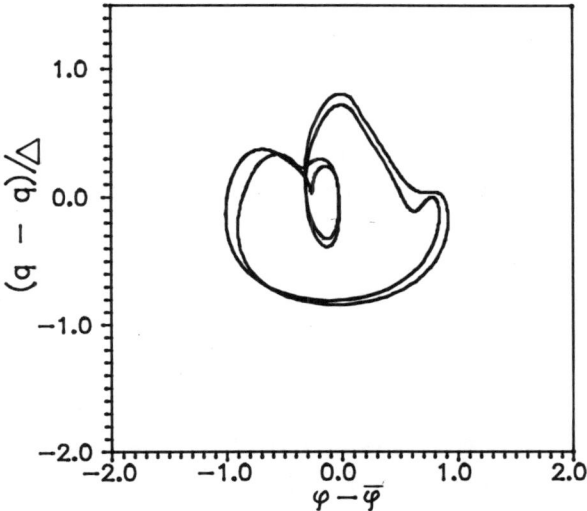

Fig.4 Phase portrait for the period-2 attractor.

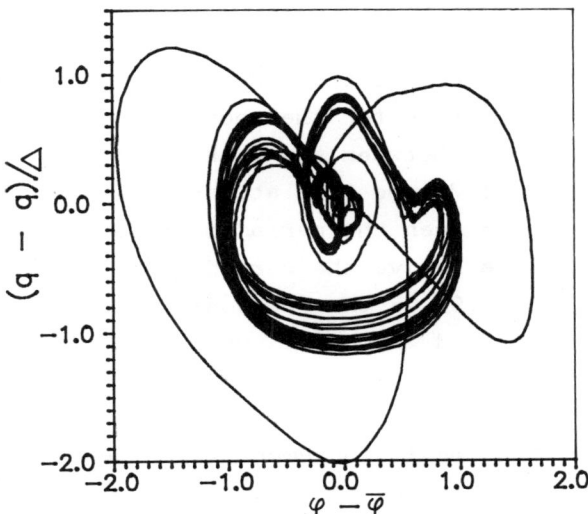

Fig.5 Strange band state for H_z = 11.825 Oe.

havior of the system was qualitatively the same: at the Walker critical drive a sharp transition occurs from the fixed point state to a strange attractor[8]. Below 3.25 μm both periodic and chaotic behavior could be obtained depending on a combination of the size of the system and on the magnitude of the drive used in the calculations[9].

As the size of the system was decreased below h = 3.25 μm and the drive field magnitude held constant at 12 Oe a period-2 attractor appeared[9]. At h = 1.1 μm a period-1 attractor was found with a spiral form of trajectory at point (0,0). Such a spiral form usually indicates the existence of a repeller. The fixed point of the system at (0,0) for drive fields higher than the critical value is a weak attractor so that it is concluded that the repeller separates the fixed point from the periodic attractor. Finally, the period was doubled (the attractor developed an additional loop) for h = 1.0 μm and no period doubling was found with a further decrease of the size of the system[9].

The dependence of the shape of the attractor on the drive field magnitude for the chosen value of h = 1.2 μm is discussed below. Between the Walker critical field of H_w = 10.92 Oe and 11.75 Oe the attractor shown in fig.2 was found. It can be seen the period of this attractor is 1. At 11.85 Oe of the drive the period-3 attractor of fig.3 was found and at 11.9 Oe the period-2 attractor depicted in fig.4 was obtained. However, when the transitions between these periodic attractors were followed through by starting at one of them and slowly varying the magnitude of the drive field in small increments as the calculations progressed, narrow ranges of the drive were found for which strange band states were found. The strange band state obtained for 11.825 Oe is depicted in fig.5 and the one obtained for 11.875 Oe in fig.6. It can be seen that in the

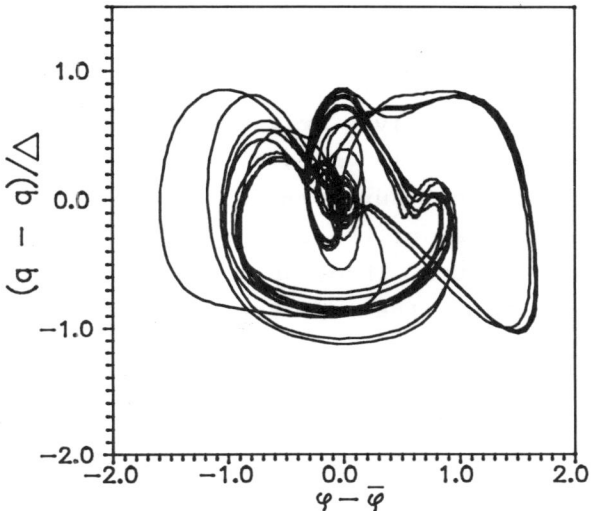

Fig.6 Strange band state for H_z = 11.875 Oe.

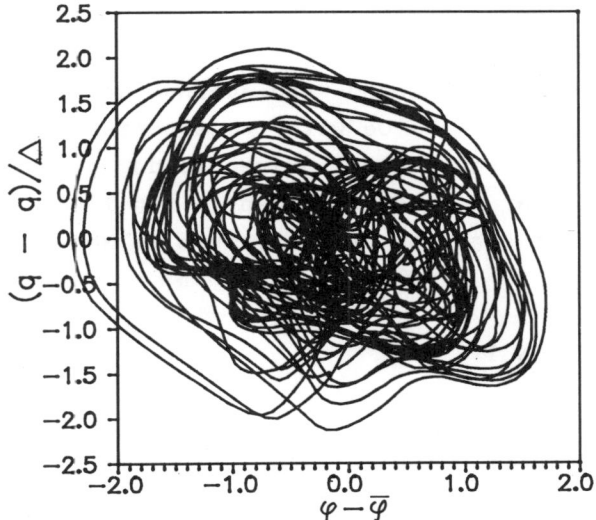

Fig.7 Chaotic state at H_z = 40 Oe.

strange band state the trajectories follow to a certain extent the shape of the trajectories of the attractors they border with in the parameter space (H_z,h). Also, it would seem that for the given drive field the trajectory seems to adhere preferentially more times to the shape of one of these attractors. Furthermore, once in a number of times around the center of the figure, a wildly different trajectory shape occurs. This is probably caused by the trajectory approaching the repeller more closely than it does at other times.

The character of the transitions found for h = 1.2 µm is the same as discussed in ref.10, where the transitions between periodic attractors for h = 1.1 µm was studied as a function of the drive field magnitude just above the Walker critical field.. There the periodic attractors were also found to be separated by strange band states.

The period-2 attractor in fig.4 was stable up to about H_z = 20 Oe. Next, at H_z = 21 and up to H_z = 23. Oe a strange band was found and again a different period-2 attractor at H_z = 25 Oe with trajectories very close to each other. At large enough drive fields a strongly chaotic state similar to the one shown in fig.1 was found. An example of such behavior of the system is shown in fig.7 for H_z = 40 Oe

3. SECTION II

Recently, preliminary investigations of chaotic states of Bloch walls containing vertical Bloch lines were also performed[11]. The results are similar to those obtained for walls containing horizontal Bloch lines and discussed in Section I: if the drive field is smaller than a critical value, the fixed point attractor, corresponding to the stationary motion state, occurs. However, when the drive field exceeds the critical magnitude, the fixed point undergoes a transition to the attractor shown in fig.8 which corresponds to the chaotic motion of the wall.

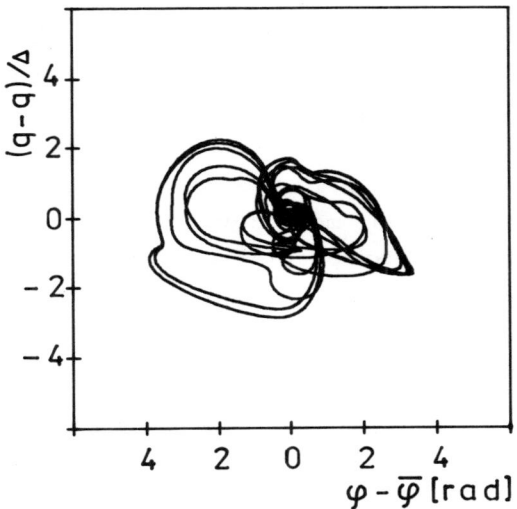

Fig.8 Chaotic state of a Bloch wall containing a pair of vertical Bloch lines

4. SECTION III

For completeness of the presentation it is necessary to mention that the chaotic behavior of the Bloch wall was also obtained numerically when the wall was treated as a one dimensional, continuous chain of magnetic moments the axis of which is perpendicular to the plane of the wall[12,13]. The drive field applied to the chain was time dependent. It would seem that the chaotic solutions are obtained in this case because the magnitude of the anisotropy is assumed low enough not to be able to keep the y-dependence of the angle 0 in the form usual for the Bloch wall (and assumed in Section I and Section II). Thus, a chaotic solution appears for just the 0(y,t) dependence.

5. CONCLUSIONS

In magnetic systems described by the Landau-Lifshitz equation of motion the route to chaos seems to be simple: a fixed point attractor at a critical value of the applied field undergoes a transition to a strange attractor with a very complicated structure. If the size of the system is limited, the route becomes more complicated: periodic attractors are found which may overlap in parameter space. At the overlap, strange band chaotic states may be found. The strange attractor of infinitely large systems reappears at large magnitudes of the drive field applied to the system.

6. REFERENCES

1. E.K.Sklyanin,"On complete integrability of the Landau-Lifshitz equation" preprint LOMI, E3, Leningrad 1979, p.32.
2. N.Bekki, K.Nozaki, in Dynamical Problems in Soliton Systems, ed.S.Takeno, Springer Series in Synergetics, vol.30 (1985) 268.
3. A.R.Bishop, K.Fesser, P.S.Lomdahl, W.C.Kerr, M.B.Williams, S.E.Trullinger, Phys. Rev. Lett.50, 1095 (1983).
4. A.P.Malozemoff, J.C.Slonczewski,"Magnetic domain walls", American Press, N.Y. 1979.
5. J.C.Slonczewski, J.Appl.Phys.44, 1759 (1973).
6. N.L.Schryer, R.L.Walker, J.Appl.Phys.45, 5406 (1974).
7. J.J.Żebrowski, A.Sukiennicki, Acta Physica Polonica A72, 299 (1987).

8. J.J.Żebrowski, A.Sukiennicki, Springer Proc. Phys.23, 130 (1987).
9. J.J.Żebrowski, to be published in Phys.Scripta.
10. J.J.Żebrowski, accepted to the ICM88 program.
11. R.A.Kosinski, A.Sukiennicki, to be presented as a poster at this conference.
12. F.Waldner, J.Magn.Magn.Mat.31-34, 1015 (1983).
13. H.Suhl, X.Y.Zhang, J.Appl.Phys. 61, 4216 (1987).

NONLINEAR RESPONSES IN MAGNETIC MATERIALS

P.E. Wigen and R. D. McMichael
The Ohio State University
Department of Physics
Columbus, Ohio 43210-1106 USA

ABSTRACT

Magnetic materials display a rich variety of nonlinear dynamic phenomena which manifest themselves as auto-oscillations, period multiplication, irregular period oscillations, intermittency, chaos and periodic windows. These materials are of particular interest because the system can be modeled by a microscopic Hamiltonian with well known parameters. The microscopic nonlinear equations provide the theoretical framework for modeling the system to explain the experimentally observed chaotic dynamics. This system is particularly relevant to nonlinear studies as most other nonlinear systems cannot be described by theoretical models based on microscopic parameters.

INTRODUCTION

"Physicists like to think that all you have to do is say 'these are the conditions, now what happens next ?'" (R.P. Feynman).

Our inquiry into the laws of nature normally start with an approximation of linear equations of motion whose solutions can often be readily determined. However, as non-linear terms become larger the analytic solution is more complicated and under extreme cases the system displays a highly disordered or chaotic behavior. "Where classical science stops chaos begins". However, the response of the system is usually deterministic in that there exists a time dependent set of equations that can be used to calculate the future behavior of the system for a given set of initial conditions. The study of such systems is a relatively new field of "deterministic chaos" that has become important in recent years. It attempts to describe the irregular motion generated by the nonlinear systems whose dynamical laws are uniquely determined but whose time evolution is unpredictable as it depends on the precise knowledge of its previous history.

Our inquiry into the laws of nature often suffer from our inability to predict the time evolution of common systems like the atmosphere, the turbulent sea, the fluctuations of wild life populations or the oscillations of the heart or the brain. A general knowledge about the behavior of such systems can have important implications to society. How general is the chaotic response to systems observed in nature and does each system have a unique "route to chaos", or are there general classes of routes?

Several fundamental questions can be asked about chaotic systems for which some answers have been determined:

Can one predict whether or not a given system will display deterministic chaos? That is, can the differential equation that describes the behavior of the system be developed? With the aid of computer simulations, some systems have been quantitatively analyzed.

Can one specify the notion of chaotic motion more mathematically or develop quantitative measures for it? Chaotic motion can be described in terms of strange attractors which have characteristic fractal dimensions, Kolmogorov entropy and Lyapunov exponents.

What is the impact of these findings on other branches of physics as well as other fields of science and technology? Non-linear mechanics is quite interdisciplinary finding applications in fields such as physiology and ecology.

Does the existence of deterministic chaos imply the end of long term predictablity for some nonlinear systems or is it possible that one can still learn something from a chaotic response? That is, can the principle of predictability still be used in our efforts to understand these complicated systems? While the precise path may be unpredictable, the motion within the basin of attraction has some predictable features.

HIGH POWER EFFECTS IN MAGNETIC RESONANCES

In ferromagnetic and antiferromagnetic resonances, a number of geometrial arrangements of the pumping field are possible. In the normal configuration the pumping field is perpendicular to the external field (perpendicular pumping). In anisotropic materials a parallel configuration of the pumping field and the external field will parametrically excite the resonance when driven at twice the precession frequency (parallel pumping). A third resonance condition occurs under perpendicular pumping at twice the precession frequency. This parametrically excited resonance is known as subsidiary resonance.

At the normal resonance configuration, a linear response of the detected signal is observed at low pumping powers. At increased powers a periodic variation in the amplitude of the precession of the magnetization vector is observed. At subsidiary resonance or parallel pumped resonance a power threshold must be exceeded beyond which a steady state response is observed. At higher drive power, the periodic and aperiodic oscillations in the amplitude of the signal are observed.

Oscillations in the detected signal at the subsidiary resonance configuration were first reported in yttrium iron garnet (YIG) as "relaxation oscillation" by Hartwick, et. al. in 1961 [1]. See Figure 1. These auto-oscillations set in at power levels that exceed those associated with the critical power required for "Suhl instabilities" [2] a mechanism that breaks up the spatial uniformity of the magnetization by coupling the uniform precession mode to degenerate 'Cooper pair' spinwaves having wavevectors ±k. The nonlinear equations give rise to the "butterfly" curve shown in Figure 2 which plots the critical power versus magnetic field for high power effects in the magnetic garnets. The auto-oscillations are observed in regions I and II of Figure 2.

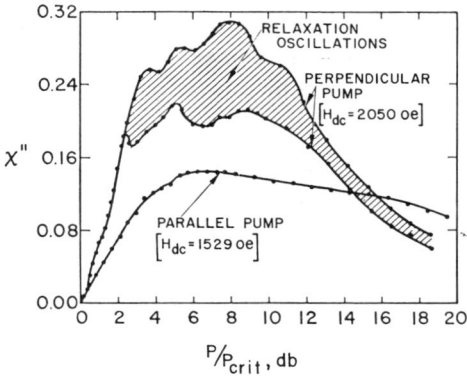

Figure 1. χ'' above the instability threshold in a YIG sphere under parallel and perpendicular pumping. (From reference 1.)

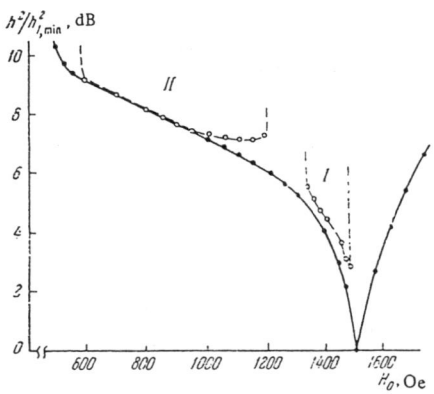

Figure 2. Critical power vs. static field for a YIG sphere (M∥<100>). Auto-oscillation occur in zones I and II. (From reference 4.)

With the progress in the study of the field of chaos in the last decade, interest in the nature of the "relaxation oscillations" shown in Figure 1 have stimulated renewed interest in these instabilities.

Spinwave instabilities in magnetic materials display a rich variety of nonlinear dynamic phenomena such as auto-oscillations, period multiplication, irregular period oscillations, intermittency, chaos and periodic windows. Chaotic behavior has been observed in magnetic systems at all of the resonance conditions for ferromagnetic materials and at parallel pumped resonance for antiferromagnetic materials.

The initial attempt to understand the auto-oscillatory behavior of the magnetization vector in YIG was made by L'Vov and Zakharov [3, 4]. Their results from experiments on YIG and theoretical models based on a microscopic Hamiltonian for the system form the foundations for the present theories used to explain the auto-oscillatory response.

Figure 3 shows the first report of relaxation oscillations in an antiferromagnetic material observed in 1974 [5] suggesting that antiferromagnetic materials would also show chaotic behavior.

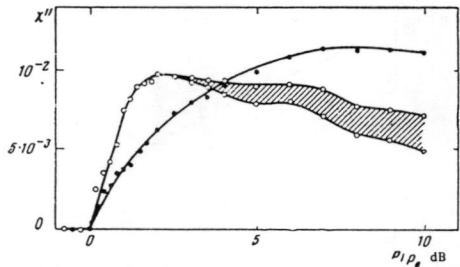

Figure 3. Beyond-threshold susceptibility for parallel pumped antiferromagnetic $MnCO_3$ for phenomena attributed to magnon pair generation (•) and nondegenerate magneto-elastic wave generation (o) showing auto-oscillations. (From reference 5.)

Magnetic systems have a number of advantages as a model system for investigating nonlinear dynamics.

Experimentally, the resonances are generally investigated in the gigahertz range and the relaxation times associated with the magnetization vector are generally in the microsecond to nanosecond range. Consequently, the relaxation oscillations are observed to occur in a frequency range between 0.1 MHz and 10 MHz. This range of frequencies is ideal for the use of electronic instruments and computers in the detection, storage and analysis of data and for studying various routes to chaos that might exist in nature. In addition, there are many material variables that can be adjusted in magnetic systems. It is easy to control and maintain the sample temperature; it is possible to vary the response frequency over 1 to 2 orders of magnitude in range; with a variation of the orientation of the sample in the magnetic field, the influence of various anisotropy energies can be investigated. The geometry of the sample is another important variable as the demagnetization factor influences the distribution of the spinwaves. Thus, it is possible to investigate "routes to chaos" in magnetic systems under a wide variety of conditions that can be readily controlled in the experiment.

In addition to these experimental advantages, detailed models can be developed from the microscopic Hamiltonian which describes the system. The resulting nonlinear equations can be used to simulate the experimental results.

EXPERIMENTAL STUDIES OF ROUTES TO CHAOS IN MAGNETIC SYSTEMS

Antiferromagnetic Materials

In the early work reported by L'Vov, et. al. [3], both periodic and stochastic behavior of the resonance signal at subsidiary resonance in YIG were reported as a function of the rf driving field. However, the early attempts to study auto-oscillations in terms of a theory of chaos were made for parallel pumped magnons in the antiferromagnet, $CuCl_2 \cdot 2H_2O$ [6]. These measurements were made at 8.9 GHz and at a temperature of 1.4°k in a TE_{101} cavity. Periodic oscillation phenomena were observed in the microwave power when the rf power was approximately 12 db above the threshold power observed for the spinwave instability. The fundamental frequency of the auto-oscillations occurs in the frequency range of 3 to 4 MHz and is observed to increase with the microwave power. This paper contains the first report of subharmonics attributed to period doubling type bifurcations in a magnetic system.

This work was followed by that of Waldner and his colleagues [7-8] on the antiferromagnetic material $(NH_3CH_2)_2CuCl_4$ at 9.36 GHz and a temperature of 1.7 K. These experiments also used the parallel pumping geometry but showed a transition to chaos by "irregular periods".

The next study was done in the antiferromagnetic material $CsMnF_3$ by Smirnov who reported an inhomogeneous distribution in the density of the excited magnons [9, 10]. As the absorbed power was increased, the density of spinwaves increased at the center of the sample. This increase was a result of a shift of the magnon dispersion relation with the increase of the precession angle of the magnetic vector. At sufficient power, a periodic redistribution of the density of the spinwaves was observed in the range of 2.5 MHz at 10 GHz drive frequencies.

Ferromagnetic Materials

The study of routes to chaos is much more flexible in the case of ferromagnetic systems. Using the very narrow linewidth garnet materials, it is possible to use spheres or thin film geometries, to investigate at either parallel or perpendicular pumping and to look either at the uniform mode conditions or the subsidiary resonance condition at twice the resonance frequency. In a magnetic sphere the uniform precession mode is degenerate with a large number of spinwaves which readily couple to the fundamental mode in the system. In the case of the thin film with the field oriented perpendicular to the plane, the resonance condition occurs at the bottom of the spinwave band and the coupling to spinwave modes is somewhat more restricted. The observations can be made either by changing the power at a given magnetic field or sweeping the magnetic field through the resonance at a given high power [11-14].

The initial chaotic analysis in the resonance of the garnets was obtained in spheres of gallium doped YIG investigated at 1.3 GHz by Gibson and Jeffries [11]. The Suhl instability was observed and the details of the onset of auto-oscillations in the magnetization were investigated at room temperature with a perpendicular pumping configuration. The auto-oscillations were found to be very dependent on the sample crystal axis orientation. As shown in Figure 4 the results of this initial experiment in YIG shows beautiful auto-oscillations of period 1, period 2 and period 4 in a bifurcation process with the system eventually ending up in a chaotic state.

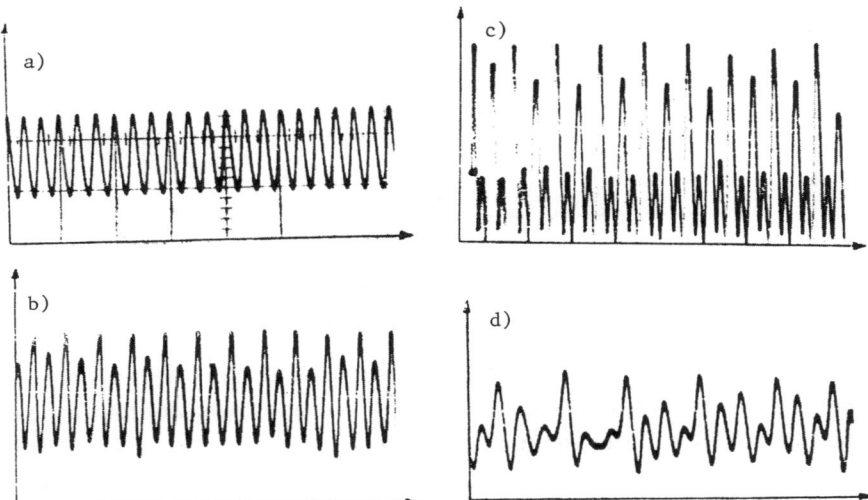

Figure 4. Auto-oscillations observed at 16 kHz in a 0.047 cm Ga-YIG sphere at the main resonance for increasing pumping power showing a period doubling route to chaos. (From reference 11.)

The observation of a subharmonic route to chaos in parallel pumped spinwaves in a YIG sphere was reported by Rezende et. al. [15-18]. The oscilloscope traces show period 1, period 2 and period 4 oscillations increasing to a chaotic response in the system with the fundamental frequency of the auto-oscillations in the range of 100 kHz.

Details of the regions of auto-oscillation for perpendicular pumped spinwave instabilities in an YIG sphere at subsidiary resonance were reported by Bryant, et. al. [12-14] and shown in Figure 5.

A very thorough investigation of the chaotic nature of the response in ferromagnetic platelettes of $(CH_3NH_3)_2CuCl_4$ under parallel pumping was

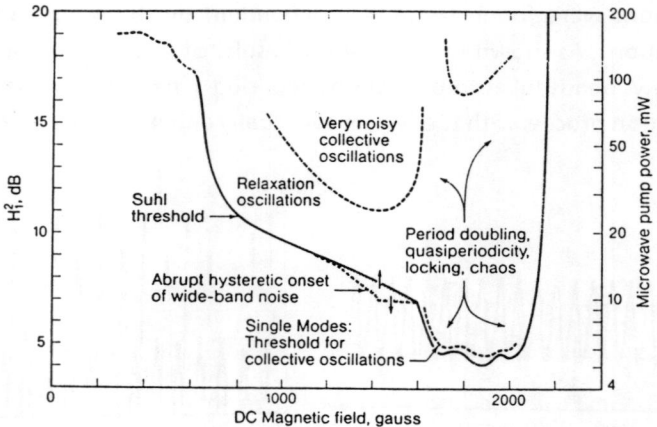

Figure 5. Region of oscillation at the subsidiary resonance in a YIG sphere. (From reference 14.)done by Yamazaki et. al. [19, 20].

Irregular spiking behavior similar to that reported by Waldner et. al. in an antiferromagnet [7, 8] was reported as well as chaotic behavior following a period doubling. The chaotic behavior was shown to be deterministic through the use of return maps and the signal was analyzed in terms of the fractal dimension and an estimate of the Kolmogorov entropy of the attractor. The largest Lyapunov exponent was found by iterating a one-dimensional function fit to a return map.

The first experiments in YIG films involved discs of approximately 1 micron thick and 3 mm diameter with observations at 1 to 2 GHz and 9.24 GHz [21, 22]. The experimental conditions involve perpendicular pumping of the main resonance at the perpendicular orientation. At the low frequency range, a slot line system was used to observe the response as a function of the power and the frequency of the system. The onset of relaxation oscillations is shown in Figure 6 and the power spectra indicatng a bifurcation process in the response on its route to chaos are shown in Figure 7.

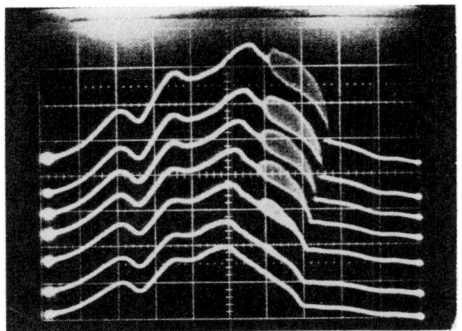

Figure 6. Onset of oscillation in a YIG film at the main resonance. The center frequency is 983 MHz the sweep span is 16 MHz and the static magnetic field is 598 G. The pumping power increases by 0.1 dB from the bottom trace to the second trace and increases at 0.5 dB intervals for the remaining traces.

The high frequency experiments were performed in a microwave cavity where the response of the system was investigated as a function of power versus the magnetic field. The oscillations observed at 9.24 GHz show a periodic oscillation with period doubling bifurcations on the route to a chaotic behavior. Observations were made in the temperature range from helium temperatures to room temperature. In Figure 8, the shift of the resonance condition to lower d.c. fields due to the "fold over" effect is shown as well as a map of regions of periodic and chaotic response in the parameter space of magnetic field vs. driving power. Figure 9 is another example of the bifurcation process observed in the film at constant power but as a function of the applied magnetic field.

THEORY

Two approaches are used to characterize the chaotic oscillations. The first approach is in terms of the strange attractors of the system. From the time-series data, a multidimensional phase space is developed from which it may be possible to determine the fractal dimensions and the metric

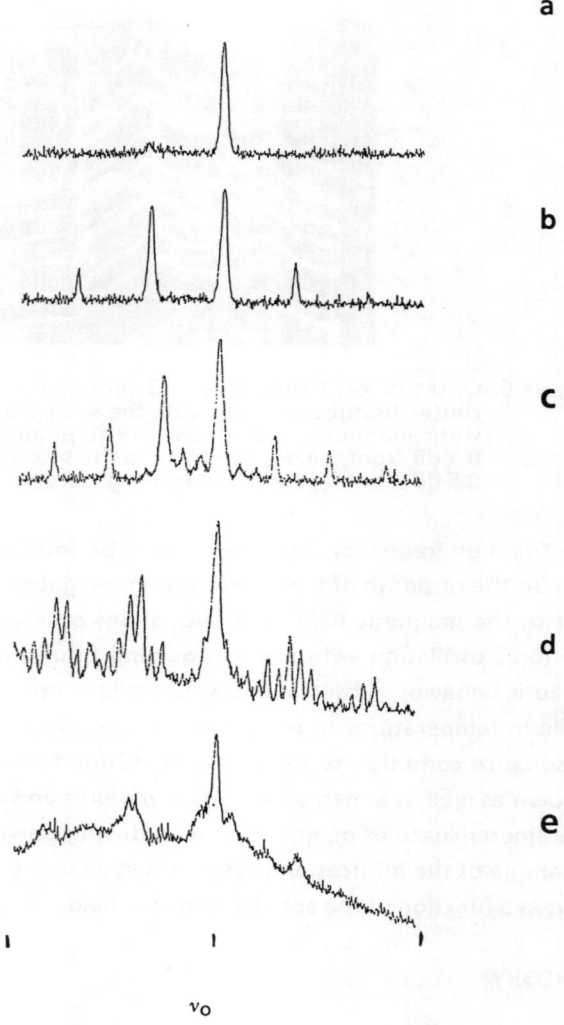

Figure 7. Power spectra for auto-oscillations in a YIG film at several power levels; a) 2.0 dB, b) 2.4 dB, c) 4.0 dB, d) 6.0 dB and e) 12.0 dB. The center frequency is $v_o = 1.02$ GHz and the width of the sweep is 20 MHz. $H_o = 623$ Oe.

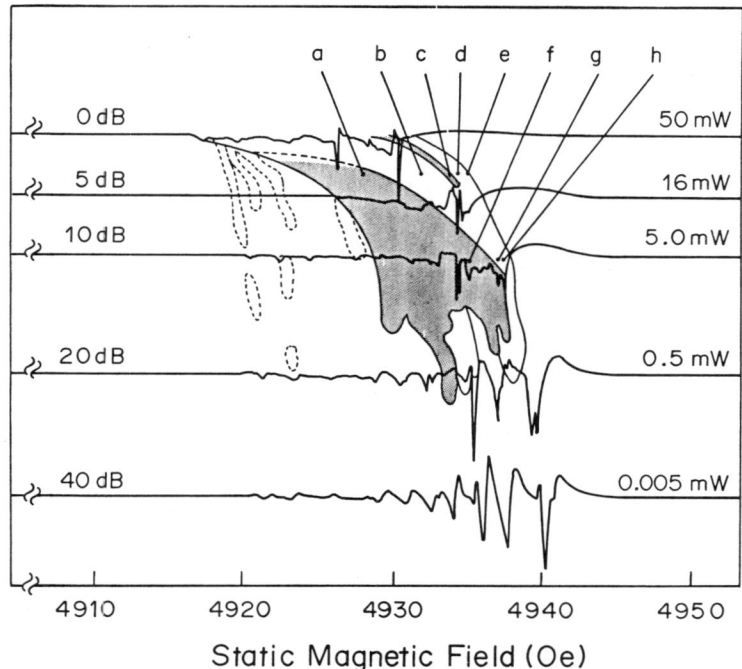

Figure 8. Regions of auto-oscillation and derivative FMR spectrum at several pumping powers in a YIG film at 9.24 GHz. The shaded areas indicate the presence of broadband noise in the power spectra. (From reference 22.)

entropy [10, 20, 23]. The second approach is the establishment of an equation of motion obtained from the microscopic Hamiltonian that describes the dynamical behavior of the magnetization vector [3, 4, 24-27].

STRANGE ATTRACTORS

To analyze the behavior of a system in terms of attractors, one must construct a phase space in which the attractor can exist. Ideally, this phase space would have axes corresponding to the amplitude and phase of each

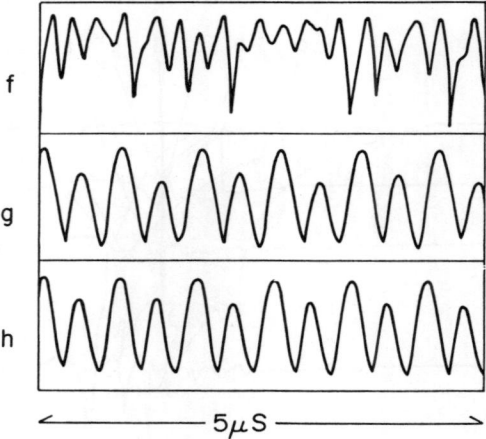

Figure 9. Oscillations observed at the points labelled f, g and h in Figure 8. (From reference 22.)

of the modes involved in the nonlinear behavior. Usually, however, one only has one experimental signal V(t). To imbed this signal in a multidimensional phase space, n-vectors can be created using the method of time delays where $x_1(t) = V(t)$, $x_2(t) = V(t + \tau)$,... $x_n(t) = V(t + (n-1)\tau)$. The time evolution of the system can be described by a point which moves along a curve in this phase space. A periodic oscillation will have a trajectory that forms a closed curve and is described as having a normal attractor. Phase portraits of unstable period 1 and period 2 signals are shown in Figure 10. For a chaotic response, the trajectory does not return unto itself, and indeed two paths having arbitrarily close initial condition will diverge exponentially from each other with time. Since the trajectory lies within a bounded region of phase-space the system has an attractor, but it is called a strange attractor due to the divergence of nearby trajectories.

Strange attractors, while imbedded in a phase space of integer dimension, have fractal characteristics which are characterized by a non-integer fractal dimension. The fractal dimension can be calculated from the experimental attractor using a method outlined by P. Grassberger and

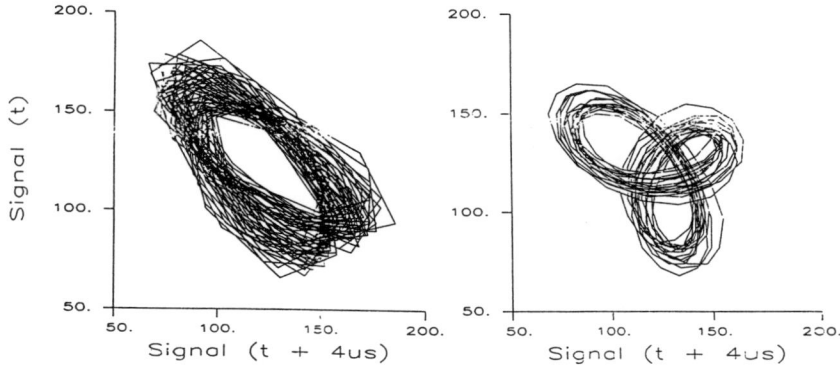

Figure 10. Phase portraits for period 1 and period 2 oscillations made by the method of time delays. (From reference 21.)

I. Procaccia [23]. This procedure was used by Yamazaki et. al. [20] to determine the fractal dimension of chaotic signals in parallel pumped ferromagnetic $(CH_3NH_3)_2CuCl_4$ as a function of pumping power. The fractal dimension was observed to saturate at high power as shown in Figure 11. The fractal dimension is related to the number of degrees of freedom that are required to describe the magnetization vector in real time.

Another quantitative measure of a strange attractor is the average rate of exponential divergence of nearby trajectories within the attractor. This average positive exponent is known as the Luyapunov Exponent. A negative exponent will characterize the average rate of exponential convergence of trajectories onto a normal attractor.

The sum of the Luyapunov exponents for a deterministic chaotic system is bounded above by a finite value, K, which is a measure of Kolmogorov entropy. For a regular periodic trajectory, $K = 0$ and for completely random noise, $K = \infty$. The magnitude of K is thus another quantitative measure of the degree of chaos associated with a strange attractor.

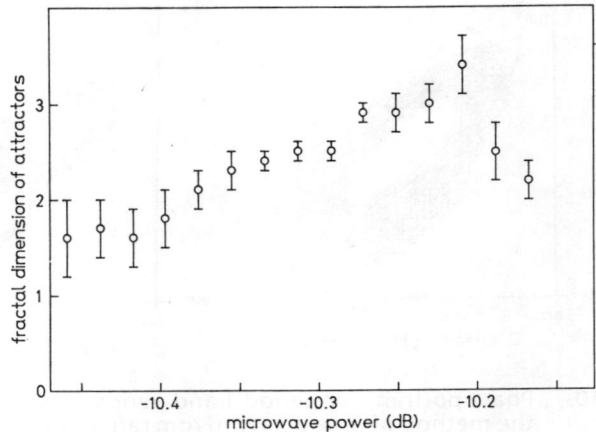

Figure 11. Fractal dimension of attractors as a function of driving power in parallel pumped ferromagnetic $(NH_3CH_3)_2CuCl_4$. (From reference 20.)

Numerical Methods

As part of the investigations of L'vov and Zakharov [3, 4] in yttrium iron garnet in the 70's, Zakharov developed an S-matrix theory to describe the behavior of the magnetization vector [4]. This theory involves a Hamiltonian which is diagonal in the ±k wavevectors or Cooper pair like spinwaves. This theory was later extended by Nakamura [24-26] to predict periodic doubling bifurcations which could be characterized by a strange attractor in the chaotic response of the system.

Two concerns remain about the application of the S-matrix theory to magnetic systems. 1) Can the S-matrix theory be developed by using the classical equations of motion of the spins and by passing the elaborate method of second quantization. 2) In the second quantization procedure, higher order terms in the spinwave operators are neglected. What is the validity of the S-matrix theory for strong excitations where large precession angles of the resonating spins are implied? In consideration of these concerns, Waldner has developed a stroboscopic approach which assumes that the static field terms are much larger than the internal effective fields

[27]. That is, that the magnetic resonance frequency is large in comparison to the aperiodic oscillations of the magnetization vector. As a result, the slow evolution of the standing waves are valid even for very large precession angles.

These theories give a set of coupled differential equations that simulate the nonlinear response of the magnetization vector due to the coupling between the uniform mode and the various spinwave modes in the magnetic system. The technique usually involves a two-mode system where the coupling to different spinwave pair modes can be varied and compared with the experimental results. Using this approach, many features of the observed variations in the magnetization vector have been obtained. A good example is given in Figure 12. However, since the system consists of many modes, limiting the system to the "dominant" nonuniform modes in the spinwave manifold is a severe limitation of the analysis. As a result, it has been argued that an entire manifold of modes pumped above threshold should be considered in the model.

Analytical Approach

Recently, Suhl [28-30] has developed an approximate analytical treatment for the succession of higher bifurcation regions in the response of the magnetization vector. The model includes a coupling not to one spinwave mode, but to all of the degenerate spinwave modes in the manifold. In an abstract space in which coordinates are given by the number of magnons in each mode, the linear resonance state is described by a fixed point along the axis of the uniform mode. As the pump power increases, that fixed point moves continuously along the uniform mode axis. When the critical threshold is reached a finite level of nonuniform spinwave densities are excited. At increasing powers, the increased density of the $k \neq 0$ spinwaves will cause the fixed point to trace out a path in the abstract space. The stability of this path with increasing signal power is examined by subjecting the points to infinitesimal deviations and examining the time development of these deviations, i.e., finding new linear modes. When the real part of the eigenvalues of the new modes

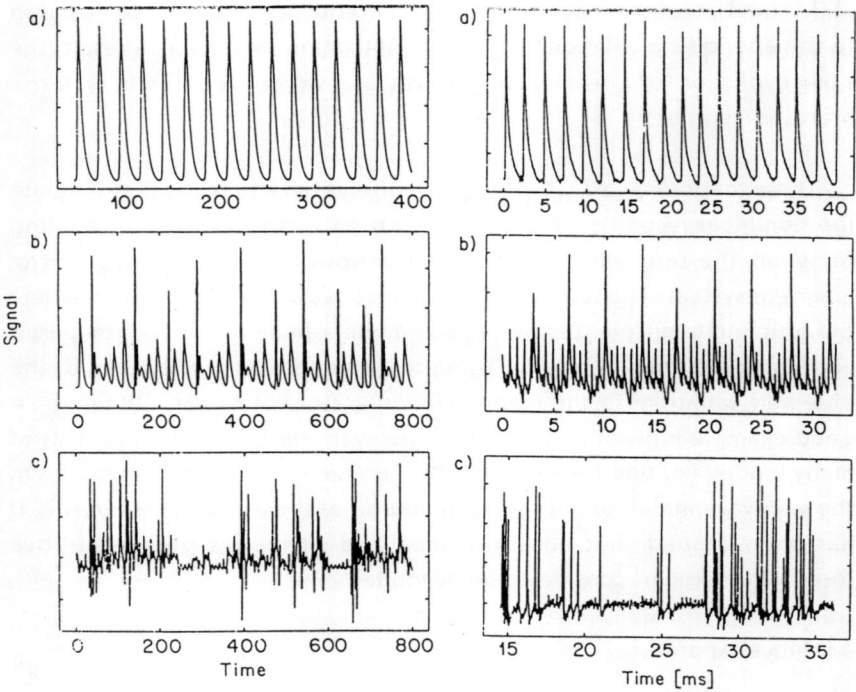

Figure 12. Two-mode simulations (left) and experimental signals (right) from M. Warden, Ph.D. Thesis, U of Zurich, (1987).

become positive, an instability occurs. This crossing is associated with a Hopf bifurcation and a new set of linear modes is required to describe a new path of the point in this phase space.

CONCLUSION

Spinwave instabilities in magnetic materials produce a rich variety of nonlinear responses of the magnetization vector such as auto-oscillations, period multiplication, irregular period oscillations, intermittency, chaos and periodic windows. The spinwave instabilities can be modeled by equations derived from a microscopic Hamiltonian with well known parameters. These nonlinear equations provide the basis for a theoretical framework to explain the dynamics observed in these systems.

The availability of these equations in magnetic systems make their study particularly relevant as most other physical systems cannot be so readily described by theoretical models. Based on numerical calculations for which the large number of spinwave modes have been truncated to a two-mode basis, qualitative agreement with observed results have been obtained. However, a reliable theory should also be able to predict the critical power and the frequency of the auto-oscillations. This may not be possible in a truncated model but the more comprehensive models become much more cumbersome.

Experimentally it would be desirable to obtain direct access to the magnon modes that are involved in the dynamics. This may be possible by using thin ferromagnetic films for which the number of degenerate modes is limited and/or by using Brillouin scattering techniques in conjunction with the magnetic resonance.

Clearly, magnetic materials are excellent systems for nonlinear studies of routes to chaos and chaotic responses. But even here, further theoretical and experimental effort is needed.

REFERENCES

1. Hartwick,T.S., Peressini, E.R. and Weiss, M.T., "Subsidiary Resonance in YIG", J. Appl. Phys. 32, 223S (1961).

2. Suhl, H., "The Theory of Ferromagnetic Resonance At High Signal Powers", J. Phys. Chem. Solids., 1, 209 (1957).

3. L'vov, V.S., Musher, S.L. and Starobinets, S.S., "Theory of Magnetization Self-Oscillations on Parametric Excitation of Spin Waves", Zh. Eksp. Teor. Fiz. 64, 1074 (1973), [Sov. Phys. JETP, 37, 546 (1973)].

4. Zakharov, V.E., L'vov, V.S. and Starobinets, S.S., "Spin-Wave Turbulence Beyond the Parametric Excitation Threshold", Usp. Fiz. Nauk, 114, 609 (1974), [Sov. Phys.-USP., 17, 896 (1975)].

5. Ozhogin, V.I. and Yakubovskii, A. Yu., "Parametric Pairs in Antiferromagnet With Easy Plane Anisotropy", Zh. Eksp. Teor. Fiz., <u>67</u>, 287 (1974), [Sov. Phys. JETP, <u>40</u>, 144 (1975)].

6. Yamazaki, H., "Oscillations and Period-Doubling of Magnon Amplitude Under Parallel Pumping in Antiferromagnetic $CuCl_2 2H_2O$", J. Phys. Soc. Japan, <u>53</u>, 1155 (1984).

7. Waldner, F., Barberis, D.R. and Yamazaki, H., "Route to Chaos By Irregular Periods: Simulations of Parallel Pumping in Ferromagnets", Phys. Rev. A, <u>31</u>, 420 (1985).

8. Waldner, F., Badii, R., Barberis, D.R., Broggi, G., Floeder, W., Meier, P.F., Stoop, R., Warden, M. and Yamazaki, H., "Route to Chaos by Irregular Periods", J. Mag. Magn. Mtls., <u>54-57</u>, 1135 (1986).

9. Smirnov, A.I., "Periodic Redistribution of the Density of Parametrically Excited Spin Waves in an Antiferromagnet", Zh. Eksp. Teor. Fiz. <u>88</u>, 1369 (1985), [Sov. Phys. JETP, <u>61</u>, 815 (1985)].

10. Smirnov, A.I., "Onset of Chaos in the Redistribution of Parametrically Excited Magnons", Zh. Eksp. Teor. Fiz. <u>90</u>, 385 (1986), [Sov. Phys. JETP, <u>63</u>, 222 (1986)].

11. Gibson, G. and Jeffries, C., "Observation of Period Doubling and Chaos in Spin-Wave Instabilities in Yttrium Iron Garnet", Phys. Rev. A <u>29</u>, 811 (1984).

12. Bryant, P., Jeffries, C. and Nakamura, K., "Spin-Wave Nonlinear Dynamics in an Yttrium Iron Garnet Sphere", Phys. Rev. Letts. <u>60</u>, 1185 (1988).

13. Bryant, P., Jeffries, C. and Nakamura, K,. "Spin-Wave Turbulence", Conference Proceedings to be published in Nuclear Physics B, (1987).

14. Bryant, P.H., Jeffries, C.D. and Nakamura, K., "Spinwave Dynamics in a Ferrimagnetic Sphere", 1-38, Phys. Rev. A (to be published).

15. Rezende, S.M., de Aguiar, F.M. and de Alcantara Bonfim, O.F., "Order and Chaos in Ferromagnetic Spin Wave Instabilities", J. Mag. Mag. Mats., <u>54-57</u>, 1127 (1986).

16. de Aguiar, F.M. and Rezende, S.M., "Observation of Subharmonic Routes to Chaos in Parallel-Pumped Spin Waves in Yttrium Iron Garnet", Phys. Rev. Letts., <u>56</u>, 1070 (1986).

17. Rezende, S.M., de Alcantara Bonfim, O.F. and de Aguiar, F.M., "Model for Chaotic Dynamics of the Perpendicular-Pumping Spin-Wave Instability", Phys. Rev. B <u>33</u>, 5153 (1986).

18. de Aguiar, F.M., Azevedo, A. and Rezende, S.M., "Strange Attractors in Spin-Wave Chaotic Dynamics", Phys. Rev. B (submitted).

19. Yamazaki, H. and Warden, M., "Observations of Deterministic Chaos of Parallel-Pumped Magnons in Ferromagnetic $(CH_3NH_3)_2CuCl_4$", J. Phys. Soc. Japan 55, 4477 (1986).

20. Yamazaki, H., Mino, M., Nagashima, H. and Warden, M., "Strange Attractor of Chaotic Magnons Observed in Ferromagnetic $(CH_3NH_3)_2CuCl_4$", J. Phys. Soc. Japan 56, 742 (1987).

21. Wigen, P.E., Doetsch, H., Ming, Y., Baselgia, L. and Waldner, F., "Chaos in Magnetic Garnet Thin Films", J. Appl. Phys. 63, 4157 (1988).

22. McMichael, R.D. and Wigen, P.E., "Field and Power Dependence of Auto-Oscillations in YIG Films", J. Appl. Phys. (to be published).

23. Grassberger, P. and Procaccia, I. "Measuring the Strangeness of Strange Attractors", Physica 9D, 189 (1983).

24. Nakamura, K., Ohta, S. and Kawasaki, K., "Chaotic States of Ferromagnets in Strong Parallel Pumping Fields", J. Phys. C: Solid State Phys. 15, L143 (1982).

25. Ohta, S. and Nakamura, K., "Power Spectra of Chaotic States in Driven Magnets", J. Phys. C: Solid State Phys. 16, L605 (1983).

26. Nakamura, K., Okazaki, Y. and Bishop, A.R., "Periodically Pulsed Spin Dynamics: Scaling Behavior of Semiclassical Wave Functions", Phys. Rev. Letts. 57, 5 (1986).

27. F. Waldner, "A 'Stroboscopic Model' For Non-Linear Ferromagnetic Resonance Phenomena", J. Phys. C: Solid State Phys. 21, 1243 (1988).

28. Suhl, H. and Zhang, X.Y., "Spatial and Temporal Patterns in High-Power Ferromagnetic Resonance", Phys. Rev. Lett. 57, 1480 (1986).

29. Zhang, X.Y. and Suhl, H., "Theory of Auto-Oscillation in High Power Ferromagnetic Resonance", Phys. Rev. B, (to be published).

30. Suhl, H. and Zhang, X.Y., "Spin-Wave Instabilities and Their Revival by Nonlinear Mechanics", preprint (to be published).

PHYSICS OF MATERIALS FOR VERTICAL BLOCH LINE DATA STORAGE DEVICES

Warwick Clegg
Department of Electrical Engineering
University of Manchester, M13 9PL
UK

ABSTRACT

Vertical Bloch line memory is an emerging technology which encodes information in the micromagnetics of a domain wall. The concepts of the storage method are outlined and some simple calculations made to determine the likely density and speed of memory operation. Experimental results are also presented to show the effects of planar fields on VBL velocities.

1 INTRODUCTION

High density memory devices which combine solid state ruggedness with other attributes such as non-volatility and radiation hardness are an attractive proposition for a number of uses. At present the practical respresentative of this class of memory is the magnetic 'bubble' device which can be bought from a small number of manufacturers in Japan, the USA and France. Despite its early promise the bubble device has never achieved a mass market, basically because of its bit cost. It is also somewhat slow, a raw bit rate before multiplexing of 100 kHz being typical. It should be noted, however, that this latter drawback is not related to physical limitations of the wall mobility in the epitaxial magnetic garnet layer supporting the bubbles, but rather to the practical problems of driving the inductive coils which provide 60 Oe or so of rotating field over a volume of few cm^3. Using current-drive propagation through apertured conductor layers [1], data rates of 20 MHz have been attained [2], though this drive method is inappropriate over a full device because of the power it consumes.

Bit cost is related to many factors such as the cost of starting materials, the number and complexity of process steps, packaging and testing. Other things remaining equal then increases in storage density should reduce the bit cost proportionally. At the time of writing 4 Mbit single chip bubble devices are available and at least one manufacturer has the serious intention to bring ion-implanted 16 Mbit single chip devices to production status within the next year. The prospect of eventually another four-fold increase in chip capacity cannot be entirely ruled out but it seems most unlikely. Densities beyond this latter are inconceivable without a fundamental change in the mechanism of storage. This is precisely what vertical Bloch line (VBL) memory offers [3] and explains why there is growing interest to understand the complex relationship between basic physics, materials and technology which could eventually enable successful manufacture of devices with densities of a few hundred Megabits per cm².

2 DEVICE CONCEPTS

Since the operating principles of VBL devices are not yet widely known, this Section aims to provide a suitable background of understanding.

2.1 Data Storage

The essential advance in storage density lies in the encoding of data not at the one-bit-per-domain level, but rather at the sub-domain level within the micromagnetics of the domain wall structure. This concept is not entirely new, Schwee [4] having proposed in 1972 that cross-ties in domain walls in thin Permalloy films could represent binary data. Also in the mid-1970's the bubble lattice file [5] was the subject of an intense development effort, and since the lattice would not sustain vacancies it was necessary to use different bubble wall states to encode binary information. The encoding scheme still used, however, one domain per bit, even though such domains could be packed significantly more closely together in a self-stabilised hexagonal array and gain a 10:1 packing advantage over the normal requirement to space bubbles by 4 diameters to avoid mutual interaction.

Figure 1 shows in simplified form an arrangement of vertical Bloch lines within the wall of an elongated stripe domain in an epitaxial garnet such as is used for bubble devices. (The nomenclature of the line is somewhat disputed and it should more correctly be termed a Néel line, particularly when viewed through the centre of the line perpendicular to the plane of the wall. However, for historical reasons, in what follows it will be called a Bloch line).

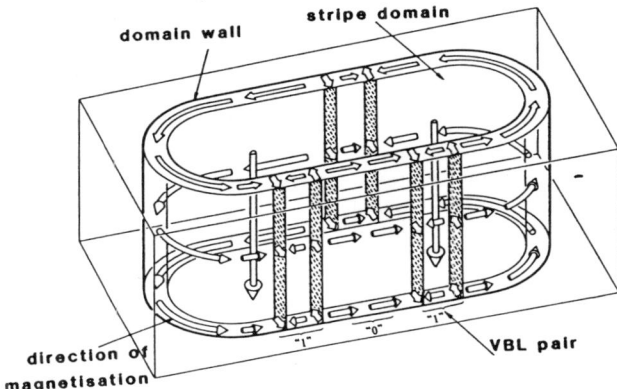

Figure 1   Schematic of the vertical Bloch line memory concept [6]

The VBL's shown are said to have negative handedness since the azimuthal angle of magnetisation at the wall centre, Φ, rotates anticlockwise through the VBL as the domain wall is followed round in a clockwise sense. A pair of negative VBL's (or a pair of positive VBL's) constitute a stable entity since they form a winding pair with a total rotation angle of 2π. At long range they exhibit a mutual magnetostatic attraction resulting from their opposite so-called σ-charges which arise because of the head-on or tail-on arrangement of the wall magnetisation at the VBL as shown in Figure 2. At short distances the exchange interaction provides a stabilising repulsion for a winding pair, though not of course for an unwinding pair (a positive and a negative VBL) so the latter will tend to coalesce and mutually annihilate. The choice of a pair of negative VBL's, rather than positive VBL's, as the preferred means of information storage is because of the nature of the Bloch line injection process as explained in Section 2.3.

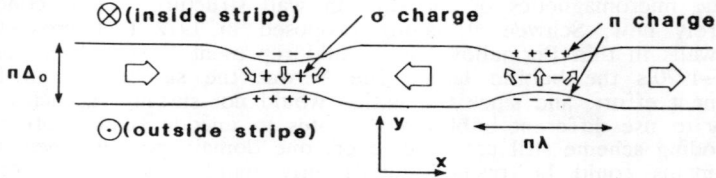

Figure 2 Two negative VBL's will attract at long range because of their opposite σ-charge. Note constriction in the wall width at the VBL arising from the π-charge.

The concept of data storage is now evident, ie that the presence or absence of a pair of VBL's at defined locations around the domain wall denotes binary information. The stripe domain is thus analogous to the 'minor loop' in a bubble memory. Among the means under consideration to confine the stripe domains the most favoured at present is a partial or full groove in the epilayer [7]. Partial grooves orthogonal to the stripes could provide stable bit positions for VBL pairs, as could ion-implanted strips [8] or overlaid strips of planar-magnetised high coercivity film such as CoPt [9]. The effect of the latter is to produce a periodic in-plane field potential well structure, within which to trap the VBL pair. As well as these passive stripe and bit stabilisation methods, conducting overlays providing local fields are also extensively used in test devices.

An illustration of the arrangement described so far is seen in Figure 3. Note the positioning of an isolated negative VBL at the stripe head and the presence of a small (1-2 Oe) planar field directed as shown. The stripe is easier to initialise in this configuration (sigma stripe) and also the single VBL aids transfer of data around the stripe head. The planar field weakly stabilises the isolated Bloch line at the head and also controls the distance between VBL's in a pair since it adds an extra term to the energy of the configuration.

2.2 Propagation

The favoured method for moving VBL pairs from one potential well to the next uses a pulsed vertical bias field. The primary effect of the pulse is to cause motion of the stripe domain wall perpendicular to the wall plane. This motion give rise to a gyrotropic force on each VBL in

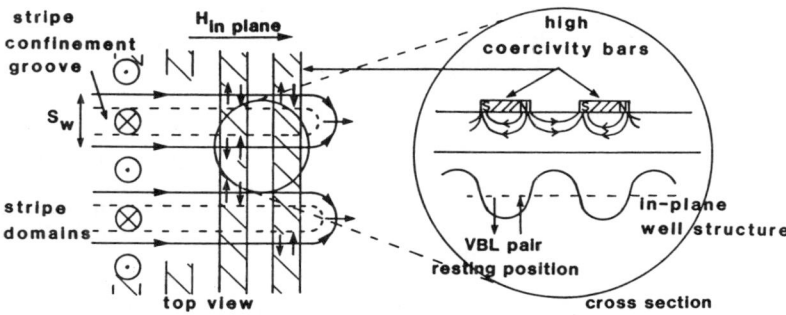

Figure 3 Stripe confinement by grooving and a periodic in–plane field potential well structure to define VBL pair bit positions.

the pair having a magnitude per unit length of [10]

$$F_g = \frac{2\pi M}{\gamma} \dot{y} \qquad (1)$$

where the symbols have their usual meaning. To achieve unidirectional propagation, rather than oscillation, the pulse must be shaped so that $\dot{y}$ (forward) is greater than $\dot{y}$ (backward) and therefore F_g (forward) will be sufficient to move the pair over the potential barrier to the next well whereas F_g (backward) will be insufficient to reverse the motion.

A pulse of the form shown in Figure 4 is appropriate for the 5 μm stripe width garnet material in current use for VBL device development. A relatively long constant drive field is found necessary to allow the wall to reach an equilibrium state [11]. Reverse propagation could be achieved by using a slow risetime followed by a fast fall time, or by reversing the drive field polarity. The former method would be preferred since the domain wall displacement would remain unidirectional from its equilibrium position.

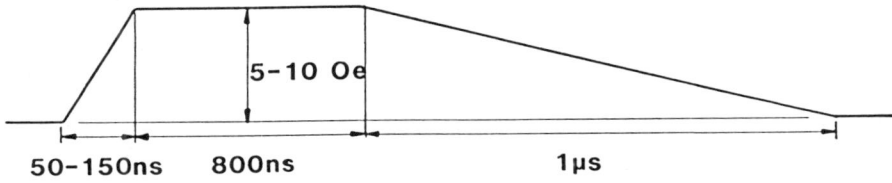

Figure 4 Typical perpendicular pulse field shape for propagation of VBL pairs in 5 μm stripe width garnet films with 1–2 Oe potential well depths.

Other methods could be used for propagation, for instance a travelling well structure produced by multiple overlaid conductors. However, for reasons of fabrication complexity and on-chip power dissipation these are less attractive than a small externally-pulsed bias field.

2.3 The Write Operation

To write data requires the controlled insertion of a pair of negative Bloch lines. One method to accomplish this is to create a Bloch curve at the stripe head and cause it to punch through to the opposite film surface. This process may be understood with reference to Figure 5 which shows the rotation of the azimuthal angle of magnetisation, Φ, through the film thickness for a Bloch wall. The form of the curve arises from the effects of the wall's demagnetising field and stray fields from the magnetisation on both sides of the wall. Near the film surfaces the latter is sufficiently strong ($|H_y| > 8M$) that the wall takes on a Néel form beyond the critical points

$$Z_1 = \frac{h}{1+e^2} \quad \text{and} \quad Z_2 = \frac{he^2}{1+e^2} \qquad (2)$$

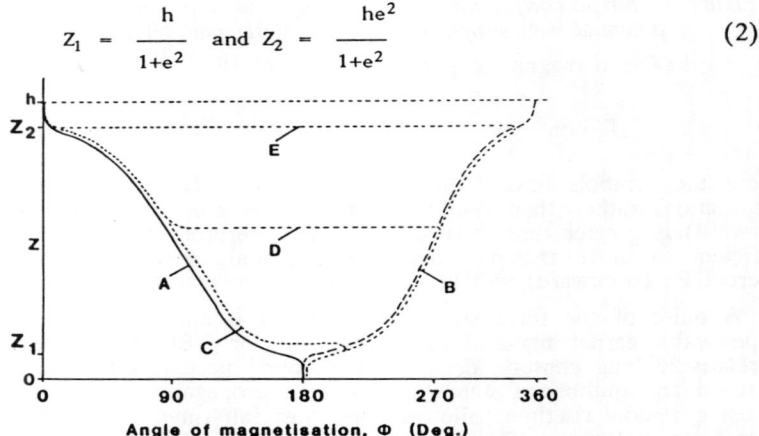

Figure 5 Azimuthal angle of magnetisation Φ at centre of wall through the epilayer thickness. A and B show opposite wall chiralities, C shows Bloch curve nucleation and D and E propagation of the curve through the film.

As the stripe head is caused to expand in a perpendicular field H_z, then Φ will precess around H_z and if the wall chirality (sense of twist) and direction of H_z is such that Φ increases then the wall configuration will change to that shown at (C) in the Figure. This corresponds to the nucleation of a Bloch curve bounding a small region of reversed wall chirality. If the head continues to expand then the Bloch curve moves through the film (D) and for a critical wall velocity [12]

$$V_p \triangleq 24 \, \gamma \, A \, / \, \left[hK^{\frac{1}{2}} \right] \qquad (3)$$

it can reach the opposite critical point (E) and 'punch through' leaving the stripe head with a statically-stable region of reverse chirality wall flanked by two vertical Bloch lines. The latter, however, comprise one positive and one negative VBL and this unwinding pair could recombine. It is therefore necessary to move the negative line round to the flank of the stripe and 'chop' the positive line from the head by causing the two side walls to merge. These operations can be accomplished by local fields from overlaid conductors. The chopping operation itself produces a further negative VBL at the stripe head and therefore reproduces the original starting condition. This is the reason for the choice of negative VBL's as the information carriers. The sequence of operations is shown in Figure 6.

Control of the data pattern written into the stripes can be achieved by positioning a 'major line' adjacent to the stripe heads, which can carry a pattern of bubble domains and spaces as determined by a conventional bubble generator. Wherever a bubble domain on the major line is positioned adjacent to a stripe head it will block the latter's expansion and so prevent injection of new VBL's. Another method of writing data, which injects VBL's into the flank of the stripe domain, has also been demonstrated experimentally [13].

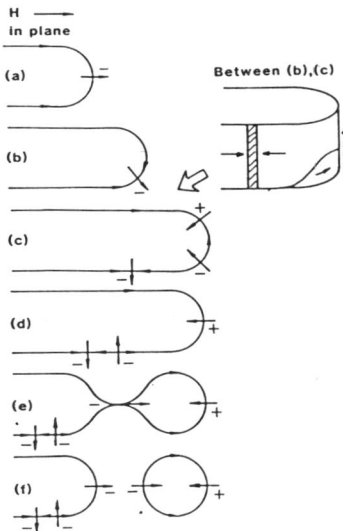

Figure 6 The VBL write process by expansion and chopping the stripe head. Stripes can be packed with a centre–to–centre spacing of 2 S_W, but bubbles on the major line need 4 S_W separation. This mis–match could be solved by an interleaved arrangement where alternate stripe domains communicate with a major line at opposite ends.

2.4 The Read Operation

This operation is based on the relative ease with which the end of a stripe domain may be chopped to produce a bubble, depending on the presence or absence of a VBL pair at the storage position nearest the stripe head. A number of manipulations are necessary, carried out by overlaid conductors, the principal ones of which are shown in Figure 7.

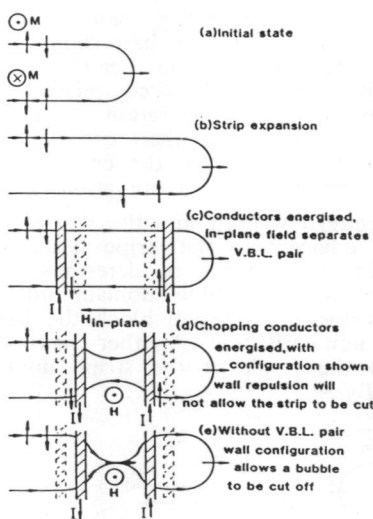

Figure 7 The read process can be accomplished by testing whether the stripe sidewalls will merge to allow a bubble domain to be cut off.

In the sequence of operations shown, the stripe head is first expanded to position only the lower VBL pair between the conductors. When energised the outer conductors provide a small local planar field which separates the Bloch lines. The inner conductors are then pulsed in such a sense as to cause a local perpendicular collapsing field. If a VBL pair were present then the sidewalls will form a 'winding' pair with a total rotation of 2π. In this case exchange energy resists the merging of the walls and so if the chopping current is carefully controlled, no bubble is produced. On the other hand, with no VBL pair present the sidewalls form an 'unwinding' pair with zero overall rotation and no exchange barrier to chopping. Note that in this case the situation corresponds to (f) in Figure 6 which automatically leaves a negative Bloch line at the stripe head. In either case, therefore, the read operation is non-destructive.

The pattern of bubbles and gaps replicating the (inverse of the) stored data is then moved along the major line to a bubble detector for electrical readout. With slight modification the above operation could be made destructive so as to erase existing data to allow an overwrite.

Having outlined the way in which VBL memory devices are likely to operate, we may now consider some of the physical aspects of their operation in more detail.

3 PHYSICAL CONSIDERATIONS

Of particular importance to a memory device are the likely density of storage and the speed of access. The former is governed by the interactions between Bloch lines and will be considered first.

Unlike bubbles, between which there is a repulsive interaction, the force between neighbouring pairs of vertical Bloch lines arising from the σ-charges is attractive. The spacing between VBL pairs should be the minimum possible whilst ensuring that propagation margins are insensitive to bit patterns. Bit potential wells should provide sufficient restoring force to overcome attractive effects but should also be minimised otherwise drive power will be excessive.

We begin by considering a simple point–charge model for a VBL to determine more easily the effect of material parameters on density, and then the effect of σ-charge distribution through the film thickness will be accounted for. Assume first that the Bloch line separates wall regions whose magnetisation is oppositely directed, then its magnetic moment is 2M. Given that the wall width is $\pi\Delta_0$ and film thickness is h, then we have a total magnetic charge m, of

$$m = 2 M \pi \Delta_0 h \qquad (4)$$

3.1 Attractive Force Between VBL Pairs

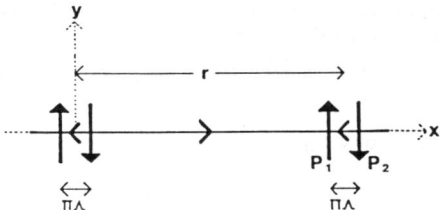

Figure 8   VBL pairs in adjacent bit positions

With reference to Figure 8, the effective field resulting at a point P, distance x from the centre of the VBL pair is

$$H_p = 2M\pi\Delta_0 h \cdot \left[\frac{1}{\left[x + \frac{\pi\Lambda}{2}\right]^2} - \frac{1}{\left[x - \frac{\pi\Lambda}{2}\right]^2} \right] \qquad (5)$$

$$\cong - \frac{4\pi M\Delta_0 \pi\Lambda h}{x^3} \qquad \text{if } \pi\Lambda \ll x \qquad (6)$$

The attractive force on a pair of VBL at distance r (bit spacing) from the initial pair is

$$F_{att} = 2 \pi M \Delta_0 h (H_{p1} - H_{p2}) \qquad (7)$$

$$= 2 \pi M \Delta_0 h \pi \Lambda \frac{\delta H}{\delta x}\bigg|_{x = r} \qquad (8)$$

which from (6) is

$$F_{att} = \frac{24\pi^4 M^2 \Delta_0^2 \Lambda^2 h^2}{r^4} \qquad (9)$$

Using standard substitutions of

$$\Delta_0 = \frac{\ell}{2Q} \quad , \quad \Lambda = \frac{\ell}{2\sqrt{Q}} \quad , \quad \ell = \frac{\sqrt{AK}}{\pi M^2} \quad \text{and} \quad Q = \frac{K}{2\pi M^2}$$

we are left with

$$F_{att} = 3\pi^3 \frac{A \ell^2 h^2}{Q^2 r^4} \tag{10}$$

3.2 Potential Well Restoring Force

Consider the potential wells produced, for example, by a set of strips of high coercive field material magnetised in the plane above the film thickness. The wells have the form

$$H_w = H_{max} \cos\left[\frac{2\pi x}{r}\right] \tag{11}$$

where H_{max} is a constant and r is the bit period. The restoring force on a VBL pair in the potential well is thus

$$F_R = 2 \pi M \Delta_0 h (H_w (P_1) - H_w (P_2)) \tag{12}$$

which is equivalent to

$$F_R = 2 \pi M \Delta_0 h \pi \Lambda \left. \frac{\delta H_w}{\delta x} \right|_{x = r} \tag{13}$$

$$= 2 \pi^2 M \Delta_0 \Lambda h \frac{2\pi}{r} H_{max} \sin\left[\frac{2\pi x}{r}\right] \tag{14}$$

giving the maximum values of F_R at $x = r/4$, $x = 3r/4$, of value

$$F_{Rmax} = 2 \pi^2 M \Delta_0 \Lambda h 2 \pi \frac{H_{max}}{r} \tag{15}$$

which after standard substitutions reduces to

$$F_{Rmax} = \frac{\pi^2}{Q} \sqrt{2\pi A} \frac{H_{max} \ell h}{r} \tag{16}$$

3.3 Drive Field Gyrotropic Force

The gyrotropic force on a VBL has the form [10]

$$F_G = \frac{2\pi}{\gamma} M h \dot{y} \tag{17}$$

Now $\dot{y}$ is difficult to determine, in that a hard domain wall containing many VBL's will move more slowly than a soft wall containing widely-spaced VBL's or none at all. Taking a worst case of a hard wall, to establish a lower limit for F_G, the appropriate wall velocity is

$$\dot{y} = \alpha \, \Upsilon \, \vartriangle_o H_z \qquad (18)$$

where H_z is the sum of perpendicular components of effective fields, ie demagnetising, wall curvature and applied fields. The lower limit for gyrotropic force is therefore

$$F_{Gmin} = 2\pi M \alpha \vartriangle_o h H_z = \sqrt{\frac{2\pi A}{Q}} \; \alpha \, h \, H_z \qquad (19/20)$$

3.4 Material Parameters

In comparing the three forces shown in Eqns 10, 16 and 20 we wish to minimise F_{att} so that the potential well magnitude H_{max} is minimised, so reducing the drive field while maintaining the bit period r low. Of the parameters in question ϱ can be taken as $S_w/8$ and the exchange constant A can be assumed as $\approx 2 \times 10^{-7}$ erg/cm. From the form of the equations it can be seen that a low h ($\sim 3-4 \, \varrho$), high Q (~ 4) material will reduce the relative effect of F_{att} and that a high α ($\sim 0.1-0.2$) increases F_{Gmin}. This latter would be expected as large damping decreases domain distortion and VBL oscillatory effects during propagation. The damping should not be too large since [14)]

$$v_{VBL} \; \alpha \; \frac{1}{\alpha} \; v_{wall} \qquad (21)$$

ie large damping increases the required wall displacement and thus H_z.

In the following estimation of density we take material parameters: $\alpha = 0.1$, $Q = 4$, $h = 4 \, \varrho$, $\varrho = S_w/8$, $A = 2 \times 10^{-7}$ erg/cm.

3.5 Storage Density

It is clearly desirable to have $F_{Rmax} \gg F_{att}$ for stable bit positions and successful propagation of random bit patterns. If we decide to make $F_{Rmax} = 10 \, F_{att}$ we can combine Eqns 10 and 16 to obtain a relationship between potential well depth (H_{max}) and bit period : stripe width ratio (r/S_w) as shown in Fig 9.

The broken line shows the choice of potential well depth which corresponds to the 'knee' of the curves. (The position of this line is derived from previous numerical calculations of VBL propagation, Ref [11)] and unpublished work). This enables an estimation of the bit period and the corresponding storage density (= $1/(r \times S_w)$ bits/cm^2) varies approximately as $1/S_w^2$ as plotted in Fig 10.

The simple calculation may be refined in a number of ways to give an indication of whether the densities shown are optimistic or pessimistic. For instance, the charge at the Bloch line may be re-calculated to take account of wall narrowing and the distribution of magnetisation directions in the line through the material thickness as follows.

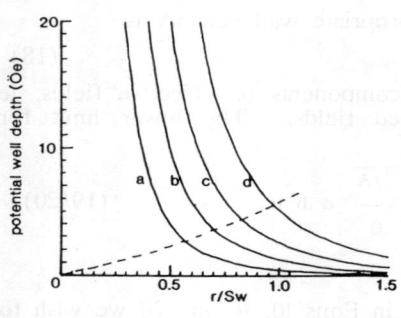
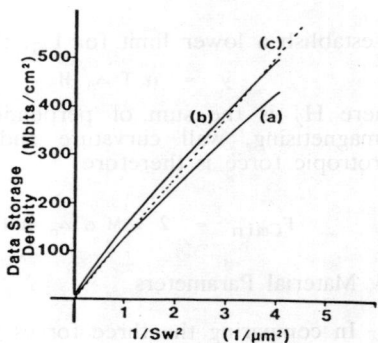

Figure 9 Potential well depth as a function of bit period : stripe width ratio for stripwidths of a) 5µm, b) 2µm, c) 1µm, d) 0.5µm.

Figure 10 Density as a function of $1/S_W^2$ for a) simple charge model, b) 2-D VBL model, and c) $r = 0.8\ S_W$.

In the presence of a Bloch line the wall width contracts due to π-charge demagnetising effects and others. The wall width parameter is given by [15]

$$\Delta = \Delta_o \left[1 + \Delta_o^2 \left[\frac{\delta\Phi}{\delta x}\right]^2 + \frac{\sin^2\Phi}{Q} \right]^{-\frac{1}{2}} \quad (22)$$

The azimuthal angle of wall magnetisation, Ψ, measured from the wall plane has the form

$$\Psi(x) = \pm 2 \arctan \exp \left[\frac{x}{\Lambda_o}\right] \quad (23)$$

hence

$$\frac{\delta\Phi}{\delta x} = \frac{\delta\Psi}{\delta x}(x) = \frac{1}{\Lambda_o} \quad (24)$$

Noting that at the Bloch line centre $\sin^2\Phi = 0$, we find

$$\Delta = \Delta_o \left[1 + \frac{\Delta_o^2}{\Lambda_o^2} + \frac{1}{Q}\right]^{-\frac{1}{2}} = \Delta_o \left[1 + \frac{2}{Q}\right]^{-\frac{1}{2}} \quad (25/26)$$

$$\cong 0.82\ \Delta_o \text{ for } Q = 4 \text{ which is reasonable} \quad (27)$$

Now consider the variation of Ψ through the thickness and the corresponding effect on the σ-charge magnitude. The magnetic moment at a given height z is

$$m = 2M \cos\Psi = 2M \sqrt{1 - \frac{1}{4}(\ln(Z/(h-z)))^2} \quad (28)$$

The thickness-averaged moment is therefore

$$m = \frac{2M}{h} \int_{\frac{h}{1+e^2}}^{\frac{he^2}{1+e^2}} \sqrt{1 - \frac{1}{4}\left[\ln\frac{z}{h-z}\right]^2} \, dz \qquad (29)$$

Note that the integral need only be taken between the limits of the two critical points since beyond these points there is no moment. This has solution

$$m = 1.28 \, M \qquad (30)$$

These results have been verified by a 3-Dimensional numerical simulation of a pair of VBL's [16]. Including wall-narrowing effects, then

$$m \doteq \pi M \Delta_0 \text{ per unit line length} \qquad (31)$$

This reduced value will give a corresponding reduction in the attractive force between VBL pairs as follows.

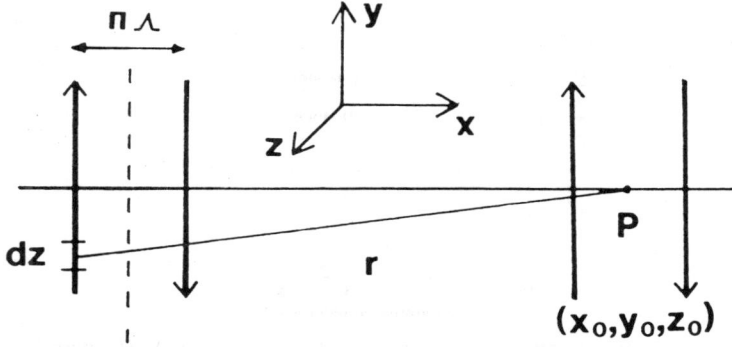

Figure 11 Geometry for calculation of 2-D line model of VBL pairs in adjacent bit positions.

From the geometry of Figure 11, then the in-plane field at a distance x_0 from the centre of a VBL is

$$H_x = m \int_0^h \left\{ \frac{x_0 + \frac{\pi \Lambda}{2}}{\left[\left(x_0 + \frac{\pi \Lambda}{2}\right)^2 + (z_0 - z)^2\right]^{3/2}} - \frac{x_0 - \frac{\pi \Lambda}{2}}{\left[\left(x_0 + \frac{\pi \Lambda}{2}\right)^2 + (z_0 - z)^2\right]^{3/2}} \right\} dz \qquad (32)$$

The thickness-averaged field at P is

$$\langle H_x \rangle = \frac{1}{h}\int_0^h H_x \, dz_o = \frac{m}{h}\left[\sqrt{1 + \frac{h^2}{\left[x_o + \frac{\pi\Lambda}{2}\right]^2}} - \sqrt{1 + \frac{h^2}{\left[x_o - \frac{\pi\Lambda}{2}\right]^2}}\right] (33/34)$$

From Eqn 8, the force of attraction on an adjacent VBL pair is

$$F_{att} = m \pi \Lambda \left. \frac{\delta H}{\delta x} \right|_{x_o} \quad (35)$$

which may now be calculated as

$$F_{att} = (\pi\triangle_o Mh)^2 \pi\Lambda \left[\frac{1}{\left[x_o - \frac{\pi\Lambda}{2}\right]^2 \left[\left[x_o - \frac{\pi\Lambda}{2}\right]^2 + h^2\right]^{\frac{1}{2}}} - \frac{1}{\left[x_o + \frac{\pi\Lambda}{2}\right]^2 \left[\left[x_o + \frac{\pi\Lambda}{2}\right]^2 + h^2\right]^{\frac{1}{2}}}\right] (36)$$

This result is plotted as a function of separation distance in Figure 12, along with the result of the point-charge model for comparison.

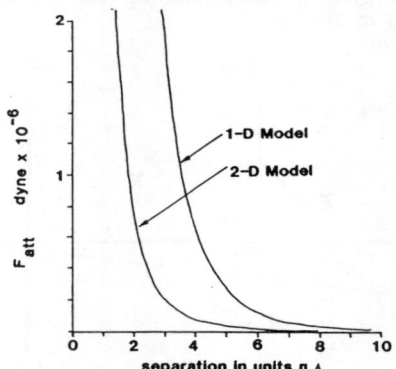

Figure 12 Comparison of the attractive force between VBL pairs for 1–D and 2–D models.

The reduced attractive force indicates that data bits may be rather more closely packed together than the original calculation indicated, and gives a corresponding improvement in density as marked on Fig 10. Also marked is a line corresponding to $r = 0.8 \, S_w$ which is seen to be a reasonable approximation to the bit spacing at any density. This is somewhat analagous to the required bit spacing in a bubble device though in this lattter case the interaction is repulsive. Note that in comparison to a bubble device memory cell of area 16 S_w^2, the VBL device of cell area 0.8 S_w^2 should show a twenty-fold improvement in density.

3.6 Speed of Operation

In estimating the speed of operation of a VBL memory, two areas need to be considered, the minor loop region where data is propagated as VBL pairs circulating around a stripe domain wall, and the input/output track where data is transfered to the stripe domains in the form of conventional bubbles propagating in a current driven circuit. Kryder [17] has calculated the time to access a current driven circuit as being proportional to track length (independent of bit density) and so we need only study VBL propagation in a domain wall. We assume VBL motion is by pulsed perpendicular bias field with fast rise and slow fall times as previously described and calculate the relaxation time of the domain wall to show the likely speed of propagation.

Figure 13 illustrates the domain model used in the calculation, where each domain wall is represented by a line current of magnitude $2Mh$ and the domain array of equilibrium spacing S_W, is disturbed by a wall displacement of $\pm y$.

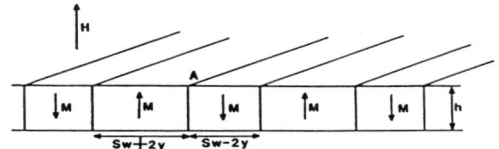

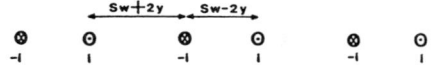

$i = 2Mh$

Figure 13   Equivalent current model of a domain array of equilibrium spacing S_W perturbed by a wall displacement of $\pm y$.

For small wall displacement, y, the wall restoring force, F_R, is:

$$F_R = -2M \left[\frac{4Mh}{S_w + 2y} - \frac{4Mh}{S_w - 2y} \right] = -\frac{32 M^2 h}{S_w^2} y \qquad (37/38)$$

Wall velocity, $\dot{y}$, is given by; $\dot{y} = \mu H_{eff}$ \hfill (39)

which implies a damping force, F_D, of:

$$F_D = \frac{2M}{\mu} \dot{y} \qquad (40)$$

In dynamic equilibrium ($\ddot{y} = 0$)

$$\frac{2M}{\mu} \dot{y} = - \frac{32 M^2 h}{S_w^2} y \qquad (41)$$

therefore,

$$y = Y_{max}\, e^{-t/\tau_C} \qquad (42)$$

which is an exponential decay of time constant τ_C where,

$$\tau_C = \frac{S_w^2}{16 Mh\, \mu} \qquad (43)$$

where $M = \sqrt{\dfrac{2QA}{\pi}} \dfrac{8}{S_w}$, $h = S_w/2$, and $\mu = \dfrac{\alpha \gamma S_w}{16Q}$ for a wall containing Bloch lines
and taking $Q \triangleq 4$, $A \triangleq 2.63 \times 10^{-7}$ erg/cm, $\alpha \triangleq 0.1$,
$\gamma \triangleq 1.77 \times 10^7$ Oe^{-1} s^{-1}
Therefore $\tau_c = 6.9 \times 10^{-4}\, S_w$ (44)
where the stripe width, S_w, is to scale from 5 μm to 0.5 μm. A time of $3\tau_c$ is allowed for the system to reach equilibrium.

For no VBL motion in the reverse direction, wall motion is quasi-static, hence

$$\dot{y} = \text{constant} = Y_{max}/T \qquad (45)$$

Taking into account the VBL gyrotropic force, Eqn (17), and the potential well restoring force, Eqn (15),
we can say that for no VBL motion, $F_G < F_R$
Therefore

$$T > \dfrac{Y_{max}\, \Gamma}{2\pi^2\, \gamma\, \Delta_0 \Lambda\, H_{max}} > 5.8 \times 10^{-6} \left[\dfrac{Y_{max}}{S_w\, H_{max}}\right] \qquad (46/47)$$

We know that, $Y_{max} \propto S_w$, therefore, $T \propto 1/H_{max}$ (48/49)
where H_{max} is the depth of the VBL pair potential well.
Typically, $T \triangleq 1.5$ μs/Oe, and for this discussion assume $T \triangleq 3\, \tau_c$ (50)
noting that this is greater than that which will cause reverse motion.

Substituting the material parameters we find the clock period, 6τ, scales from 400 kHz ($S_w = 5$ μm) to 4 MHz ($S_w = 0.5$ μm). As the number of bits per unit length of minor loop is proportional to S_w, the time taken to circulate VBL pairs once around a loop is proportional to the loop length and is independent of density. Coupled with the results of Kryder [17] we can therefore conclude that the access time of a VBL memory is independent of device density and depends only on architectural considerations (size and number of major and minor tracks and physical dimensions of the device).

4 EXPERIMENTAL MEASUREMENTS OF VBL VELOCITY

4.1 Procedure

A single vertical Bloch line in a straight domain wall may be directly driven by an in-plane field. In this case the gyrotropic coupling acts in the inverse sense and causes the wall to kink slightly. This local wall deflection provides a means of observing the position of the VBL during dynamic experiments. If the position of the VBL is reset to a consistent starting point between applications of the pulsed planar field, then the motion during the pulse can be observed stroboscopically and the VBL velocity measured [18].

The test structure and waveforms are shown in Figure 14. An S=0 stripe domain is held in the elongated potential well generated by a confinement current. The in-plane field coils 1 and 2 generate a field of 6.2 Oe/A parallel to the stripe domain with rise and fall times of 60 ns. A field in the positive x direction (H_{ip} reset) moves VBL1 and VBL2 together while a field in the negative x direction (H_{ip} move) separates them. When current is passed down the VBL hold conductor in the direction indicated, an in-plane field is generated beneath the conductor in the negative x direction. This is sufficient to prevent VBL2 passing beneath the hold conductor during the reset pulse.

The rise and fall times of the current in the VBL hold conductor are longer than that required to generate VBLs and are typically of the order of a few μs. The in-plane field reset pulse is approximately twice the length of the move pulse so VBL2 always returns to the VBL hold conductor.

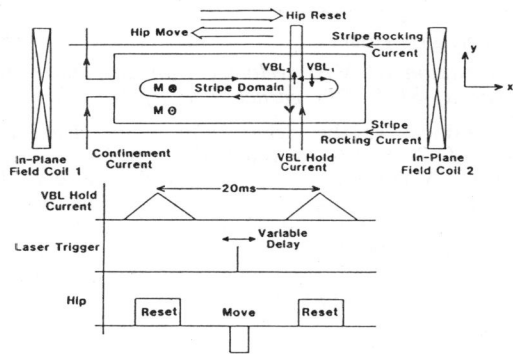

Figure 14 Schematic of test structure and test pulses

Illumination of the sample is by a 90 kW Rhodamine 6G dye laser, pumped by a pulsed 600 kW flowing N_2 laser. The 5ns FWHM pulses at 580 nm are coupled fibre-optically to an Orthoplan microscope and a SIT camera records the events. Image processing by frame grabber and pc improve the visibility of the small wall deflections and overall the accuracy of wall velocity measurements is 2m/s.

Two materials have been examined, their parameters being:

h = 4.15 μm, ℓ = 0.61 μm, Q = 5.5, α = 0.116, $4\pi M$ = 174 Gauss, γ = 1.79 x 10^7 1/Oe.s, A = 1.98 x 10^{-7} erg/cm (SEW 303 # 22, kindly supplied by NEC Ltd) and h = 2.03 μm, ℓ = 0.66μm, A = 1.0 x 10^{-7} erg/cm, $4\pi M_s$ = 111 Gauss, α = 0.12, Q = 5.4, K_u = 2650 erg/cm³, γ = 1.77 x 10^{-7} 1/Oe.s (SMYG # 33, kindly supplied by LETI).

4.2 Results

Figure 15a shows the variation in VBL velocity with in-plane drive field amplitude for the SEW 303 material. The VBL did not move at drive fields below 2.0 Oe. This is therefore the VBL coercive field. At drive fields between 2.0 Oe and 3.0 Oe the VBL did not always move and therefore the velocity could not be measured in this range. The velocity between these drive fields is found by extrapolating the linear increase in velocity with drive field amplitude. The initial VBL mobility is 10 m/s/Oe.

The plot shows a peak velocity of 41.3 m/s at a drive field of 5.6 Oe.

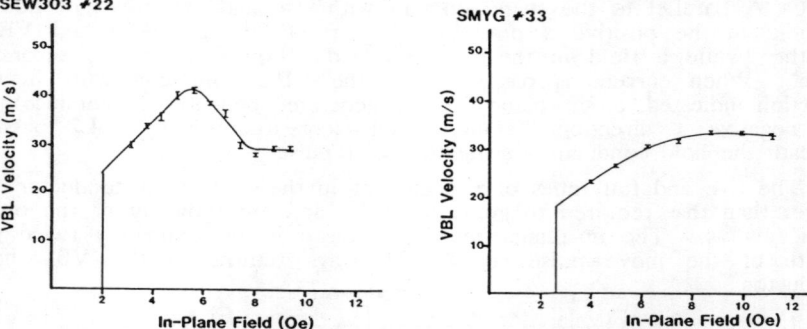

Figure 15 The variation of VBL velocity with amplitude of in-plane drive field for a) SEW 303 garnet, and b) for SMYG garnet

The velocity reduces as the drive field is increased above this until a saturation velocity of 29.5 m/s is reached at 8 Oe. Figure 15b shows a similar plot for the SMYG material. The VBL coercive field in this case is 2.6 Oe and the initial mobility is 5.5 m/s/Oe. The peak velocity for this material is 34 m/s which occurs at 10 Oe drive field. The VBL mobility first begins to drop at 6 Oe drive field. No peak in the VBL velocity was observed. In neither of the materials was there evidence of overshoot after the termination of the drive pulse.

4.3 Discussion

The stripe rocking method was used to determine the in-plane field necessary to nucleate horizontal Bloch lines (HBLs). In the SEW 303 material an S=1 strip domain begins to reduce in length and rocking amplitude, indicating that HBLs have been nucleated, when the in-plane field is increased above 5.6 Oe. This value of in-plane field corresponds to the drive field which gives peak VBL velocity. In the SMYG material it becomes apparent that HBLs have nucleated when the in-plane field is increased above 6.2 Oe. This field corresponds to the drive field above which the VBL velocity departs from a linear increase with drive field.

The likely stripe domain wall structure for the three cases $H_{ip} < H_{crit}$, $H_{ip} = H_{crit}$ and $H_{ip} > H_{crit}$ is shown in Figure 16 where H_{ip} is the in-plane drive field and H_{crit} is the critical field necessary to nucleate HBLs. The arrows in the Figure show the direction of magnetisation at the centre of the wall. In Figure 16a the in-plane drive field is less than the field necesary for HBL nucleation. The VBL moves with a velocity V_1. This is the wall structure which will occur in the linear part of the VBL velocity vs in-plane drive field graph. Figure 16b shows the wall magnetisation structure when the in-plane drive field is equal to the HBL nucleation threshold. Bloch loops have formed in the wall to the right of the VBL to reduce the Zeeman energy of the wall. The VBL has unwound with the lower Bloch loop which accounts for its distorted appearance. The VBL moves with a velocity V_2. The wall structure accounts for the departure of the VBL velocity from a linear increase

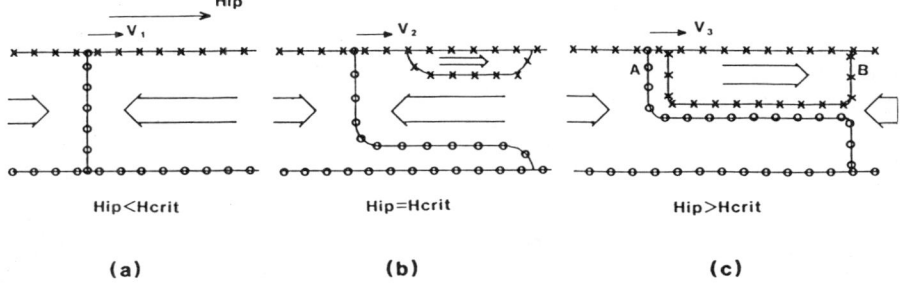

Figure 16 Likely wall structures for increasing in—plane field.

with drive field. As the upper Bloch loop expands a retarding force will act on the VBL which results in the drop in VBL mobility. As the drive field is increased above H_{crit} (Figure 16c) the horizontal Bloch lines move to the middle of the film thickness, reaching equilibrium at $\frac{1}{2}(h \pm \pi \Lambda)$ where $\pi\Lambda$ is the Bloch line width. The VBL retarding force will increase as the horizontal Bloch lines get nearer to the centre. This explains why the velocity breakdown observed in the SEW 303 material occurs. When the horizontal Bloch lines reach the centre of the film thickness, velocity breakdown stops as the VBL retarding force is then at its maximum value. It is expected that if the drive field is increased above this then the velocity may begin to increase, although at a much reduced rate. The approximate theoretical field required to move the horizontal Bloch lines to the centre of the film thickness (excluding exchange energy between HBL's) is given by [15].

$$H_{ip} = 4(2\pi A)^{\frac{1}{2}}/h \qquad (51)$$

For the SEW 303 material this gives a field of 10.75 Oe which is in reasonable agreement with the drive field at which velocity breakdown stops, 8.5 Oe. No velocity breakdown was observed in the SMYG material. This is due to the sample being thin, with Bloch line width being a significant fraction of the material thickness. It is estimated that the Bloch line width is 0.5 μm, the material thickness being 2.03 μm. In the SEW 303 material the Bloch line width is 0.4 μm and the material thickness 4.15 μm. Velocity breakdown will therefore be much more noticeable in the thicker material.

With reference to Figure 16c, the thickness averaged total rotation of the wall's azimuthal angle of magnetisation at A is equal to π and at B equal to 0. Gyrotropic deflection of the wall as the Bloch lines move is therefore only seen at A. All measurements of VBL position are therefore of position A. When the drive field was terminated, position A was not seen to move. The nucleated HBLs must therefore unwind to their respective surfaces leaving a VBL at A.

The high velocities observed in both materials are promising for fast operation of a memory device. However, the planar fields parallel to the domain wall must be maintained below H_{crit} or an instability in the vertical form of the VBL will result.

5 PERFORMANCE SUMMARY AND CONCLUSIONS

From the physical considerations presented in Section 3, there is every reason to anticipate that VBL memory will offer very high storage density together with a better data transfer rate than bubble memory. At the practical level, however, then assuming continued successful progress in developing the necessary on-chip control functions with operating margins having acceptable overlap, the limiting aspect of the technology will be the lithographic requirements for bit-position definition.

If the minimum bit spacing calculated in Section 3.6 is relaxed somewhat to $r = S_W$, and if the bit-defining structure (e.g. overlaid strips of high-coercivity film) has unity mark-space ratio, then the minimum lithographic feature size is seen to be $S_W/2$. A very welcome aspect of the bit lithography is its extremely simple nature - there are no complex geometries as for a bubble device, only a highly regular linear grating structure. With efficient planarisation of the stripe grooves and the use of extra capacity on-chip to allow for a degree of redundancy to cope with defects, there are good grounds for inferring that fabrication yields should be acceptable even with sub-micron bit spacing.

For a byte-organised chip using interleaving (9 read/write lines) then the likely operational characteristics, with strip width as a parameter, are as follows.

Table 1 VBL memory, likely characteristics

$S_W(\mu m)$	5	2	1	0.5
Min. Lithography (μm)	2.5	1	0.5	0.25
Density (Mbits/cm²)	4	25	100	400
Clock Rate (MHz)	0.4	1	2	4
Data Transfer Rate (MHz)	3.2	8	16	32

A design study [19] of a prototype chip of this general layout concludes that with on-chip power dissipation of 1W, its average access time (with unidirectional propagation) would be around 1ms. This latter figure is an order of magnitude faster than bubble memory and nearly two orders of magnitude faster than rotating magnetic memory.

6 REFERENCES

1) Bobeck A H, Blank S L, Butherus A D, Ciak F J and Strauss W, "Current-Access Magnetic Bubble Circuits", Bell Syst Tech J $\underline{58}$, No 6, pp 1453-1540, 1979.
2) Breed D J and Verhulst A G H, "High Frequency Magnetic Bubble Devices", Microelectronics J $\underline{12}$, No 5, pp 15-18, 1981.
3) Konishi S, "A New Ultra-High Density Solid State Memory: Bloch Line Memory", IEEE Trans Mag, MAG-$\underline{19}$, No 5, pp 1838-1840, 1983.
4) Schwee, L J, "Proposal on Cross-Tie Wall and Bloch Line Propagation in Thin Magnetic Films", IEEE Trans Mag, MAG-$\underline{8}$, No 5, pp 405-407, 1972.
5) Voegeli O, Calhoun B A, Rosier L L and Slonczewski J C, "The Use of Bubble Lattices for Information Storage", AIP Conf Proc $\underline{24}$, pp 617-621, 1974.
6) Konishi S and Hidaka Y, "1 Gbit/cm² High Density Solid State Memory is Realised: Bloch Line Memory", Nikkei Electronics No $\underline{323}$, pp 141-167, 1983 (in Japanese).
7) Klein D and Engemann J, "Bloch Line Memory: Dams for Stripe Domain Confinement", J App Phys, $\underline{58}$, No 8, pp 4071-4073, 1985.
8) Hidaka Y and Matsutera H, "Bloch Line Stabilisation in Stripe Domain Wall for Bloch Line Memory", IEEE Trans Mag, MAG-$\underline{20}$, No 5, pp 1135-1137, 1984.
9) Suzuki T et al, "Chip Organisation of Bloch Line Memory", IEEE Trans Mag, MAG-$\underline{22}$, No 5, pp 784-789, 1986.
10) Slonczewski J C, "Theory of Bloch Line and Bloch Wall Motion", J App Phys $\underline{45}$, No 6, pp 2705-2708, 1974.
11) Matsuyama K and Konishi S, "Computer Simulation of Domain Wall and Vertical Bloch Line Motion in a Bubble Garnet Film", IEEE Trans Mag, MAG-$\underline{20}$, No 5, pp 1141-1143, 1984.
12) Malozemoff A P, Slonczewski J C and De Luca J C, "Translational Velocities and Ballistic Overshoot of Bubbles in Garnet Films", AIP Conf Proc $\underline{29}$, pp 58-64, 1976.
13) Hidaka Y, "A Bloch Line Pair Generator Using the Flank Wall Near the Stripe Domain Head for the Bloch Line Memory", Jap J App Phys $\underline{25}$, No 3, pp L228-L231, 1986.
14) Konishi S, Matsuyama K, Chida I, Kubota S, Kawahara H and Ohbo M, "Bloch Line Memory, and Approach to Gigabit Memory", IEEE Trans Mag, MAG-$\underline{20}$, No 5, pp 1129-1134, 1984.
15) Malozemoff A P and Slonczewski J C, "Magnetic Domain Walls in Bubble Materials", Chapter 4, Academic Press, 1979.
16) Ronan G, Clegg W and Farrow G S D, "Bloch Line Interaction and its Consequences for VBL Memory Operation", IOP Solid State Physics Conf, Reading, Conf Proc Paper MG10, 1985.
17) Kryder M, "Current Accessed Magnetic Bubble Devices: Projections of Performance", IEEE Trans Mag, MAG-$\underline{17}$, No 5, pp 2392-2400, 1981.
18) Heyes N, Ronan G and Clegg W, "Stroboscopic Observation of High Speed VBL Motion Using an Optical Sampling System", IEEE Trans Mag, MAG-$\underline{24}$, No 2, pp 1741-1743, 1988.
19) Ronan G, Farrow G and Clegg W, "The Effects of Domain Wall Relaxation Times on VBL Memory Operation", IOP Solid State Physics Conf., London, Conf. Proc. Paper MG12, 1986.

MULTILAYERS : PAST, PRESENT AND FUTURE

I.B. PUCHALSKA and H. NIEDOBA

CNRS, Laboratoire de Magnétisme et d'Optique des
Solides, 92195 Meudon Cedex, France

ABSTRACT

The twolayered magnetic structure is the basis of multilayered system. Domain structure and magnetization process in strongly and weakly coupled two identical 80Ni-20Fe films separated by C are shown as an example. The interesting characteristics of strongly coupled films like low coercivity, regular ripple and high velocity of the walls are stressed.
Depending on components, multilayers can display modulated or sharp superlattice structure. Magnetic properties depend on type of the elements, their thickness and arrangement. The main magnetic characteristics : magnetization, anisotropy and coercivity in the different multilayered systems are discussed.

1. INTRODUCTION

The idea of the multilayers as a new material for information storage was suggested by Oakland and Rossing[1] and next by Broadbend[2] and Fuller and Sullivan[3]. The first observations of interaction of domains walls in superimposed films were made by Puchalska and Spain[4] using electron microscope in the Lorentz mode.

The use of electron microscope for the observation of magnetic domains and walls in thin films relies upon deflection of the electrons by magnetization within the sample and the image of the wall corresponds to the boundary between regions which are magnetized differently. Observations in a single film are characterized by an alternation of light and dark wall images in accordance with alternation of the magnetization orientation within domains (Fig. 1).

Thus, in the case of superimposed films adjacent wall images which are both light or dark indicate that such walls are within different films (Fig.2).

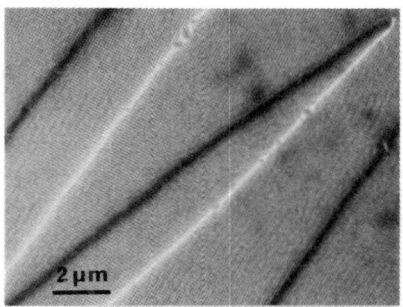

Fig.1. Lorentz micrograph of domain walls in 80Ni-20Fe single layer

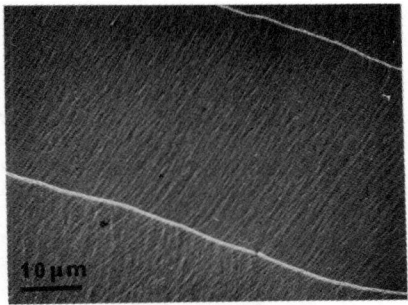

Fig.2. Lorentz micrograph of domain walls in twolayered 80Ni-20Fe film

The theoretical model of the magnetostatic coupling between two magnetic layers (A_1 and A_2) separated by a nonmagnetic material (B) is due to Louis Néel[5]. Discussing the model of the interaction between the two magnetic layers Néel introduced the terms of the coupling field H_m and superficial coupling energy $E_s = H_m \cdot M \cdot d$ where M is magnetization, d thickness of the layer A_1. The coupling field H_m originates from the layer A_1 and acts on the layer A_2 (Fig.3) giving rise in some cases to an asymmetric hysteresis. The coupling originates from exchange and

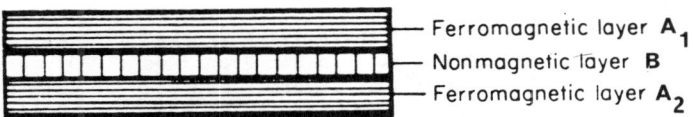

Fig.3. Schematic representation of twolayered film.

magnetostatic effects. According to the author the second one is the principal reason of the phenomemon. It may originate from :
- demagnetizing field of the two layers
- interactions of magnetic walls
- coupling through a non-magnetic layer (B) due to magnetic poles at the films surfaces.

The interest in the multilayers was twofold : 1) fundamental - concerning magnetostatic and exchange coupling, interactions of domains and walls and 2) industrial - concerning memory devices.

In the sixties single permalloy films were considered to be a good medium for memories based on magnetization reversal. However this process was inappropriate due first to wall creep provoked by demagnetizing field (e.g.[6]) and second to coinciding current reversing (wiping of) the information bits. It was found that in the strongly coupled films wall-creep could be solved (see e.g.[7]) which led to the idea of replacing the single layers by multilayers. More sophisticated applications[3] (more "sophisticated" in those days but realised at the present time) were supposed to use the weakly coupled films composed of soft and hard magnetic layers to realise a non destructive readout memory in which the magnetization of a "readout film" can be oriented by magnetostatic field (coupling field H_m) of a "storage film" after every readout cycle.

In those years many interesting papers were written on this subject (see e.g. Yelon review[8]) and possible applications were proposed but few were ever realized. In the seventies, the interest for ferromagnetic multilayers diminished considerably since the bubble materials became fashionable as to be the solution for memories. Some interesting papers appeared on garnet sandwich films (e.g.[9] and Fig. 4) however due to difficult technology this material had never been used for applications.

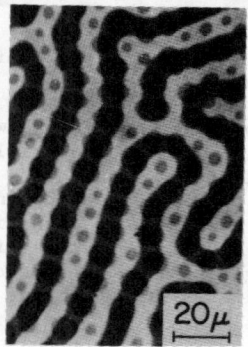

Fig.4. Stripe domains and bubbles in garnet sandwich films (Eu Ga YIG 4μm/GGG 4μm/ EuGa YIG 8μm). Courtesy of P.J. Grundy

At the beginning of the eighties there was a renewed interest in the multilayers and during the ICM session in San Francisco (1985) more than twenty papers were presented on several aspects of layered structures. At present the multilayers are one of the most interesting subjects in physics of magnetic films and surfaces. Multilayers composed of many thin (on atomic scale) layers, coupled films (metallic or amorphous) several hundreds Angströms thick, surfaces, interfaces are subjects of numerous investigations.

The magnetic properties of multilayers depend strongly on the preparation technology, substrates, magnetic components and their thickness, quantity of layers as well as on the thickness and material of the intermediate layers.

In the next paragraphs we will show several examples of two layers and multilayers systems.

2. TWO LAYERS SYSTEM

2.1. Sandwich films based on 80Ni-20Fe (Permalloy)

As an example we will show some recent results of strongly and weakly coupled films of 80Ni-20Fe/C/80Ni-20Fe[10]. The samples were deposited by electron beam evaporation in the presence of magnetic field = 200 Oe in a vacuum of about 3×10^{-6} Torr. Thickness of both Permalloy layers is D = 300Å, thickness of carbon layer is respectively d_c= 20Å, 50Å, 150Å and 270Å. The temperature of deposition of the first and second magnetic layer was 50°C and 100 - 150°C respectively.

Sample of d_c= 20Å is strongly coupled. It is difficult to know if in the doublelayers film of d_c= 20Å discussed here the coupling originates only from exchange (due to the discontinuities of C) or if it exists some contribution of magnetostatic effect. Swiatek et al.[11] showed recently the exchange couplings of 80Ni-20Fe films across several non-magnetic interlayers, but not across C. It is possible that exchange coupling across carbon films is similar to that in Pd, Au and Bi (see Fig. 5).

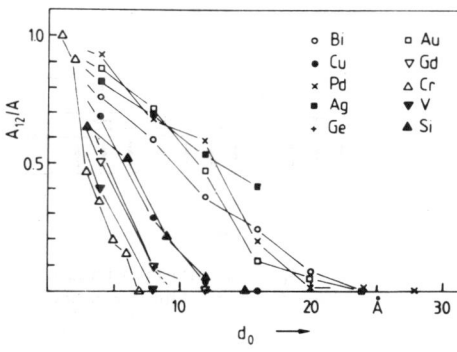

5. Interlayer exchange A_{12} as a function of interlayered thickness do for 80Ni-20Fe double layers with various interlayered materials (after[11])

In the strongly coupled films, magnetic domains are superimposed and are identical in both layers (Fig.6a,b and 9). Magnetization in two layers is paralell which is a typical magnetization arrangement in the case of the strong coupling exchange effect (see e.g.[8]). Fig.6a,b show

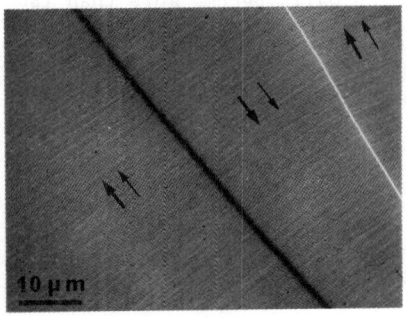 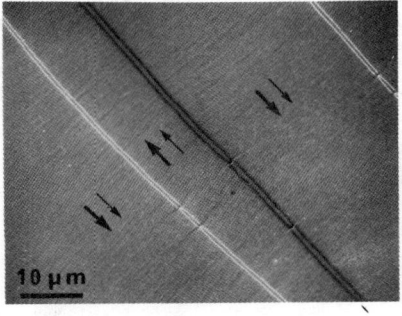

(a) (b)

Fig.6. Magnetic domains in the strongly coupled 80 Ni-20 Fe films d_c= 20Å a)after saturation in easy axis. b) after saturation in hard axis. Two different arrows indicate magnetization direction in the bottom and top layers.

domain structure in the sample with d_c = 20Å after saturation in easy
and hard direction respectively. After applying the field in the easy
axis and then reducing to zero the alternating Néel walls are light and
dark and may be either 180°/180 or 180°/-180°[7]. After applying the
field in the hard direction (Fig.6 b) the alternating light and dark
twin walls of the same contrast appear. This phenomenon was observed
before by Birgagnet et al. [12] and was interpreted by existence of
so called quasi-walls[13] which appear to close the magnetic flux over
two layers as is shown in Fig.7. The twin walls consist of the segments
linked by some sort of lines. It is possible that the segments of the
Néel walls are of different chirality and the lines have the Bloch
character. Magnetization distribution of the segments of twin walls is
proposed in Fig.8. Some dynamical investigations concerning this
structure are in progress.

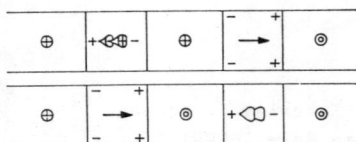 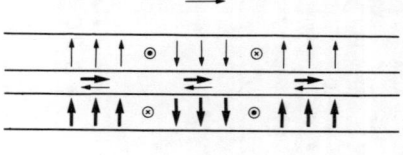

Fig.7. Magnetization distribution in wall and quasi wall (after[12])

Fig.8. Proposed interpretation of the lines in the twin walls

The meander-shape wall (Fig.9) appears in the sample after
saturation in the hard direction and was interpreted by the effect of
converting the wall from Q/Q wall into $Q/Q-2\pi$[7]. The reason for
conversion is that if a field is applied antiparallel to the
magnetization of the wall centers, the walls become more than 180°
walls and their stray field cannot be compensated by corresponding
quasiwalls. By motion of a Bloch line in one layer a more than 180°
wall is converted into minus - less - than 180° wall for which a stray
field compensation by a quasi wall in the neighbouring layer is again
possible.

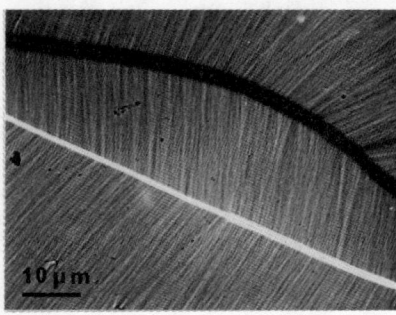

Fig.9. Meander-shape wall in the strongly coupled 80Ni-20Fe films after saturation in hard direction, d_c = 20Å

It is to be noticed that in the strongly coupled specimen the magnetization ripple is very regular which is an interesting feature since this structure plays an important role in many magnetic properties like magnetization reversal, susceptibility and hysteresis. Hysteresis in the easy and hard directions of this specimen are shown in Fig.10a and b respectively.

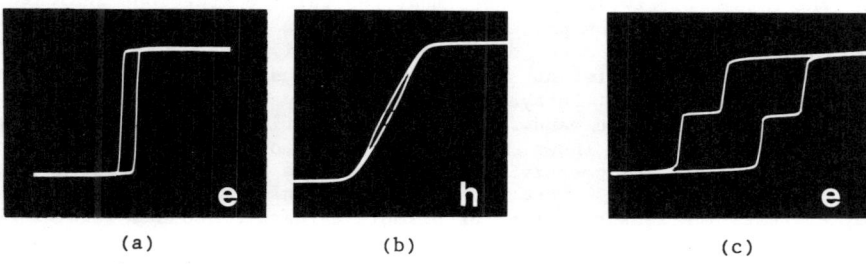

(a) (b) (c)

Fig.10. Magnetization hysteresis in the easy (e) and hard (h) directions of the sample with d_c= 20Å (a and b) and d_c= 270Å (c)

The form of loops are similar to those obtained in an uniaxial perfect single Permalloy film, however in the two layered strongly coupled films (d_c= 20Å) coercivity is low (H_c= 0,2 Oe). The coercivity H_c of the single Permalloy film of the same composition and thickness (600Å) is 3 Oe. The same value of coercivity has been found in the specimen where d_c= 50Å.

Fig.11 and 12 show the Lorentz pattern of domains and walls in the weakly coupled specimens of d_c= 150Å and 270Å respectively. Magnetic structure in these films differ considerably from that observed in the strongly coupled films : domains are not superimposed, ripple is not regular, the distance between the walls of the same contrast is large and the cross-ties appear. This is really significant because cross-tie cannot exist in the strongly coupled specimen where the walls

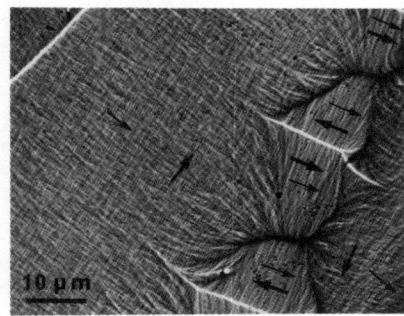

Fig.11. Magnetic domains in the 80Ni-20Fe double-layered films. d_c= 150Å

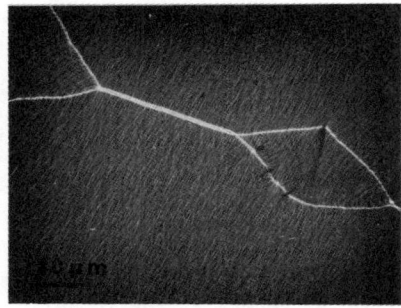

Fig.12. Magnetic domains in 80Ni-20Fe doublelayered films. d_c= 270Å

are magnetostaticly strongly coupled[7]. Similar structure was observed by Tsukahara et al[14] in amorphous doublelayered films. In some regions of the samples described here ripple of one layer is perpendicular to ripple of the second which means that in this region direction of magnetization in two layers is perpendicular to each other (Fig. 10). The hysteresis of the specimen d_c= 270Å is shifted (Fig. 9c). In the case of twolayered specimens the shift of the loop indicates that the magnetization process does not occur in the same way in the two layers. In the case described here this effect was casual and possible due to different temperatures of the film deposition (t_{A1}= 50°C and t_{A2}= 150°C). Kobayashi et al[15] have observed similar shift of hysteresis in the sandwich ferrimagnetic films with perpendicular anisotropy and have shown that the shift of hysteresis is due to different values of coercivity. This effect is important for magnetic and magneto-optical recording applications[16] and for the magnetoresistive sensors based for example on hard Pt Co and soft Ni-Fe films[17].

3. MULTILAYERS.

The Renaissance of old/new system : multilayers has started at mid-eighties. The search for new materials for information storage and recent development of ultra high vacuum techniques and characterization of systems have resulted in an extremely rapid growth of the this field of magnetism. The basic magnetic layers are in general Co, Ni, Fe and Ni-Fe. For the intermediate non-magnetic layers, several materials can be used : Cu, Mg, Nn, Au, Ag, C, etc.. Depending on the magnetic and non-magnetic components and their coupling, thethicknesses and quantity of layers, the system displays different magnetic properties often quite different from bulk.

3.1. Structures.

For some artificial multilayers, the chemical modulation mode is sinusoidal due to interdiffusion. These multilayers are often called compositionally modulated films. The example can be Fe/V which are similar 3d elements with the same bcc structure and Ni/Cu which have both fcc structure and are soluble with each other[18,19].
On the contrary, the combination of the bcc and fcc metals like Fe/Mg has chemically sharp interfaces (square wave modulation mode); these elements are insoluble with each other and form the system which is called "metallic superlattice". However Fe/Cu which are also bcc and fcc metals and are easy to form in the multilayered system, have not exactly "superlattices" magnetic properties[20].

3.2. Coupling effect

In general, the coupling mechanism in multilayers is similar to that observed in the doubleayered systems and depends on the length scale (thickness) of the layers. At the short length scale there exist: 1) direct exchange coupling and, 2)RKKY (Ruderman-Kittel-Kasuya-Yosida) interaction due to electronic mechanism via conduction electrons. At the long length scale (above 50Å) the coupling is magnetostatic due to dipolar interaction.

3.3 Magnetization

Magnetization in monolayers and multilayers has been investigated by many authors. Fu and Freeman[21] showed that magnetic moments in monolayers of chromium, iron and even vanadium which is not magnetic in bulk - may be very important ("giant magnetic moments" - authors say) if the monolayers are deposited on the noble metal. On the contrary nickel in the form of monolayer Au/Ni/Au and Cu/Ni/Cu (d_{Ni} <5 Å) is non-magnetic[21,22].

Multilayers with sufficiently thin magnetic layers and sharp interfaces (superlattices) may also display an important magnetic moment. However, technology of such a system needs sophisticated apparatus and high precision. Depending on components and number of the layers and their thickness, magnetization is strong, weak or non-existing. Here are some examples : the Fe(15Å)/V(16Å) multilayers are ferromagnetic but Fe(d_{Fe})/V(16Å) with d_{Fe} <6 Å are non-magnetic[18]. Fe and V are similar 3d elements with the same bcc structures and interdiffusion is not negligible. If Fe layer is thinner than 6Å, Mössbauer spectra give an appearance of non-magnetic fraction of the atoms[19]. The authors conclude that Fe/V system is not suitable to construct a model system to investigate on two-dimensional ferromagnet.

On the other hand, Fe/Mg superlattices are ferromagnetic even if Fe layer is 2Å thick[18]. The local magnetic moment of Fe atoms in the interface sites in contact with Mg has been found close to the bulk value[19]. Interesting results are reported on the system Fe/Cr[24,25]. If the chromium layer is thicker than 30Å (d_{Fe}= 60Å and 30 Å) the multilayers are ferromagnetic, however if d_{Cr} < 30Å the system is antiferromagnetic.

Draaisma et al[26] investigated Pd/Co and Pd/Fe multilayers and showed that saturation magnetization M_s in the systems may exceed that of the bulk ferromagnetic elements. This is thought to be due to polarization of the Pd atoms at the interface.

3.4. Anisotropy

Magnetic anisotropy is one of the most amazing features in multilayers. In 1954 Louis Néel[27] predicted the existence of magnetic surface anisotropy, caused by reduced symetry in the surroundings of the surface atoms. Such an anisotropy may also be present at the interface between a magnetic and a nonmagnetic metal. In multilayers, surface anisotropy plays a crucial role and in many cases provokes an important change of magnetic properties. The results are sometimes spectacular, first of all if it is concerned with the appearance of a perpendicular anisotropy. For example in the multilayers Fe(15Å)/V(16Å) and Fe(8Å)/Mg(16Å) the easy axis is in the plane[18,19], however in Fe(2Å)/Mg(16Å) the easy direction is along the film normal[19]. Another example is the system of Pd/Co and Pd/Fe multilayers[26,28,29] where the anisotropy at the surface tends to orient the magnetization perpendicular to the film. In the case of Pd/Co this leads to a perpendicular anisotropy for Co layers below a thickness of 7.2Å. The Co layers can even be made as thin as 2Å in which case it is possible to obtain the rectangular hysteresis loop with a high remanence when the field is applied along the film normal[26]. It is remarkable that even with the large shape anisotropy, typical for the thin films which tend to keep the magnetization in the plane of the film expected for the thinnest films, the total magnetic easy axis is in the perpendicular direction.

The origin of perpendicular anisotropy in such a thin films is not very clear at present. Some speculations lead to a magnetostrictive effect due to the lattice mismatch at the interfaces (e.g.[28]. On the other hand, calculations of the interface magnetic anisotropy for the monolayer of Fe, Ni and V using a self-consistent local-orbital method give no negligible value of this anisotropy[33]. However further work is needed.

3.5 Coercivity.

Depending on applications, authors are searching either for materials with a low coercivity (e.g. for magnetic heads) or with a high coercivity (e.g. for magnetic or magnetooptical recording). Jubb et al[31] investigated the 80Ni-20Fe/Al multilayers and found very low coercivity (H_c= 0.1-0.2 Oe) in the structure with five layers of 400Å Ni/Fe and six layers of Al of the thickness d_{Al}= 10-100Å. Minimum coercivity was obtained for the Permalloy layer of the thickness 400-500Å (four to five layers of Permalloy). As the Permalloy-layer thickness is decreased below 300Å, the coercivities began to increase. It is possibly due to discontinuities in the Ni/Fe layers. For Fe/C multilayered system in the case of the period = 50Å, H_c= 0.3 Oe[32].

High coercivity H_c= 550 Oe has been found in the Pd(10.3Å)/Co(4.7Å)multilayered system with a perpendicular anisotropy which may be a good candidate for a vertical magnetic recording medium[28].

SUMMARY

In the past, the investigations of the twolayered systems were focused on the origin of coupling effect, domain structure and magnetization process. Many magneticians were involved in this field and their theoretical and experimental results serve these days for understanding the more complex structures.

At the present time, the twolayered systems based on Permalloy films and other soft layers, or on hard layers are investigated in view to use these materials for information storage, magnetic heads and propagation elements.

Since mid-eighties, multilayers have became the main topic in the magnetic thin films laboratories. Technology of these materials is complicated and needs sophisticated apparatus for the sample fabrication and characterization. Theoretical investigations show that some magnetic multilayers may have important magnetic moments and perpendicular anisotropy in the case of very thin layers. Many experimentalists work on these problems and often their results are satisfactory.

The future will show whether the multilayers would remain only the intellectual problem or will also become the medium for applications.

REFERENCES

1. OAKLAND, L.J. and ROSSING, T.D., J. Appl. Phys., **30**, 54S (1959)

2. BROADBENT, K.D., Proc. Inst. Radio Engrs., **48**, 1728 (1960)

3. FULLER, H.W. and SULLIVAN, D.L., J. Appl. Phys., **33**, 1063 (1962)

4. PUCHALSKA, I.B. and SPAIN, R.J., Compte Rendus, Acc.Sc., **254**, 2937 (1962)

5. NEEL, L., Compte Rendus, Acc. Sc., **255**, 1676 (1962)

6. DOYLE, W.D., Physics Bulletin, **22**, 645 (1971)

7. FELDKELLER, E., J. Appl. Phys., **39**, 1181 (1968)

8. YELON, A., Physics of Thin Films, Academic Press, 205 - 299 (1971)

9. LIN, Y.S. and GRUNDY, P.J., J. Appl. Phys., **45**, 4084 (1974)

10. NIEDOBA, H., PUCHALSKA, I.B., Proceeding of 12-ICMFS Le Creusot (France), 180 (1988)

11. SWIATEK, P., SAUREBACH, F., PANG, Y., GRUNBER, P., ZINN, W., Proceedings of the 3rd International Conference on Physics of Magnetic Materials, Szczyrk-Biala (Poland), 1986, ed. World Scientific, 389 (1987)

12. BIRAGNET, F., DEVENYI, J., CLERC, G., MASSENET, O., MONTMORY, R. and YELON, A., Phys. Stat. Sol., **16**, 569 (1966)

13. SLONCZEWSKI, J.C. and MIDDELHOEK, S., Appl. Phys. Lett., **6**, 139 (1965)

14. TSUKAHARA, S., MORITA, H., YAMAMOTO, M. and FUJIMORI, H., IEEE Trans. Mag., **MAG-22**, 775 (1986)

15. KOBAYASHI, T., TSUJI, H., TSUNASHIMA, S. and UCHIYAMA S., Jap. J. Appl. Phys., **20**, 2089 (1981)

16. TSUNASHIMA, S., CHOE, Y.J., ITOH, K. and UCHIYAMA, S., Proceeding of 12-ICMFS, Le Creusot (France), 444 (1988)

17. HILL, E.W. and McCULLOUGH, A.M., IEEE Trans. Mag., **24**, 1707 (1988)

18. SHINJO, T., Proceedings of International Conference on the Application of the Mössbauer Effect, Leuven (Belgium), 1985, in Hyperfine Interactions

19. SHINJO, T., HOSOITO, N., KAWAGUCHI, K., NAKAYAMA, N., TAKADA, T. and ENDOCH, Y., J. Mag. Mag. Materials, 54-57, 737 (1986)

20. KOZONO, Y., KOMURO, M., NARISHIGE, S., HANAZONO, M. and SUGITA, Y., J. Appl. Phys., 61, 4311 (1987)

21. FU, C.L. and FREEMAN, A.J., J. Mag. Mag. Materials, 54, 777 (1986)

22. RENARD, J.P. and BEAUVILLAIN, P., Physica Scripta, T19, 405 (1987)

23. CHAPPERT, C., RENARD, D., BEAUVILLAIN, P., RENARD, J.P. and SEIDEN, SEIDEN, J., J. Mag. Mag. Materials, 54-57, 795 (1986)

24. BAIBICH, M.N., BROTO, J.M., CREUZET, G., ETIENNE, P., FERT, A., FERT, A.R., HADJOUDJ, S. and NGUYEN van DAU, F., Proceeding of 12-ICMFS, Le Creusot (France), 256 (1988)

25. GRUNBERG, P., SCHREIBER, R., PANG, Y., BRODSKY, M.B. and SOWERS, H., Phys. Rev. Letters, 57, 2442 (1986)

26. DRAAISMA, H.J.G., de JONGE, W.J.M. and den BROEDER, F.J.A., J. Mag. Mag. Materials, 66, 351 (1987)

27. NEEL, L., J. Phys. Radium, 15, 225 (1954)

28. CARCIA, P.F., MEINHALDT, A.D. and SUNA, A., Appl. Phys. Lett., 47q 178 (1985)

29. Den BROEDER, F.J.A., DONKERSLOOT, H.C., DRAASIMA, H.J.G. and de JONGE, W.J.M., J. Appl. Phys., 61, 4317 (1987)

30. SCHULLER, I.K., NATO Advances Study Institute on Physics, Fabrication and Application of Multilayered Structures, June, Ile de Bendor (France) 1987

31. JUBB, N.J., DAVIS, R.E., REITH, T.M., KOLAR, H.R. and LEAVITT, J.A., J. Appl. Phys., 57, 4192 (1985)

32. KOBAYASHI, T., NAKATANI, R., OOTOMO, S. and KUMASAKA, N., J. Appl. Phys, 63, 3203 (1988)

33. GAY, J.G. and RICHTER, R., Phys. Rev. Lett., 56, 2728 (1986)

COHERENT SPIN WAVES IN MAGNETIC FILMS

A.G.Gurevich

A.F.Ioffe Physico-Technical Institute, Academy
of Sciences USSR
Leningrad 194021, USSR

ABSTRACT

Properties and means of excitation of coherent spin waves are reviewed paying the most attention to propagating coherent spin waves in magnetic films and in particular to the parametric excitation of partly coherent spin waves by magnetostatic-wave pumping.

1. INTRODUCTION

The idea of spin waves (delocalized excitations of a ferromagnet) was suggested in 1930 by Bloch [1]. Microscopic theory of such excitations was given in 1940 by Holstein and Primakoff [2]. In 1946 the notion of spin waves or magnons was used by Akhiezer [3] in founding the theory of ferromagnetic relaxation. Dyson [4] and many others studied the thermodynamics and relaxation processes in ferromagnets on the basis of more precise spin-wave theories, taking into account the spin-wave interactions. In all the above mentioned investigations spin waves have been regarded as non-coherent.

After the discovery of ferromagnetic [5] and antiferromagnetic [6] resonances it became evident that in these phenomena coherent magnons or spin excitations with wave vector $k = 0$ are created with phases determined by the phase of field that excites them.

Spin waves with $k \neq 0$ also can be coherent. It became quite clear after Herring and Kittel in 1951 created the macroscopic spin-wave theory [7]. Spin waves have been regarded in this theory as electromagnetic waves in a medium for which the material equations are the Landau - Lifshitz equations of motion of magnetization [8].

Kittel has shown in 1958 [9] that coherent standing spin waves with rather high k values can be excited in thin films by uniform ac magnetic field. It has been immediately confirmed experimentally [10], and a new direction in experimental magnetism, spin-wave resonance has been born.

Some years earlier a new class of nonlinear phenomena in magnetically ordered substances was discovered [11]. These phenomena were explained [12,13] on the basis of an idea of partly coherent spin waves parametrically excited by uniform ac magnetization or directly by ac magnetic field.

It is clear that quite coherent propagating spin waves can also exist. Such waves were observed first in rods of single-crystal yttrium iron garnet (YIG) and in early 70-th in epitaxial YIG films. They were called magnetostatic waves as their dispersion relations can be derived from shortened, magnetostatic Maxwell equations. These waves became an object of intense practical interest.

All the above mentioned questions concerning coherent spin waves are briefly reviewed below. Most attention is payed to propagating coherent spin waves in films including parametric excitation of partly coherent spin waves under the influence of entirely coherent spin waves in films.

2. STANDING SPIN WAVES IN FILMS

The spin wave dispersion law in isotropic ferromagnet on condition that ka 1 (where k is the wave number and a is the magnetic lattice constant) is given by

$$\left(\frac{\omega}{\gamma}\right)^2 = (H_{i0} + Dk^2)(H_{i0} + Dk^2 + 4\pi M_0 \frac{k_x^2 + k_y^2}{k^2}), \quad (1)$$

where ω is the circular frequency, γ is the gyromagnetic ratio, H_{i0} is the internal dc magnetic field (directed, so as the dc magnetization, along z axis) and D is the exchange stiffness constant. To obtain the ω or H_{i0} eigenvalues in particular for a film it is necessary to take into account the boundary conditions: the usual electrodynamic conditions and complementary ones for ac magnetization $\vec{m}$. Kittel [9] suggested that on the boundaries of a ferromagnet, i.e. on the film surfaces

$$\vec{m} = 0. \quad (2)$$

An alternative condition

$$\frac{\partial \vec{m}}{\partial n} = 0 \quad (3)$$

was given before by Ament and Rado [14]. A more general condition can be written as follows [15]:

$$\begin{aligned}\frac{\partial m_x}{\partial n} + \xi\, m_x \cos 2\theta &= 0, \\ \frac{\partial m_y}{\partial n} + \xi\, m_y \cos^2\theta &= 0,\end{aligned} \quad (4)$$

where $\vec{n}$ is a normal to the surface, is the angle between $\vec{n}$ and $\vec{M_0}$, y axis is parallel to the surface and ξ is the pinning parameter ($\xi = 0$ corresponds to the case of Ament and Rado and $\xi = \infty$ to the case of

Kittel). From these conditions for normally magnetized film ($\theta = 0$) with thickness d the equation for $k_z \equiv k$ can be easily obtained:

$$\operatorname{ctg} kd = \frac{k^2 - \xi_1 \xi_2}{k(\xi_1 + \xi_2)}, \qquad (5)$$

where ξ_1 and ξ_2 are the pinning parameters on the surfaces of the film.

If $\xi_1 = \xi_2 = \infty$ (the case investigated by Kittel) we deduce from Eq.5

$$k_m = \frac{m\pi}{d}, \qquad (6)$$

and using Eq.1 obtain the eigenvalues of frequency:

$$\frac{\omega_m}{\gamma} = H_{10} + D\left(\frac{\pi m}{d}\right)^2. \qquad (7)$$

In Fig.1 the distributions of ac magnetization (with circular polarization) corresponding to these eigenvalues are given. It is clear that uniform ac field will excite the modes with odd m's and their intensities will decrease with growing m. Experiments are usually carried out at ω = const. and the exchange stiffness D can be obtained from the positions of absorption vs dc field maxima.

The parameter ξ is the ratio of surface anisotropy and exchange energies. Its negative value corresponds to the easy anisotropy plane. The case $\xi_1 = -\xi_2 = \xi$, i.e. antisymmetrical boundary conditions is of some interest. In this case Eq.6 is fulfiled and the constant D can be obtained regardless of the ξ value [16]. Such boundary conditions can be realized [17] by means of covering the surfaces of the film under investigation with thin films having M_0 values higher and lower than M_0 of the main film.

Standing spin waves in thin magnetic films have been for many years an object of great interest. Different more complicated cases have been studied theoretically, in particular the cases of tangential magnetization and of nonuniform magnetization. A lot of experiments have been made. Now spin-wave resonance (SWR) is a common method for exchange stiffness measurement. It should be noted that this method can be used and is widely used for metal films, their conductivity exerting no influence if $d \ll \delta$, where δ is the skin depth.

3. PROPAGATING SPIN WAVES IN FILMS

Obtaining the dispersion law of spin waves with a wave-vector component k_t parallel to film surface is a more complicated problem than in the above discussed case of SWR where $k_t = 0$. The reason for this is that one wave with a certain value of $\vec{k}$ normal component can not satisfy all the boundary conditions and therefore be an eigenmode of the problem.

Situation becomes more simple in the case of thick ($d \gtrsim 5 \mu m$) films and sufficiently small ($k_t \lesssim 10^4$ cm^{-1}) wave vectors when the influence of exchange interaction on the dispersion law can be neglected and complementary boundary conditions ignored. Such "non-exchange" spin waves are called magnetostatic waves (MSW). As it has been mentioned above they became an object of great practical interest.

In Fig.2 the dispersion characteristics of these waves are shown in three important particular cases. In first two cases: of normal magnetization and of tangential magnetization and propagation along the magnetization the waves are volume ones. That means the trigonometric dependence of all the ac components on the coordinate in the direction of the normal to the film. In both cases an infinite number of modes can exist. But these cases differ substantially in the character of $\omega(k_t)$ dependence. For normally magnetized film this dependence is a growing one, i.e. the group velocity is positive. If the film is tangentially magnetized (and the wave propagates along dc magnetization) frequency decreases with growing k_t, i.e. the group velocity is negative.

In the third case only one mode exists with exponential dependence of ac components on coordinate in the direction of normal to the film not only outside the film but also inside it. Such wave is called a surface one. The dispersion law for it derived by Damon and Eshbach [18] is the following:

$$(\frac{\omega}{\gamma})^2 = (H_{i0} + 2\pi M_o)^2 - (2\pi M_o)^2 e^{-2k_t d}. \qquad (8)$$

The group velocity is positive here and decreases with growing k_t. The peculiarity of this wave is its non-reciprocity: for given direction of dc magnetization it propagates along a given film surface only in one direction.

Dispersion laws of non-exchange magnetostatic waves have been derived for arbitrary directions of magnetization and propagation and for more complicated structures including ferromagnetic and dielectric layers and metal surfaces. In Fig.3 such laws are shown for surface waves in some structures.

Magnetostatic waves in films are effectively excited by means of metal conductors (antennae) located on or near film surface and fed from an appropriate waveguide. Transversal dimensions of these conductors must be sufficiently small, of the order of or less than $1/k_t$.

The single-crystal films of ittrium iron garnet (YIG), stoichiometric or Ga-substituted are commonly used for MSW propagation. They are grown, so as garnet films for bubble devices and magnetooptics, by the liquid epitaxy on thin and precisely finished slabs of high-quality gadolinium gallium garnet (GGG) single crystals.

The excange interaction that must be taken into account for thin films and (or) large k_t values complicates materially the dispersion laws of spin waves in films. It was mentioned above that for standing waves (i.e. $k_t = 0$) an infinite number of eigenfrequencies exists. These frequencies (or in the case ω = const. the resonance H_{i0} values) are the initial points of dispersion-characteristic branches for $k_t \neq 0$. Some of these branches repulse as it is shown in Fig.4. The repulsion so as the initial (for $k_t = 0$) splitting depends on the value of pinning parameter ξ. The calculation of such dispersion characteristics (spectra) comes to finding the simultaneous solutions of electrodynamic equations (in magnetostatic approximation) and equations of motion satisfying both electrodynamic and complementary boundary conditions.

Two ways of solving this problem are known. The first one used first by Gann [19] consists in that the magnetostatic potential of the wave satisfying all the boundary conditions is found as a sum of three partial solutions that, taken separately, do not satisfy them. Following the second way used first by Vendik, Chartorizhskii and Kalinikos [20,21] the interdependence of ac field and magnetization resulting from electrodynamic equations and boundary conditions is first found; then taking this interdependence into account one finds the solutions of equation of motion satisfying the complementary boundary conditions.

The two ways are equivalent and both can be realized only using a computer.

It should be noted that to derive the electric field $\vec{e}$ of magnetostatic wave and the energy flux it is necessary to overstep the limits of magnetostatic approximation and use the full Maxwell equation connecting $\vec{e}$ with magnetic induction $\vec{b}$. But the velocity of energy flow coinsides practically with group velocity that can be determined from dispersion equation found in magnetostatic approximation.

The magnetostatic waves in films have the following unique features that provoke the interest to them:
1) broad frequency range limited from above only by the increase of dc magnetic field and from below by the rise of domains,
2) possibility of frequency tuning by changing the dc magnetic field,
3) range of wave numbers $k = 10^2 - 10^3$ cm^{-1} convenient for applications and not depending on frequency,
4) low and controllable in broad range ($\sim 10^6 - 10^8$ cm sec^{-1}) group velocity,
5) possibility of changing the dispersion law by simple means, e.g. by choice of wave type, thicknesses of layers ets.,
6) nonreciprocity (for the surface wave),
7) low decrement,
8) simple excitation ,
9) planar technology of devices.

A large number of microwave devices using MSW has been proposed and designed (see e.g. [22]): delay lines, non-dispersive or with given dispersion law, fixed or controllable; filters with different bandwidths; resonators, in particular for microwave semiconductor generators etc.

Devices analogous to some of the above mentioned were formerly designed using surface acoustic waves (SAW). MSW have great advantages as compared with SAW the most important of which are the possibility to work at higher frequencies, the possibility of tuning, nonreciprocity and simlicity of excitation.

All the above mentioned devices are linear ones. Their dynamical range is limited by nonlinear phenomena which arise sometimes at rather low power levels. On the other hand, these nonlinear phenomena can be used to design some new, nonlinear devices.

4. PARAMETRIC EXCITATION OF SPIN WAVES

The idea of parametric excitation of spin waves was born more than 30 years ago to explane the unexpected results of experiments of Bloembergen and Damon [11] and many others' on ferromagnetic resonance in high microwave fields. In these experiments the saturation of resonance began in fields much less than those at which it should begin according to the solution of equation of motion for a given uniform magnetization mode. Anderson and Suhl [23] suggested, and soon it was brilliantly corroborated by Suhl theory [12], that the reason of the observed phenomena is the parametric excitation of spin waves under the influence of uniform oscillation of magnetization. The basis of it is of coarse again the nonlinearity of Landau Lifshitz equation but containing now not only uniform mode but also the manifold of spin waves that always exist at the thermal level in magnetically ordered substances. The nonlinearity of equation of motion leads to coupling of different modes and under some conditions to energy transmission from the pumping mode (excited by the external field) to spin waves. At a certain (threshold) pumping amplitude the energy transmitted to some group of spin waves exceeds the energy losses due to relaxation, and the amplitudes of this group begin to rise. Of course this rise is limited by other nonlinear processes, and at every pumping amplitude a certain level of spin-wave excitation sets in.

As Suhl has shown spin-wave pairs arise here, the waves in a pair having the same frequency ω_k and wave vectors (in the case of uniform pumping) $\vec{k}$ and $-\vec{k}$. The pumping frequency ω_p as always at parametric excitation satisfies the following condition:

$$\omega_k = n \frac{\omega_p}{2}, \qquad n = 1,2 \ldots \qquad (9)$$

The threshold should be lower at lesser n, but the condition (9) is not always fulfiled for $n = 1$, and then the case $n = 2$ is realized. Suhl has found that for $n = 1$ (the first order parametric process) the threshold magnetization value is given approximately by

$$m_{1\,th} \sim \frac{\Delta H_k}{4\pi}, \qquad (10)$$

where ΔH_k is the linewidth of unstable (parametrically excited) spin waves. These waves have $k \sim 10^4$ cm^{-1} and propagate at an angle $\theta_k = 45°$ with $\vec{M}_0$. The threshold value of magnetic field turns out very small if the first order process is allowed at ferromagnetic resonance that is the case (for spherical samples) at frequencies

$$\omega_p < \frac{2}{3} \gamma \, 4\pi M_0, \qquad (11)$$

i.e. for YIG at room temperature at frequency $f_p = \frac{\omega_p}{2\pi} = 3.27$ GHz. But if the first order process is allowed only far from resonance the threshold field gets two orders higher. Threshold fields for the second order process are of the same order ($\sim 10^{-2}$ Oe in the case of YIG) as for the first order one far from resonance. At second order process spin waves with $k \sim 10^5$ cm^{-1} and $\theta_k = 0$ are excited.

Schlömann, Green and Milano [13] have shown theoretically and experimentally that ac magnetic field $\vec{h}_z$ parallel to dc magnetization $\vec{M}_0$ can also parametrically excite spin waves (parallel pumping). Spin waves with k values varying with changing dc magnetic field in the range, approximately, $10^4 - 10^6$ cm^{-1} are excited here. It gives an opportunity to derive the k dependence of spin-wave relaxation parameter ΔH_k from measured h_z threshold values.

In terms of corpuscular theory the above mentioned processes can be treated (Fig.5) as splitting of, accordingly, one magnon, two magnons or a photon in a pair of magnons with wave vectors $\vec{k}$ and $-\vec{k}$. These magnons (or the corresponding spin waves) are only partly coherent: only sum of their phases is determined by the phase of pumping field, the difference of them being quite accidental.

Coherence of pumping is not a necessary condition of parametric excitation. Non-coherent magnons accumulating in a certain region of k-space due to relaxation of primary spin waves (parametrically excited by coherent pumping) can also play the role of pumping.

The processes occurring in spin system above the threshold are very complicated. They have been studied thoroughly by Zakharov, L'vov and Starobinetz [24] in the case of parallel pumping. It was found that at not very high (less than 10 dB above threshold) pumping levels the spin-wave packet in k-space remains narrow, and the limiting of parametric-spin-wave number occures mainly due to deviation of spin-wave-pair phases from their optimum value. Furthermore in the system of parametrically excited magnons low-frequency oscillations can arrise.

At higher pumping levels the spin-wave packet becomes unstable relative to arising of another packets in k-space, and at still stronger pumping fields transition to chaos begins.

Parametric excitation of spin waves gets much more complicated when the pumping magnetization or field becomes nonuniform. In particular, it has been shown theoretically [25] and experimentally [26] that the threshold becomes higher if the pumping field exists only in a small region. A qualitative explanation of this is as follows: the parametric spin waves leave the region of pumping before their amplitudes encrease. It tells materially on threshold values if the mean path of spin waves l_k exceeds the dimension d of the region of pumping. In the case of parallel pumping the following expression takes place:

$$\frac{h_{z\,th}}{(h_{z\,th})_0} = \sqrt{1 + (\eta \frac{l_k}{d})^2} = \sqrt{1 + (\eta \frac{2 v_{gr}}{d \gamma \Delta H_k})^2} \qquad (12)$$

where $(h_{z\,th})_0$ is the threshold field for uniform pumping, v_{gr} is the group velocity of parametric spin waves and η is a multiplier of the order 1. Eq.12 is valid qualitatively even in the general case of nonuniform pumping, d being now the distance at which pumping magnetization or field decreases materially, i.e. $d \sim \frac{|m|}{|\nabla m|}$. In the opposite case when $l_k \ll d$ the parametric excitation can be considered as local one: with growing pumping power it begins at the point where magnetization or field reaches first its threshold value - the same as for uniform pumping.

5. PARAMETRIC SPIN WAVE EXCITATION IN FILMS

Interesting phenomena connected with parametric excitation of spin waves take place in films in which coherent spin waves propagate. Let us consider for example a device (Fig.6) in which excitation and receiving of surface magnetostatic wave take place. At low level of input power P_1 this device is a linear delay line. At a certain (threshold) value of $P_1 = P_{th}$ nonlinearity arises: the losses N_{12}, N_{13} and N_{14} (determined as $N_{1i} = -10 \lg (P_i/P_1)$) stop to be constant. The nonlinearity is a result of parametric excitation of spin waves, and its threshold should be low ($\sim 1 \mu W$ in YIG films) if the first order process is allowed. The parametrically excited spin waves in this case, according to the result of direct experiment [27] on light scattering, have $k \sim 10^4 cm^{-1}$ and can be regarded as uniform plane waves. Then the boundaries of the region where the first order process exists can be easily found by comparing the dispersion law (1) of these waves and the dispersion law (8) of pumping wave.

In Fig.7 these boundaries and experimental frequency dependence [28] of the threshold power are shown. One can see that P_{th} values are really small only in the region of first order processes. These values depend strongly on k (as one can see e.g. in Fig.8) and on film thickness.

Transition from P_1 threshold values to threshold magnetization has been carried out [29] on the supposition that the parametric excitation is local and pumping magnetization doesn't differ materially from the surface wave magnetization extrapolated on antenna axis. The derived in such way m_{th} values are almost constant in the whole region of first order processes (Fig.8) and exceed 1.5 - 2 times the values calculated according to Suhl theory [12]. The difference can be due to non-locality of excitation. As regards the strong $P_{th}(k)$ dependence it correlates with the k-dependence of the pumping-wave group velocity. Thus, although (in terms of local model) the pumping is the magnetization in the near zone (where the magnetostatic wave is not yet formed) the value of it differs only slightly from the magnetization of the wave extrapolated on the antenna axis. In this sence one can speak of parametric excitation of spin waves by MSW pumping.

The non-uniformity of pumping complicates strongly all the phenomena above threshold. The region of parametric excitation broadens with growing input power, and the ratio of pumping field or magnetization to its threshold value becomes a function of coordinates. In Fig.9 the dependences of the losses of energy transmitted in different ports of the device shown in Fig.6 on the input power are given[30]. As one can see the powers transmitted in ports 3 and 4 at first continue to increase above threshold. It can be the result of broadening of the parametric excitation region. Then the increase stops and the P_3 and P_4 values remain approximately constant in a broad ($\sim$ 20 dB) dynamic range. This is used for designing a nonlinear device, a power limiter. From Fig.9 one can also see that the losses of energy transmitted into port 2 decrease above threshold. This is utilized in another nonlinear device, a weak-signal suppresser or signal-to-noise enhancer[31].

Especially complicated nonlinear phenomena arise at simultaneous propagation of coherent spin waves with different frequencies. Let us consider a case when one of these waves is a strong, above-threshold wave and the other is a weak one. Then the first wave has an influence on the losses and phase of the second one. It turned out to be substantial[32] when the weak-wave frequency (see Fig.10) lies near not only to the frequency f_s of the strong wave but also to its satellites $f_s \pm 2\delta f$ where δf is the frequency of oscillations in spin-wave system.

6. CONCLUSION

Collective excitations of exchange-ordered magnetic system, spin waves or magnons can be, so as the other quasi-particles, coherent or non-coherent. Non-coherent magnons play a great role in the thermodynamics of magnetically ordered substances, determine their equilibrium magnetization and contribute to the specific heat and heat conductivity.

Coherent magnons with $k = 0$ are nothing else that the oscillations of magnetization excited by ac magnetic field and reaching their maximum values at ferro-, antiferro- or ferrimagnetic resonance. Coherent standing spin waves are excited in thin films by uniform ac field, and coherent propagating spin waves can be excited in films by means of thin conductors. The relaxation of coherent magnons occures due to their interaction with non-coherent magnons, phonons, charge carriers and localized excitations of atoms or ions.

Partly coherent spin waves are parametrically excited by coherent ones (oscillations or waves of magnetization), by ac magnetic field and also by non-coherent magnons.

Coherent magnons, not only with $k = 0$ but also propagating spin waves with $k = 10^2 - 10^3$ cm^{-1} are of great practical interest. The use of such waves is the basis of spin-wave electronics, a new technology in microwave engineering.

REFERENCES

1. Bloch F., Zs. f. Phys. 61, 206 (1930).
2. Holstein T. and Primakoff H., Phys. Rev. 58, 1098 (1940).
3. Akhiezer A., J. Phys. USSR 10, 217 (1946).
4. Dyson F.J., Phys. Rev. 102, 1217, 1230 (1956).
5. Griffiths G.H.E., Nature 158, 670 (1946).

6. Ubbink J., Poulis J.A., Gerritsen H.J. and Gorter C.G., Physica 18, 361 (1952).
7. Herring C. and Kittel C., Phys. Rev. 81, 869 (1951).
8. Landau L.D. and Lifshitz E.M., Phys. Zs. d. SU 8, 153 (1935).
9. Kittel C., Phys. Rev. 110, 1295 (1958).
10. Seavy M.H.,Jr. and Tannenwald P.E., Phys. Rev. Lett. 1, 168 (1958).
11. Bloembergen N. and Damon R.W., Phys.Rev.85,699 (1952).
12. Suhl H., J. Phys. Chem. Sol. 1, 209 (1957).
13. Schlömann E., Green J.J. and Milano U., J. Appl. Phys. 31, 386 S (1960).
14. Ament W.S. and Rado G.T., Phys. Rev. 97, 1558 (1955).
15. Soohoo R.,"Magnetic Thin Films", Harper Row Publ., N.Y., Evanstone and L. (1965).
16. Korchagin Yu. A., Khlebopros R.G. and Chistyakov N.S., Fiz. Tverd. Tela 14, 2121 (1972).
17. Wigen P.E., Kooi C.F. and Shanabarger M.R., Phys. Rev. Lett. 9, 206 (1962).
18. .Damon R.W. and Eshbach J.R., J. Phys. Chem. Sol. 19, 308 (1961).
19. Gann V.V., Fiz. Tverd. Tela 8, 3167 (1966).
20. Vendik O.G. and Chartorizhskii D.N., Fiz. Tverd. Tela 12, 1538 (1970).
21. Kalinikos B.A., Izv. VUZ, Ser. Fiz., N 8, 42 (1981).
22. Castera G.P., J. Appl. Phys. 55, 2506 (1984).
23. Anderson P.W. and Suhl H., Phys. Rev. 100, 1788 (1955).
24. Zakharov V.E., L'vov V.S. and Starobinetz, Usp. Fiz. Nauk 114, 609 (1974).
25. L'vov V.S. and Rubenchik A.M., Preprint Inst. Yad. Fiz. SO AN SSSR N 1 (1972).
26. Melkov G.A. and Sholom S.V., Fiz. Tverd. Tela 29, 3257 (1987).

27. Srivanasan G., Patton C.E. and Emtage P.E., J. Appl. Phys. <u>61</u>, 2318 (1987).
28. Gusev B.N., Gurevich A.G., Anisimov A.N., Chivilyova O.A., Vinnik M.A. and Berezin I.L., Fiz. Tverd. Tela <u>28</u>, 2969 (1986).
29. Chivilyova O.A., Anisimov A.N., Gurevich A.G., Gusev B.N., Vugal'ter G.A. and Sher E.S., Fiz. Tverd. Tela <u>29</u>, 1774 (1987).
30. Chivilyova O.A., Anisimov A.N. and Gurevich A.G., Zh. Tech. Fiz. <u>58</u>, 1204(1988).
31. Adam J.D., IEEE Trans. <u>MAG-16</u>, 1168 (1980).
32. Chivilyova O.A., Anisimov A.N., Gurevich A.G., Yakovlev S.V. and Averin A.N., Pis'ma v Zh. Tech. Fiz. <u>13</u>, 1497 (1987).

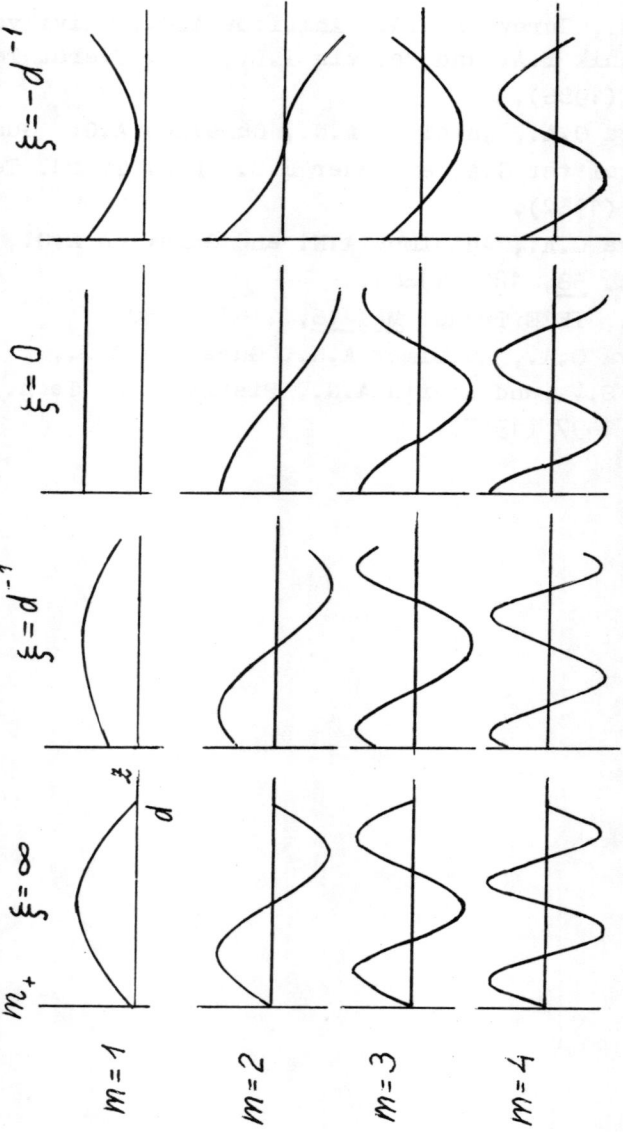

Fig.1. Standing spin-wave magnetizations in normally magnetized ferromagnetic film with symmetric boundary conditions.

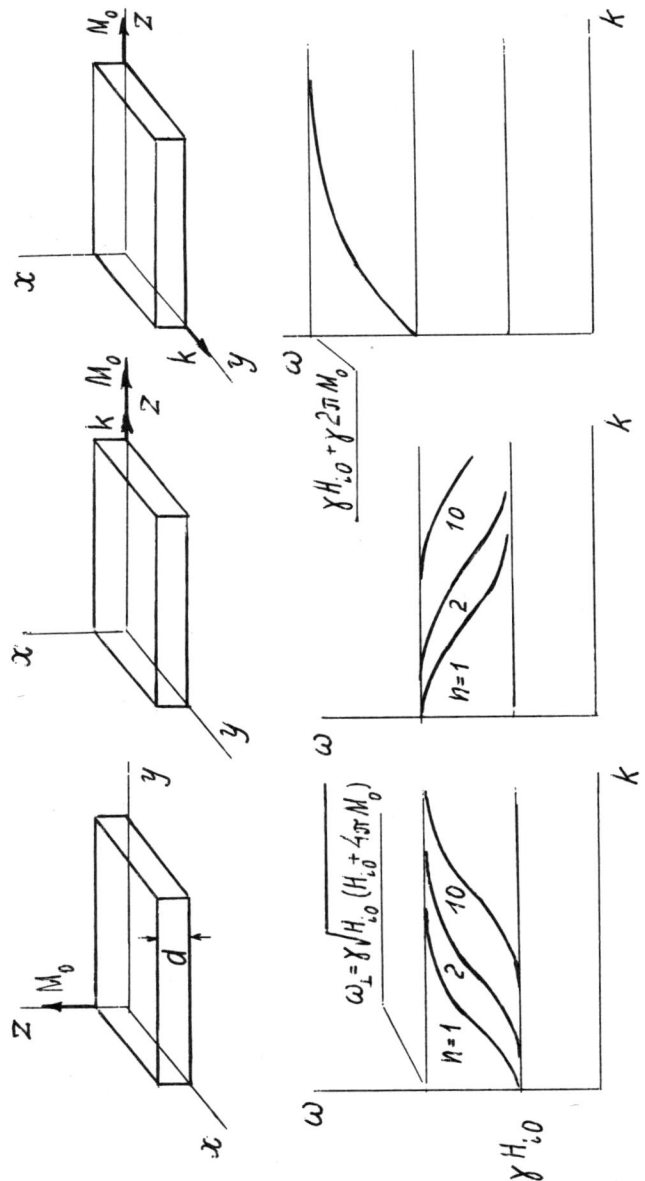

Fig.2. Dispersion characteristics of non-exchange spin waves in a film.

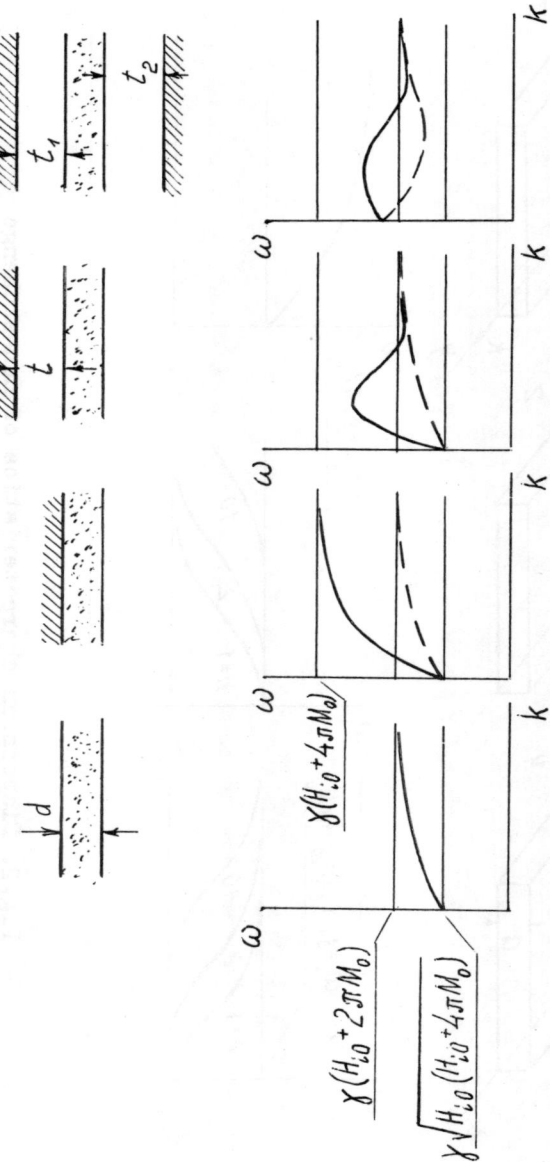

Fig.3. Dispersion characteristics of surface non-exchange spin waves in planar structures.

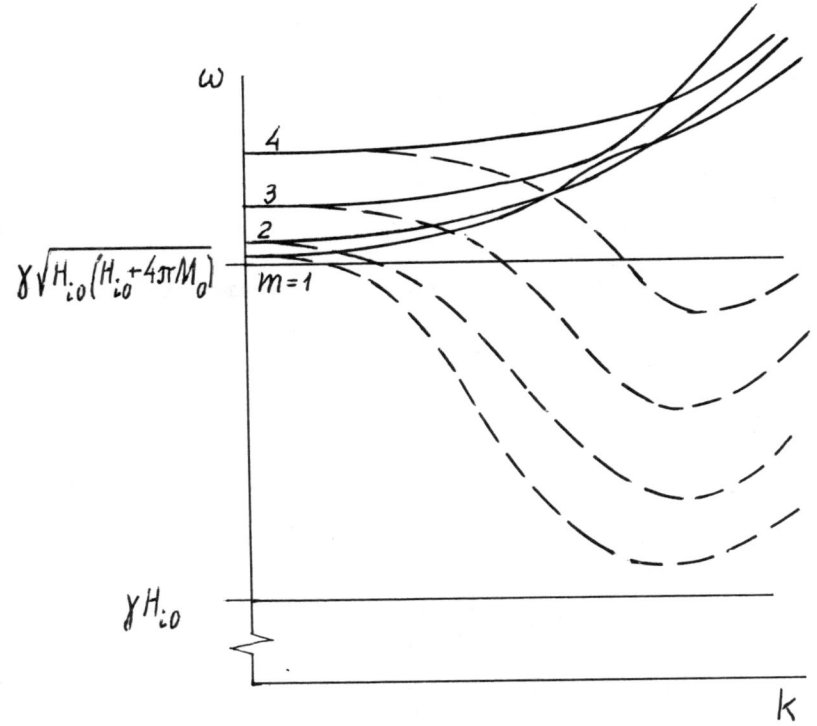

Fig.4. Spectrum of exchange spin-waves in tangentially magnetized film ($\theta = 0$). Solid lines correspond to $\vec{k} \perp \vec{M}_0$, dashed lines — to $\vec{k} \parallel \vec{M}_0$.

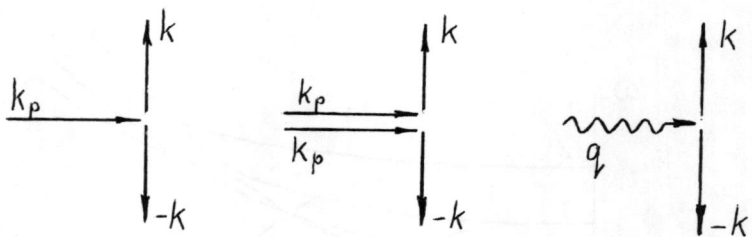

Fig.5. Elementary processes at parametric excitation of spin waves.

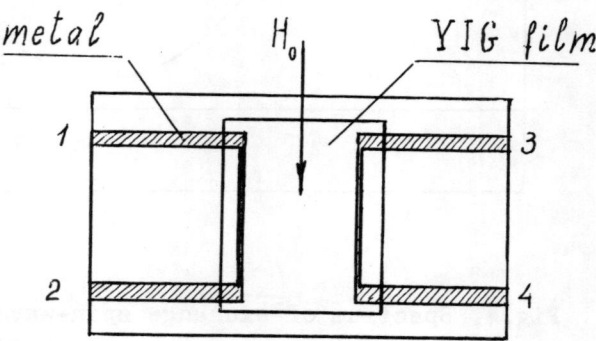

Fig.6. A device for excitation and receiving of surface magnetostatic wave.

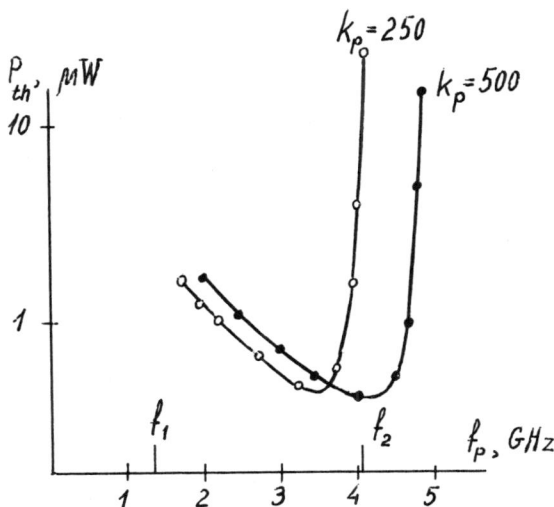

Fig.7. Frequency dependence of threshold power for parametric excitation of spin waves by surface magnetostatic wave. YIG film, $d = 7.2\,\mu m$. Boundaries of first-order-process region are shown for $k = 250$ cm^{-1}.

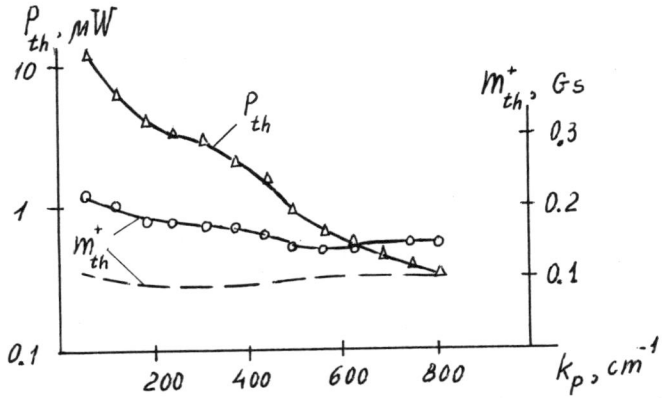

Fig.8. Dependences of threshold power and magnetization on k of surface magnetostatic pumping wave. YIG film, $d = 15.2\,\mu m$, frequency 3.2 GHz. The dashed line has been calculated using Suhl theory.

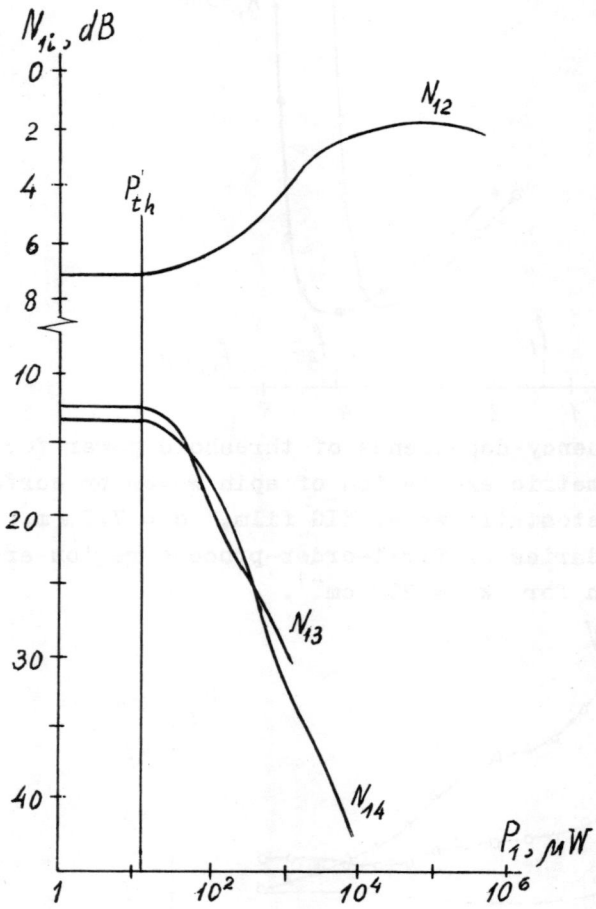

Fig.9. Energy losses for transmission from port 1 into ports 2, 3 and 4 of the device shown in Fig.6. Thickness of the film and frequency are the same as in Fig.8. $k = 200$ cm^{-1}.

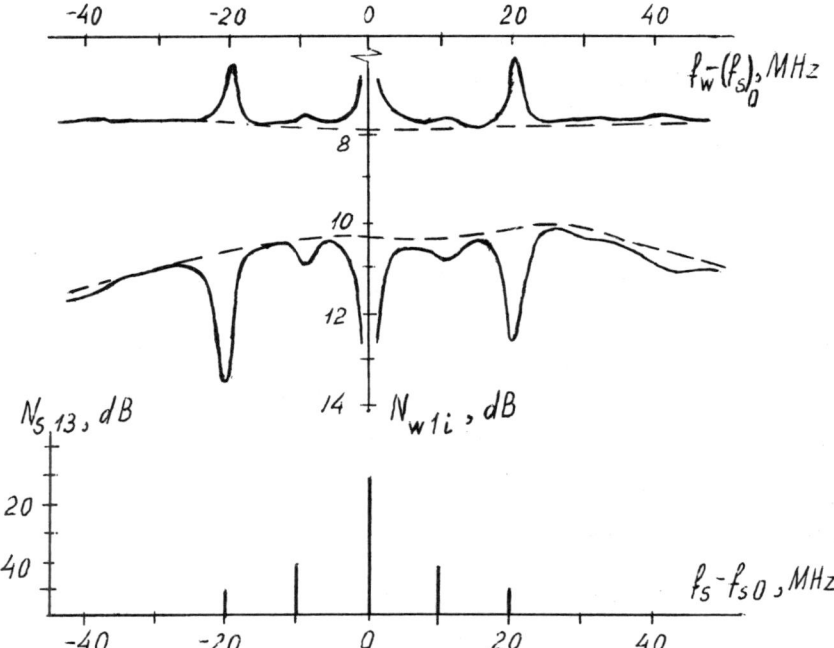

Fig.10. Losses of energy of weak magnetostatic wave vs the difference between its frequency and frequency of the strong wave. The strong-wave spectrum is also shown. YIG film, $d = 26 \mu m$, $H_{10} = 297$ Oe. Solid lines correspond to $P_{1s} = 270 \mu W$, dashed lines – without strong wave.

COUPLED FERROMAGNETIC RESONANCES OF IRON FILMS

M. Pomerantz, J. C. Slonczewski, and E. Spiller
IBM Research Division, T. J. Watson Research Center,
Yorktown Heights, N.Y. 10598
USA

ABSTRACT

Samples of two films of Fe, separated by a layer of carbon, were studied by ferromagnetic resonance (fmr). For Fe thicknesses less than 4 nm the resonance fields depended on the film thicknesses. This allowed the resonances of such films to be resolved when their thicknesses were sufficiently different. For a given pair of Fe films the thickness of the C spacer was varied. A variety of behavior was observed, which we attribute to coupling between the films. Calculations of the fmr when the films are coupled by an isotropic exchange interaction were made, for both ferromagnetic and antiferromagnetic coupling. The calculated fmr have similarities to the observed, including a dramatic 'splitting' of one absorption line into two. Some differences between the theory and the experiments also remain.

1. INTRODUCTION

The exchange interaction arises from the overlap of wavefunctions, and thus has a range of the order of atomic sizes. It is estimated that the range of the exchange field is perhaps 1 nm in metals[1] and 0.1 nm if the separation is non-metallic[2]. For many years[3] there have been attempts to observe the exchange interaction between separated macroscopic magnets. This requires separations and smoothness on the scale of 1 nm or better. The difficulty in achieving this has limited the success of the early work. With the improvement of thin-film preparation techniques it seems feasible to make magnetic layers and spacers that have the necessary properties, particularly in the case of metallic spacers where the spacing may be larger. Recently there have appeared a number of works involving metallic spacers[4-6] where exchange couplings have effects. The present authors have been studying Fe films separated by spacers of amorphous carbon (a-C). In a previous paper[7] we showed that there was no exchange coupling > 100 Oe. for C thicknesses ≥ 1nm. Thus, the absence of exchange coupling indicates that the C was non-metallic. This is expected from the observations[8] that the optical band gap of a-C is 0.4 - 0.7 eV. Carbon is a poor conductor and in the thicknesses used in our work,(less than 2 nm), the carriers may be so sparse as to be ineffective, or they may be localized. In the present paper we present results on the experiments and theory of films in which the C is so thin that magnetic coupling is observed. Some background information from our earlier work is reviewed first, and then some experimental data are presented. Calculations of the angle dependence of fmr for various strengths of exchange coupling are presented. Both ferromagnetic and antiferromagnetic couplings are considered. These model calculations have some, but not all of the features seen in experiments.

    The preparation and characterization of these films has been described previously[9, 10]. The strong reason for choosing the materials Fe and C is that these have been shown[11] to form layers whose smoothness is of the order of 0.3 nm. This value is determined from x-ray reflectivity, which is affected by waviness of the layers over distances of mms. Thus the uniformity of the thicknesses of the layers may be even better than 0.3nm. Since the range of the exchange interaction is expected to be of this scale, it is necessary to have such smoothness. Otherwise the situation becomes more complicated because of other effects, such as dipolar fields and pinholes. The films are made by e-beam evaporation at pressures of about 10^{-6}torr. The procedure followed was to coat a set of Si wafers with a 3 nm layer of C, which actually tends to smooth the already very smooth surface. Then a layer of Fe of about 1.7 nm was deposited on all the substrates. The spacer layer of C was deposited next, but during the evaporation a shutter was moved across the wafers so that different thicknesses of C covered each Fe film. Then a second Fe film of different thickness, about 1.2 nm, was evaporated on all the samples. Thus the Fe films are as nearly identical as possible on all the samples. A final overcoating of 3 nm of C was deposited to protect the Fe. In this paper we shall discuss the results for only this particular set of Fe thicknesses. Other samples were prepared and some variability of results was found, which shall be detailed elsewhere.

The choice of Fe and C may seem strange if one recalls that C can be an important impurity in Fe, necessary for the formation of steels. However, the diffusion of C in Fe is only substantial at elevated temperatures[12], which our samples do not experience. At ordinary temperatures C hardly diffuses or reacts with Fe. Our samples are another evidence of these facts, since the Fe films of thickness about 2 nm or less have retained their magnetic properties for more than three years. If there were substantial diffusion between Fe and C the films would have deteriorated by now. However, there may be reaction and diffusion between the Fe and the C at the time of evaporation. Mossbauer experiments[13] give evidence of a layer of iron carbide about 0.5 nm thick at the interface of evaporated Fe and C. This may explain, in part, our observation[9] that the effective magnetization of these films decreases as their thickness decreases. Since carbides have lower magnetizations than pure Fe, the relative amount of carbide affects the average magnetization. A 0.5 nm interface of Fe carbide would represent a larger fraction of the total magnetic layer as the layer was made thinner.

The statement that the effective magnetization varies with the thickness is another way of expressing the observation[9] that the fmr fields depend on the thicknesses of the films. We use this property to prepare two films whose resonances are resolved by making their thicknesses different enough. This is illustrated in Fig. 1, which shows the fmr of films separated by about 4 nm of C. The resonances were observed at various angles in order to show the well known fact that the resonances cross, which happens at about 82 °. We expect that the films are magnetically uncoupled at this spacing, so that the spectrum is the simple superposition of the resonances of the two films. The results of a calculation of the superposition of resonances are illustrated in Fig. 2. (The method of calculation will be described below.) The parameters of the two films that gave a good fit were $4\pi M$ = 0.73 T and 0.87 T, thickness 1.2 and 1.6 nm, g = 1.95 and 1.97, respectively.) The good agreement of theory and experiment supports the assumption that the resonances are uncoupled.

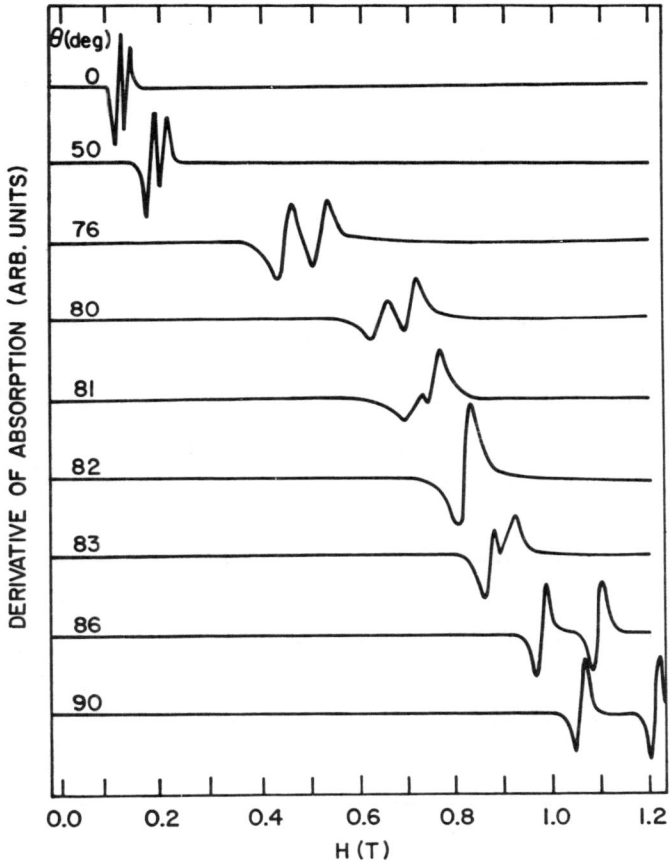

Fig. 1. Fmr of Fe films of thicknesses about 1.7 nm and 1.2 nm, separated by about 4 nm of C. θ is the angle between the dc magnetic field and the plane of the film. Frequency = 9.4 GHz. The amplitudes at angles 0 and 50° are reduced by factors 5 and 3, respectively, compared to the data at other angles.

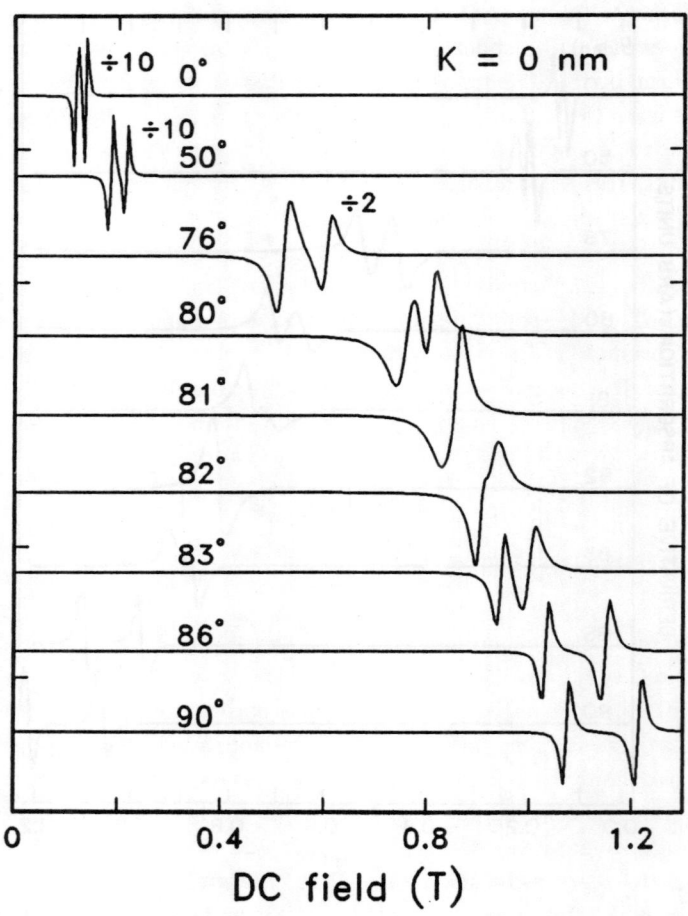

2. Calculation of the fmr, with parameters chosen to give a good fit (see text). The total absorption is the simple superposition of the fmr of the two films, i.e., coupling, K = 0.

In contrast to the uncoupled case, we also observe intermediate and strong coupled regimes as the spacing is reduced. The strong coupled regime is characterized by there being one resonance line at all angles. It was found[10] that there were two different behaviors in this case. In one, the resonance corresponded to a single film of larger magnetization than either of the component films. This we explained as the physical amalgamation of the two films into one thicker one. As mentioned above, thicker films have larger effective magnetizations. In the second strongly coupled case, the single resonance occurred at fields intermediate to the component films. This was interpreted as the resonance of two distinct but strongly coupled films. The resonances then were predicted to occur at a particular weighted average of the separate resonances. The occurrence of one case or the other was related to the vacuum system pressure which affects the amount of impurity at the junction of the two films. We wish to consider in more detail now the intermediate coupling case.

2. THE INTERMEDIATE COUPLING REGIME

By 'intermediate coupling' we mean the kind of behavior illustrated in Fig. 3. Here the C spacer is about 0.8 nm thick. We observe a single resonance at an angle of 0°, i.e., when the external field is parallel to the plane of the films. Recall that these Fe films were made simultaneously with those shown in Fig. 1 where two resonances were seen at this angle. As the angle is increased toward the perpendicular we continue to observe one line, until an angle of about 82°. Then a line emerges from the high field side of that resonance. At perpendicular there are two distinct lines. Thus 'intermediate' coupling will denote the case in which there is *one* line at low angles and *two* lines at perpendicular.

One possible explanation of this observation is that the lines at low angles overlap, and merely give the appearance of being a single line. We check this by calculating the resonance shapes, with parameters of the films chosen to get a reasonable fit at 90°, and superposing the two lines. As shown in Fig. 4, the shapes of the resonances clearly are that of two poorly resolved lines below 80°, rather than the single line observed at these angles. The assumption of simple superposition fails, so we are lead to consider the effects of coupling between the films.

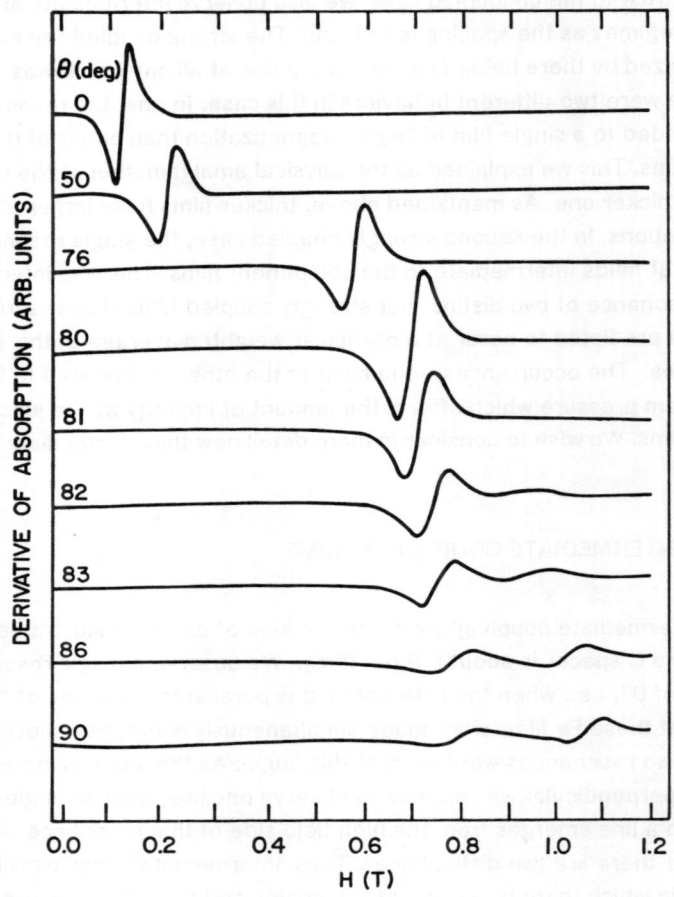

Fig. 3. Fmr of Fe films evaporated simultaneously with those of Fig. 1. The Fe film separation is about 0.8 nm. The amplitudes at angles 0, 50 and 76 ° are reduced by factors 4.6, 3.7, and 2.6, respectively, compared to the data at other angles.

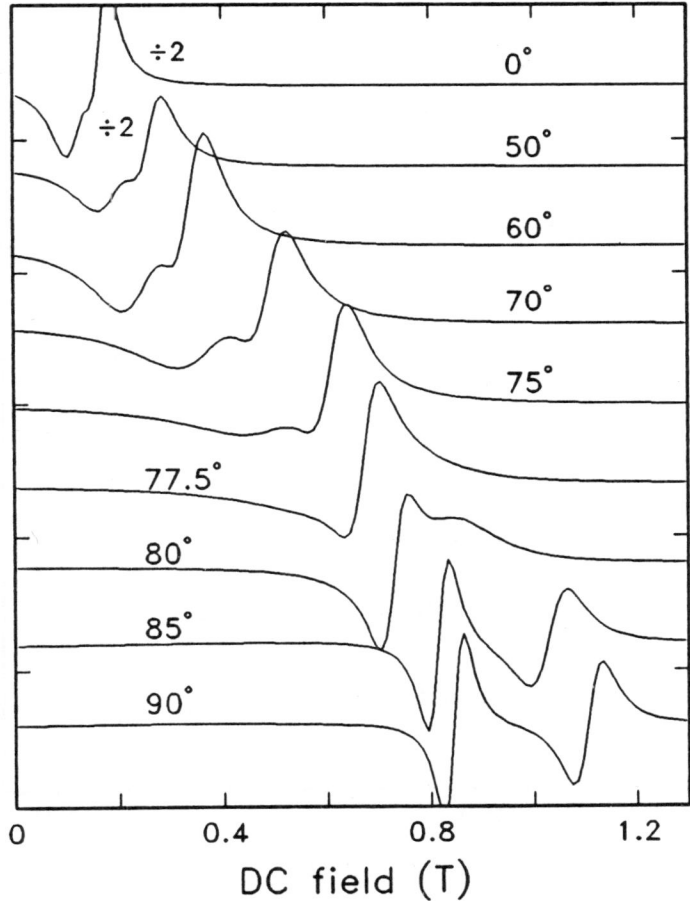

4. Calculation of the fmr, with parameters chosen to give a good fit at perpendicular resonance. The resonances of the two films are assumed uncoupled ($K = 0$).

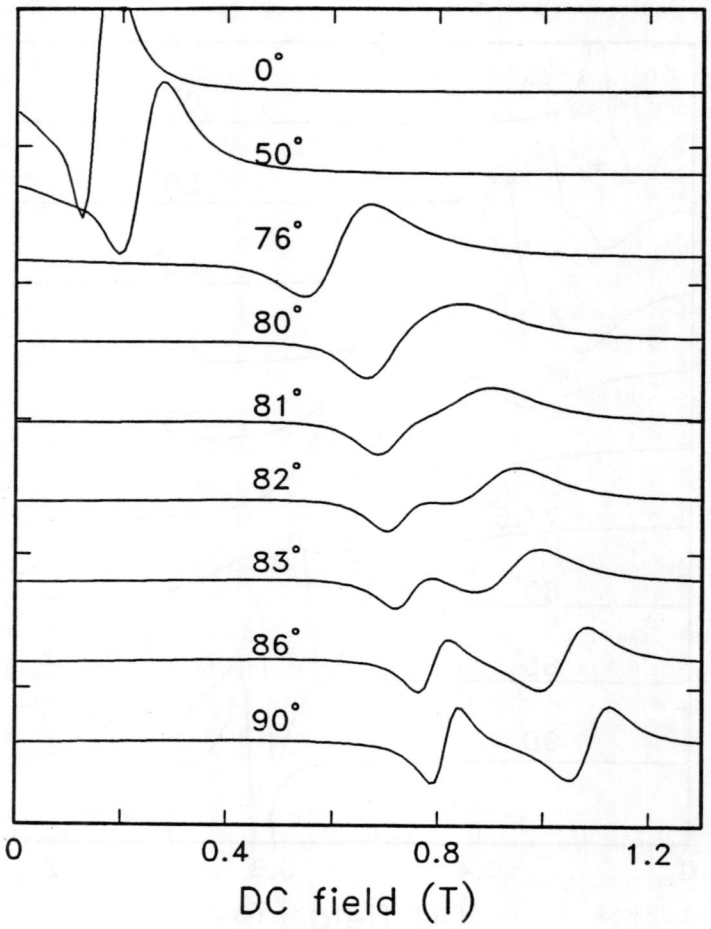

5. Calculation of the fmr, with parameters chosen to give a good fit at perpendicular resonance. The resonances of the two films are assumed coupled by about 300 Oe. (K = 0.8 nm).

In modeling the coupled magnetic films it is important to know the mechanism of coupling. We are not sure whether, in addition to exchange coupling, there are significant dipolar fields, or pinhole connections between the films. (Pinholes are sufficiently few that the films do not act as a single thick film.) We have calculated the fmr under the assumption that the coupling is by exchange fields acting through the barrier. This model is formulated by adding to the equations of motion of the magnetizations of the separate films, $\vec{M}_a$ and $\vec{M}_b$, a term $Ka^{-1}\vec{M}_b \times \vec{M}_a$ representing the torque on $\vec{M}_a$ due to exchange interaction with $\vec{M}_b$, where K is an average exchange parameter which is a steep function of the separation between the Fe films. (The exchange field is roughly $H_e = 370 \times K(Oe.)$.) The film thicknesses are a and b. This gives the coupled Landau-Lifshitz equations:

$$d\vec{M}_a/\gamma dt = \vec{H}_a \times \vec{M}_a + Ka^{-1}\vec{M}_b \times \vec{M}_a \qquad (1)$$

Here $\vec{H}_a = \vec{H}_{ext} - 4\pi M_{az}\hat{z} + Re\vec{h} \exp(i\omega t) - \lambda d\vec{M}_a/dt$, where $\vec{H}_{ext}$ and $\vec{h}$ are the applied static and microwave magnetic fields, respectively, and λ is a damping parameter. The anisotropy or other fields could be included, with an increase in complexity but no new physics. The equation for $\vec{M}_b$ is obtained by interchanging a and b. These equations were solved for the equilibrium positions of $\vec{M}_a$ and $\vec{M}_b$. Then the frequencies of small oscillations about these positions were found.

This was done for several angles of the external field with respect to the surface. The results for the cases of no coupling were already shown as Figs. 2 and 4. With K = 0.8 nm (corresponding to an exchange interaction of about 300 Oe.) the resonances appear as in Fig. 5. At angles less than 80° there seems to be only one resonance. At about this angle a second resonance appears. At 90° the lines are quite distinct. These results show the qualitative behavior observed in Fig. 3, but are not quite satisfactory in detail. The theory predicts that for ferromagnetic coupling the higher field line is stronger than the low field line for all angles. This is a property of the symmetric mode of a *ferromagnetically* coupled system. In the experiment, however, it appears that the lower field line is stronger; the upper field line seems weak as it pulls out of the lower line. This can also be seen in Figs. 6 and 7, which show the resonances at perpendicular and parallel fields, respectively, for a set of samples. The C spacer thickness in curve (a) is about 4 nm., and is decreased in (b) to

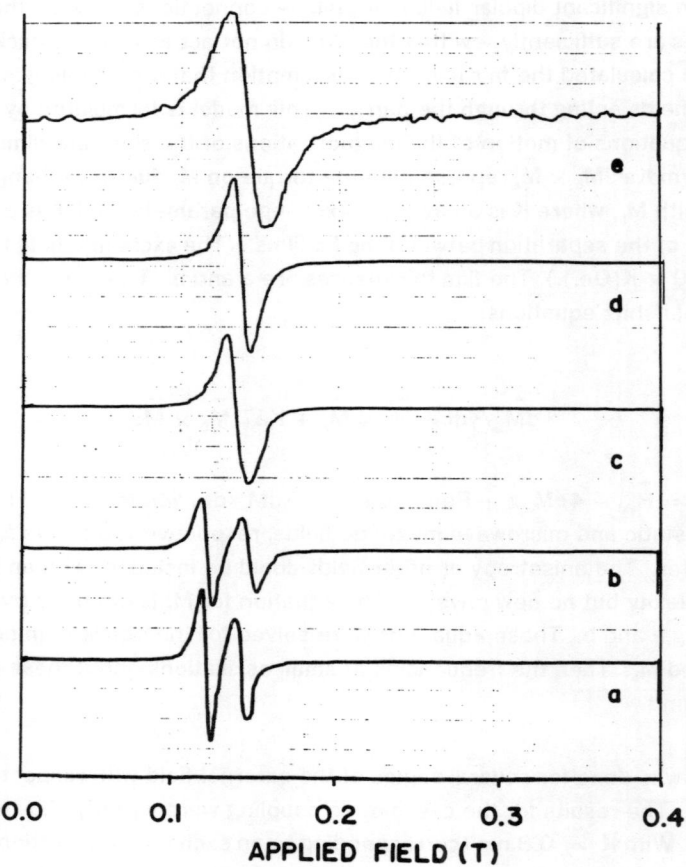

6. Measured fmr in parallel orientation, for C spacer thicknesses (nm) of a) 4 ; b) 2 ; c) 1.2 ; d) 1.0 ; e) 0.8.

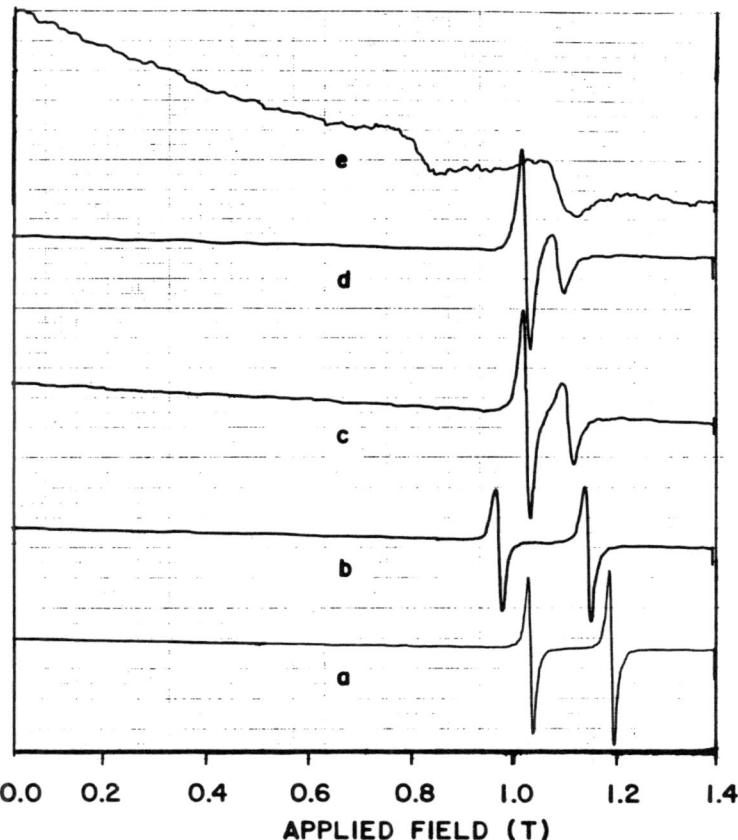

7. Same as Fig. 6, except that the external field is perpendicular to the film.

2.0 nm, (c) 1.2 nm (d) 1.0 nm (e) 0.8 nm. Samples c, d, and e are cases of intermediate coupling, presumably of progressively greater coupling because the C spacing is decreasing. In samples c and d the high field line seems to have less intensity than the lower field line.

Thus we are lead to consider the possibility of *antiferromagnetic exchange interaction*. By general reasoning, the lower field line would be relatively more intense with this kind of coupling. We attempted to calculate the angle dependence of the coupled resonances, but for antiferromagnetic exchange interaction at a general angle there are calculational problems we have not yet solved. We can, however, properly calculate the resonances in the special cases of $\theta = 0°$ and $90°$. Some results are shown in Figs. 8 and 9. where the resonances are shown at several magnitudes of coupling. The expectation that the lower field line tends to be stronger is confirmed. One also observes that for couplings $\geq$ -0.8 nm (about 300 Oe.) there is only one strong line at parallel field but two lines at perpendicular. Thus the qualitative features of the measurements are obtained.

A problem that remains is that the coupling does not quantitatively give the observed resonance fields. If we assume that the films of Fig. 3 are the same as Fig. 1, then we would expect that the parameters of Fig. 2 could be used for calculations including coupling. When this is done, as in Figs. 8 and 9, the resonance positions do not agree with those seen in Fig.3. In the coupling range up to 300 Oe. the coupling does not tend to cause a lowering of the resonance fields toward the values seen in Fig. 3, or sample e of Fig. 7.

The problem may not be a theoretical one, however. Experimentally, there seems to be some irregular behavior as the C thickness is varied in this range of very thin layers. For example, we observe differences between samples a and b in Figs. 6 and 7, although both appear to have uncoupled resonances. Thus there seems to be some variability of the effective magnetization of the samples despite that they were made simultaneously and adjacent to each other. This could arise because the C spacer layer is reacting with the Fe, and the quantity of C available for reaction is different for the different samples. Also background impurity that attaches to the C spacer may depend on the the precise nature of the surface which might depend on the amount of carbide formed.

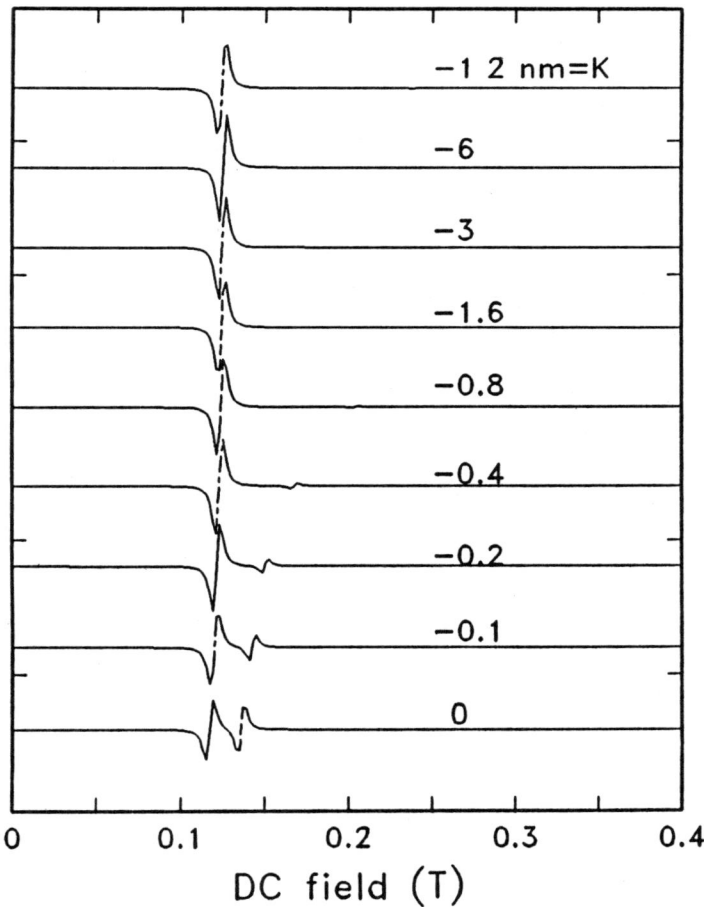

8. Calculated fmr for various values of the coupling parameter, K. The negative sign represents antiferromagnetic coupling. The Fe film parameters are the same as used in Fig. 2. Fmr in parallel orientation.

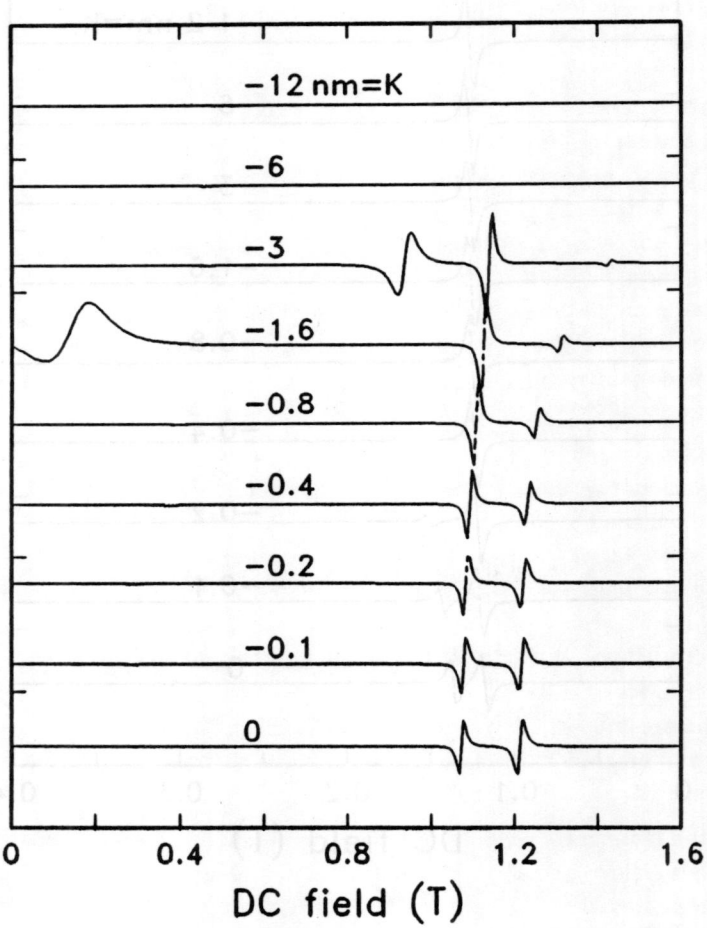

9. Same as Fig. 8, except for the perpendicular orientation.

3. CONCLUSIONS

We have prepared samples of Fe films in the thickness range of less than 2.0 nm., and separated them with layers of C. The effective magnetization of the films depends on their thicknesses, possibly because of the formation of carbides when the films are deposited. Once formed, however, the magnetic properties are stable for years. As the thickness of the C spacer is varied we observe changes in the fmr which are not explainable by the superposition of the two resonances. Calculations that include the exchange interaction as a coupling term in the equations of motion give better qualitative agreement with the experiments. Calculations of a somewhat similar model by Cochran, et al,[14] for the parallel resonance case show similar changes in the intensities of coupled resonances as the coupling is changed in magnitude and in sign.

The discrepancies that remain between theory and experiment are likely due to lack of control of the formation of the interface of the Fe layers and the C spacer. Since the exchange interaction is such a sensitive function of distance, small variations can have large effects. One of us (J. C. S.) has shown that the exchange effect can be treated as a tunneling phenomenon[15]. Thus, the details of the potential barriers caused by the spacer are important. One of the predictions of this theory is that the exchange interaction may be either ferro- or antiferromagnetic depending on the barrier properties. This may explain the previously puzzling observation[7] that the coupling in our Fe/C layers appears to be antiferromagnetic.

As the techniques of multilayer film fabrication improve it will become possible to eliminate effects due to chemical and structural uncertainty. The angle dependence of the fmr should then be a most useful method for the quantitative measurement of the exchange interaction between magnetic films. We believe we have seen the kinds of phenomena that should be found in all exchange coupled films.

4. REFERENCES

1. Grunberg, P., J. Appl. Phys. 57, 3673-3677 (1985).
2. White, R. M. and Friedman, D. J., J. of Magn. and Magn. Mat. 49, 117-123 (1985).
3. Yelon, A. in "Physics of Thin Films", ed. Francombe, M. H., and Hoffman, R. W., (Academic Press, New York,1971) 205.   reviews early work on interacting magnetic films .
4. Salamon, M. B., Sinha, S., Rhyne, J. J., Cunningham, J. E., Erwin, R. W., Borchers, J., and Flynn, C. P., Phys. Rev. Lett. 56, 259-262 (1986).
5. Grunberg, P., Schreiber, R., Pang, Y., Brodsky, M. B., and Sowers, H., Phys. Rev. Lett. 57, 2442 - 2445 (1986).
6. Majkrzak, C. F., Cable, J. W., Kwo, J., Hong, M., McWhan, D. B., Yafet, Y., Waszczak, J. V., and Vettier, C., Phys. Rev. Lett. 56, 2700-2703 (1986).
7. Pomerantz, M., Slonczewski, J. C., and Spiller, E. in "Proceedings of the International Symposium on Physics of Magnetic Materials", ed. Takahashi, M., Maekawa, S., Gondo, Y., Nose, H., (World Scientific Publ. Co, Singapore,1987) 64-71.
8. Robertson, J., Adv. in Phys. 35, 318 (1986).   reviews the properties of a-C .
9. Pomerantz, M., Slonczewski, J. C., and Spiller, E., J. of Magn. and Magn. Matls. 54-57, 781-782 (1986).
10. Pomerantz, M., Slonczewski, J. C., and Spiller, E., J. App. Phys. 61, 3747 (1987).
11. Spiller, E., Mat. Res. Soc. Symp. Proc. 56, 419-433 (1986).
12. Bozorth, R. M. "Ferromagnetism", (Van Nostrand, New York,1951).
13. Shinjo, T., Kawaguchi, K., Yamamoto, R., Hosoito, N., and Takada, T., Sol. State Comm. 52, 257 (1984).   and private communication from Prof. T. Shinjo .
14. Cochran, J. F., Heinrich, B., and Arott, A. S., Phys. Rev. B 34, 7788 (1986).
15. Slonczewski, J. C., Phys. Rev. B, submitted (1988).

ENHANCED MAGNETO-OPTICAL EFFECTS OBSERVED IN
COMPOSITIONALLY MODULATED FILMS

Tadataka Morishita

NHK Science and Technical Research Laboratories,
Setagaya, Tokyo 157, Japan

Compositionally modulated films, which are composed of magnetic and dielectric layers, are suitable for interference enhancement of the magneto-optical effects. A Kerr rotation angle of more than 25 degrees has been observed for the sample made of bilayers of 100 A FeTb / 150 A SiO.

1. INTRODUCTION

It is still left as a challenging problem to find a new material showing large magneto-optical effects. Because the magnetic Kerr rotation angle is usually not large enough to obtain a satisfactory signal-to-noise ratio for optical readout of magnetically recorded informations.
On the other hand, interference enhancement of the Kerr rotation is also very attractive since it is realized by using usual materials. Conventional interference techniques coating a dielectric layers on a film surface require certain relations between refractive indices of dielectric and magnetic materials, and also between a thickness of

dielectric layer and the wavelength of light used.

A compositionally modulated (CM) structure consisting of magnetic and transparent dielectric layers enables us to enhance the Kerr effect by a novel method. In a CM film the enhancement is caused by a combination of multi-reflections in the protective surface layer and inner dielectric layers sandwiched by magnetic layers, as schematically shown in Fig. 1. The latter becomes more important for contribution to a resultant Kerr rotation as the layer thickness decreases.

The CM films, however, show very often an anomalous Kerr rotation; the maximum Kerr rotation is attained not always in the magnetic state where magnetizations in all the magnetic layers align along an applied magnetic field but in the state where magnetizations point antiparallelly in some of layers. So the magneto-optical properties in a CM film including interference effects should be completely investigated for a guiding principle to obtain a large remanent Kerr rotation.

In this paper, first, Kerr and Faraday hysteresis loops observed in CM films made of three FeTb layers with alternating SiO_2 layers are interpreted through the calculation for the model shown in Fig. 1, and it is demonstrated that interference effects are essential in determining magnitude and sign of the magneto-optical rotations[1]. Secondly, a strongly enhanced Kerr rotation is presented for CM films of an FeTb/SiO system of which magnetic properties have been examined taking consideration of their application to magneto-optical recording[2,3,4].

2. EXPERIMENTAL

Samples were prepared on glass substrates by two methods: an alternate rf magnetron sputtering with a SiO_2 and a mosaic FeTb targets, and an alternate evaporation from Fe, Tb and SiO sources. The deposition rates were 1 A/sec or less for all constituents. The FeTb layer thickness is limited to less than 200 A, which is thin enough to transmit visible light through several layers. The SiO_2 and SiO layers ranged from 10 A to 200A in thickness. The protective layer (500 to 700 A) was also deposited on the top with a corresponding silicon oxide in order to prevent oxidation and to enhance the Kerr rotation. The Kerr rotation spectra were measured from the air side of a sample for light of wavelengths from 4000 to 9000 A in a magnetic field of 4 kOe. Kerr and Faraday hysteresis loops were traced out in a field up to 10 kOe for several wavelengths. The magnetization and coercivity were measured with a vibrating sample magnetometer. The uniaxial magnetic anisotropy constant was determined with a torque magnetometry.

3. RESULTS AND DISCUSSION

3.1 Interference Effects in $FeTb/SiO_2$ CM Films[1]

Figure 2 shows Kerr hysteresis loops which are observed ; (a) from the film side, (b) through the substrate, for the film consisting of three FeTb layers separated by intermediate SiO_2 layers by using three light sources of 4000, 6000 and 8000 A. The thickness of FeTb layer is 100 A and that of SiO_2 layer is 30 A. The thickness of the top layer is 500 A. The shape of hysteresis loops strongly depends on the side from which they are observed. It should be noticed that none of hysteresis loops show a maximum value of Kerr rotation at the remanent state. This makes a remarkable contrast with Faraday hysteresis loops presented in Fig. 3. All the loops, however, show a characteristic shape having steps whose number is equal to that of FeTb layers. Each step occurs at a fixed magnetic field independent of the wavelength of light and the side of a sample. We, therefore, have attributed the step to reversal of the magnetization belonging to one of FeTb layers. The existence of the steps, at the same time, means that coercivity differs layer by layer.

The observed hysteresis loops are decomposed into three component loops which enable us to estimate the Kerr rotation and Faraday rotation angles, including sign, of each layer. Decomposition of the observed Kerr (Fig.2 a) and Faraday (Fig.3 a) loops is shown in Fig.4 (a) and (b), respectively. From results of the decomposition of the Faraday hysteresis loops, it is clear that an individual FeTb layer has an almost equal Faraday rotation angle with the same sign. The observed loops, therefore, show the maximum at the remanent state. While, in the Kerr hysteresis loops, each layer has a Kerr rotation angle with a different magnitude and sign although FeTb layers were deposited under the same conditions. Moreover the sign of the Kerr rotation angle changes depending on the wavelength of light. For example, the FeTb layer corresponding to the second step has the rotation angle of the positive sign at 4000 and 8000 A, but the negative sign at 6000 A. These results suggest that the Kerr rotation in CM films cannot be explained by simply adding contributions from FeTb layers.

Calculations have been made for the magneto-optical effect in CM films using the method which takes interference effects into account[5]. First, using Fresnel coefficients for the boundary between the substrate and the first FeTb layer we have calculated the reflection and transmission coefficient for the second boundary between the first FeTb layer and the adjacent SiO_2 layer, considering a phase shift due to traveling in the FeTb layer. Using the coefficients for the second boundary those for the third boundary have been calculated in the same way. Successive calculations have given the coefficients for the boundary between the top SiO_2 layer and air, which are the reflectivity and transmissivity for a CM film. The results involve the effects of multi-reflection and transmission at all the interfaces of the film. We have calculated the Kerr and Faraday rotation angles using wavelength dependent refractive indices for the right- and left-handed circularly polarized light estimated from the reported dielectric constants of FeTb alloy[6]. Optical constants reported by Philipp[7] were used for SiO_2. As is seen from Fig.5 the calculated loops well reproduce the observed Kerr and Faraday hysteresis loops in Figs. 3 and 4.

We next calculated the dependence of the Kerr rotation of each

FeTb layer in the FeTb/SiO$_2$ CM film on the FeTb layer thickness. The thicknesses of SiO$_2$ and protective layers are fixed at 30 A and 800 A, respectively. The wavelength of light is 6328 A. As is seen from results in Fig. 6(a), all the FeTb layers have the negative sign when the FeTb layer is thinner than 25 A. However such FeTb layers are too thin to show magnetization perpendicular of the film. The same kind of calculation, in Fig. 6(b), has been made for FeTb/ SiO CM films composed of dielectric having a larger refractive index. In this case the FeTb layers show the Kerr rotation of the same sign in a wider range of thickness.

3.2 Strongly Enhanced Kerr Rotation in FeTb/SiO CM Films[2,3,4]

Magnetization perpendicular to the film is necessary for the polar Kerr rotation. We should, first, investigate magnetic anisotropy of FeTb/SiO CM film. It should be noted, in Fig. 7, that perpendicular anisotropy is still dominant in a CM film with a FeTb layer thickness of less than 50 A. The uniaxial anisotropy constant (K_u) of FeTb/SiO CM film is larger than that of Fe/Tb CM film as shown in Fig. 8. The perpendicular anisotropy seems to be additionally induced at the interfaces between FeTb and SiO layers. Coercivity (H_c) and magnetization (M) are also shown as a function of the FeTb layer thickness. Thermal stabilities were examined by annealing samples at 150 oC, for 1 hr., in $1 * 10^{-5}$ Torr, and results are presented in Fig. 8.

The dependences of the Kerr rotation and reflectivity on the FeTb layer thickness at a wavelength of 6328 A are presented in Fig. 9. The FeTb layer thickness at which a sharp increase in the Kerr rotation occurs, strongly depends on the wavelength of light. The enhancement of the rotation angle is inevitably accompanied by a reduction of the reflectivity because enhancement by means of multi-reflection becomes important under the conditions for like an anti-reflection.

Figure 10 is observed Kerr rotation spectra for typical three samples with a constant FeTb layer thickness. The wavelength at maximum in a spectrum strongly depends on the thickness of the protective layer. In order to investigate an effect of the CM structure on the Kerr rotation spectrum, calculations were carried out in consideration of summing up multiple beams, as shown in Fig. 1, which are multi-reflected and pass through FeTb layers in the CM film[8]. The calculated Kerr rotation spectra are shown in Fig. 11. The spectrum(a) is for the film with a protective SiO layer of 650 A on a thick(500 A) FeTb layer. The spectra (b) and (c) are for CM films composed of five FeTb layers alternating with four SiO layers and a protective layer. To obtain the spectrum(d) we have varied the protective layer by 20 A in thickness for the case (c). As is evident from comparing the spectrum(a) with others, the CM structure is suitable for the enhancing the magneto-optical effects.

The CM structure has an advantage of satisfying the conditions for interference enhancement by adjusting a thickness ratio of the magnetic to dielectric layer, because it is possible to look on the dielectric constant, that is the square of the refractive index, averaged over both kinds of layers as an effective dielectric constant of the films. It is very important to notice that the magnetic properties remain unchanged as far as the CM structure exists.

As mentioned above the interference enhancement usually goes with the reduction of the reflectivity. This is crucial for conventional applications as magneto-optical recording. We, however, expect that the giant Kerr rotation will develop a new application by itself.

4. CONCLUSION

The magnitude and sign of the Kerr and Faraday rotation of each FeTb layer in a CM film have been determined through decomposition of observed hysteresis loops into component loops. The experimental results have been reproduced by calculations including interference effects in the CM structure. The guiding principle for obtaining a large remanent Kerr rotation has been described. A kerr rotation angle of more than 25 degrees has been observed for FeTb/SiO CM films. Calculations have demonstrated that the Kerr rotation angle runs up to 80 degrees.

ACKNOWLEDGMENT

I am grateful to Professor K. Sato of Tokyo University of Agriculture and Technology, Messrs. R. Sato and N. Saito and Ms. M. Kajiura as coworkers during the course of this study ,and also would like to thank Professor K. Tsushima of The Royal Institute of Technology and Professor T. Ando of Tokyo University for suggestive discussion.

References

1) Saito, N., Sato, R., Morishita, T. and Kajiura, M., to be published in Proc. of MRS Symp.K, (1988) Tokyo.

2) Sato, R., Saito, N., Morishita, T. and Kajiura, M., to be published in Proc. of JIMIS-5, (1988) Kyoto.

3) Sato, R., Saito, N. and Morishita, T., presented at Joint MMM-INTERMAG, (1988) Canada, BE-07

4) Morishita, T., Sato, R., Sato, K. and Kida, H., to be published in Proc. of ICM, (1988) Paris.

5) Crook, A.W., J. Opt. Soc. Am. $\underline{38}$, 954 (1948).

6) Allen, R. and Connell, G.A.N., J. Appl. Phys.,$\underline{53}$ 2353 (1982).

7) Philipp, H.R., "Handbook of Optical Constants of Solids", (Academic Press, New York) 1985, p.765.

8) Ohta, K., Takahashi, A., Deguchi, T., Hyuga, T., Kobayashi, S. and Yamada, H., Proc. of SPIE, $\underline{382}$, 252 (1983).

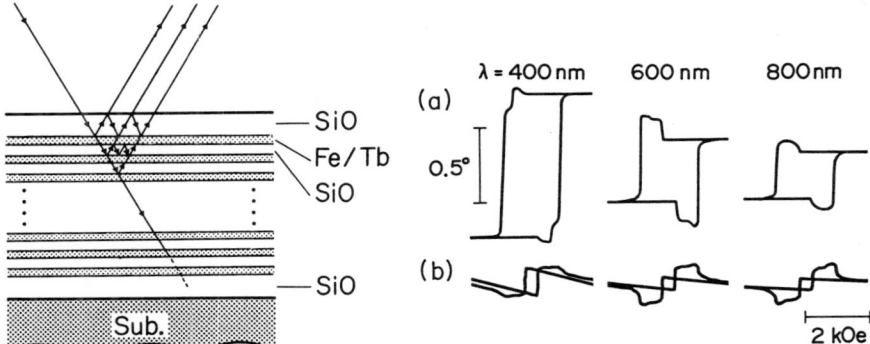

Fig.1: Schmatic structure of the compositionally modulated film with a protective layer.

Fig.2: Kerr hysteresis loops observed (a) from the film side and (b) through the substrate for a FeTb/SiO$_2$ CM film.(Ref.1)

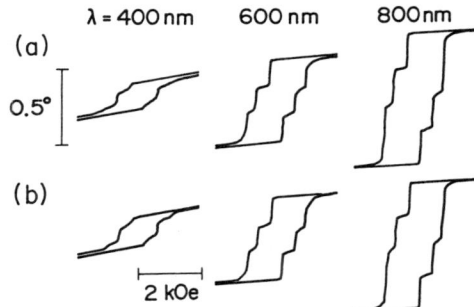

Fig.3: Faraday hysteresis loops observed (a) from the film side and (b) through the substrate for the same sample as in Fig.2.(Ref.1)

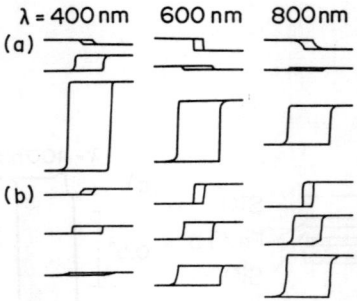

Fig.4: Decomposition of the Kerr(a) and Faraday (b) hysteresis loops observed from the film side.(Ref.1)

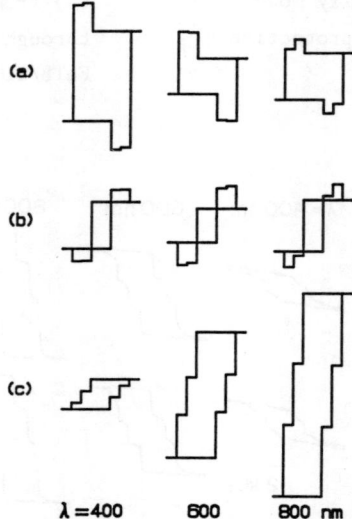

Fig.5: Calculated hysteresis loops: (a) and (b) are Kerr loops observed from the film side and through the substrate, respectively. (c) is Faraday loops.(Ref.1)

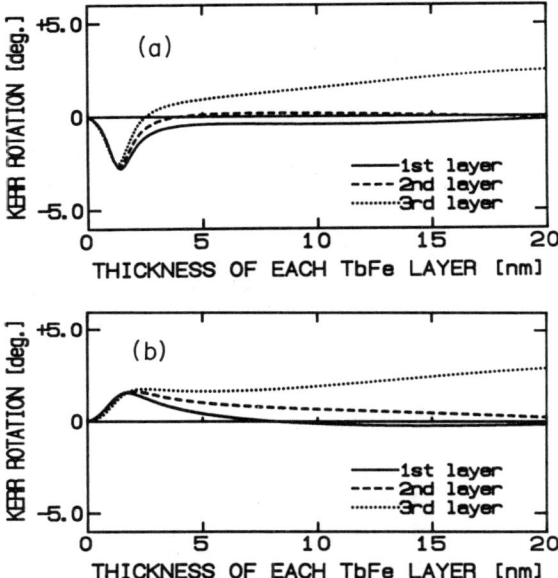

Fig.6: Calculated dependence of the Kerr rotation on the layer thickness for each FeTb layer in (a) FeTb/SiO$_2$ and (b) FeTb/SiO CM films.(Ref.1)

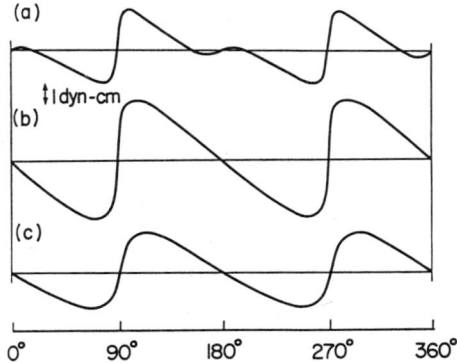

Fig.7: Magnetic torque curves for CM films with a thickness of FeTb layer of (a)41 Å,(b)52 Å and(c)75 Å.(Ref.2)

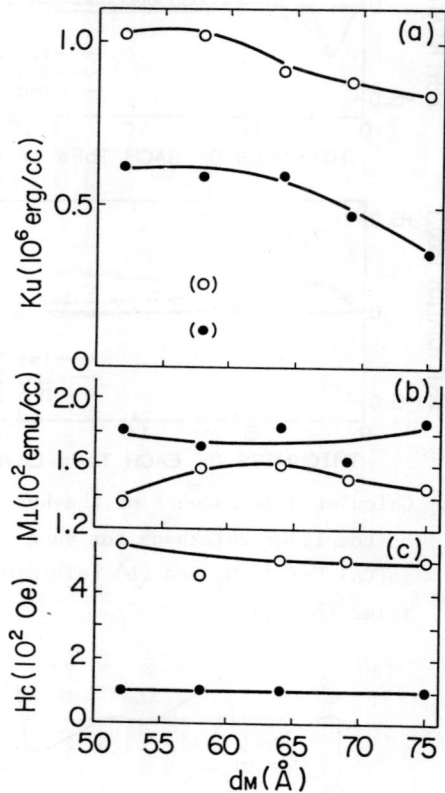

Fig.8: FeTb layer thickness dependences of (a) K_u, parentheses for an Fe/Tb CM film (b) M (c) H_c. Open circles: as deposited, full circles: annealed.(Ref.2)

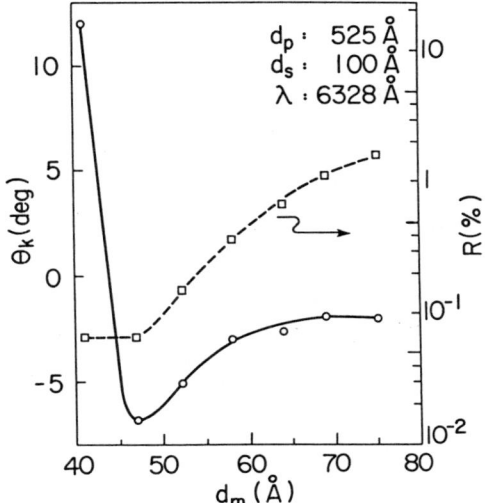

Fig.9: Dependences of Kerr rotation angle (solid line) and reflectivity (broken line) on the thickness of FeTb layer. The thickness of SiO layer is 100 Å and that of protective layer is 525 Å.(Ref.3)

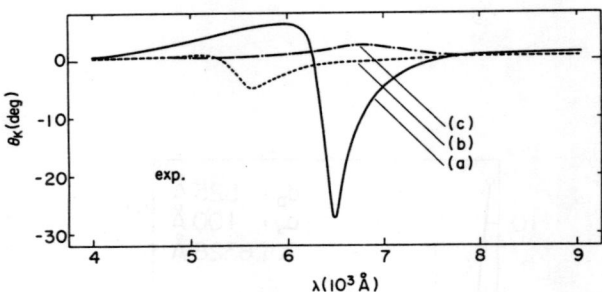

Fig.10: Observed spectra of the Kerr rotation for typical three samples: a) dt = 650 A, dm = 100 A, ds = 150 A b) dt = 560 A, dm = 100 A, ds = 150 A c)dt = 670 A, dm = 100 A, ds = 50 A, where dt, dm and ds are the thickness of the protective, FeTb and SiO layers, respectively.(Ref.4)

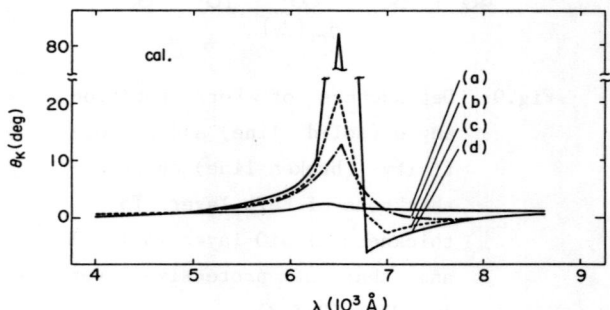

Fig.11: Calculated Kerr rotation spectra: a) dt = 650 A, dm = 500 A, ds = 0 b) dt = 650 A, dm = 100 A, ds = 250 A c) dt = 650 A, dm = 100 A, ds = 300 A d) dt = 630 A, dm = 100 A, ds = 300 A.(Ref.4)

EFFECT OF LOCAL ANISOTROPY FLUCTUATION ON THE PERMEABILITY OF SPUTTERED Fe-Si-Al ALLOY FILMS *

Migaku TAKAHASHI

Electronic Engng., Faculty of Engng.,
Tohoku University, Sendai JAPAN **

Structure and magnetic properties of Fe-Si and Fe-Si-Al (Sendust) sputtered alloy films are systematically examined. For the structure, non equilibrium phase like as disordered α was formed by changing the sputtering parameters and this α phase transforms to ordered DO_3 phase by annealing. Magnetic anisotropy and magnetostriction were determined for each phases. Evaluation of the magnetic inhomogeneities represented as magnetic ripple and skew and also of the induced uniaxial magnetic anisotropy, H_k, was carried for the films as a function of composition. Based on these fundamental results, influence of magnetic inhomogeneities on the permeability, μ_{eff}, in a high frequency is discussed in connection with the structural inhomogeneities.

*----------------

* Presented at the 4th International Conference on Physics of Magnetic Materials, Szczyrk-Bila, Poland, September 4-10, 1988
** Mailing address : Dept. of Electronic Engng., Tohoku University, Sendai, 980 JAPAN

1. Introduction

Fe based binary and ternary alloys with Si and Al represent the peculiar magnetic properties and are still interested in the field of the metalphysics and also the technical application [1-23]. With the recent progress of the high density recording, soft magnetic materials with the excellent permeability, μ, around a couple of 10 MHz and also high saturation magnetic moment, M_S, is required to realize a digital VTR devices. In particular, much attention is now repaid to the Fe-Si-Al alloys for the most useful potential candidate for the thin film VTR and also the MIG heads [1-9]. Especially for the Sendust alloy, 9.6wt%Si, 5.5wt% Al, bal.Fe in composition which shows the excellent soft magnetic properties as a maximum permeability, μ_m=162000, and initial permeability, μ_i=35100 [24]. However, from the structural view point, metallurgical phases of Fe-Si-Al alloys are fairly complex and the two types of ordered structure, DO_3, B2 and the several types of two phase field of DO_3 and B2 with disordered α are existed in equilibrium. Furthermore, it has been systematically clarified by present authors [16-23] that the intrinsic magnetic properties like as magnetocrystalline anisotropy, K_1, and the magnetostriction constants, λ_{100} and λ_{111} change drastically by the slight differences of the metallurgical structure and alloy compositions. Therefore, in the research of the magnetic films in this alloy system, experiment should be developed based on the exact knowledge for the various metal physical properties of the bulk specimens. In the present research, Fe-Si-Al alloy films with different alloy compositions are fabricated systematically by d.c. magnetron sputtering and

investigated
1) Magnetic anisotropy and magnetostriction for each metallurgical phases formed by the different sputtering condition.
2) Evaluation of the magnetic inhomogeneities represented as magnetic ripple and skew and the induced uniaxial magnetic anisotropy, H_K.
3) Influence of magnetic inhomogeneities on the permeability, μ, in a high frequency.

Within the frame work in this paper, the precise mechanisms of the frequency dependence of permeability, μ, is discussed in connection with the magnetic inhomogeneities caused by the structural defects in the films, refering to the difference of the phases formed between bulk and films.

2. Experimental Procedure

Films were prepared by d.c. magnetron sputtering onto crystallized glass substrates under an Ar pressure 5 mTorr, a current of 300 mA and a gas flow rate of 5.5 ccm. The thermal expansion coefficient of the substrates is about 130×10^{-7} /°C. The film thickness is fixed at 1.5 μm for the composition dependence experiment, and is varied from 0.3 to 1.5 μm for the thickness dependence. All films were step-wise annealed in vacuum up to 600 °C for 2 hours.

The magnitude of effective permeability is determined by the inductive method using a ferrite core. The composition of the films was determined by E.P.M.A. and the magnetostriction for the films was measured by means of an optical cantilever method.

For the quantitative analysis of the local anisotropy fluctuation, the dynamic differential susceptibility measurement proposed by H. Hoffmann [25-26] was used. The dynamic differential susceptibility, χ, experimentally obtained is represented by the following equation.

$$\chi = k \frac{Ms}{Hk} \left\{ h(\alpha) + \frac{1}{4\pi\sqrt{2}} S^2 \frac{Ms\sqrt{d}}{(A\,Ku)^{\frac{5}{4}}} h(\alpha)^{-\frac{1}{4}} + 3 <\theta^2> h(\alpha)^{-1} \right\}^{-1} \quad (1)$$

Here, $h(\alpha)$ is a normalized effective single domain field ($h(\alpha)=(H_{dc}-H_k)/H_k$). The second and third terms are effective intrinsic demagnetizing fields caused by the existence of the local anisotropy fluctuation. By fitting the experimentally obtained curves to this equation, we determined the magnitude of the structure constant, S, and of the skew angle, θ, from the coefficients of second and third terms. The structure constant, S, corresponds to the magnitude of microscopical local anisotropy fluctuation indirectly, and is a function of local magnetic anisotropy constant, grain size, and the number of grains of films along the normal axis to film plane. In crystalline films, S is represented phenomenologically as [25-27]

$$S_{D\cdot F} = \frac{K_s \sigma_1 D}{\sqrt{n}} = \frac{D}{\sqrt{n}} \sqrt{\frac{8}{105}} \sqrt{\left[K_1 + \frac{3}{8}(\lambda_{100}-\lambda_{111})\sigma_{in}\right]^2 + \frac{7}{16}\left[\frac{3}{2}[\lambda_{100}-\lambda_{111}]\sigma_{in}\right]^2}$$

$$= K_1\, f(\psi,\phi) + \lambda\, \sigma_{in} \quad (2)$$

Here, $f(\psi,\phi)$ corresponds to the distribution of the crystal orientation of each grain, σ_{in} the internal stress in the films, and λ the magnetostriction constant of the films. In the present study, S is multiplied by a coefficient equal to power of the exchange stiffness constant, $A^{-5/8}$ (the value of A is almost constant in the present narrow composition region).

3. Results and Discussion

3-1 Structure

In Fig.1, the unit cell model of B2 and DO_3 ordered structure normally found in this alloy system is shown.

As seen in this figure, the lattice is composed of two kinds of sublattice. One is simple cubic lattice of site (I), the other is NaCl type cubic lattice of site (II).

Namely, in the case of DO_3 ordered structure, Fe atoms occupy lattice site (I) and also lattice site (II).

For B2 type of ordered structure, Fe atoms occupy only lattice site (I) and the rest atoms are distributed randomly in the site (II).

Here, disordered, α, means no order of Fe atomsite in the cell, which corresponds to the usual b.c.c. structure.

In Fig.2, the phase diagrams of Fe-Si, and Fe-Al, and the isothermal section of Fe-Si-Al alloys at 600 °C is shown [28]. In Fe-Si-Al ternary alloys, two types of ordered structure, B2 or DO_3, and disordered, α, furthermore, two phases mixtures like as $B2+DO_3$, $\alpha+DO_3$ and $\alpha+B2$ are found. As clearly seen in this figure, these phases are strongly sensitive to the alloy composition and also to the heat treatment.

Therefore, the instability of the soft magnetism for these alloys mainly arises from these phase instabilities of the alloys.

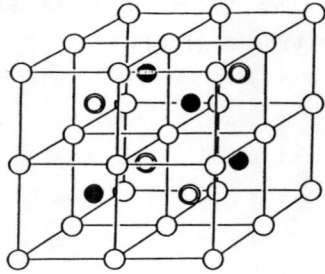
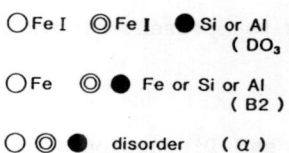

○ Fe I ◎ Fe I ● Si or Al
                 (DO$_3$)

○ Fe   ◎ ● Fe or Si or Al
                 (B2)

○ ◎ ● disorder   (α)

Fig.1 Unit cell model of B2 and DO$_3$ ordered structure

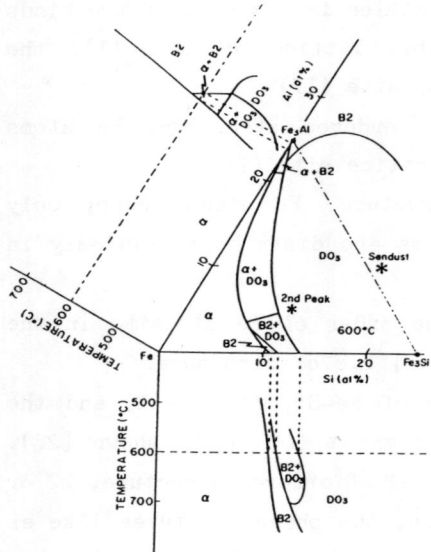

Fig.2 Phase diagram of Fe-Si-Al ternary alloys [28]. (✻) corresponds to Sendust and Second peak alloys.

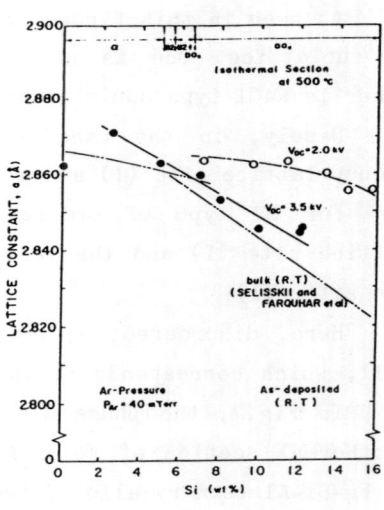

Fig.3 Lattice constant, "a", against Si concentration for Fe-Si films sputtered under two different sputtering conditions. Two — dot-dash line (———) corresponds to that of the bulk crystals reported by I. P. Selisskii[32] and M. C. Farquhar et al.[33].

While, in the case of thin films, it is easily assumed that the phase of films formed by sputtering or evaporation may be different from that of bulk samples expected from the equilibrium phase diagram. As an example, in Fig.3, the dependence of the lattice constant, "a", on Si concentration for the Fe-Si alloy films sputtered by the two different conditions is shown.

In the same figure, concentration dependence of "a" for bulk samples (two-dot-dash line) and also the isothermal section of Fe-Si binary alloy phase diagram at 500 °C are also shown.

As seen, the concentration dependence of "a" for the film sputtered under the applied d.c. Voltage, V_{DC}=2.0 kV, (O) is different from that of V_{DC}=3.5 kV (●). Namely, for the films sputtered under V_{DC}=3.5 kV, the concentration dependence of "a" is very similar to that of bulk from 5 to 12wt%Si concentration. However, for the films sputtered under V_{DC}=2.0 kV, between 5 and 16wt%Si, "a" is just on the extrapolated line from pure Fe to about 5wt%Si. These results mean that the disordered α as non equilibrium phase is formed for the films sputtered V_{DC}=2.0 kV and that the DO_3 ordered structure as equilibrium phase is formed for the films sputtered V_{DC}=3.5 kV from 7.0 up to 16wt%Si concentration. These assumption is well supported by the results of electron diffraction.

In Fig.4, the change of X-ray diffraction by annealing for the Sendust alloy film composition is shown. As seen in this figure, only fundamental reflection line is observed for the as-deposited film. However, for the film annealed at 600 °C, super lattice reflection of 111 and 200, which characterize the existence of DO_3 ordered structure, becomes to be clearly observed. This fact means that the disordered α is formed in an as-deposited state and that this non equilibrium phase, α,

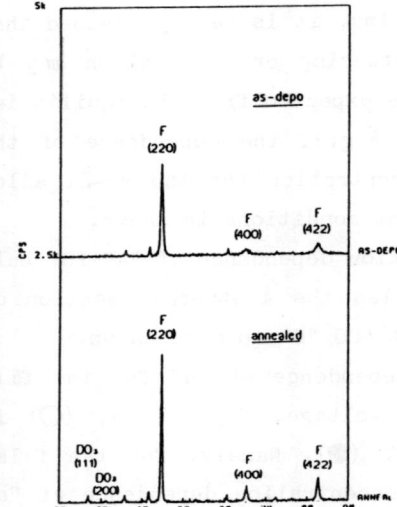

Fig.4 Change in the X-ray diffraction patterns resulting from annealing in Sendust sputtered films.

transforms to DO_3 phase as equilibrium by annealing.

As seen these examples, in the case of sputtered films of Fe-Si-Al alloy, non equilibrium phase is formed by changing the sputtering parameters and also the heat treatment.

Therefore, to realize the soft magnetic properties like as bulk samples, we must pay careful attention to the metallurgical phase of films, which are strongly connected to the sputtering condition.

3-2. Magnetostriction

Based on the structural results shown in 3-1, in this section, attention is paid to the difference of the magnetostriction between bulk and films.

In Fig.5, the concentration dependence of the magneto-

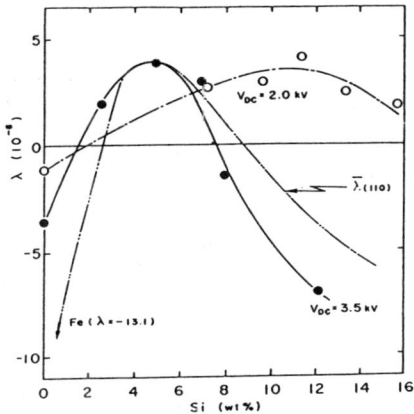
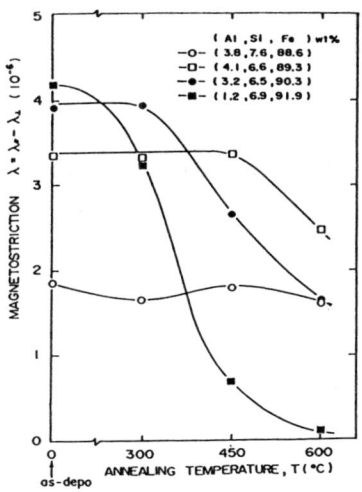

Fig.5 Magnetostriction, λ, against Si concentration for Fe-Si films sputtered under two different sputtering conditions. Two — dot-dash line corresponds to the magnetostriction, $\lambda_{(110)}$, composed from the values of the magnetostriction constants λ_{100} and λ_{111} [34] in the case of the (110) sheet texture.

Fig.6 Changes of magnetostriction, λ, against annealing temperature, T, for the Fe-Si-Al alloy films with various compositions.

striction, λ, for Fe-Si sputtered films is shown. The sign of λ of the film sputtered under V_{DC}=3.5 kV (●), is negative at pure Fe. λ becomes zero around 2.0wt%Si and takes a positive maximum around 5.0wt%Si. With further increasing Si cocentration, λ crosses zero again around 8.0wt%Si and changes sign from positive to negative. While, in the case of the film sputtered under V_{DC}=2.0 kV (○), concentration dependence of λ is completely different from that of the films sputtered under V_{DC}=3.5 kV, especially in the concentration region more than 6.0wt%. Namely, λ crosses zero around 2.0wt%Si and retains positive sign up to 16wt%Si. In the same figure, calculated composition dependence of $\lambda_{(110)}$ by taking into account the preferred orientation of each grain [(110) Sheet texture], is shown by two-dot-dash line. In this calculation, magnetostriction constants, λ_{100} and λ_{111} determined by the single crystal experiment [19-23] for each α and DO_3 phase is used. As seen, the experimentally obtained concentration dependence of λ for films (V_{DC}=3.5 kV) agrees fairly well with that of calculated one. This fact means that the each grain of films sputtered under V_{DC}=3.5 kV consists of stable DO_3 ordered structure in the concentration range more than about 6.0wt%Si and that the preferred orientation of grains is realized in the films. It should be here noted that λ of non equilibrium α phase realized from 6.0 to 16wt%Si takes positive sign and the magnitude of λ is nearly about 3×10^{-6}.

In fig.6, changes of λ by annealing for the various film composition of Fe-Si-Al alloys are shown. As seen in this figure, λ changes and decreases remarkably around 300 °C for the films with Fe concentration more than 88wt%. This sudden change of λ by annealing is well explained by the phase transformation in each grain of the film from the disordered α-phase (as -deposition) to ordered DO_3 phase (annealed). The temperature at which the sudden decrease of λ happens well corresponds to the ordering temperature in the films. A similar annealing dependence is also observed in the case of the coercive force, H_c.

In Fig.7, the dependence of λ on concentration in as-deposited films in the Fe-Si-Al ternary alloy system is shown. In this figure, the main central composition of the bulk Sendust alloy (5.5wt%Al, 9.6wt%Si, bal.Fe) is represented by asterisk mark (*), and the equi-value lines of λ determined from the

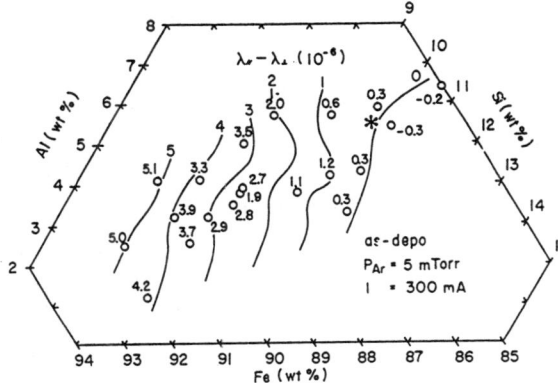

Fig.7 The values of magnetostriction, λ, for the films in an as-deposited state as a function of Fe, Si and Al concentrations. (✻) corresponds to the composition of Sendust alloy.

Fig.8 The values of magnetostriction, λ, for the films annealed at 600°C for 2 hrs as a function of Fe, Si and Al concentrations. (✻) corresponds to the composition of the Sendust alloy.

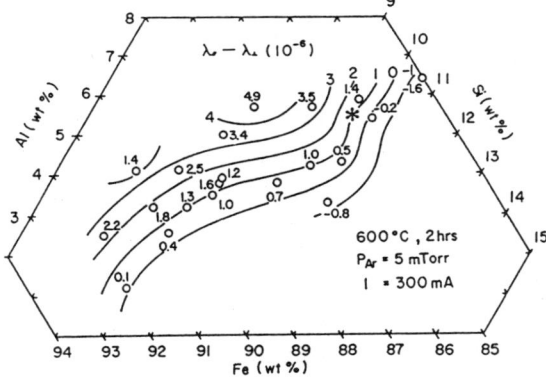

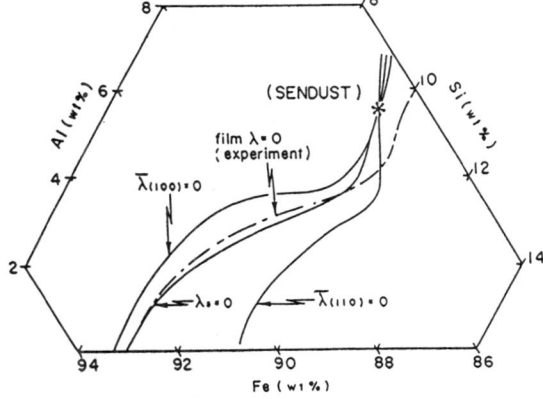

Fig.9 Concentration dependence of the calculated zero magnetostriction in the case of $\overline{\lambda}(100)$, $\overline{\lambda}(110)$ and λ_s. Dashed line corresponds to zero magnetostriction of the sputtered films (—·— , present data).

value of λ at each concentration of the film are also drawn. As seen in this figure, λ shows a zero value around Fe concentration about 85wt%, and gradually increases (positive sign) with the increment of Fe concentration.

In Fig.8, the concentration dependence of λ for the films anealed for 2 hours at 600°C is shown. Comparing the dependence of λ on concentration for the films after annealing to that of the films in an as-deposited state (Fig.7), we found that a big difference in the concentration dependence of λ is caused by annealing. It should be noted that the compositions at which zero magnetostriction is realized are changed from 85wt% (as-deposited) to 87∿93wt% (annealed at 600 °C) in Fe content by annealing.

Generally in polycrystalline material, by taking into account the preferred orientation of each grain (sheet texture), the apparent magnetostriction can be calculated using the magnetostriction constants λ_{100} and λ_{111} in the cubic case[11]. Therefore, by comparing the experimental results with that of the calculated one, we can obtain the information on the preferred orientation of grains and also on the ordered structure.

In Fig.9, as an example , the calculated concentration dependence of the zero magnetostriction is shown by solid line in the Fe-Si-Al alloy system. For the calculation, the crystal orientation of each grain was assumed to be (i) random, (ii) (100) sheet texture and (iii) (110) sheet texture, respectively. The values of the magnetostriction constants, λ_{100} and λ_{111}, as determined by present authors for Fe-Si-Al [19-23] and by M. Goto for Fe-Si [29] alloy single crystals with DO_3 ordered structure, were used. As seen in this figure, composition dependence of zero magnetostriction depends strongly on the difference of the preferred orientation of grains. By comparing these concentration dependences of $\bar{\lambda}$ and λ_s obtained by calculation with that of experiment, one can find a nice agreement for the concentration dependence of λ experimentally obtained with that of $\lambda_s=0$ in the Fe content region more than 87wt%. This fact suggests that each grain of the films annealed at 600 °C consists of stable DO_3 ordered structure and that the crystal orientation of the grains in films is nearly random. This conclusion is well supported by the experimental fact that the diffracted X-ray intensities from each crystal plane of the films is the same as that of the powdered samples (random orientation).

It is worth while noting that the informations for the phase formed and the preferred orientation of grains of the films are obtained through the magnetostriction measurement, refering to the magnetostriction constants of each ordered phase.

3-3 Magnetic Anisotropy

3-3-(i) Magnetocrystalline Anisotropy

Generally, in polycrystalline films, it is fairly difficult to determine exactly the intrinsic magnetic anisotropy constant, K_1 of each grain. Therefore, in present case, K_1 of each different non equilibrium phase is estimated from the results of single crystals with the concentrations near the phase-boundary concentrations (Fig.2). In Fig.10, temperature dependence of K_1 for the Fe-Si-Al alloy single crystals with various different compositions, (2∿5wt%Al, 5∿6wt%Si, bal Fe) is shown. The magnitude of K_1 ($\simeq 10^5$ erg/cc) in two phase mixture (B∿D) like as $B2+DO_3$ and $\alpha+DO_3$ is found to be remarkably larger than that of K_1 ($\simeq 10^3$ erg/cc) in DO_3 single phase (A). Furthermore, K_1 in two

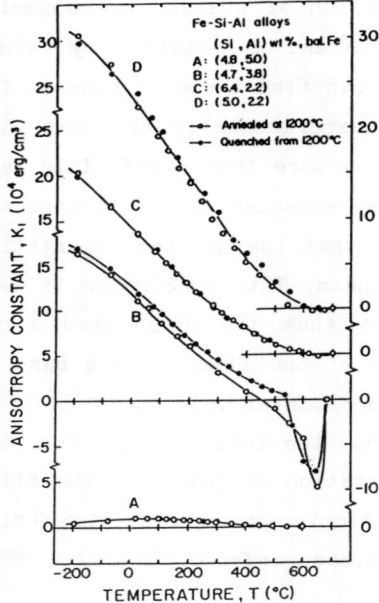

Fig.10 Temperature dependence of K_1 for Fe-Si-Al alloys.

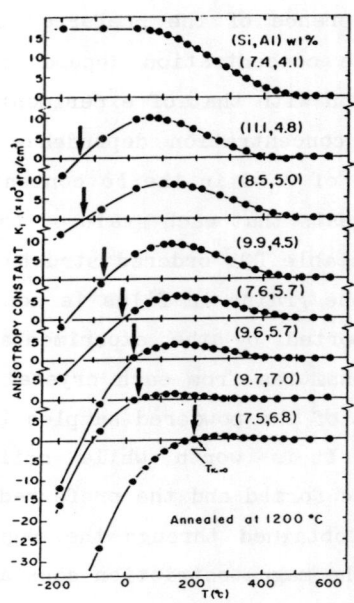

Fig.11 Temperature dependence of K_1 for the Sendust alloys annealed at 1200°C with various compositions. Arrows indicate the temperature, $T_{K1=0}$, at which K_1 crosses zero.

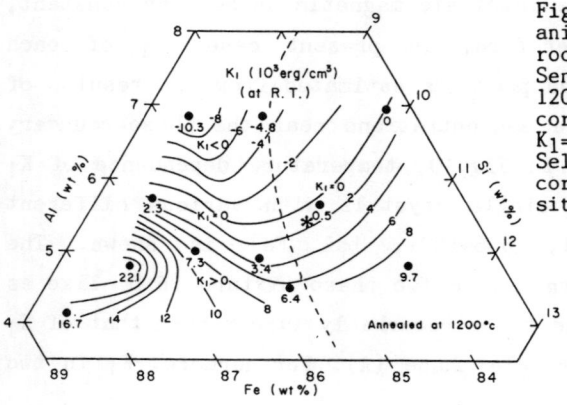

Fig.12 Magnetocrystalline anisotropy constant, K_1, at room temperature for the Sendust alloys annealed at 1200°C. Broken line corresponds to the line of $K_1=0$ determined by I. P. Selisskii[32]. (✻) mark corresponds to the composition of the Sendust alloy.

phase mixture decreases monotonically with increasing temperature, however, K_1 in DO_3 single phase takes a broad maximum at a infinite temperature as seen in Sendust alloys (Fig.11). Therefore, K_1 in α or B2 phase near Sendust composition which is formed as non equilibrium phase is estimated to be much larger than that of DO_3 phase.

As a typical representative of DO_3 single phase, precise concentration dependences of K_1 is presented for the Fe-Si-Al alloys with various compositions near Sendust. In Fig.11, the temperature dependence of K_1 for the alloys with various composition near Sendust is shown. The temperature dependence of K_1 for each alloy is found to be different. For most alloys examined, the magnitude of K_1 with negative sign ($\simeq 10^4$ erg/cc) becomes gradually small and crosses zero with increasing temperature. With further increasing temperature, K_1 changes its sign from negative to positive and takes a broad miximum. It should be noted that the temperature at which K_1 becomes zero, $T_{K1=0}$, differs remarkably for each alloy composition. Namely, for the alloy (7.5wt%Si, 6.8wt%Al, bal.Fe), $T_{K1=0}$ is found to be 200 °C, however, for the alloy (11.1wt%Si, 4.8wt%Al, bal.Fe), $T_{K1=0}$ is found to be -140°C. No sign change of K_1 took place in the alloy with the concentration 7.4wt%Si, 4.1wt%Al, bal.Fe. The appearance of $T_{K1=0}$ at which K_1 becomes zero mainly concern for the soft magnetism in Sendust. The anomalous temperature dependence of μ_i [30] is well explained by taking into account this temperature dependence of K_1. Based on the results shown in Fig.11, the concentration dependence of K_1 at room temperature is shown in Fig.12. In this figure, the equi-value lines for K_1 are also drawn. From the figure, Sendust alloy (✻) is located between the line of $K_1=0$ and $K_1=2$ ($\times 10^3$ erg/cc). The composition dependence of K_1 obtained for the specimens slowly cooled from 1200 °C agrees fairly well with that of the earlier results [19] (slowly cooled from 900 °C). This fact means that the degree of order in DO_3 structure of the alloys is not changed by these two different annealing temperature. While in the same figure, the line of $K_1=0$ previously reported by Zaimovsky and Selissky is also shown. As seen, the line of $K_1=0$ determined by the present experiment is completely different from that of the results obtained by Zaimovsky and Selissky.

As seen in these examples, it is easily understood that K_1 in Fe-Si-Al alloy change drastically by the difference of ordereded phase and also the composition.

3-3-(ii) Uniaxial and Local Anisotropy

In the case of sputtered Fe-Si-Al alloy films near Sendust composition, uniaxial anisotropy, H_k, is induced in-plane in an as-deposited state and this H_k is not diminished by the successive annealing process up to 600 °C.

In Fig.13, the composition dependence of H_k is shown. In the same figure, equi-value lines of K_1 in DO_3 phase (one-dotted line) and of magnetostrition, λ, of the films (two-dotted line) are also drawn. The magnitude of H_k has a minimum at about 2 Oe in the composition region where $K_1 \simeq 0$ ($\times 10^3$ erg/cc) and $\lambda \simeq 2 \sim 3$ ($\times 10^{-6}$). It should be here noted that the magnitude of H_k in this concertration region corresponds to a magnitude of the uniaxial magnetic anisotropy constant, K_u, of about 1×10^3 erg/cc, and which is nearly comparable to that of K_1. The equi-value lines of H_k are fairly independent of the magnitude of λ, and run almost along the equi-value lines of K_1. The magnitude of H_k increases with increasing the absolute value of K_1 for the grains

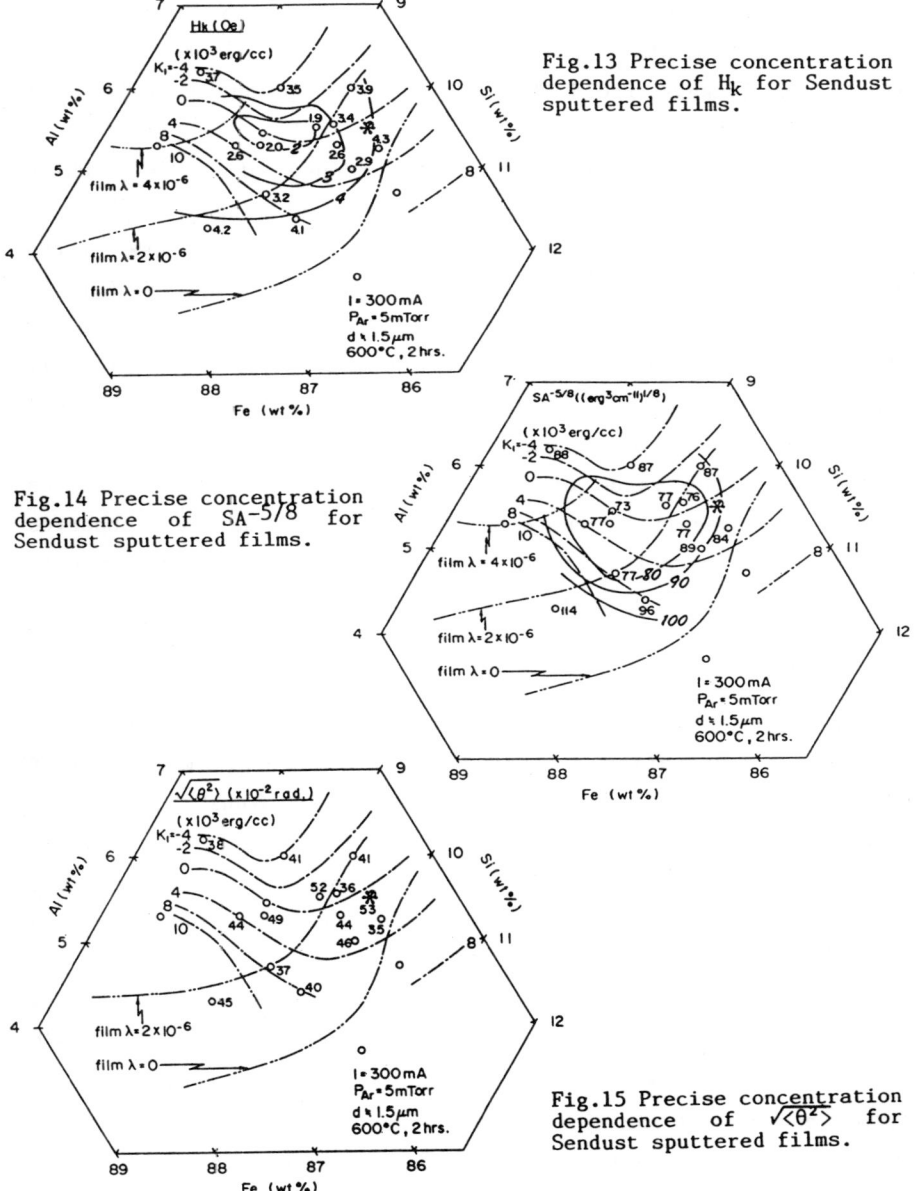

Fig.13 Precise concentration dependence of H_k for Sendust sputtered films.

Fig.14 Precise concentration dependence of $SA^{-5/8}$ for Sendust sputtered films.

Fig.15 Precise concentration dependence of $\sqrt{\langle\theta^2\rangle}$ for Sendust sputtered films.

in the films with DO_3 ordered structure. For the origin of H_k, preferred alignment of the Anti Phase Boundary in DO_3 structure along the leakage field direction, which produces Fe-Fe atom pairs, may mainly concern.

In Fig.14, the composition dependence of $SA^{-5/8}$, which represents the anisotropy fluctuation in short range (ripple), is shown. The values of $SA^{-5/8}$ show a shallow minimum peak in the broad composition region where $K_1 \approx 0$ (x10^3 erg/cc) and $\lambda \approx 3$ (x10^{-6}). Furthermore, the composition dependence of equi-value lines of $SA^{-5/8}$ is very similar to that of μ_{eff} at 5 MHz (Fig.21). Equi-value lines of $SA^{-5/8}$ run almost along the equi-value lines of K_1, which is very similar to the case of H_k. With an increment of the value of K_1, the magnitude of $SA^{-5/8}$ becomes large. This result corresponds to the fact that the observed microscopical magnetic inhomogeneities are mainly concerned to the magnetocrystalline anisotropy constants, K_1.

In Fig.15, the composition dependence of $\sqrt{\langle\theta^2\rangle}$, which represents the long range anisotropy fluctuation (skew), is shown. The magnitude of $\sqrt{\langle\theta^2\rangle}$ is about $0.35 \sim 0.53$ (x10^{-2} rad) and no clear concentration dependence was found. This magnitude of $\sqrt{\langle\theta^2\rangle}$ in Fe-Si-Al (Sendust) alloy films is found to be much larger than that of Co based amorphous films [31].

In order to make clear the origin of anisotropy fluctuation, thickness dependence of H_k and $SA^{-5/8}$ is investigated for three different composition of Fe-Si-Al alloy films.

In Fig.16, the thickness dependences of $SA^{-5/8}$ and H_k is shown. Microscopic anisotropy fluctuations represented as $SA^{-5/8}$ increase with decreasing film thickness, and the magnitude of $SA^{-5/8}$ becomes two or three times larger than that of the films of 1.5 μm with decreasing the film thickness for each alloy

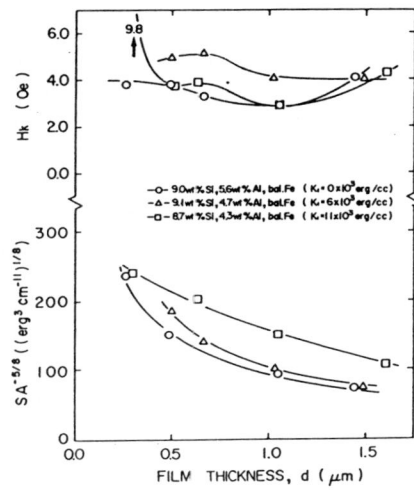

Fig.16 Change of the values of H_k and $SA^{-5/8}$ against film thickness, d.

composition. This fact means that the microscopic magnetic inhomogeneities of the films due to structural inhomogeneities increase remarkably with the decrement of the film thickness. On the other hand, H_k is independent of the film thickness and shows nearly constant value from 3 to 4 Oe.

While, the results for the metallographical observation of grains by SEM indicates that all films examined (d=1.5∿0.3 μm) were composed of fine grains about 300 Å without columnar structure. And no peculiar difference was found with increasing the film thickness. From the X-ray diffraction analysis, two structural informations were obtained.

(i) ; the (110) sheet texture is mainly realized at thick films of 1.5 μm. However, with decreasing the film thickness, preferred orientation of each grains changes from (110) to (110)+(100) mixed texture,

(ii) ; the lattice constant of the film has a fairly strong thickness dependence.

Namely, at the thick film of 1.5 μm, lattice constant, "a", is nearly the same as that of bulk samples with DO_3 ordered structure (a=2.853 Å). However, with decreasing the film thickness, the lattice constant gradually decreases (a=2.835 Å).

This fact means that the well orderd DO_3 structure could not be easily formed in each grain of the films, especially in a thin films less than 1.0 μm. Therefore, in Fe-Si-Al (Sendust) alloy films, local anisotropy fluctuation is mainly caused by the inhomogeneous preferred orientation of grains, rather than magnetostrictive interaction.

3-4 Coercive force and Permeability

In Fig.17, change of H_c by annealing is shown for various composition of alloy films. As seen, H_c decreases discontinously around 300 °C and the values of H_c suddenly becomes small from 20∼30 Oe (as-deposited state) to 0.4 Oe. This temperature at which sudden decrease of H_c took place well corresponds to Fe concentration of films and shifts to relatively low temperature with increasing Fe concentration. Origin of this discontinious change of H_c by annealing is considered to be the phase transformation of each grain from disordered α (as-deposited) to ordered DO_3 structure. While, in the case of the film with about 95wt%Fe, disordered α phase is stable in equilibrium. Therefore, no sudden decrease of H_c was not found.

In Fig.18, concentration dependence of H_c for the films annealed at 600 °C for 2 hours is shown. Asterisk mark (✷) corresponds to the Sendust alloy. Magnitude of H_c shows a shallow minimum peak of about 0.4 Oe near the Sendust concentration. From the figure, magnitude of H_c shows a big change by

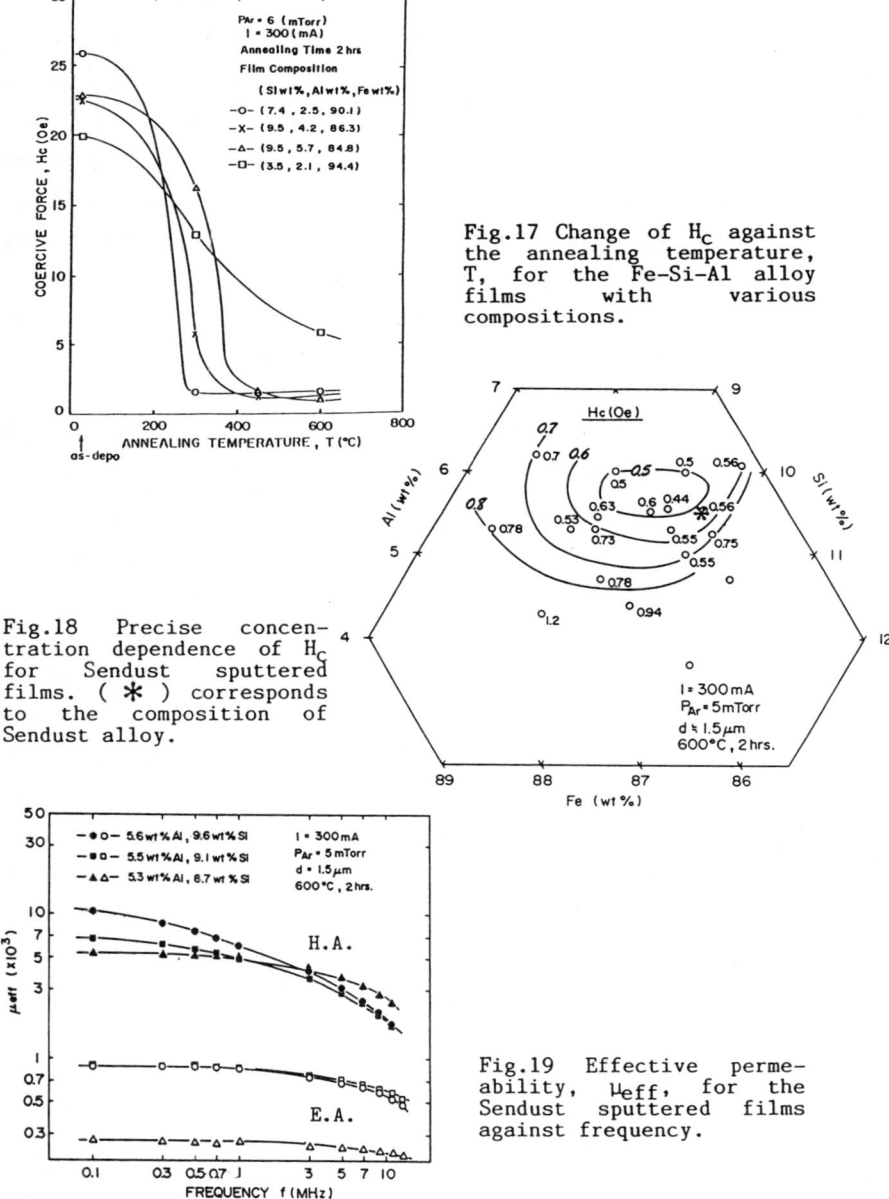

Fig.17 Change of H_C against the annealing temperature, T, for the Fe-Si-Al alloy films with various compositions.

Fig.18 Precise concentration dependence of H_C for Sendust sputtered films. (✻) corresponds to the composition of Sendust alloy.

Fig.19 Effective permeability, μ_{eff}, for the Sendust sputtered films against frequency.

a slight difference of the film concentration by about 1∼2wt%.

In Fig.19, as an example, the effective permeability along the magnetically easy (E.A.) and hard (H.A.) axis is shown as a function of frequency for the three different compositions of the films near Sendust. From the figure, the values of μ_{eff} along H.A. are found to be about one order higher than that of E.A.. This observed anisotropy for μ_{eff} is caused by the strong uniaxial anisotropy in plane as seen in Fig.13. The frequency dependence of μ_{eff} changes remarkably depending on the film concentration even in the Sendust concentration.

In Figs.20 and 21, the concentration dependence of μ_{eff} along H.A. at 100 kHz and 5 MHz is shown, respectively. ✶ mark corresponds to the bulk Sendust concentration. In this figure, equi-value lines of magnetocrystalline anisotropy constants, K_1, determined by the single crystal experiments (Fig.12) and of magnetostriction constant of films, λ, (Fig.8) determined by the previous experiment are shown. As seen in Fig.20, μ_{eff} shows a sharp peak of about 10000 at the Sendust concentration at which $K_1 \approx 0$ (x10^3 erg/cc) and $\lambda \approx 1$ (x10^{-6}) are realized. Equi-value lines of μ_{eff} run almost along the equi-value lines of K_1 and μ_{eff} gradually increases with decreasing K_1. Furthermore, on the equi-value line of $K_1 \approx 0$ (x10^3 erg/cc), μ_{eff} gradually increases with decreasing λ. While, in the case of 5 MHz, the values of μ_{eff} show a shallow maximum peak of about 4000 in the broad composition region at which $K_1 \approx 0$ (x10^3 erg/cc) and $\lambda \approx 3$ (x10^{-6}). Furthermore, the equi-value lines of μ_{eff} run also almost along the equi-value lines of K_1. These results suggest that the magnitude of μ_{eff} up to about 10MHz strongly depends on the magnitude of K_1 for each grain in the case of Fe-Si-Al (Sendust) alloy films and that in order to realize the high value of μ_{eff},

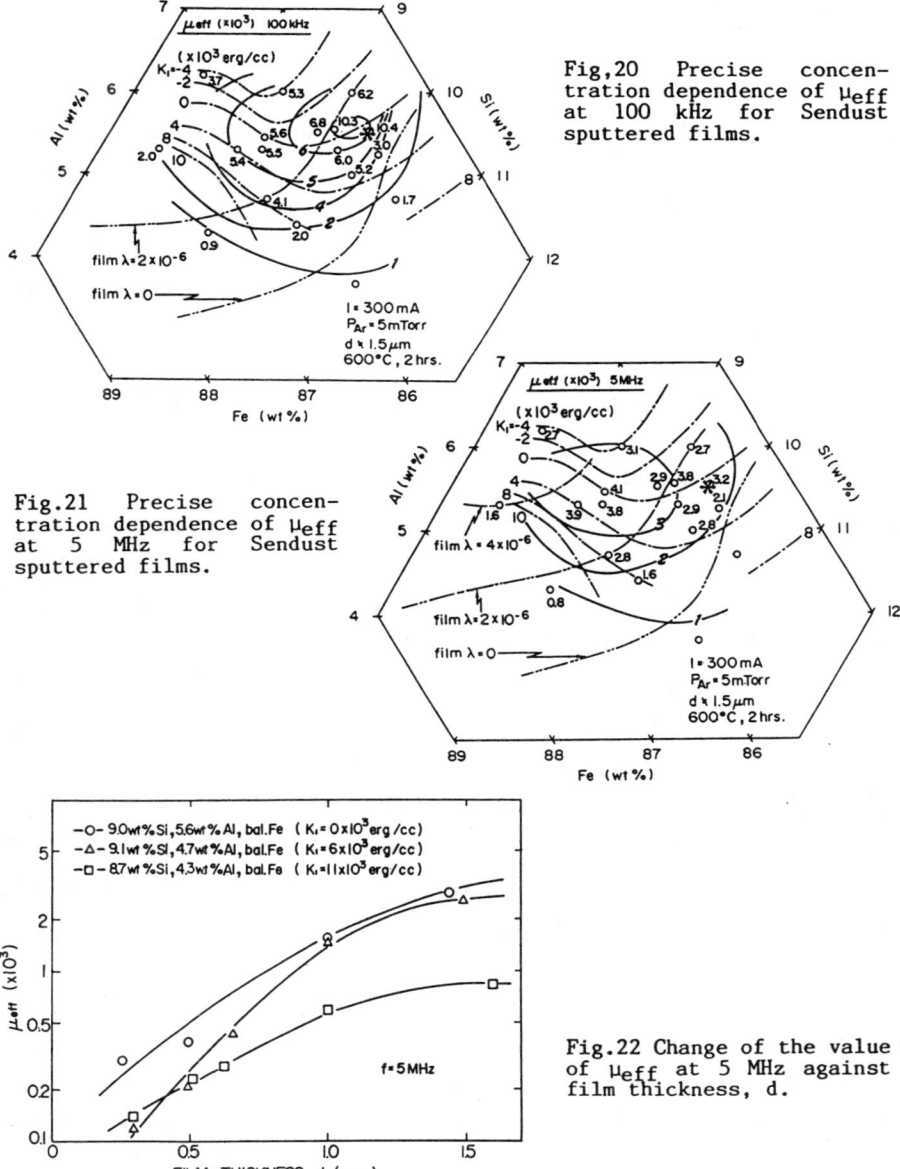

Fig.20 Precise concentration dependence of μ_{eff} at 100 kHz for Sendust sputtered films.

Fig.21 Precise concentration dependence of μ_{eff} at 5 MHz for Sendust sputtered films.

Fig.22 Change of the value of μ_{eff} at 5 MHz against film thickness, d.

adjustment of exact concentration with $K_1 \approx 0$ ($\times 10^3$ erg/cc) is important.

Furthermore, by comparing the concentration dependence of μ_{eff} at 5 MHz to that of SA-5/8 (Fig.14), we can easily find the composition region where the maximum peak of μ_{eff} at 5 MHz were realized and see that peak form is very similar to that of SA-5/8. Therefore, it is supposed that the magnitude of μ_{eff} in high frequency region is strongly influenced by the magnitude of SA-5/8 rather than H_k. This assumption is well supported by the results of thickness dependence of μ_{eff}.

In Fig.22, the thickness dependence of μ_{eff} at 5 MHz is shown for the films of three different compositions. K_1 for each alloy composition corresponds to 0, 6 and 11 ($\times 10^3$ erg/cc), respectively. The value of μ_{eff} decreases remarkably with decreasing film thickness for each composition of the films. Furthermore, the value of μ_{eff} decreases with increasing the value of K_1 for each alloy.

By comparing the results shown in Fig.22 with that of Fig.16, we can find well relationships between effective permeability and the local anisotropy fluctuation of the films. Here, the structure constant, S, which characterize the microscopic local anisotropy fluctuation is caused phenomenologically by two terms (Eq.(2)). One is due to the inhomogeneous distribution of grain orientations (concerning the magnitude of K_1) and the other, the magnetoelastic effect.

The composition dependence of the magnetostriction of the polycrystalline specimen under (100), (110) and random orientation of grains have been already shown in Fig.9. Using these calculated results, the magnetostriction for these alloy compositions in the present case must decrease from about 2∿1 ($\times 10^{-6}$)

to a value less than 1 ($\times 10^{-6}$) with the decrement of the film thickness, because of the change of preferred orientation of grains. This fact implies that the contribution of the magneto-elastic effect to the magnitude of $SA^{-5/8}$ is small. Therefore, the remarkable decrement of μ_{eff} with decreasing film thickness is caused by the change of the preferred grain orientation itself in films.

3-5 Relationships between permeability and anisotropy.

Based on all of the results obtained, relations between the permeability, μ_{eff}, and the microscopical anisotropy fluctuation is discussed.

In Fig.23, μ_{eff} at 5 MHz obtained for the various compositions of Fe-Si-Al alloy (Fig.21) is plotted against the uniaxial magnetic anisotropy induced during sputtering (Fig.13). In the same figure, the calculated results of μ_{eff} (●) by assuming the ideal uniform rotation process along hard axis in an uniaxial ferromagnet is also shown. As clearly seen, the magnitude of μ_{eff} is lower than that of calculated value and no clear relationship between μ_{eff} and K_u was found.

While, in Fig.24, μ_{eff} at 5 MHz (Fig.22) normalized by saturation magnetization, M_s, is plotted against the microscopical local anisotropy fluctuation S (Fig.16). As seen, normalized μ_{eff} systematically changes against S and is changed nearly proportional to $1/\sqrt{S}$. This fact means that the magnitude of μ_{eff} at high frequencies is strongly related to the magnitude of microscopic local anisotropy fluctuation caused by inhomogeneous preferred grains orientation. Therefore, to realize an excellent high μ_{eff} at high frequencies for the sputtered Fe-Si-Al films, the homogeneous preferred grain orientation is important.

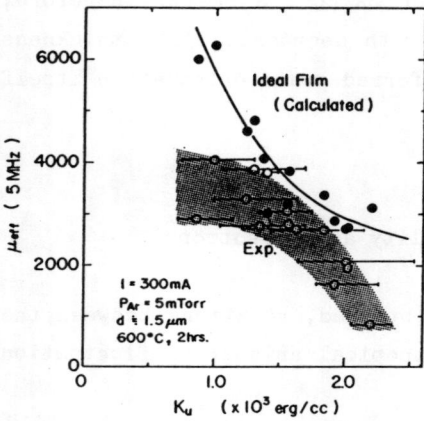

Fig.23 Change of the value of μ_{eff} along hard axis at 5 MHz against the uniaxial magnetic anisotropy, K_u. (**) corresponds to the calculated results of μ_{eff} by assuming the ideal uniform rotation process in an uniaxial ferromagnet.

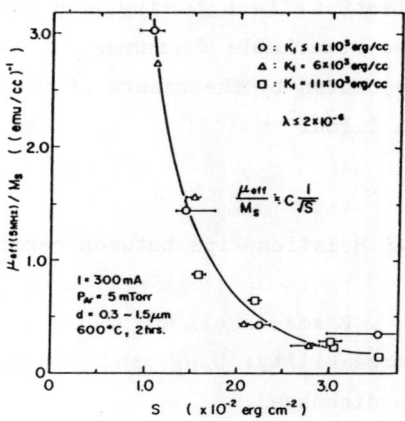

Fig.24 Change of the value of μ_{eff} along hard axis at 5 MHz normalized by saturation omagnetization, M_s, against the microscopical local anisotropy fluctuation, S.

4. Summary

Structural and Magnetic properties of Fe-Si and Fe-Si-Al (Sendust) sputtered films are systematically examined. As a result, it was clarified that

1) non equilibrium phase like as disordered α can be formed at room temperature by changing the sputtering parameters and this disordered α transforms to DO_3 phase as equilibrium by annealing,
2) magnetocrystalline anisotropy, K_1, and magnetostriction, λ, changes drastically by affecting the difference of the phase,
3) uniaxial magnetic anisotropy, K_U, is induced in plane. For the origin, K_U is considered to arise from a preferential alignment of Anti Phase Boundary (APB) in DO_3 phase parallel to the leakage field direction of magnetron sputtering,
4) anisotropy fluctuation in short range is found. For the origin, inhomogeneous preferred orientation of grains in DO_3 is mainly concern, rather than the magnetoelastic coupling energy,
5) μ_{eff} at high frequency is strongly related to the magnitude of microscopical local anisotropy fluctuation, rather than H_k. To realize a excellent high μ_{eff}, the homogeneous preferred grain orientation is important.

REFERENCE

1) K. Kajiwara, M. Hayakawa and K. Aso ; IEEE Trans. Magn., January (1989, in press)
2) C. W. M. P. Sillen, J. J. M. Ruigrok, A. Broese van Groenou, and U. Enz ; IEEE Trans. Magn., VOL.24, No.2, March 1802-1804 (1988)
3) K. Saito, T. Shimizu and H. Ishida ; IEEE Trans. Magn., VOL. MAG-23, No.5, September 2925-2927 (1987)
4) T. Nishiyama, K. Noguchi, K. Mouri, H. Iwata and T. Shinohara ; IEEE Trans. Magn., VOL. MAG-23, No.5, September 2931-2933 (1987)
5) Y. Hashimoto ; IEEE Trans. Magn., VOL. MAG-23, No.5, September 3167-3172 (1987)
6) H. Tomiyasu, I. Sato, K. Kanai ; J. Magn. Soc. Jpn., Vol.11, No.2, 105-108 (1987)
7) M. Kadono, T. Yamamoto, K. Nago, M. Michijima, T. Muramatsu, T. Kira and H. Kyotani ; J. Magn. Soc. Jpn., Vol.12, No.2, 97-102 (1988)
8) T. Kumura, Y. Kunito, H. Sato and T. Ishida ; J. Magn. Soc. Jpn., Vol.12, No.2, 103-106 (1988)
9) H. Iwata, T. Nishiyama, K. Noguchi, K. Mouri, S. Suwabe and T. Shinohara ; J. Magn. Soc. Jpn., Vol.12, No.2, 107-110 (1988)
10) R. Minakata ; IEEE Trans. Magn., MAG-23, No.5, September 3236-3238 (1987)
11) M. Takahashi, S. Suwabe, T. Narita and T. Wakiyama ; J. Magn. Soc. Jpn., Vol.10, No.2, 307-310 (1986)
12) M. Takahashi, T. Sato, T. Narita, R. Goto and T. Wakiyama ; J. Magn. Soc. Jpn., Vol.11, No.2, 299-302 (1987)
13) M. Takahashi, T. Sato, N. Kato and T. Wakiyama ; J. Magn. Soc. Jpn., Vol.11, No.2, 303-306 (1987)

14) M. Takahashi, N. Kato, T. Sato and T. Wakiyama ; IEEE Trans. Magn., MAG-23, No.5, September 3068-3070 (1987)

15) M. Takahashi, N. Kato, T. Shimatsu, H. Shoji and T. Wakiyama ; J. Magn. Soc. Jpn., Vol.12, No.2, 305-310 (1988)

16) M. Takahashi, T. Tomitani, H. Arai and T. Wakiyama ; J. Magn. Soc. Jpn., Vol.10, No.2, 225-228 (1986)

17) M. Takahashi, H. Arai, T. Tanaka and T. Wakiyama ; J. Magn. Soc. Jpn., Vol.10, No.2, 221-224 (1986)

18) M. Takahashi, H. Arai, T. Tanaka and T. Wakiyama ; IEEE Trans. Magn., MAG-22, No.5 638-640 (1986)

19) M. Takahashi, S. Nishimaki and T. Wakiyama ; J. Magn. Magn. Mater., Vol.66, 55-62 (1987)

20) M. Takahashi, H. Arai and T. Wakiyama ; J. Magn. Soc. Jpn., Vol.11, No.2, 251-254 (1987)

21) M. Takahashi, H. Arai, T. Kobayashi and T. Wakiyama ; J. Magn. Soc. Jpn., Vol.11, No.2, 255-258 (1987)

22) M. Takahashi, H. Arai and T. Wakiyama., IEEE Trans. Magn., MAG-23, No.5, 3523-3525 (1987)

23) T. Tanaka, M. Takahashi, H. Matsushima, T. Wakiyama and D. Watanabe ; J. Magn. Soc. Jpn., Vol.12, No.2, 269-272 (1988)

24) H. Masumoto and T. Yamamoto ; Trans. Jpn. Inst. Metals, Sendai, Vol.1, 127 (1937)

25) H. Hoffmann ; Phys. Stat. Sol., 33, 175 (1969)

26) K. Kempter and H. Hoffmann ; Phys. Stat. Sol., 34, 237 (1969)

27) W. D. Doyle and T. F. Finnegan ; J. Appl. Phys., 39, 3355 (1968)

28) T. Miyazaki, T. Tsuduki, T. Kosakai and Y. Fujimoto ; Trans. Jpn. Inst. Metals, 46, 1111 (1982)

29) M. Goto and T. Kamimori, ; J. Phys. Soc. Jpn., 52, 3710 (1983)

30) M. Takahashi ; Jpn. J. Appl. Phys., Vol.56, No.10, 1289-1306 (1987)

31) M. Takahashi, T. Shimatsu, Y. Shimada and T. Wakiyama ; J. Magn. Soc. Jpn. Vol.11, No.2, 275-278 (1987)

32) I. P. Selisskii ; Z. Fiz. Khim. SSSR, 20, 597 (1946)
33) M. C. Farquhar, et al. ; J. Iron Steel Inst., 152, 457 (1945)
34) T. Kamimori, M. Shida, M. Goto ; proceeding

MULTILAYERED MAGNETIC FILMS - PROPERTIES AND APPLICATIONS

P J Grundy* and M Ohkoshi+*
*Department of Physics, University of Salford,
Salford M5 4WT, U.K.
+*Faculty of Engineering, University of Hiroshima,
Higoshi-Hiroshima 724, Japan.

ABSTRACT

Ferromagnetic multilayers are composite materials with interesting and, in some cases, potentially applicable properties. This paper briefly reviews the field and describes, in particular, the fabrication and characterisation of some types of multilayer which could, in principle, find application in magnetic or magneto-optic media or in devices. Systems highlighted are Co/TM multilayers which can exhibit varying anisotropy orientations and RE-TM and Heusler alloy multilayers which may provide more stable and sensitive media for magneto-optic applications.

1. INTRODUCTION

Thin film multilayer structures are a novel composite material and provide opportunities to explore and exploit unusual magnetic behaviour [e.g.1]. The system usually contains many wavelengths or periods, defined as the summed thickness of the two (usually) components, i.e $\lambda = d_x + d_y$. Such systems provide opportunities to explore the fundamental magnetic interactions across spacings approaching atomic dimensions and also to investigate extrinsic properties related to compositional, growth and microstructural features. Investigations of this kind compliment and benefit, of course, from studies of single and bi-layer films.

Sequential deposition of individual films from two sources by evaporation or sputtering is a convenient method of fabricating periodic structures with nominal individual layer thicknesses down to the monolayer level or below. It should also be noted that layered deposition of compositionally modulated systems can also be effected by electrodeposition from solution [2].

Lattice matching between oriented layers can produce a true superlattice with epitaxial, coherent interfaces involving various degrees of strain which can be relieved by the nucleation of misfit dislocations. Polycrystalline and amorphous films allow the fabrication of three distinct systems with a wide variety of atomic arrangements, (sometimes involving symmetry changes), microstructures and grain sizes. This variety is embellished by the possibility of chemical disorder and mixing at the layer interfaces. The wide variety of structural and microstructural possibilities can be built up from ferro-, antiferro-, ferri- or non-magnetic elements or alloys, thus providing a broad spectrum of interactions and differing magnetic properties.

In general, deposition onto "cold" substrates, either in UHV evaporation or molecular beam systems (MBE), promotes the formation of clean and physically and chemically sharp interfaces. Methods involving a greater energy transfer to a film growing on a "warm" substrate in poorer vacuum, i.e. in comnventional evaporation or sputtering, can lead to diffuse physical and chemical modulations. Obviously, mutual insolubility of the components will also encourage sharp and well defined multilayers.

This paper will briefly review a selected literature on multilayers and will concentrate on some systems which suggest a possible application, however tenuous as yet. The work of others will be discussed where relevant to the themes of this paper, but the choice is selective and many excellent and important investigations have, of necessity, been neglected.

2. FABRICATION AND CHARACTERISATION

The systems we shall discuss have been prepared by sputtering (r.f., d.c. or ion beam sources) or evaporation onto a variety of substrates. Multilayers may be prepared either by deposition onto a stationary substrate through programmed shutters over two sources, or onto a substrate that rotates to a position above each source in turn. The "open" interval of the shutters or the "transit" time of the substrate over the two sources determine the individual film thicknesses and the multilayer wavelength. Sputtering is more easily controlled and well defined structures down to very small periods can be obtained using differing sputtering powers and small transit times [3].

Many experimental techniques can be used to characterise multilayers. Here we are concerned with magnetic, magneto-optical and structural measurements. Of particular interest are magnetization and hysteresis loop parameter measurements, some torque investigations, and electron microscopy and X-ray diffractometry in structural investigations. Diffractometry is particularly useful for structural identifications at large angles and for measurement of the multilayer period and quality of the interfaces at low angles. Rutherford back-scattering nuclear spectroscopy is also useful in studying compositional modulations and film thicknesses for individual layer thicknesses greater than about 50Å.

There is a growing list of measurement techniques that are being applied to in-situ investigations. Of particular importance are those using electron, atom, ion or photon beams giving structural, compositional and spin polarisation information. These clearly have a particular relevance in MBE and UHV studies of ideal systems [4] but they will not be of concern in this paper.

3. RESULTS AND DISCUSSION

In investigating the magnetic properties of multilayers it is necessary to recognise the relevance of results from very thin single layer films and their surfaces. Although a conclusive picture is perhaps not yet clear, moment changes of the order of 10-30% are predicted and observed [5,6] at surfaces and strong anisotropies are also measured [7]. Related effects are expected at the interface between layers where changes in local coordination number also occur. Added to this will be the extra factors in a multilayer of the nature of the interface and the components of the adjacent layers. Reductions in moment are also measured and predicted at interfaces [8] and particular interactions can occur, i.e ferro-ferro or ferro-antiferro [9,10], between nearest and next nearest layers.

3.1 Co/TM and related multilayers

Such interesting effects have been observed in Co/Cr [1,3,11], Co/W [12], Co/Pd and Co/Pt [13,14] multilayers. However, some of these systems have contrasting structural properties. This is especially the case for the Co/W system prepared by sputtering [12]. Figure 1 describes the dependence of structure on the individual layer thicknesses, d_{Co} and d_W, for a series of multilayers. The four regions,

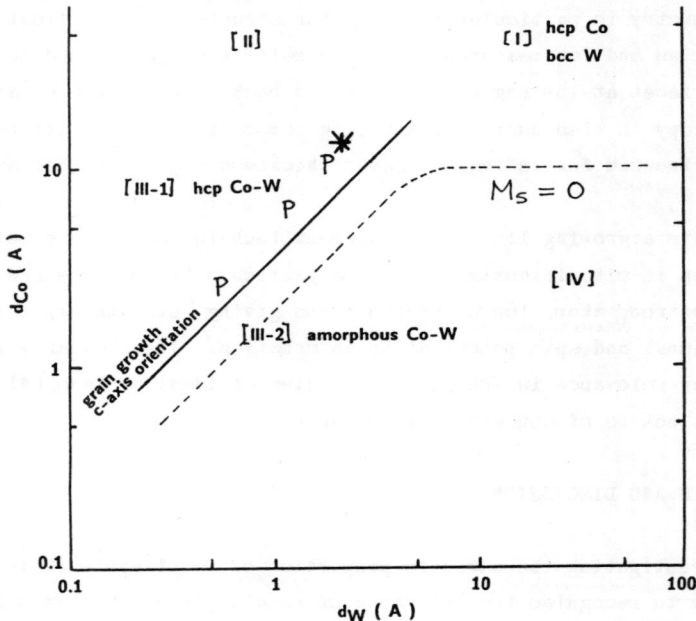

Figure 1. A description of the structure of Co/W multilayers in terms of layer thicknesses.

one subdivided into two parts, show differing structures (atomic arrangements) within the individual layers and also different microstructures. Low angle X-ray diffraction (LXRD) showed the multilayers to be well formed and periodic down to a wavelength of λ = 6Å. LXRD patterns are shown in Figure 2 for multilayers representative of some of the regions of Figure 1. High angle (HXRD) patterns are also included to define the atomic arrangements observed within the individual layers. The strong dependence of the LXRD maximum on d_W rather than d_{Co} suggested that the surface of the cobalt layers is less well defined than that of the tungsten films. This may be the result of a complex mixing effect or some preferential diffusion. The first order LXRD reflection was of maximum intensity for 3.5Å $\langle$ d_W $\langle$ 4Å corresponding to the formation of a tungsten monolayer.

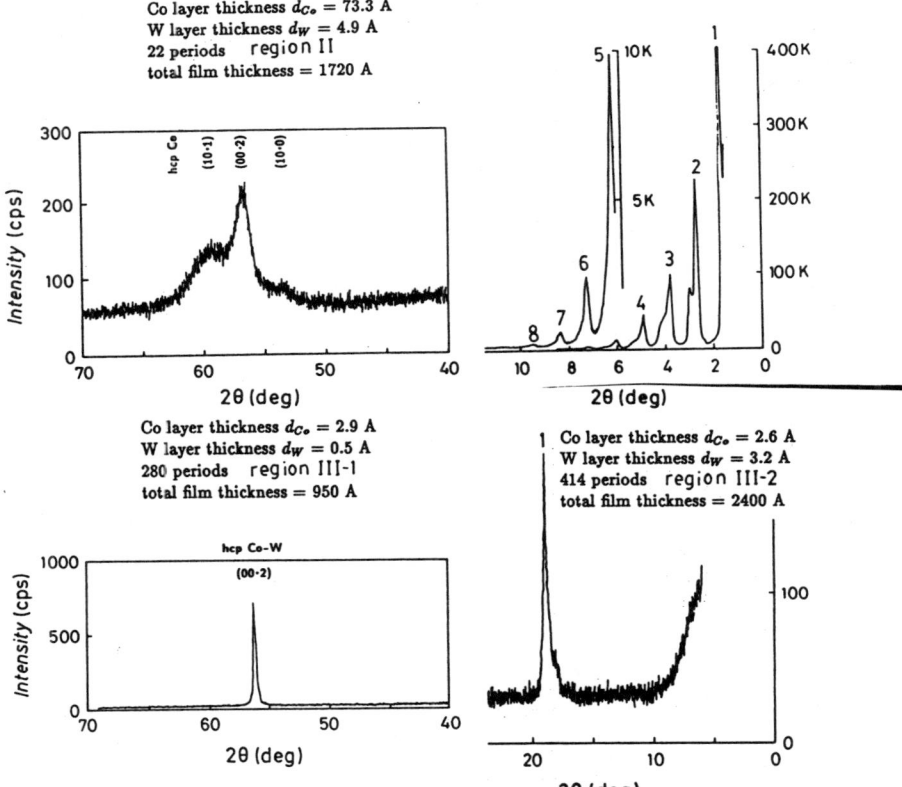

Figure 2. LXRD and HXRD diffraction data from multilayers in Figure 1.

A good representation of the compositional modulation can be given by an RBS spectrum, as shown in Figure 3a, but resolution limitations preclude the application of the technique to systems with individual layers less than about 50Å thick. The cobalt and tungsten signals are well separated, i.e. their atomic numbers are sufficiently different, and the concentration changes are certainly clear for the tungsten layers with their more efficient back-scattering. Figure 3b gives the calculated, relative proportions of cobalt and tungsten as a function of depth in the multilayer. The mean values of the concentration oscillations suggest the effective composition of the multilayer as $Co_{84}W_{16}$. This corresponds well to the ratio of thicknesses in this particular example (from region I) of ~5.3/1, Co/W.

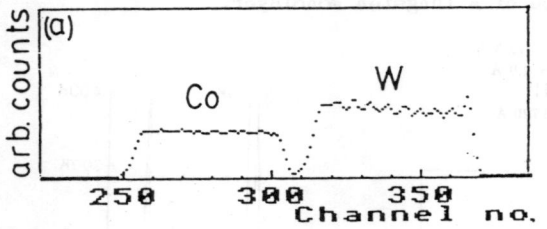
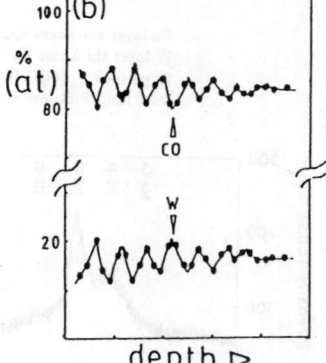

Figure 3. (a) RBS spectrum for a 9 period Co/W multilayer. (b) Concentration of Co and W as a function of depth.

In regions I and II the cobalt layers were polycrystalline with a random orientation, separated by (110) textured tungsten layers in I but amorphous tungsten layers in II. Region III is sub-divided into two parts. To the left of the boundary line, which corresponds to an effective tungsten concentration, C_W, of about 15 at%, the multilayers are crystalline with a preferred c-axis texture, see Figures 1 and 2. Small periods produce a more definite [0001] c-axis orientation and correspond to significant grain growth. Indeed, electron microscopy (TEM) of such multilayers, Figures 4a and b, shows the presence of pseudo-single crystal layers with [0001] perpendicular to the layer plane and stripe domain structures typical of films with perpendicular anisotropy and a large component of magnetization oriented normal to the plane of the multilayer.

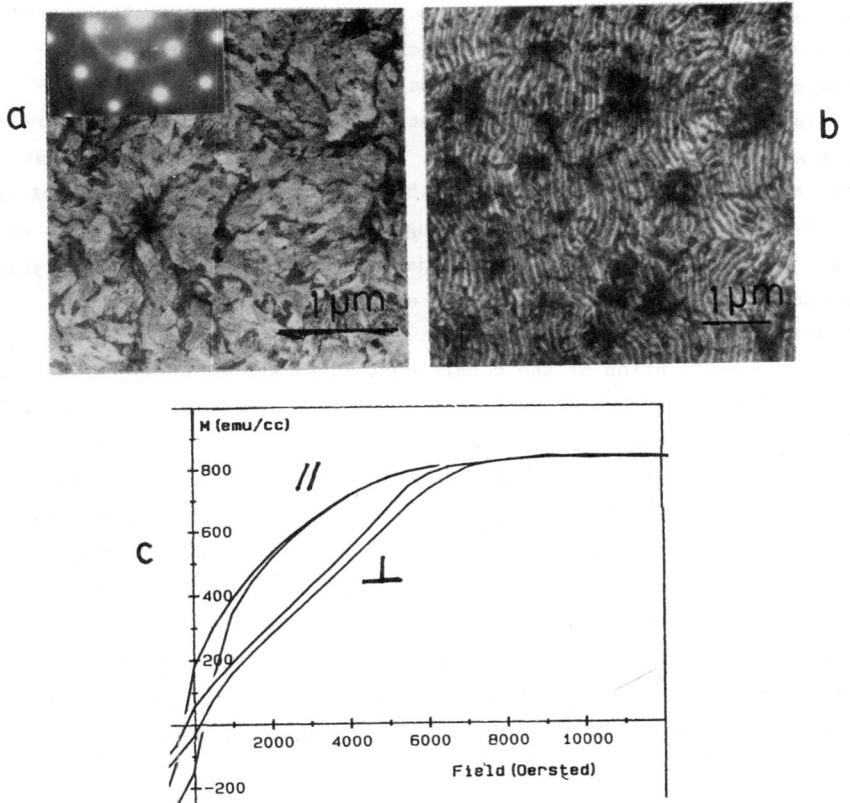

Figure 4. (a) TEM micrograph and TED pattern, (b) Lorentz TEM micrograph showing domains and (c) Hysteresis loops for P* in Figure 1.

The stripe domains in this system, which is representative of the multilayers marked by P in Figure 1, indicate strong exchange coupling through the tungsten spacing layers. At P* the cobalt layers are about 10Å thick and the tungsten is at or below monolayer thickness. Of course, as the period decreases further the tungsten films will become discontinuous and the system will eventually approximate to a homogeneous alloy. The hysteresis loops of Figure 4c point to the development of perpendicular anisotropy, and torque measurements on such multilayers suggest an effective anisotropy constant, K_{eff} of between 10^5 and 10^6 ergs cm^{-3}. The maximum K_{eff} is obtained at the boundary with region III-2 as indicated by the individual multilayers P on Figure 1. The characteristics of the perpendicular loop of Figure 4c, such as the "shoulder" on the demagnetizing curve point to the presence of a well defined stripe domain structure in the multilayer and demagnetization by domain wall motion.

In region III-1 the lattice constants of the hexagonal structure increase with increasing d_W until at the boundary with III-2 they are well above the bulk values at c = 2.55Å and c = 4.12Å (cf a = 2.50Å and c = 4.06Å for bulk cobalt). This suggests some interfacial strain or alloying. To the right of the boundary in region III-2 the multilyers are completely amorphous but with a still very obvious periodic structure (Figure 2).

The magnetization of the cobalt content of the multilayers, M_{Co}, decreases with increasing C_W for d_{Co} less than about 10Å. For greater thicknesses of cobalt it is independent of d_W and C_W, as shown in Figure 5. The values of M_{Co} at small layer thicknesses can be explained if a notionally mixed or alloyed, non-magnetic interface about 6.4Å thick forms between the the cobalt and tungsten layers. The "pure" cobalt central part then has the moment of bulk cobalt down to a thickness of ~3.6Å. This "dead" layer at the interface contains tungsten at a concentration greater than about 25 at% and, as mentioned above, its width agrees well with the observation of clear LXRD patterns down to a period of about 6Å and also with computations of the multilayer profiles.

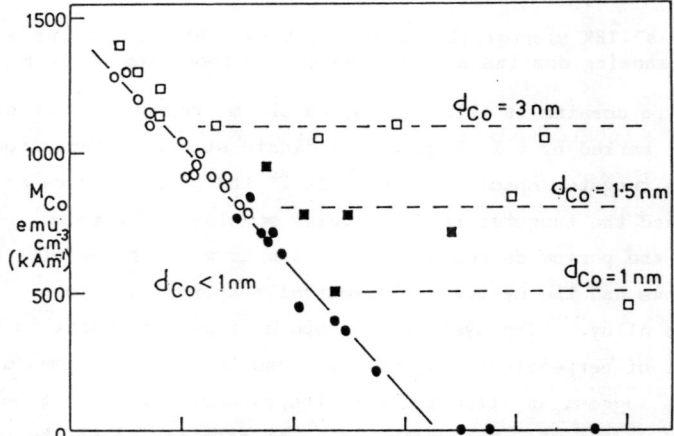

Figure 5. M_{Co} of Co/V multilayers as a function of C_W. The abscissa scale is defined in Figure 6. Open symbols for crystalline and full symbols for amorphous multilayers.

Coercivity in the multilayers is related to the structure of the cobalt layers and the degree of preferred orientation. As shown in Figure 6, H_c peaks at the same value of C_W as I_{0002}, clearly reflecting the contrast between the relatively hard crystalline layers in region III-1 and the magnetically soft multilayers in region III-2.

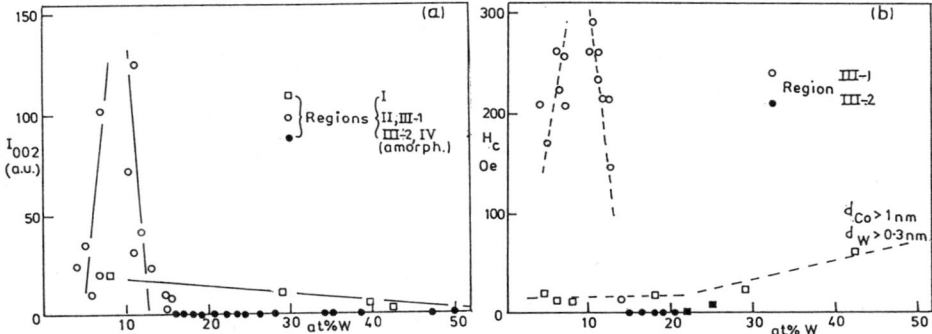

Figure 6. (a) Intensity of the (0002) HXRD maximum and (b) Coercivity as a function of C_W in Co/W multilayers.

Similar magnetic and structural results have been obtained for Co/Cr multilayers [3], except for the fact that, as in the case of homogeneous CoCr alloys, no amorphous layers are formed. Transmission electron diffraction (TED) patterns and Lorentz TEM micrographs of the domain structure in multilayers prepared with d_{Co}/d_{Cr} = 20Å/6Å and 3.6Å/1Å are shown in Figure 7. The TED patterns show the presence of b.c.c. chromium and h.c.p. cobalt and illustrate the development of a c-axis texture in the cobalt pattern for the 3.6Å/1Å multilayer with strong hkil reflections for l = 0. The stripe domains in the 20Å/6Å multilayer extend throughout the system and again suggest exchange coupling across the chromium and a perpendicular component of magnetization. This component and the associated anisotropy is developed further in the 3.6Å/1Å system as shown by the maze and bubble-type domain patterns. The contrast is low in this micrograph as a result of the imaging mode used in Lorentz microscopy which utilises any remaining horizontal components of magnetization. The hysteresis loops for such multilayers

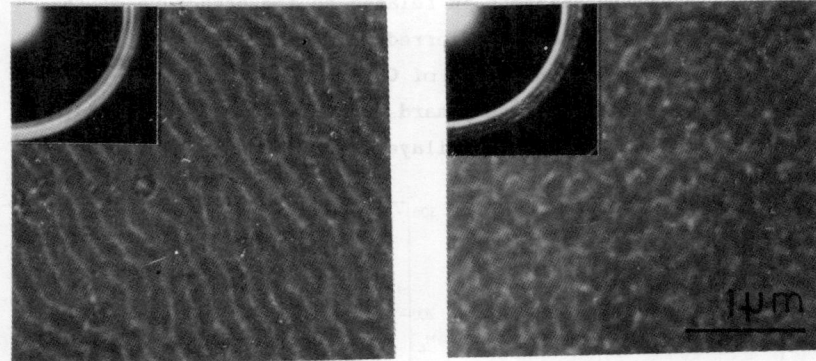

Figure 7. Lorentz TEM micrographs and TED patterns for Co/Cr multilayers with (a) 20Å Co/6Å Cr and (b) 3.6Å Co/1Å Cr.

prepared in the ratios 20Å/6Å and 3.6Å/1Å are given in Figure 8. The gradual steepening of the perpendicular loops with the reduction of the cobalt layer thickness shows the development of an effective perpendicular anisotropy in agreement with the domain observations.

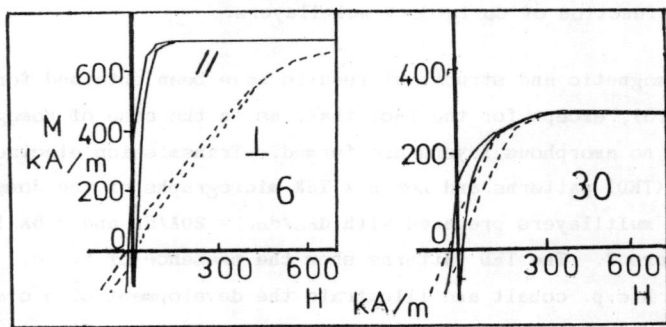

Figure 8. VSM hysteresis loops for the multilayers of Figure 7.

Our results on Co/Cr are in agreement with those from more exrtensive investigations [1] in which perpendicular anisotropy is developed in sputtered Co/Cr multilayers if the Co/Cr thickness ratio is less than about 3/1 and the the cobalt layer thickness is less than 15Å. These multilayers have properties that make them suitable for perpendicular recording applications and it would be interesting to investigate their recording behaviuor and their corrosion and tribological properties.

Turning to work on other Co/TM systems, particularly evaporated Co/Pd [13] and Co/Pt [14], we find much stronger anisotropies which are again associated with thin cobalt layers of 10Å or less. Values of the effective anisotropy, i.e a surface anisotropy related to the number, and perhaps quality, of the interfaces plus any volume anisotropies, such as magnetocrystalline or magnetoelastic, is greater than the shape term. If K_{eff} is positive perpendicular orientation of the magnetization can be obtained. As discussed above, stripe domains have been observed in Co/V and Co/Cr multilayers, and also in Fe/SiO$_2$ systems [15] where they suggest that surface anisotropies may exist in some Fe-based multilayers and that they may be great enough to overcome the increased magnetostatic term.

In contrast to these evaporated Co/Pd multilayers, perpendicular magnetization is not observed [13] in ion beam sputtered multilayers. Energy transfer may be responsible for enforced mixing and an absence of surface anisotropy. This may, of course, be responsible for the reduced effects reported above for Co/V and Co/Cr systems. Indeed, recent work [16] on the Co/Au system has shown that on annealing the multilayers to encourage the formation of sharp interfaces through the mutual insolubility of the two components perpendicular anisotropy is increased.

Other interesting effects are observed in Co/Pd and Co/Pt. The cobalt appears to be f.c.c. with a <111> texture; this is not surprising when f.c.c. interlayers are used (in contrast to b.c.c. V and Cr). Also the cobalt moment in Co/Pd is increased above that of the bulk metal because of an induced spin polarisation of the palladium atoms at the interface.

In general Fe/TM multilayers do not exhibit perpendicular magnetization (but see Fe/SiO$_2$ above), presumably because of a lack of preferred orientation and larger magnetostatic shape energies. However, as mentioned earlier, interesting exchange interactions occur which could lead to applications in sensors. This is particularly so, for instance, in Fe/Cr multilayers [9] in which there is a field dependent change from antiferromagnetic to ferromagnetic behaviour. There is an associated change in current transmission across the multilayer through the chromium layers giving a large change in magnetoresistance.

Two further examples of the possible application of multilayers as sensors are in the development of controllable magnetostriction, magnetization and permeabiltiy. In analogy to surface anisotropy one might expect a surface magnetostriction. Indeed Ni/Ag and Ni/C multilayers do show [17] a variable λ_{eff} which is inversely related to the thickness of the nickel layers. Rare earth-transition metal based multilayers would be of interest in this context.

Magnetic recording heads, particularly video heads, require as large an M_s as possible, low magnetostriction, anisotropy and coercivity and very high permeabilities. A crystalline/amorphous multilayer of Fe/CoNbZr films provides these requirements [18]. The amorphous CoNbZr alloy spacing layer provides for the nucleation of small grains in the iron films thus minimising local anisotropies and giving low coercivites. An optimum structure appears to be a multilayer of 100Å Fe/50Å CoNbZr. Two interesting observations are that the iron films do appear to grow with a <110> texture and that the nucleation of the iron grains appears to be reflected through the CoNbZr film, i.e. the growth of the iron is somehow coherent.

3.2. RE-TM and Heusler multilayers

Multilayers have been investigated which contain either rare earth-transition metal alloys as one component or as both components. There are two main objectives in these experiments and they are both related to a particular application, that of thermo-magneto-optic (TMO) recording. The first objective is to increase the stability of the medium with respect to oxidation and the second is to improve the optical properties of the medium, i.e. the magneto-optic Kerr rotation angle, θ_K, and the reflectivity. It would be ideal, of course, if both these requirements could be satisfied simultaneously.

Compositionally modulated films of TbCo can show increased resistance to oxidation and enhanced anisotropy [19]. The modulation is achieved by r.f. sputtering layers onto a substrate held alternately at zero bias and at a negative voltage bias. In this way alternate layers of Tb-rich and Co-rich alloys are obtained with differing microstructures. The bias causes re-sputtering of the growing film, a loss of rare earth and a refinement of the grain structure.

The compositional modulation is illustrated in the RBS spectrum of
Figure 9 which shows that for this four (n = 4) layered film the cobalt
content in each layer is not altered by the bias but the terbium content
oscillates from one to the next. The magnetic properties of the
multilayers are shown via the magneto-optic and vibrating sample
magnetometer loops of Figure 10. The thickness of each layer can be
calculated from the total thickness of the multilayer (3000Å). The
systems for n = 2 and 4 show an anomalous loop associated with the
different compensation temperatures of the two different compositions
and the presence of a compensation domain wall at each interface. The
thickness, δ, of this flux adjusting wall is calculated at 150Å and
hence any multilayer where the layer thickness, d, is greater than 2δ
will exhibit a complex magnetization behaviour, i.e. for n < 10. For
large n > 50 (d < 60Å) the multilayers behave as a single film, i.e the
loop for n = 76 is characteristic of a single film of the average
composition. Significant mixing is probable in bias sputtering of
multilayers.

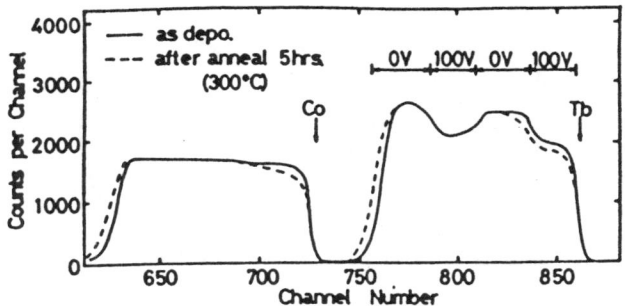

Figure 9. RBS spectrum for a 4-layer TbCo film of 0-100V bias
modulation. Solid lines as-deposited, broken lines annealed.

Stable domains could be laser written in these multilayers and "bit"
sizes as small as 2μ achieved. This shows that adequate values of
cercivity and anisotropy can be obtained in such modulated structures.
The resistance to oxidation is certainly increased in these materials.
Figure 11 plots the change of magnetization of a set of 0-150V modulated
films as a function of aging time at 50°C in a relative humidity of
100%. The magnetization of the 150V single layer changed too quickly to

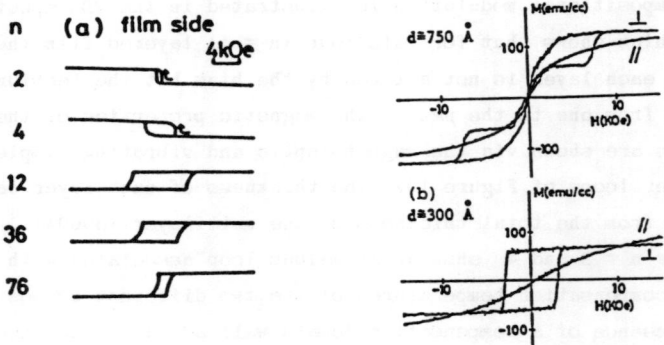

Figure 10. (a) Kerr loops for 0-150V modulated TbCo multilayers, (b) VSM loops for systems with n = 4 and 10.

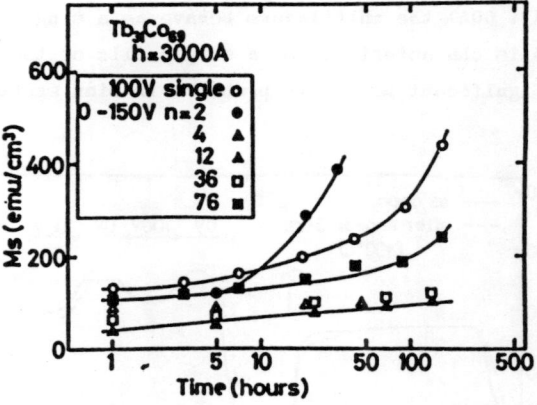

Figure 11. Results of aging TbCo multilayers at 50°C and 100% humidity.

be included in the graph but it is clear that for n = 12 and 36 excellent effective resistance to oxidation of the rare earth is obtained and M_s stays almost constant with time. Presumably at smaller layer thicknesses the mixing effects reduce the effectiveness of the columnar microstructure of the zero bias layers in combatting oxidation. Bias induced, compositional modulation is therefore able to provide media suitable for TMO applications with adequate optical properties and stabilities approaching that of single layer films intentionally protected with metallic or dielectric overlays [20].

An alternative strategy to improve the performance of TMO media is
to construct RE-TM/X multilayers, where X is a good reflector and a
chemically stable material. It is possible that such multilayer
systems, e.g. TbFeCo/Ag, could give enhanced θ_k's and reflectivities.
Multilayers of FeCo/Ag with a dielectric interference layer have given
Kerr rotations of several degrees with good reflectivity [21].

Strong candidates for TMO applications are Heusler alloy films,
typically the stoichiometric compound PtMnSb. These half-metallic
materials are ferromagnetic with a large Kerr rotation and Polar θ_k's of
over 1° [22]. There are at least two particular difficulties with this
material. The first is that as-deposited films are amorphous and hence
non-magnetic. The second is that the ferromagnetic crystalline $C1_b$
structure has cubic symmetry. Annealing and crystallization (> 150°C)
of the as-deposited films or deposition onto a heated substrate gives
the $C1_b$ structure and films with in-plane magnetization with M_s
typically 550 emu cm^{-3}. The average grain size of these films is 50nm
[22] which would produce significant noise in any optical read process.

Attempts to introduce perpendicular anisotropy, increased coercivity
and reduced grain size in PtMnSb films have used several strategies.
The simplest approach is to use a biasing scheme such as that provided
in a bilayer. Figure 12a shows polar Kerr loops taken from the
substrate (glass) and air sides of a 2000Å TbFeCo/150Å PtMnSb bylayer.
The increased coercivity and remanence due to the biasing effect is

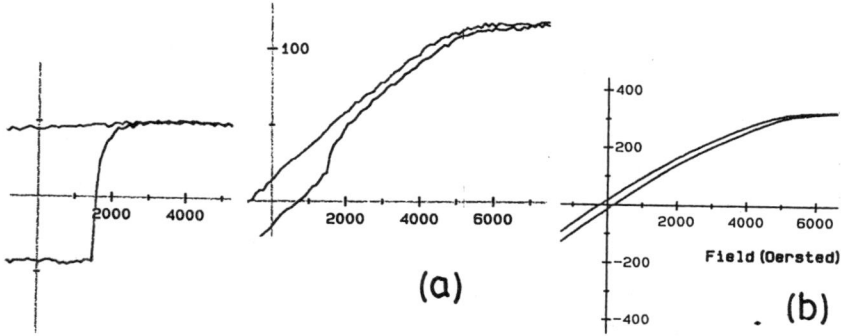

Figure 12. Kerr loops for (a) glass and air sides of a
glass/TbFeCo/PtMnSb bilayer and (b) a single PtMnSb film

clear in the PtMnSb loop. There is also some indication of a
perpendicular component of anisotropy in comparison with the loop from a
single PtMnSb film, Figure 12b.

A second approach is to make use of a multilayer system and to
explore the possibility of a surface contribution to anisotropy. At the
same time experience with Co/TM multilayers would suggest that it might
be possible to obtain a reduction in grain size, particularly if the
spacing layers were amorphous. Figure 13 compares TEM micrographs of a
single $Pt_{31}Mn_{38}Sb_{31}$ film and a PtMnSb/V, 27Å/12Å multilayer, the latter
annealed at 300°C for 2 hours. The decrease in grain size is clear.
Unfortunately with the reduction in grain size comes a decrease in
magnetization and smaller Kerr rotations. No perpendicular anisotropy
has been obtained yet. Further experiments with other biasing
arrangements and multilayer structures are in progress.

 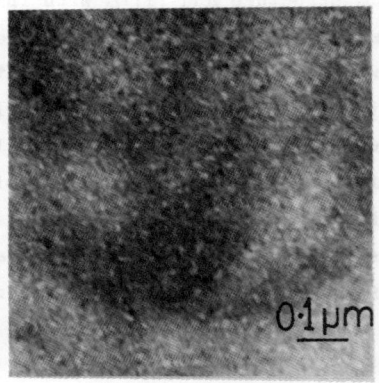

Figure 13. TEM micrographs of the microstructure of (a) single PtMnSb
film and (b) PtMnSb/V, 27Å/12Å multilayer.

Experiments with potential superlattice systems, where the multilayer
has components which are almost lattice matched have been reported [23].
PtMnSb/NiMnSb (lattice mismatch 4%) multilayers have been sputter
deposited with equal layer thicknesses between 20Å and 200Å. A textured
microstructure was obtained with a preferred ⟨111⟩ growth of the $C1_b$
phase but unfortunately no perpendicular anisotropy was detected and
reduced Kerr rotations were obtained. However an effective
perpendicular anisotropy of about 4×10^5 ergs cm^{-3} has been reported [24]
in related Pt/MnSb multilayers with a 12.5Å Pt + 50Å MnSb period.

However these multilayers were annealed at 400°C before measurement and the integrity of the layered system must be in question.

4. SUMMARY AND CONCLUSIONS

In this paper we have discussed in general terms the concept of a composite magnetic material formed by depositing many layers of, usually, two different components. Our attention has focused on particular choices and on our small contribution to the enormous and growing body of information on multilayers. We have considered only certain magnetic and structural characterisation techniques. It is obvious that the use of measurement methods at the atomic level, such as ferromagnetic resonance and Mossbauer spectroscopy has provided much complementary information. Use of in-situ characterisation techniques is particularly useful in probing the properties of the growing films and the multilayer structure.

Apart from their intrinsic interest it is possible that potential applications exist for these materials. The category and perhaps quality of a multilayer is determined by its method of fabrication; MBE techniques can give sharp and coherent crystalline interfaces while the more rapid deposition techniques of evaporation and, especially, sputtering provide polycrystalline and amorphous systems. The feasibility of any application will depend on, amongst other factors, fabrication costs. Large volume products will look to sputtering, conventional evaporation and, possibly, electrodeposition methods. More sophisticated and exacting structures will merit MBE techniques.

There is no doubt that the attention given to multilayer systems will grow, driven both by scientific curiosity and their possible exploitation in new device and materials applications.

5. REFERENCES

+*Now at Max Planck Institut fur Metallforschung, Stuttgart, F.R.G. Previously at the University of Salford.
1. Sato, N., J. Appl. Phys. 59, 2514 (1986).
2. Atzmony, U., Swartzendruber, L.J., Bennett, L.H., Dariel, M.P., Lashmore, D.S., Rubinstein, M., and Lubitz, P., J. Magn. Magn. Mat. 69, 237 (1987).

3. Babkair, S.S., and Grundy, P.J., IEEE Trans. Magn. MAG-24, 1710 (1988).
4. Korecki, J., and Gradmann, U., Phys. Rev. Lett. 55, 2491 (1985).
5. Freeman, A.J., Fu, C.L., Lee, J.I., and Oguchi, T., Proc. ISPMM, Sendai,1987, World Scientific Pub. Co. Ltd., Singapore, p221, 1987.
6. Szpunar, B.J., J. Magn. Magn. Mat. 49, 93 (1985).
7. Gradmann, U., J. Magn. Magn. Mat. 54-57, 733 (1986).
8. Gyorgy, E.M., Dillon, J.F., McWhan, D.B., Rupp, L.W., and Testardi, L.R., Phys. Rev. Lett. 45, 57 (1980).
9. Baibich, M.N., Broto, J.M., Creuzet, G., Etienne, P., Fert, A., Fert, A.R., Hadjoudj, S., and Nguyen Van Dau, F., paper Th2-03, 12th ICMFS, Le Creusot, August, 1988.
10. Sakakima, H., Krishnan, R., and Tessier, M., J. Appl. Phys. 57, 3651 (1985).
11. Stearns, M.B., Lee, C.H., and Vernon, S.P., J. Magn. Magn. Mat. 54-57, 791 (1986).
12. Ohkoshi, M., Grundy, P.J., and Babkair, S.S., paper at ICM, Paris, 1988, to be published in J. de Phys.
13. den Broeder, F.J.A., and Draaisma, H.J.G., Proc. ISPMM, Sendai, 1987, World Scientific Pub. Co. Ltd., p234, 1987.
14. Carcia, P.F., ibid p240.
15. Tsukahara, S., Matsuoka, S., Sekino, K., Itoh, A., Togami, Y., Morishita, T., and Tsushima, K., ibid p287.
16. den Broeder, F.J.A., Kuiper, K., van de Mosselaer, A.P., and Hoving, W., paper ThP-27, 12th ICMFS, Le Creusot, August,1988.
17. Zuberek, R., Szymczak, H., Krishnan, R., Youn, K.B., and Sella, C., IEEE Trans. Magn. in press and paper Th3-03, ibid.
18. de Wit, H.J., Witmer, C.H.M., and Dirne, F.W.A., IEEE Trans Magn. MAG-23, 2123 (1987).
19. Honda, S., Ohkoshi, M., and Kusuda, T., IEEE Trans. Magn. MAG-22 22 (1986),
20. Grundy, P.J., Lacey, E.T.M., and Wright, C.D., IEEE Trans. Magn. MAG-23, 2632 (1987).
21. Reim, W., and Weller, W., paper 4D-4, ICM, Paris, July, 1988, to be published in J. de Phys.
22. Grundy, P.J., and Attaran, E., to be published in J. Magn. Magn. Mat.
23. Takanishi, K., Fujimori, H., Shoji, M., and Nagai, A., Jpn. J. Appl. Phys. 26, L1317 (1987).
24. Kawanake, T., and Naoe, M., paper ThP-36, 12th ICMFS, Le Creusot, August, 1988.

Magneto-Optical Materials for Data Storage Applications

R Carey, D M Newman and B W J Thomas
Department of Applied Physical Sciences
Faculty of Applied Science
Coventry Polytechnic
COVENTRY CV1 5FB
England

Abstract

The search for suitable materials for thermo-magnetic recording (TMR) and the concurrent development of thermo-magneto-optic (TMO) erasable disk systems has led to impressive improvements in the CNR in the latter over the past four to five years (~45dB in 1983 to ~60dB in 1987).

The progress has in large part been empirical, with not nearly as significant progress being made in the basic understanding of the magnetic properties (eg the origin of K_u) or the magneto-optic properties (eg the origin of the Kerr rotation θ_k) of the materials used.

Of particular interest to date have been various homogeneous RE-TM alloys - binary, ternary and quaternary - and a general approach to the empirical categorisation of these different alloys is now fairly well established. Progress has also been made in the development of alternative TMO materials, some of which are beginning to show considerable promise.

In this paper the development of TMO materials is discussed, with particular reference being made to reported work on the magnetic and magneto-optic properties of amorphous RE-TM alloy films, as well as some more recent developments that are likely to challenge the dominance of these homogeneous amorphous alloys in the future.

1. AMORPHOUS HEAVY RE-TM ALLOY FILMS

Ferrimagnetic behaviour is anticipated only in RE-TM alloys in which the RE is one of the heavy RE metals (Gd, Tb, Dy, Ho, Er, Tm, Yb and Lu); the light RE metals (La, Ce, Pr, Nd, Pm, Sm and Eu) are expected to couple ferromagnetically with the TM in

RE-TM alloys. Thus, to date, the heavy RE-TM alloys have been studied in far greater detail as potential TMO storage materials, and of these the interest in Gd and Tb alloys has been most intense. Of the transition metals, only Co and Fe, when alloyed with these RE metals, produce ferrimagnetic behaviour at or above room temperature. Major interest has therefore centred on amorphous alloy films of binary or ternary combinations of Gd, Tb, Fe and Co as potential TMR media.

Ferrimagnetic behaviour is not essential for TMR, indeed 'Curie point writing' was first used for TMR in MnBi, a ferromagnetic alloy. However, the possibility of magnetic compensation, and the consequent magnetic and thermomagnetic properties of the heavy RE-TM films, is very attractive for TMO applications. The more recent interest in light RE-TM alloys and both light and heavy RE-TM alloys will be discussed later.

The development of different RE-TM alloy films is based on the observed magnetic and magneto-optic properties of the possible binary combinations of Gd, Tb, Co and Fe, ie GdCo, TbFe, GdFe and TbCo, when compared with the desirable material requirements shown in Table 1. A considerable volume of data is now available on the magnetic properties of these alloys. All four can be prepared in amorphous form by vacuum evaporation and/or sputtering under different conditions. The resultant microstructure and hence some of the magnetic and, to a lesser extent, the magneto-optic properties are

TABLE 1 Important material requirements that must be considered when selecting a material for a thermo-magneto-optic recording medium.

1. MAGNETIC REQUIREMENTS

	Requirement	Corollary
(i)	M_s normal to film plane	$K_u \gg 2\pi M_s^2$
(ii)	Magnetization distribution stable against stray magnetic fields and small temperature variations	$H_c (\sim K_u/M_s)$ large at room temperature ($\sim$ 5KOe);
(iii)	Small domains, diameter $\sim 1\mu m$	Large $M_s H_c$ product. Minimum stable diameter $d_0 \sim \sigma/M_s H_c$
(iv)	Regular and reproducible domains	Defect free films. σ large c.f. demagnetizing energy

2. THERMOMAGNETIC REQUIREMENTS

Requirement — **Corollary**

(i) TMR possible with available solid state lasers (pulsed rapidly) without film damage
— At data rates required with up to ~10 mW available at film, temperature changes up to ~100°C possible
H_c (T) therefore gives
H_c (Room Temp) ~5kOe
H_c (Elevated Temp) ~500Oe

(ii) No crystallization of amorphous alloy
— The critical temperature (compensation or Curie) << crystallization temperature

(iii) Magnetization distribution unaffected by small temperature excursion during read process
— Read power (~1 to 2 mW) << Threshold for writing or erasing

(iv) Complete erasure without degradation of SNR
— H_c large

(v) Noise level after write process ≈ noise level before write process (ie N_{mod} ≈ 0)
— M_s small close to critical temperature

(vi) Direct overwrite with same laser
— Only demonstrated in simple RE-TM alloys with T_{comp} ~60°C

3. MAGNETO-OPTIC REQUIREMENTS

Requirement — **Corollary**

(i) $M_s \perp$ to film plane
— Polar Kerr effect readout

(ii) SNR as large as possible but at least >25dB
— Maximum $\sqrt{R}\, \theta_k$ (=K)
(RE-TM, θ_k~0.3°, R~0.6 leads to SNR ~28dB)

(iii) Reduce media noise to minimum
— θ_k and R both uniform over disk surface

(iv) $\frac{d\theta_k}{dT}$ small at elevated temperature
— Compensation point writing to maintain SNR

influenced by the preparation conditions. However, ferrimagnetic behaviour with compensation temperatures below, at and above room temperature has been reported for all four binary alloys. The range of compositions over which compensation is observed above room temperature for GdFe and TbFe is very narrow and compensation temperatures significantly higher than room temperature are not observed. On the other hand, the Curie point for these two alloys varies little with composition about the room temperature compensation composition and is typically ~120° C and 210° C for TbFe and GdFe respectively. Thus they may be used for Curie point writing, whereas GdCo and TbCo with relatively high Curie temperatures are only suitable for compensation point writing. The necessary large value of K_u for perpendicular magnetization is also observed in all four alloys, with square hysteresis loops and high coercivity close to compensation. These apparently attractive overall features of these alloys are not identical for all four. In particular, although the Gd alloys (GdCo and GdFe) exhibit very strong temperature variations of coercivity close to compensation, the coercivity at temperatures away from compensation is not large enough to support stable submicron diameter domains in either alloy ie the product $M_s H_c$ is too small. TbFe and TbCo alloy films have significantly greater H_c values and the $M_s H_c$ product is generally large enough for stable submicron domains to exist.

Thus the Tb binary alloys would seem to be more attractive than the Gd alloys for high density TMR. Moreover, whilst the origin of K_u is not clearly understood, it is generally believed that, as the symmetrical 'S' state of the Gd ion does not produce any local anistropy, the development of the uniaxial anistropy in the Gd alloys may well arise from less favourable sources than in the Tb alloys. In this context, however, it should be noted that a large thermal expansion anomaly is known to occur in TbFe amorphous alloys at compositions useful for TMR [78]. This may create local variations in K_u which, if the latter is wholly or in significant part strain induced, leads to possible loss of data or even damage to the film when it is thermally cycled during the write and erase process, depending on the precise choice of substrate.

A considerable amount of empirical data is also available on the magneto-optic properties of these materials. At room temperatures, and λ ~633nm, the polar Kerr rotation is not large in any of the binary alloys :-

GdCo ~0.33°, TbCo ~0.29°, GdFe ~0.29°, TbFe ~0.23°

Of the binary alloys it would seem that TbCo films are most satisfactory for high SNR, with suitably high K_u, H_c and $M_s H_c$ product for TMR using compensation point writing, whilst TbFe has good magnetic and thermomagnetic properties for Curie point writing but relatively poor magneto-optic and thermo-magneto-optic properties.

The poor read-out signal available from TbFe may be overcome by using a second layer as the read layer provided it faithfully reproduces the data in the TbFe storage layer. TbFe has recently been used very effectively for TMO disks using both GdFeCo(i) and GdTbFe(ii) as the overlayer, both possessing significantly higher K values than TbFe, and are well coupled to the TbFe through exchange coupling, see (i) Asaca ADS-5000, [60] and (ii) Nippon Optic.

Unfortunately, some early reports on TbCo indicated that a large K_u could only be developed by sputtering the films in the presence of significant bias voltage. This is not desirable when using plastic substrates. However, evaporated films were reported to suffer from crystallization problems above 390K [24]. More recently, sputtered TbCo films with large K_u, prepared in zero bias conditions look quite promising, with compensation temperature ~220° C and no reported crystallization problems for temperatures >410° C [31] & [32].

Although TbFe is currently being used as a Curie point storage material (in conjunction with a read over-layer) and TbCo is still a promising material for compensation point writing, the search for more efficient single layers for TMR has led to a number of studies of ternary alloys based on the established properties of the binary alloys. (Indeed, most of the disks currently available for evaluation utilize ternary alloys.)

Although it might be misleading to attempt to simplify the general properties of the ternary alloys based on Gd, Tb, Co and Fe, a few general points can be made. It is quite likely that alloys rich in Fe will have relatively low Curie points and will probably be used as Curie point materials, whilst those rich in Co, with relatively high Curie points are more likely to be used as compensation point materials. The addition of Co to alloys containing Fe is likely to raise the Curie point and will therefore lead to a larger room temperature Kerr rotation that may be virtually temperature independent over the range of temperatures used for TMR. Alloys containing Gd are likely to have greater Kerr rotations and the addition of Tb will probably produce significant increases in K_u and H_c at room temperature. Since Tb is both very expensive and comparatively rare, it is not surprising that attempts have been made to use other non 'S' state heavy rare earths, such as Dy or Ho, as alternative sources of perpendicular anisotropy. Finally, since the Kerr effect is due almost entirely to the TM subnetwork at room temperature for the wavelengths of interest, studies of the magneto-optic properties of evaporated Fe-Co alloy films [67] leads one to anticipate that in alloys containing both Fe and Co, the Kerr rotation will vary both with Fe and Co content and will be a maximum when the Fe and Co contents are roughly equal.

Bearing in mind the significance of Tb to the desirable magnetic properties of the possible ternary alloys, it is also not surprising that the three ternary alloys to emerge as useful TMR materials are GdTbFe, TbFeCo and GdTbCo. Although, as expected from above, GdFeCo films produce a relatively high Kerr rotation (eg Masui et al [84]), these ternary alloys tend to exhibit small perpendicular anisotropy or even in-plane anisotropy [98]. To be able to take advantage of the large rotation for high density TMR, increased perpendicular anisotropy is essential. This may be achieved either by forming the quaternary alloy GdTbFeCo (or GdDyFeCo or GdHoFeCo) [98] or, by again designing an exchange coupled double layer system, incorporating the GdFeCo as an effective read layer.

Of the ternary alloys of interest, TMR in GdTbFe is achieved using Curie point writing [23], whereas GdTbCo exhibits a wide range (-10° C to 240° C) of compensation points as the Co content is varied from 78 to 72 at %, so that compensation point writing is possible [44] & [20]. The Curie point for the TbFeCo alloys of current interest (~200° C) makes them suitable for Curie point writing, but the precise composition selected for a disk storage layer (eg RE dominant or TM dominant at room temperature) can have a significant effect on the final CNR.

The choice of material and its composition is often discussed simply in terms of the mode of 'writing' to be adopted and the likely effect on problems associated with routine disk production. The basic requirements for a TMO disk material include a large room temperature coercivity and a temperature variation of H_c that permits TMR with available writing energies. The first of these implies that the composition for many alloy systems is fairly close to that for room temperature compensation. The latter requirement further implies that either the Curie point (or compensation point) is not significantly above 200° C. Within these limitations the choice of alloy and its composition is still extensive. The final decision may be determined by the ease of mass production of a disk. (From this point of view it seems almost axiomatic that a single layer is preferable to a complex multiple layer structure and a binary alloy is preferable to both ternary and quaternary alloys.) It will also depend on the chemical stability of the available alloys. There can be no doubt, however, that the composition of a given alloy system must be optimised for maximum CNR.

For some compensation point writing materials, the Curie temperature is well removed from the compensation temperature. This leads to a Kerr rotation, for the write/read process, that is both greater and less temperature dependent than Curie point writing materials and a significantly improved CNR. On the other hand, it has often been pointed out that, since the Curie point for a given alloy is much less sensitive to the RE-TM ratio than the compensation point, accurate control of the Curie point over the whole disk area is likely to be easier. This makes Curie point materials appear more attractive for large scale production. High coercivity TbFeCo, with a Curie point ~210° C, is at the moment the most popular of the ternary alloys for TMR. Recent reports (eg [72]) on the optimization of the CNR for high density storage using TbFeCo do indicate, however, that accurate control of the composition is absolutely essential; a view supported by Takahashi [93] and Okada et al [64]. A detailed study of TbFeCo disks [97] suggests that the noise difference measured before and after writing data (N_{mod}) for RE dominant and TM dominant TbFeCo alloys, and the associated difference in the CNR, arises mainly from the more irregular shape of the domains in the TM dominant material.

Reports on domain regularity and the domain nucleation processes in TMR materials often refer to the work of Huth [30] showing clear irregularities in recorded bits in GdCo, and that of Tanaka and Imamura [101] on TbFe, TbGdFe and TbCoFe, showing superior writing and erasing properties for RE dominant films. Shieh and Kryder [81] and [82], studying the factors governing domain regularity and reproducibility in GdTbCo films, also conclude that thermally written domains in this alloy are more regular in materials with high coercivity and a RE dominance at room temperature. They point out that domain shape is more strongly affected by film magnetic parameters than the writing parameters. Improvements are observed in films with large domain wall stiffness which increases with domain wall energy σ (and therefore K_u), low magnetization and smaller aspect ratio d/h (h = thickness of the film). They further show that domain regularity depends crucially on the temperature dependence of both H_c and M_s at the elevated temperatures used in the write and erase process.

Recently, Takayama et al [98] reported comparisons between disks using materials exhibiting a large Kerr rotation (Gd(RE)FeCo, with RE = Tb, Dy and Ho), and a TbFeCo disk with similar Curie and compensation temperatures. Despite the large θ_k values of the quaternary alloys compared with for TbFeCo, the poor domain regularity in the quaternaries produced significantly larger N_{mod} values and hence smaller CNRs (12dB less in the case of GdTbFeCo). Moreover, the carrier signal neither increases sharply nor saturates, even in higher external fields (~600Oe), indicating the poor writing (and erasing) characteristics of these quaternary alloys.

Irregular domain shape is attributed to the increased role of the wall energy in domain formation in Gd based alloys. This is in agreement with conclusions reached by Shieh and Kryder, [81] and [82], on the significance of wall stiffness in GdTbCo alloys, and indicates that a good deal of experimental work is still needed to provide optimum materials for TMR.

In the case of TbFeCo, detailed investigation has produced TbFeCo disks with CNR ~60dB at 1MHz, falling only to 55dB at 5MHz (Okada et al [65]). A CNR in excess of that possible with a single layer TbFeCo disk has been reported for a double layer TbFeCo disk in which the read layer is just TM rich (giving a higher θ_k) and the storage layer just RE rich (Fujii et al [14]). The reports to date on high coercivity GdTbCo are also encouraging; of the simple ternary alloys, it is probably the only other on a par with TbFeCo for applications in TMR.

Clearly, the selection of a RE-TM alloy and its composition for TMR must include consideration of domain regularity. The magnetic properties (H_c and M_s) at elevated temperatures, notwithstanding comments made here and elsewhere concerning the importance of high Kerr rotation, are therefore extremely important. Control of these magnetic parameters in a homogeneous alloy film demands very close control of the composition over the usable disk area for the CNR to be maximized.

Finally, although both the successful incorporation of pulsed magnetic fields at increased frequencies and two layer systems utilizing the different thermomagnetic properties of the component films, have been reported to produce direct overwrite, successful overwrite with a pulsed laser in a single homogeneous layer has only been demonstrated to date in materials with compensation points some 50° C higher than room temperature [79] and [84].

2. DOPANTS IN RE-TM FILMS

There has been a large number of investigations into the enhancement of the Kerr effect by the addition of dopants to RE-TM alloys. It has been well known for some time, for example, that the addition of the Bi^{3+} ion to the rare earth garnets can produce enhanced Faraday rotations, and it is therefore not surprising that a wide range of additives to RE-TM alloys have been tested for increased magneto-optic interaction. Whilst both the magnetic and magneto-optic properties of these alloys are usually affected by additives, in general the effect on the Kerr rotation is not nearly so marked as the effect on the compensation temperature (and to a lesser extent the Curie temperature). The addition of 10at% Bi to GdFe increases θ_k (but in GdCo decreases) from 0.33° to 0.4° [71] whereas 11at% Bi decreases T_{comp} from 300K to 100K and T_c from 460K to 440K [22]. Although the study of the effect of dopants is of considerable value to the development of the understanding of magnetic and magneto-optic effects in amorphous RE-TM alloys films, the major function of additives to date in practical disk systems is in the passivation of the alloys. When possible the selected additive (eg Pt, Cr, Al or Ti) should also produce an increase in both θ_k and the reflectance (rather than a decrease) so that K is increased.

3. LIGHT RE-TM ALLOYS

It was noted earlier that ferrimagnetic behaviour is only anticipated in heavy RE-TM alloys, and that the magneto-optic interaction in these, at room temperature and at the wavelengths of interest, is predominantly due to the TM subnetwork. Ferromagnetic coupling on the other hand is anticipated (and observed) in the light RE-TM alloys.

Since the 4f electrons of the light RE elements such as Nd and Pr are located within ~2eV of the Fermi level of these alloys, the magneto-optic interaction due to the RE subnetwork is not negligible, as is the case in the heavy RE elements with 4f electrons ~8eV below the Fermi level. The magneto-optic contribution of both the light and heavy RE elements add to that of the TM so that an increased θ_k is anticipated in the light RE-TM alloys.

This increase in θ_k is indeed observed in binary and ternary light RE-TM alloys such as NdFe, PrFe and NdFeCo. The surprisingly large perpendicular anistrapy developed in some of these alloys, which have Curie temperatures suitable for TMR, has led to the addition of some light RE material to the heavy RE-TM alloys in an attempt to increase θ_k. The reported results on Tb(PrCo) (T_{comp} ~120° C, T_c ~400° C [83]) and TbNd(FeCo) [68] are quite encouraging. Replacement of Tb by Nd in (NdDy)(FeCoTi) alloy films [15] may also prove advantageous. Films of this alloy are reported to exhibit positive anistropy for very thin films of even 5nm thickness. The room temperature coercivity, H_c ~7.6kOe [85] and Curie temperature, T_c ~300 C, make this alloy quite suitable for TMR.

4. COMPOSITION MODULATED (CM) RE-TM AMORPHOUS ALLOY FILMS

One of the most promising recent developments in the study of TMR materials is the production of CM amorphous RE-TM alloy films. These novel materials, made of ultra thin layers of alternately deposited RE and TM, (eg Tb/Fe [74] and Tb/Co [28]), exhibit magnetic properties that make them well suited to the TMR application. Large perpendicular anisotropy and coercivity combined with relatively large magnetization and low Curie temperature indicates that small, stable domains may be written into these materials using TMR. The measured values of θ_k also suggest the attainment of acceptable CNR's when these materials are used in a disk. It is interesting that Sato et al [75], studying a range of different RE-(FeCo), CM disks with a CNR >52dB (at 1800 rpm and 1MHz) for all the compositions studied, show that the best CNR is obtained for Dy (FeCo) - 57dB at 1MHz (falling to only 38dB at 10MHz) - rather than for Tb/TM despite the latter having a larger θ_k.

Perhaps even more significant for TMR systems is the fact that the magnetization and high coercivity of CM films are both relatively insensitive to the thickness of the RE layers [54]. This implies that the control of the magnetic properties over the whole of the disk surface required for regular and reproducible domains may well be easier with these films than with the homogeneous RE/TM films already discussed.

5. OTHER MATERIALS

Although a large part of the effort to date has been concentrated on various RE-TM alloys, much progress has also been achieved with other materials that do not have the same corrosion problems. Of these, CrO_2 is of passing interest only since its magnetization is always in plane. However, the fact that the thermomagnetic properties of this conventional magnetic recording material are suitable for TMR suggests that other magnetic recording materials may also be suitable for TMR. Thus, the magneto-optic properties of other materials for perpendicular magnetic recording have been widely investigated eg Co-CoO [66] and alumite layers [1].

Studies of the magnetic and magneto-optic properties of corrosion resistant magnetic fluorides are not particularly rewarding, whereas those on oxide films are much more promising. These films are also highly corrosion resistant and have a strong magneto-optic interaction. Oxide films of (a) Bi-substituted garnets, (b) M-type barium ferrite, and (c) Co-containing spinels such as cobalt ferrite, have been reproducibly grown on glass substrates by sputtering (a, b and c), pyrolysis (a and c) and ion plating (c). The films are polycrystalline and possess a perpendicular uniaxial anistropy. Reverse domains (~1 μm diameter) have been successfully written into each of these materials using a focussed laser beam. Unfortunately, for both ferrite-type materials, attempts to increase the magneto-optic interaction by introducing more Co leads to a decrease in the perpendicular anistropy with an associated undesirable deterioration in the squareness of the perpendicular hysteresis loop. In the garnets, however, Bi substitution leads to both an increased perpendicular anistropy and a stronger magneto-optic interaction. Early work on these polycrystalline materials indicates that one of the main problems concerns the media noise produced by grain boundaries. Krumme et al [43] and more recently Gomi and Abe [19] and Shono et al [87] report, however, that garnet films prepared in pure Argon (and pure Argon and Hydrogen) are of improved quality. Because these garnets are weakly absorbing, it is necessary to provide a backing Cr film to achieve TMR with a semiconductor laser. Measured CNRs (~45 dB) are not yet comparable with the best achieved by RE-TM alloy disks, but further improvement of the medium should lead to an increase in the near future. Although not ideal as a single layer storage medium, the good transparency of these oxides in the near infra red, makes them well suited to multi-layer designs [19]. The ability to produce a good quality garnet film with a large Kerr rotation does suggest its use as a highly efficient 'read layer' to be coupled with a suitable storage layer such as TbFe as demonstrated by Yokoyama et al [114], producing an order of magnitude improvement in the figure of merit.

A similar approach may be considered with the semi-metallic ferromagnet PtMnSb and the semimagnetic semiconductor CdMnTe. Bulk PtMnSb has been shown [112] to have the largest known room temperature polar Kerr rotation of any magnetic material. The rotation produced by thin evaporated films of PtMnSb on glass substrates is even larger (θ_k~2°) than the bulk value [86]. Unfortunately, these films possess an in-plane anistropy. However, some early studies of PtMnSb/TbFe layered structures [49] indicate a significant increase both in the remanence and coercivity perpendicular to the film surface. CdMnTe films prepared by the ionized cluster beam (ICB) method are known to possess a large Verdet constant [41]. A more recent report on multi-layer structures of this material, shows how the Faraday rotation spectrum may be controlled to produce energy shifts in the peak rotation.

6. SUMMARY

The development of RE-TM amorphous alloy films suitable for TMR has been quite impressive over the recent four or five years and a number of disks are now available for evaluation from different suppliers (see Table 2). The medium used for most of these disks is based on the ternary TbFeCo, although TbFe, TbCo, GdTbFe, GdTbCo and some quaternary alloys have also been used successfully. It seems unlikely that, with such a relatively short history of development, the alloys studied so far will be the best possible for the TMR application.

With more understanding of the origin of the magnetic and magneto-optic properties of amorphous alloys, further improvements are certain. At present, great interest surrounds the design of the CM structures and the development of improved garnet layers. Both activities provide real challenges to the present domination of the RE-TM alloys for TMR media. The search for new very reliable and stable media producing high SNR will undoubtedly continue for some years after the first generation of RE-TM disks are introduced into the market.

7. ACKNOWLEDGEMENTS

The authors wish to thank the Department of Trade and Industry, London for permission to publish this report.

8. REFERENCES

The references in this paper (listed below) are extracted from a more detailed report on 'Magneto-optic Data Storage Systems' prepared by the authors for the Department of Trade and Industry, London. Only the primary author is listed for each reference.

TABLE 2 Magneto-optic Disks Available

Manu-facturer	Diam-eter (mm)	Substrate	Material	Capacity (MBytes)	CNR (dB)	Status
3M	130	Plastic	RETM	300/300	57	Samples Available
Bull	130	Glass	GdTbFe or TbFe		~50	Under Development
Daicel	130	PC	TbFeCo	500/500	>45	Pilot Plant
Fujitsu	130	Glass	TbFeCo	500/-		Announced June 1987
Hewlett Packard	90	Glass	GdTbFe		55/6	Prototype
Hitachi	130		TbFeCo	225/225	50	Available 1988
KDD		PMMA	TbFeCo		50	Technology Available
Matsushita	130	Glass/Plastic	TbFeCo	200/200	54	Under Development
Maxell	130	PC	TbFeCo+	250/250		Available 1988
NHK	305	Glass	TbFe/GdFeCo		47	In-House Only
Philips	120	PC	GdTbFe	450/450	55	Samples Available
Plasmon	120	PC	RETM	330/330	45	Prototype
Ricoh	89	Glass	Barium Ferrite	250/-	50	Available 1990
Sanyo	130	PMMA/PC	TbFeCo	326/326	56	Samples Available
Sharp	130	Glass	GdTbFe	150/150	>48	Pilot Plant
Sony	130	PC	TbFeCo	325/325		Samples Available
Sumitomo	130	PMMA or PC	TbDyFeCo		55	Prototype
TDK	130	PC	TbFeCo +	300/300	>50	Samples Available
Toshiba	130	PMMA	TbCo	300/300		Samples Available
Verbatim	90	PMMA or Glass	TbFeCo		40/-	Prototype

Author	Title	Reference	Year	No Ref
Abe M	Magneto-optical Effect in Anodized Al(2) O(3) Film with Micropores Electrodeposited with Co	J.Appl.Phys. Vol 57 3909-3911	1985	1
Fujii Y	Exchange-Coupled Double-Layer TbFeCo Films	Paper 22A-06 Int Symp Magneto-optics Kyoto	1987	14
Funada S	Magnetic Properties of New Magneto-optic Recording Media - NdDyFeCoTi	Paper CG-02 Intermag Tokyo Japan April 14th-17th	1987	15
Gomi M	Bi-Substituted Garnet Films Crystallized during RF Sputtering for M-O Memory	Paper EC-11 Intermag Tokyo Japan April 14th-17th	1987	19
Hairston D K	Magnetic Properties of CoGdTb Thin Films for Thermomagnetic/Magneto-Optic Recording	Topical Meeting on Optical Data Storage, Washington D C, Paper TuBB4	1985	20
Hartmann M	Magnetic & Magneto-Optical Properties of Amorphous GdFePb Films	J.Appl.Phys.(USA) 56 (10) 2870-2873	1984	22
Hartmann M	Erasable Magneto-optic Recording	Phil Tech Rev Vol 42 No2 37-47	1985	23
Hartmann M	Erasable Magneto-optic Recording Media	IEEE Trans Mag VolMag 20 No 5 1013-1018	1984	24
Honda S	Change of Magnetic Properties in Compositionally Modulated TbCo Sputtered Films	IEEE Trans Mag VolMag(22),No5,1221-1223	1986	28
Huth B G	Calculations of Stable Domain Radii Produced by Thermomagnetic Writing	IBM J Res Dev March 1974 100-109	1974	30
Ichihara K	Highly Reliable TbCo Film for Erasable Optical Disk Memory	Topical Meeting on Optical Data Storage, Washington D C, Paper WAA2	1985	31
Ichihara K	The Underlayer Film Treatment Effect on TbCo Film Properties in Optical Disk Memory Structure	IEEE Trans Mag VolMag 22 No5 1331- 1334	1986	32
Koyanagi T	Magneto-optical Properties of Cd1-xMnxTe Films Prepared by Ionized Cluster Beam (ICB) Deposition Technique	J Mag Soc Jap Vol 9 No2 141-144	1985	41
Krumme J P	Bismuth Iron Garnet Films Prepared by RF Magnetron Sputtering	J.Appl.Phys.(USA) 57 (2B) 3885-3887	1985	43

Kryder M H	Control of Parameters in RE-TM Alloys for Magneto-optical recording Media	IEEE Trans Mag VolMag(23) No1 165-167	1987	44
Matsubara K	Magneto-optical Kerr Effect of PtMnSb/TbFe Multilayered Films	Paper 21A-10 Int Symp Magneto-optics Kyoto	1987	49
Morishita T	Abnormal Increases in Ku and Hc without Magnetic Compensation in Compositionally Modulated FeTb Films	Paper CG-06 Intermag Tokyo Japan April 14th-17th	1987	54
Nomura T	High Data Rate Recording on Magneto-optic Disk	Paper DB-01 Intermag Tokyo Japan April 14th-17th	1987	60
Okada M	Erasing Characteristics of Magneto-Optical Recording Medium	IEEE Transl.J.Mag. Vol TJMJ-1(6) 705-707	1985	64
Okada M	High C/N Magneto-optical Disks using Plastic Substrates for Video Image Applications	Paper DB-07 Intermag Tokyo Japan April 14th-17th	1987	65
Ota Y	Kerr Effect of Co-O Films	IEEE Trans Mag VolMag 20 No5 1030-1032	1984	66
Reim W	TbxNdy(FeCo)1-x-y:Promising Materials for Magneto-optical Storage	J Appl Phys 61(8),3349-3350	1987	68
Sakurai Y	RE-TM Amorphous Films for Magneto-Optical Recording	IEEE Trans Mag (USA) VolMag 19 No 5 1724-1736	1983	71
Sato M	Write/Read Characteristics of Magneto-optical Disk	Paper CG-09 Intermag Tokyo Japan April 14th-17th	1987	72
Sato N	Magneto-optical Recording on Amorphous Rare Earth-Transition Metal Films with an Artificially Layered Structure	Paper CG-07 Intermag Tokyo Japan April 14th-17th	1987	75
Satoh T	Magneto-Optical Properties and Thermal Expansion Anomaly	Sci.Rep.Res.Inst. Tohuku Univ.Ser.A(Japan) 32 (2) 190-199	1985	78
Sheih H P	Magneto-optic Recording Materials with Direct Overwrite Capability	App.Phys.Lett. 49(8),473-474	1986	79
Shieh H P	The Effect of Wall Stiffness on Reproducibility of Thermomagnetically Written Domains	Topical Meeting on Optical Data Storage,Washington D C, Paper WAA4	1985	81
Shieh H P	Dynamics and Factors Controlling Regularity of Thermomagnetically Wrtitten Domains	J Appl Phys 61 1108-1122	1987	82

Shieh H P	Magnetic Propeerties of Amorphous Tb(Pr,Nd)Co Films	Paper FC-09 Intermag Tokyo April 14th-17th	1987 83
Shieh H P	Operating Margins for Magneto-optic Recording Materials with Direct Overwrite Capability	IEEE trans mag VolMag 23 No1 171-173	1987 84
Shimoda T	Properties of NdDyFeCoTi Magneto-optical Media Made from Cast Alloy Target	Paper 22a-08 Int Symp Magneto-optics Kyoto	1987 85
Shiomi S	Magnetic and Magneto-optic properties of PtMnSb Thin Films	Paper 21B-11 Int Symp Magneto-optics Kyoto	1987 86
Shono K	Magneto-optical Recording of Sputtered Garnet Films using Laser Diode	Paper EC-12 Intermag Tokyo Japan April 14th-17th	1987 87
Takahashi A	High Quality Magneto-optical Disk	SPIE 695,Optical Mass Data Storage,San Diego,18-22	1986 93
Takayama S	Magnetic and Magneto-optical Properties of TbFeCo Amorphous Films	J Appl Phys 61 (7) 2610-2616	1987 97
Takayama S	Magneto-optical Recording of Gd Based Amorphous Alloy Systems with High Kerr Rotation	Paper CG-05 Intermag Tokyo April 14th-17th	1987 98
Tanaka S	The Thermo-magnetic Writing and Erasing Properties and Kerr Rotation Angle of Amorphous RE-Fe Thin Films	JMMM (Neth) 35 205-207	1983 101
Van Engen PG	PtMnSb, a Material with a very High Magneto-optical Kerr Effect	App.Phys.Lett. 42(2),15,202-204	1983 112
Yokoyama Y	Garnet Film with Rectangular Hysteresis Loop and Application to Thermomagnetic Recording Medium	J.Magn.and Magn.Mat.(Neth) Vol.35 Pt(1-3) 175-177	1983 114

THEORY OF MAGNETOSTRICTION IN AMORPHOUS FERROMAGNETS

Manfred Fähnle, Jürgen Furthmüller and Reiner Pawellek
Institut für Physik, Max-Planck-Institut für Metallforschung,
Heisenbergstr. 1, 7000 Stuttgart 80, F.R. Germany

G. Herzer
Vacuumschmelze GmbH, Grüner Weg 37, 6450 Hanau, F.R. Germany

ABSTRACT

A survey is given of the calculation of the effective magnetostriction constant as well as the atomic level magnetostrictive deformations and stresses in amorphous ferromagnets within the framework of linearized theory of elasticity. The mechanism of magnetostriction is clarified, and the conditions for zero magnetostriction as well as the effects of annealing or application of external stress are discussed.

1. INTRODUCTION AND CLASSIFICATION OF THE THEORY

The soft magnetic properties of amorphous ferromagnets depend[1,2] on internal stresses introduced during the quenching process, because they exhibit an effective magnetostriction constant λ_s^{eff} which is sometimes comparable to the one of crystalline ferromagnets. For instance[3], in iron-rich (cobalt-rich) metallic glasses λ_s^{eff} is about $20 - 35 \times 10^{-6}$ ($-4 \cdot 10^{-6}$), whereas for polycrystalline iron (cobalt) a value of about $-7 \cdot 10^{-6}$ ($-10 \cdot 10^{-6}$) has been observed. Very attractive from this point of view are therefore some near-zero magnetostrictive cobalt-based metallic alloys[4-6] ($|\lambda_s^{eff}| \approx$ some 10^{-7}). The effective magnetostriction constant of these materials depends strongly on small structural changes induced by annealing[6,7] or by application of external stress[5,8]. A detailed knowledge of the magnetostriction mechanisms in amorphous ferromagnets is required to derive conditions for zero magnetostriction and to understand the effect of these small structural changes.

In crystalline ferromagnets magnetostriction is determined by the strain-derivative of the macroscopic magnetic anisotropy tensor. In contrast, in macroscopically isotropic amorphous ferromagnets the macroscopic magnetic anisotropy is zero. However, there are nevertheless strong local magnetic anisotropies originating from single-ion couplings[9-14] (or sometimes dipolar or pseudodipolar couplings[12-14]). As a result, the material may be conceived[15-19] as consisting of small structural units (basically an atom under consideration and its nearest neighbours) with strong uniaxial anisotropy and with easy axes varying randomly from site to site (Fig. 1a). In contrast, in monocrystalline ferromagnets all structural units have the same orientation (Fig. 2a). As a result of the random orientation the

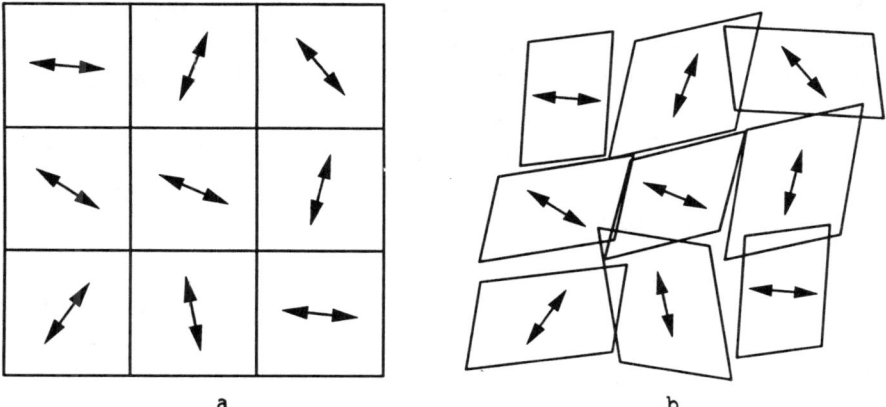

Fig. 1: Schematic representation of the structural units and the local magnetic anisotropy axes (↔) in an amorphous ferromagnet in the unstrained state (a) and in the incompatible spontaneously strained state (b). Note that Fig. 1b does not yet include the strain-induced reorientation of the easy axis directions, see Fig. 3.

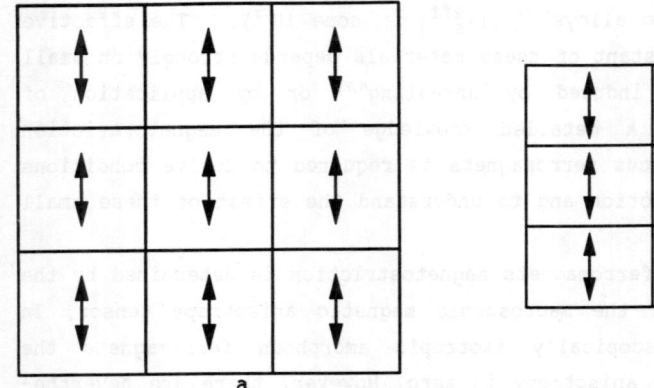

Fig. 2: Schematic representation of the structural units and the local magnetic anisotropy axes (↔) in a monocrystalline ferromagnet in the unstrained state (a) and the magnetostrictively strained state (b).

magnetic anisotropy tensor $K_{ij}(\underline{r})$ of amorphous systems is a spatially varying quantity with zero volume average, $\langle K_{ij}(\underline{r})\rangle = 0$. If we cut the material into these structural units and apply a strong magnetic field while hindering the units to rotate rigidly, each of them exhibits a spontaneous magnetostrictive strain $\epsilon_{ij}^{mag}(\underline{r})$ like a small monocrystalline ferromagnet with corresponding orientation. This spontaneous magnetostriction is described by the local magnetoelastic tensor

$$B_{klij}(\underline{r}) = \partial K_{ij}(\underline{r})/\partial \epsilon_{kl}, \qquad (1)$$

and the local magnetostriction tensor

$$\lambda_{ijkl}(\underline{r}) = - S_{ijmn}(\underline{r})\, B_{mnkl}(\underline{r}), \qquad (2)$$

(with the tensor S_{ijmn} of the elastic compliances) via

$$\epsilon_{ij}^{mag}(\underline{r}) = \lambda_{ijkl}(\underline{r})\gamma_k\gamma_l, \qquad (3)$$

where the vector $\underline{\gamma}$ denotes the direction cosines of the magnetization. The tensors K_{ij}, B_{ijkl}, λ_{ijkl} and S_{ijkl} are thereby determined by the atomic short-range order and the chemical composition.

It was assumed by O'Handley and Grant[20] that the effective macroscopic magnetostriction tensor λ_{ijkl}^{eff} of amorphous ferromagnets is in good approximation given by the simple volume average $<\lambda_{ijkl}(\underline{r})>$ of the local tensor. Because the magnetostriction tensor is a 4th rank tensor, this volume average in general is non-zero even for macroscopically isotropic systems with $<K_{ij}> = 0$. However, whereas in monocrystalline materials all structural units deform coherently (Fig. 2b), all units of the amorphous system are strained in a different way because of their different orientation (Fig. 1b). Therefore the spontaneously strained state is incompatible, i.e. there are artificial overlaps as well as free space between the units (Fig. 1b). Additional elastic deformations of the units must be performed to restore compatibility. As a result, the effective magnetostriction tensor λ_{ijkl}^{eff} is different from the volume average $<\lambda_{ijkl}(\underline{r})>$, and the correct strain field $\epsilon_{ij}(\underline{r})$ is strongly inhomogeneous.

In papers which focus on the microscopic origin of magnetostriction in amorphous ferromagnets[12-14,21-23] the inhomogeneous character of the strain field is neglected from the very beginning, relating the initial state and the magnetostrictively strained state by a homogeneous strain field. In contrast, in this paper as well as in Refs. 15-19 we consider in full detail the inhomogeneity of the strain field. The price we must pay is that we cannot say anything about the microscopic origin of magnetostriction in amorphous ferromagnets. In our phenomenological theory we simply assume the existence of a continuum tensor field $K_{ij}(\underline{r})$ with zero (or near-zero) volume average describing the local magnetic anisotropy, and we do not consider the microscopic origin of this anisotropy. We then calculate the inhomogeneous magnetostrictive deformations associated with $K_{ij}(\underline{r})$ by the linearized theory of elasticity, i.e., we consider only magnetostriction related to elastic processes and neglect phenomena related to anelastic or plastic processes[12,13].

One aim is to calculate the volume average $<\epsilon_{ij}(\underline{r})>$ of the magnetostrictive strain field, which is associated with an effective magnetostriction tensor $\lambda_{ijkl}^{eff} \neq <\lambda_{ijkl}(\underline{r})>$ by

$$\langle \epsilon_{ij}(\underline{r}) \rangle = \lambda^{eff}_{ijkl} \gamma_k \gamma_l \quad . \tag{4}$$

We consider in the following only the fully saturated state for which $\underline{\gamma}$ is constant in the whole material. From λ^{eff}_{ijkl} we obtain the effective magnetostriction constant

$$\lambda^{eff}_s = \frac{2}{3} (\lambda^{eff}_{1111} - \lambda^{eff}_{2211}) \quad . \tag{5}$$

Especially, we discuss conditions for zero macroscopic magnetostriction, i.e. $\lambda^{eff}_s = 0$. We show that λ^{eff}_s may be zero although the material is strongly magnetostrictive on a local basis. The effect of structural changes due to annealing or application of stress is discussed. Furthermore, we consider the inhomogeneities of the magnetostrictive strain field by calculating correlation functions of type $\langle \epsilon_{ij}(\underline{r}) \epsilon_{kl}(\underline{r}') \rangle$. The knowledge of these inhomogeneities is important for the question whether it is reasonable to characterize the material solely by the effective magnetostriction constant λ^{eff}_s.

2. BASIC IDEAS

We approximate the spatial variation of the anisotropy tensor $K_{ij}(\underline{r})$ by dividing the system into small structural units, for which K_{ij} is nearly constant like in small monocrystalline ferromagnets, whereas K_{ij} varies from unit to unit. For amorphous materials these units basically consist of an atom under consideration and its nearest neighbour atoms. We can apply the theory also to the case of polycrystalline ferromagnets, for which the units are much larger. To be specific, we consider units with uniaxial anisotropy and more or less randomly distributed directions of the local anisotropy axes (Fig. 1). A justification of this assumption for amorphous ferromagnets is given in Refs. 9,10. Because of the random orientation of the units all local material tensors, for instance K_{ij} and λ_{ijkl}, are spatially fluctuating quantities in a global coordinate system. The theory is of course also valid for other forms of the spatial fluctuations.

We now perform the gedanken-experiment already sketched in sect. 1. We first switch off the exchange interactions between the magnetic moments, yielding a system without magnetization and without magnetostrictive deformations. This state represents the initial state or reference state for the following calculations. It should be noted that in the case of amorphous materials this reference state is not stress-free, but there are atomic level structural stresses related inherently to the randomness of the amorphous structure[24-26].

In the second step we cut the system into the above defined structural units, not allowing for relaxation processes which would remove the atomic level structural stresses. We then switch on the exchange interactions and apply a very strong external magnetic field, so that all units exhibit a magnetization with the same direction. As a result the units will deform spontaneously to lower their magnetic anisotropy energy, and the total deformation may be decomposed into strains and rigid rotations. If we do not allow rigid rotations for the moment by exposing additional constraints, the units will exhibit the spontaneous magnetostrictive strains given by eq. (3) like small monocrystalline ferromagnets with the corresponding orientation, i.e. with the corresponding value of the local magnetostriction tensor $\lambda_{ijkl}(\underline{r})$ in the fixed global coordinate system. Because we consider isolated units at the moment, these spontaneous magnetostrictive strains are stress-free and are called quasiplastic strains[27]. As a result of these strains the magnetic anisotropy energy of the units is lowered[17] both by modifications of the local anisotropy strengths and by strain-induced reorientations of the easy axis directions (Fig. 3). Both contributions are denoted as "conventional" mechanism, because they determine also magnetostriction in crystalline ferromagnets. They

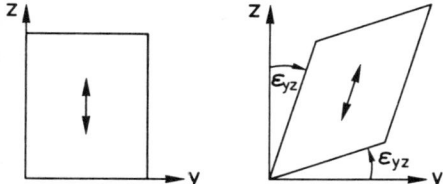

Fig. 3: Schematic illustration of the strain-induced reorientation of the easy axis direction.

are totally described by the local magnetoelastic tensor B_{ijkl}, eq. (1), or the local magnetostriction tensor λ_{ijkl}, eq. (2).

After the spontaneous magnetostrictive deformations the differently deformed units no longer fit together to a compact material (Fig. 1b). Additional elastic deformations must be performed in order to restore compatibility of the material. The total strain $\epsilon_{ij}(\underline{r})$ is given by

$$\epsilon_{ij}(\underline{r}) = \epsilon_{ij}^{mag}(\underline{r}) + \epsilon_{ij}^{el}(\underline{r}) \ . \tag{6}$$

If no additional elastic strains were needed the total magnetostrictive strain could be calculated as a simple average over the local spontaneous strains $\epsilon_{ij}^{mag}(\underline{r})$, yielding $\lambda_{ijkl}^{eff} = <\lambda_{ijkl}(\underline{r})>$ according to eqs. (3,4), as in the theory of O'Handley and Grant[20]). The elastic strains ϵ_{ij}^{el} characterize the contribution of the elastic coupling effects between neighbouring units in a real material. Because of these elastic coupling effects the units cannot deform as if they were elastically decoupled. As a result, the effective magnetostriction tensor λ_{ijkl}^{eff} is different from $<\lambda_{ijkl}(\underline{r})>$. Furthermore, the elastic strains are accompanied by stresses

$$\sigma_{ij}(\underline{r}) = C_{ijkl}(\underline{r}) \ \epsilon_{kl}^{el}(\underline{r}) \ , \tag{7}$$

where C_{ijkl} denotes the tensor of the elastic constants.

It is obvious that the existence of a more or less random strain field $\epsilon_{ij} = 1/2(\partial_i s_j + \partial_j s_i)$ with $\underline{s}$ denoting the displacement vector implies a random field of rigid rotations $\omega_{ij} = 1/2(\partial_i s_j - \partial_j s_i)$. This means that although so far the primary spontaneous deformation is a pure strain deformation (we did not allow for rigid rotations) we need a rotation field ω_{ij} concomitant to the elastic strain field ϵ_{ij}^{el} in order to restore compatibility.

We now finally allow that the units may perform rigid rotations not only when being compelled to restore compatibility but also intentionally to obtain a better alignment of the easy axis directions with the magnetization in order to lower the local anisotropy energy. I.e. we consider the rigid rotations also as a primary source for magnetostrictive deformations ("reorientation" mechanism), and it corresponds to the inclusion of a rotation dependent term in the energy functional. When neglecting the rigid rotations as source for magnetostriction we can start from an energy functional which contains only the total strains $\epsilon_{ij}(\underline{r})$, and minimize this functional with respect to the displacement field $\underline{s}$ (see sect. 3). By $\omega_{ij} = 1/2$ $(\partial_i s_j - \partial_j s_i)$ a rotation field concomitant to ϵ_{ij} is obtained although the energy functional does not contain a rotation-dependent term.

Because magnetic anisotropy implies shape anisotropy of the units, a better alignment of the easy axis directions induces a macroscopic deformation of the sample. For single crystals this mechanism is not relevant because it just represents a rigid rotation of the whole sample. In Ref. 16 it has been shown by simple energy considerations that the reorientation mechanism may possibly contribute to magnetostriction processes in near-zero magnetostrictive alloys, whereas in strongly-magnetostrictive materials the conventional mechanism is certainly dominant. For details see sect. 5.

Removing in our gedanken-experiment the constraint of zero rigid rotations would mean that all isolated units would perform rather large rotations to obtain a perfect alignment of the easy axis directions with the magnetization direction. In reality, however, there will be only rather small rotations due to the elastic coupling between neighbouring units. It is therefore not meaningful to introduce the concept of spontaneous magnetostrictive rotations ω_{ij}^{mag}, but we consider from the beginning only the small total rotations ω_{ij}.

Instead of attacking the problem by a statistical approach as in a former paper[28], we derive and solve in the following section the magnetoelastic equations for the magnetostrictive deformations in inhomogeneous ferromagnets.

3. DESCRIPTION OF THE FORMALISM

In this section we solve the problem by two methods, the balance-of-force method and the incompatibility method. Both methods yield in principle exact solutions for the at least symmetric part of the distorsion tensor $\beta_{ij} = \partial_i s_j$ in the form of two different infinite perturbation series. Because of practical reasons we can calculate numerically only a few terms in each series. We thus obtain two approximate solutions, and from a comparison of the results of the two methods for the same order of perturbation theory we get a feeling for the accuracy of the calculation. For details see Refs. 15-19.

3.1 Balance-Of-Force Method

In this method the magnetoelastic equation for the displacement field $\underline{s}$ is obtained by minimization of the total energy

$$\phi = \int d^3r \left[\phi_{el}(\underline{\underline{\epsilon}}(\underline{r})) + \phi_{mag}^{(\epsilon)}(\underline{\underline{\epsilon}}(\underline{r})) + \phi_{mag}^{(\omega)}(\underline{\underline{\omega}}(\underline{r})) \right] . \tag{8}$$

The various terms in eq. (8) have the following meaning:
(i) The elastic energy density

$$\phi_{el} = 1/2 \; c_{ijkl} \; \epsilon_{ij} \; \epsilon_{kl} . \tag{9}$$

We neglect all contributions to ϕ_{el} for instance from Grad $\underline{\underline{\epsilon}}$-terms, which are required in principle to describe accurately the very short wavelength fluctuations in our continuum theory. Furthermore, although we consider rotations of the local units we assume that the effect of torque stresses[29,30] which would contribute Grad $\underline{\underline{\omega}}$-terms, may be neglected. These torque stresses appear for example in covalent-bound crystals, but they are absent in systems with purely central forces. Although there are strong hints that metalloid atoms like Si, B, P contribute covalent binding forces in amorphous alloys, we assume in the following that the main part of the binding energy results from central forces, and we therefore neglect Grad $\underline{\underline{\omega}}$-terms.

(ii) The strain-dependent part of the magnetic anisotropy energy density giving rise to the conventional mechanism

$$\phi_{mag}^{(\epsilon)} = B_{ijkl} \; \epsilon_{ij} \; \gamma_k \gamma_l . \tag{10}$$

(iii) The rotation-dependent part of the magnetic anisotropy energy density, responsible for the reorientation mechanism

$$\phi_{mag}^{(\omega)} = 2\gamma_i \, \omega_{ik} \, K_{kj} \gamma_j \, . \tag{11}$$

This contribution can be derived in the following way: If the local structural unit rotates, the tensor K_{ij} remains unchanged in a co-rotating coordinate system. To describe the situation in a global system, we must perform a coordinate transformation between the rotated and the global system. Inserting the transformed $\underline{K}$-tensor into the expression for the anisotropy energy density, $\phi_{ani} = K_{ij}\gamma_i\gamma_j$, this leads in linear approximation to eq. (11).

We now minimize the total energy ϕ with respect to the displacement field $\underline{s}$ with $\epsilon_{ij} = 1/2 \, (\partial_i s_j + \partial_j s_i)$ and $\omega_{ij} = 1/2 \, (\partial_i s_j - \partial_j s_i)$, thereby automatically fulfilling the compatibility conditions. (Minimizing with respect to $\underline{\underline{\epsilon}}$ yields the spontaneous magnetostrictive strains, eq. (3).) A rigid rotation of the whole sample which would occur for the case $\langle K_{ij}\rangle \neq 0$ is prevented by an additional constraint. We obtain the following balance-of-force equation for the volume (surface effects are neglected):

$$\partial_j C_{ijkl}(\underline{r}) \partial_k s_l(\underline{r}) = - \partial_j \tau_{ij}^{mag}(\underline{r}) \, . \tag{12}$$

The quantity

$$\tau_{ij}^{mag} = B_{ijkl}\gamma_k\gamma_l + e_{ijk}\,[\underline{\gamma} \times (\underline{\underline{K}} - \langle\underline{\underline{K}}\rangle)\cdot\underline{\gamma}]_k \tag{13}$$

describes the effect of the conventional mechanism (first, symmetric term) and of the reorientation mechanism (second, antisymmetric term).

Eq. (12) may be solved iteratively by a perturbation approach with the fluctuations $\delta C_{ijkl} = C_{ijkl} - \langle C_{ijkl}\rangle$ of the elastic constants as perturbation parameters[15,16], yielding for the distorsion tensor β_{ij} up to second order

$$\begin{aligned}\beta_{ij}(\underline{r}) = &\int d^3 r' \, \Gamma_{ijkl}^{(0)}(\underline{r}-\underline{r}') \, \tau_{kl}^{mag}(\underline{r}') + \iint d^3 r' \, d^3 r'' \, \Gamma_{ijkl}^{(0)}(\underline{r}-\underline{r}') \\ &\Gamma_{mnpq}^{(0)}(\underline{r}'-\underline{r}'')\delta C_{klmn}(\underline{r}') \, \tau_{pq}^{mag}(\underline{r}'') + \iiint d^3 r' \, d^3 r'' \, d^3 r''' \\ &\Gamma_{ijkl}^{(0)}(\underline{r}-\underline{r}') \, \Gamma_{mnpq}^{(0)}(\underline{r}'-\underline{r}'') \, \Gamma_{rstu}^{(0)}(\underline{r}''-\underline{r}''') \, \delta C_{klmn}(\underline{r}') \\ &\delta C_{pqrs}(\underline{r}'') \, \tau_{tu}^{mag}(\underline{r}''') \, . \end{aligned} \tag{14}$$

Here $\Gamma_{ijkl}^{\{0\}}(\underline{r}-\underline{r}')$ is the strain Green's tensor for a hypothetical homogeneous system with elastic constants $\langle C_{ijkl}(\underline{r})\rangle$. For isotropic system it has the general form[31,32]

$$\Gamma_{ijkl}^{\{0\}}(\underline{r}) = E_{ijkl}\delta(\underline{r}) + F_{ijkl}(\vartheta,\varphi)/r^3 + \Gamma_{ijkl}^{\{0\}\,hom} \quad . \tag{15}$$

For large volumes V of the sample we have

$$\Gamma_{ijkl}^{\{0\}\,hom} = \frac{1}{V}\,[-E_{ijkl} - (C_{ijkl}^{\{0\}})^{-1}] \quad . \tag{16}$$

Because we consider infinitely extended systems (V → ∞), this homogeneous part plays only a role for the calculation of integrals of type $\int \Gamma_{ijkl}^{\{0\}}(\underline{r}-\underline{r}')d^3r$, which yield $-(C_{ijkl}^{\{0\}})^{-1}$.

3.2 Incompatibility Method

This method is strongly related to our gedanken-experiment of cutting the system into small units, which exhibit spontaneous but incompatible magnetostrictive deformations, and restoring compatibility by additional elastic deformations. As discussed in sect. 2 it is not reasonable to introduce the concept of spontaneous rigid rotations. The incompatibility method is therefore not able to consider the reorientation mechanism. Because in nearly all cases the reorientation mechanism yields no or a negligibly small contribution to magnetostriction (sect. 5), it is reasonable to neglect this mechanism for the moment and to deal with the conventional mechanism by use of the incompatibility method (for mathematical details see Ref. 16).

We first calculate the stress-free spontaneous magnetostrictive strains $\epsilon_{ij}^{mag}(\underline{r})$ according to eq. (3). To restore compatibility, we have to perform additional elastic deformations $\epsilon_{ij}^{el}(\underline{r})$ according to

$$\text{Ink } \underline{\underline{\epsilon}}^{el}(\underline{r}) = -\text{Ink } \underline{\underline{\epsilon}}^{mag}(\underline{r}) \quad . \tag{17}$$

The total strain then is given by eq. (6), $\underline{\underline{\epsilon}} = \underline{\underline{\epsilon}}^{mag} + \underline{\underline{\epsilon}}^{el}$. The elastic strains are related to stresses $\underline{\underline{\sigma}}(\underline{r})$ via eq. (7), yielding

$$\epsilon_{ij}(\underline{r}) = \epsilon_{ij}^{mag}(\underline{r}) + S_{ijkl}(\underline{r})\,\sigma_{kl}(\underline{r}) \quad . \tag{18}$$

Eq. (17) with $\epsilon_{ij}^{el} = S_{ijkl}\sigma_{kl}$ has to be solved with the restriction $\text{Div}\underline{\sigma} = 0$, which means that there are no external forces. A perturbation approach with the fluctuations $\delta S_{ijkl}(\underline{r})$ of the elastic compliances yields[16] up to second order a formula totally analogous to eq. (14) with β_{ij} replaced by σ_{ij}, τ_{ij}^{mag} by ϵ_{ij}^{mag}, δC_{ijkl} by δS_{ijkl} and the strain Green's tensor $\Gamma_{ijkl}^{\{0\}}$ by the stress Green's tensor $\Delta_{ijkl}^{\{0\}}$, which may be directly calculated[16] from $\Gamma_{ijkl}^{\{0\}}$.

It should be noted that the incompatibility method yields only the symmetric part ϵ_{ij} of the distorsion tensor β_{ij}. For a calculation of the antisymmetric part ω_{ij} (and for dealing with the reorientation mechanism, see above) we must use the balance-of-force method.

From the perturbation series for β_{ij} (or ϵ_{ij}) we determine the volume averages $<\epsilon_{ij}(\underline{r})>$, which are related to λ_{ijkl}^{eff} and λ_s^{eff} by eqs. (4,5), as well as correlation functions of type $<\epsilon_{ij}(\underline{r})\,\epsilon_{kl}(\underline{r}')>$, which characterize the local magnetostriction.

4. INPUT FOR NUMERICAL CALCULATIONS

For an ergodic situation the statistical properties of the system are characterized by the full set of n-point correlation functions for all material tensors, i.e. K_{ij}, B_{ijkl}, λ_{ijkl}, C_{ijkl} and S_{ijkl}. Because this statistical information is of course not fully available for amorphous ferromagnets, we model the material by a situation for which the spatial fluctuations of all material parameters are strongly correlated. As discussed in sect. 2 we assume that the material is composed of small structural units for which the material tensors are constant and exhibit for instance hexagonal symmetry. For every local unit we can define a special local coordinate system with the z-axis parallel to the local easy axis, in which the material tensors have the well-known form of tensors in hexagonal crystalline materials. For simplicity we assume that the local material parameters (defined in the local coordinate systems) do not fluctuate from unit to unit. The only fluctuating quantities are the angles ϑ and φ characterizing the orientations of the local easy axes with respect to a fixed global coordinate system. The material tensors defined in the global system, which enter all our equations, can be obtained from the tensors defined in the local systems by tensor transformation, for example

$$c_{ijkl}(\underline{r}) = a_{i\mu}(\underline{r})a_{j\nu}(\underline{r})a_{k\kappa}(\underline{r})a_{l\lambda}(\underline{r})c^{local}_{\mu\nu\kappa\lambda} \quad . \tag{19}$$

The fluctuations of the tensors in the global system then arise totally from the spatial fluctuations of the matrix $a_{i\mu}[\vartheta(\underline{r}),\varphi(\underline{r})]$ of the transformation between the local and the global coordinate system. Because an isotropic tensor yields - due to its invariance under rotations - a nonfluctuating global tensor, the spatial fluctuations of all tensors are related to the anisotropic properties of the units.

Within this simple structural model the whole statistical information about the system is given by the set of n-point correlation functions of type $<a_{i_1 j_1}(\underline{r}_1)...a_{i_n j_n}(\underline{r}_n)>$, where $<>$ denotes the ensemble average. For the following we assume that these correlation functions can be written in the form

$$<a_{i_1 j_1}(\underline{r}_1)...a_{i_n j_n}(\underline{r}_n)> = <a_{i_1 j_1}(r_1)...a_{i_n j_n}(r_1)>g^{(n)}(\underline{r}_1,...\underline{r}_n) \tag{20}$$

with $g^{(n)}(\underline{r}_1,...\underline{r}_1) = 1$ and (for statistically homogeneous systems) $g^{(n)}(\underline{r}_1,...\underline{r}_n) = g^{(n)}(\underline{r}_1-\underline{t},...\underline{r}_n-\underline{t})$, with $\underline{t}$ arbitrary. The functions $g^{(n)}$ exhibit the macroscopic symmetry of the system, e.g. invariance under rotations for macroscopically isotropic materials.

The calculations of the volume averages (e.g. $<\epsilon_{ij}>$) are performed in real space for three different situations:

(i) Macroscopically isotropic systems. Taking into account the properties of the Green's tensors it turns out that we need as input only the isotropic distribution function density $P(\cos\vartheta,\varphi) = 1/4\pi$ for the angles ϑ and φ, from which we calculate the amplitudes $<a_{i_1 j_1}(\underline{r}_1)...a_{i_n j_n}(\underline{r}_1)>$ of the correlation functions in eq. (20). The functions $g^{(n)}$ enter only via the normalization relation $g^{(n)}(\underline{r},...\underline{r}) = 1$.

(ii) Systems with isotropic correlation functions $g^{(n)}$ but anisotropic distribution function densities, with an anisotropy parameter δ:

$$P(\cos\vartheta,\varphi) = (1/4\pi)[1 + \delta (3\cos^2\vartheta - 1)] \quad . \tag{21}$$

The functions $g^{(n)}$ enter only via the normalization relation.

(iii) Systems with isotropic distribution function density $P(\cos\vartheta,\varphi) = 1/4\pi$ but anisotropic correlation functions. Considering here only first order perturbation theory, the required 2-point correlation function may be characterized by two correlation lengths, $\xi_\parallel$ and $\xi_\perp$, via

$$g^{(2)} = g^{(2)} \left[\frac{x^2 + y^2}{\xi_\perp^2} + \frac{z^2}{\xi_\parallel^2} \right] . \qquad (22)$$

It turns out that a knowledge of the parameter $\alpha = 1 - \xi_\perp^2/\xi_\parallel^2$ suffices.

Obviously for the calculation of $\langle\epsilon_{ij}\rangle$ and of λ_s^{eff} no length scale enters which characterizes the size of the units. The results are thus valid both for amorphous and polycrystalline ferromagnets.

The calculation of correlation functions of type $\langle\epsilon_{ij}(\underline{r})\epsilon_{kl}(\underline{r}')\rangle$ are performed in k-space, and we consider only isotropic systems ($P(\cos\vartheta,\varphi) = 1/4\pi$, $g^{(n)}$ isotropic) and the 0. order approximation, i.e. the first term in eq. (14). For $\underline{r} = \underline{r}'$ we need only the normalization condition for $g^{(2)}$, and the results hold both for amorphous and polycrystalline ferromagnets. For $\underline{r} \neq \underline{r}'$ we insert $g^{(2)}(r-r') = e^{-|r-r'|/r_0}$ with the structural correlation length r_0, and the results are now different for amorphous and polycrystalline ferromagnets.

To obtain physically reasonable results we insert[16] for the tensor components in the local coordinate systems the values for crystalline Co (at T = 300 K) ("Co-like materials") and Gd at T = 4 K ("Gd-like materials"). This does not mean that we have to insert the Co values when discussing the near-zero magnetostrictive Co-based alloys, or that the values obtained for Gd parameters are representative only for alloys containing rare-earth atoms, because the material parameters may be changed very effectively[4] by alloying. We assume that these values are representative for the strongly magnetostrictive alloys. To characterize near-zero magnetostrictive alloys we insert similar values for K_{ij}, C_{ijkl} and S_{ijkl}, but values of B_{ijkl} and λ_{ijkl} which are two orders of magnitude smaller[16].

5. RESULTS
5.1 Conventional Or Reorientation Mechanism?

It turns out that there is no contribution of the reorientation mechanism to $<\epsilon_{ij}(\underline{r})>$ and hence to λ_s^{eff} for macroscopically isotropic systems. This is due[16] to the antisymmetric character of the reorientation term in eq. (13). As a result, all contributions of the reorientation mechanism to the local magnetostrictive deformations (see below) average out for macroscopically isotropic systems, as well as for systems with anisotropic $P(\cos\vartheta,\varphi)$. For systems with anisotropic correlation functions there is a contribution of the reorientation mechanism to λ_s^{eff}, which is, however, several orders of magnitude smaller than the contribution of the conventional mechanism[16].

In contrast, there is an effect of the reorientation mechanism to the local magnetostrictive deformations[19], i.e. to quantities like $<\epsilon_{ij}^2(\underline{r})>$, even for macroscopically isotropic systems. However, for Co- and Gd-like materials this effect is typically 2 or 3 orders of magnitude smaller than the contributions from the conventional mechanism. The reorientation mechanism may be of some significance for the local magnetostriction of near-zero magnetostrictive alloys. By playing around with the values for the material parameters situations can be found for which the reorientation mechanism is even dominant (while still being zero for $<\epsilon_{ij}>$ and λ_s^{eff}).

To conclude, apart from the local magnetostriction of near-zero magnetostrictive alloys the magnetostriction of amorphous and polycrystalline ferromagnets is almost totally determined by the conventional mechanism, as in monocrystalline ferromagnets. As discussed in sect. 2, the conventional mechanism includes both the modification of the local anisotropy strengths and the strain-induced reorientations of the local easy axes (Fig. 3). The relative importance of these two contributions depends sensitively on the material parameters[17]. For Co-like materials 96% of λ_s^{eff} is due to the strain-induced reorientations, whereas for Gd this contribution amounts to 44% only. Similar results were obtained for the local magnetostriction[19].

5.2 Calculation Of The Effective Magnetostriction Constant.

Zero Magnetostrictive Systems

In the following we compare the results for λ_s^{eff} as obtained by the balance-of-force method (bfm) and the incompatibility method (im). It should be noted again that the latter method does not incorporate the effect of the reorientation mechanism. Because this effect is either zero or negligibly small as far as λ_s^{eff} is concerned, it is nevertheless reasonable to compare the results of the two methods.

It is easily seen for example from eqs. 4,14-16 that the 0. order approximation ($\delta C_{ijkl} = 0$) of the bfm and the im yields $\lambda_{ijkl}^{eff} = -\langle C_{ijpq}\rangle^{-1} \langle B_{pqkl}\rangle$ and $\lambda_{ijkl}^{eff} = \langle \lambda_{ijkl}\rangle$, respectively. In this order the effective magnetostriction tensor is related to the simple volume averages of the local tensors B_{ijkl} and λ_{ijkl}, as if there were no elastic couplings between neighbouring units and as in the theory of O'Handley and Grant[20]. This is by no means trivial: For a hypothetical system with magnetically anisotropic but elastically isotropic units we have $\delta C_{ijkl} = 0$. In such a system there are of course strong contributions of the elastic interactions to the local magnetostrictive deformations, because the elastic coupling guarantees compatibility. However, all these contributions average out on a macroscopical scale in the above discussed hypothetical system.

A contribution of the elastic couplings to λ_s^{eff} only arises from the higher order terms in the perturbation series originating from the fluctuations δC_{ijkl} (or δS_{ijkl}), i.e. in our model related to the elastic anisotropy of the units. We calculate λ_s^{eff} up to second order in δC_{ijkl} (δS_{ijkl}), and the difference ($\lambda_s^{eff,2.order} - \lambda_s^{eff,0.order}$) is a measure for the effect of elastic couplings.

5.2.1. Macroscopically isotropic systems. In Table 1 we compare the results of the bfm and the im for λ_s^{eff} in different orders of perturbation theory for Co- and Gd-like materials. Obviously for Co-like materials there is only a small effect of the elastic couplings between neighbouring units, so that the 0. order term of the im, i.e. the simple volume average of the local magnetostriction tensor, yields a rather accurate estimate for λ_{ijkl}^{eff} as in the theory

Table 1: Results for λ_s^{eff} according to the balance-of-force method (bfm) and the incompatibility method (im).

	Co		Gd	
	bfm	im	bfm	im
0. order	-7.04×10^{-5}	-6.47×10^{-5}	3.63×10^{-6}	1.07×10^{-5}
1. order	-6.54×10^{-5}	-6.79×10^{-5}	7.52×10^{-6}	7.52×10^{-6}
2. order	-6.71×10^{-5}	-6.85×10^{-5}	7.25×10^{-6}	7.36×10^{-6}

of O'Handley and Grant[20]. In contrast, for Gd-like materials there is a rather large contribution of the elastic coupling effects. Playing around with the material parameters we can also find situations for which $\lambda_s^{\text{eff},0.\text{order}} = 0$ although the material is magnetostrictive on a local scale ($\lambda_{ijkl}(\underline{r})$, $B_{ijkl}(\underline{r}) \neq 0$), just because the volume average of the local tensors is zero. In this case λ_s^{eff} results totally from elastic coupling effects! Similarly, we can also find situations where $\lambda_s^{\text{eff},2.\text{order}} = 0$ although $\lambda_{ijkl}(\underline{r})$, $\langle\lambda_{ijkl}(\underline{r})\rangle \neq 0$. In this case the non-zero volume average $\langle\lambda_{ijkl}(r)\rangle$ is compensated by the contributions of the elastic couplings. Altogether, zero magnetostriction may arise either because the material is not magnetostrictive on a local basis ($\lambda_{ijkl}(\underline{r})$, $B_{ijkl}(\underline{r}) \equiv 0$, no spontaneous magnetostriction) or for systems which exhibit spontaneous local magnetostriction but for which the various contributions to λ_s^{eff} average out.

5.2.2. Macroscopically anisotropic systems and effect of annealing and application of external stress.

It turns out[16] that for reasonable values of the anisotropy parameter δ (α) there is only a very small (negligibly small) contribution of the anisotropy of $P(\cos\vartheta,\varphi)$ (of $g^{(2)}$) for strongly magnetostrictive systems. The situation may be different for the near-zero magnetostrictive alloys, where small structural changes may induce considerable changes of λ_s^{eff}. For example, changes of $|\Delta\lambda_s^{\text{eff}}| \approx$ some 10^{-7} may be obtained by strong annealing[6] or by application of external stress[5] of about 1000 MPa. Effects of the same order of magnitude are obtained by our theory for material parameters of Co and Gd by inserting an anisotropic distribution function density $P(\cos\vartheta,\varphi)$ according to eq. (21) with $\delta \approx 10^{-2}$. Provided that this value of δ is realistic we thus can explain the experimental results if we assume that the near-zero magnetostrictive Co-rich alloys exhibit local λ_{ijkl}- and B_{ijkl}-tensors (i.e. local magnetostriction) of the same order of magnitude as strongly magnetostrictive alloys and that $\lambda_s^{eff} \approx 0$ results from a cancellation of various terms in the expression for λ_s^{eff}. However, if we assume that in those materials the local λ_{ijkl}- and B_{ijkl}-tensors are two orders of magnitude smaller than in strongly magnetostrictive alloys (sect. 4), we cannot explain the experimental results on this line. Furthermore, the value of $\delta = 10^{-2}$ is comparable[33] to the structural anisotropy in metallic glasses induced by mechanical creep after annealing the sample at 300°C for 24h under a tensile stress of 800 MPa. It therefore seems to be reasonable to give an interpretation of the annealing effect on this line. It is, however, hard to see how a structural anisotropy of this magnitude should arise by mechanical creep when applying the external stress at room temperature[5], especially in view of the fact that the change in λ_s^{eff} may be observed immediately after application of external stress and that the effect is reversible[5,8]. On the other hand it is unrealistic to assume that an anisotropy of this magnitude may arise when allowing only for elastic reactions on external stress[34]. We thus conclude that the experimentally observed effect of external stress on λ_s^{eff} may not be explained by an anisotropy of $P(\cos\vartheta,\varphi)$. In Ref. 16 we have given simple arguments for the conjecture that the stress effect may be explained by a stress-induced modification of the local B_{ijkl}-tensor, i.e., $\partial B_{ijkl}/\partial \sigma_{mn}$, whereas a stress-induced modification of the local C_{ijkl}-tensor does not suffice.

5.3 Atomic Level Magnetostrictive Deformations and Stresses

To get a feeling for the spatial inhomogeneity of the magnetostrictive deformation field we have calculated[19] correlation functions of type $\langle t_{ij}(\underline{r}) t'_{kl}(\underline{r}')\rangle$ with t_{ij} and t'_{ij} given by ϵ_{ij} or ω_{ij}, yielding the following results:

(i) The local deformations are also nearly exclusively determined by the conventional mechanism, whereas the contribution of the reorientation mechanism is non-zero but negligibly small in most cases.

(ii) The results for $<\omega_{ij}^2(\underline{r})>$ are comparable to those for $<\epsilon_{ij}^2(\underline{r})>$ although $<\omega_{ij}(\underline{r})> = 0$ and $<\epsilon_{ij}(\underline{r})> \neq 0$. This results from the fact (sect. 2) that a random field ϵ_{ij} implies a random field ω_{ij}.

(iii) There may be strong spatial fluctuations of the magnetostrictive deformations. For instance, for the macroscopically isotropic Gd-like material the balance-of-force method yields $<\epsilon_{33}^2> = 0.396 \times 10^{-9}$ and $<\epsilon_{33}>^2 = 0.132 \cdot 10^{-10}$. However, these fluctuations are on a scale which is given by the structural correlation length r_0. For amorphous materials r_0 is about the nearest-neighbour distance a_0, and we are dealing with atomic scale magnetostrictive fluctuations. As a result, the fluctuations average out on all scales much larger than a_0. Because the soft magnetic properties of amorphous alloys are mainly determined[1,2] by the interactions of domain walls with defect structures of extension of about 100 Å, the magnetostrictive behaviour may be still characterized by one macroscopic constant λ_s^{eff} when dealing with this problem. The atomic scale magnetostrictive fluctuations must be taken into account only when considering the interaction of domain walls with the atomic scale structural randomness.

(iii) The elastic part ϵ_{ij}^{el} of the strain field is related to stress according to eq. (7). These atomic level magnetostrictive stresses are typically 3-4 orders of magnitude smaller than the atomic level structural stresses[24-26], which are inherently related to the randomness of the amorphous structure.

Acknowledgement: The authors are indebted to Prof. E. Kröner and Prof. H. Kronmüller for helpful discussions and to Ms. Hermann for typing the manuscript.

REFERENCES
1) Kronmüller, H., Fähnle, M., Domann, M., Grimm, H., Grimm, R. and Gröger, B., J. Magn. Magn. Mat. 13, 53 (1979).
2) Kronmüller, H., J. Appl. Phys. 52, 1859 (1981).
3) O'Handley, R.C., Solid State Commun. 21, 1119 (1977).
4) O'Handley, R.C., Phys. Rev. B18, 930 (1978); J. Appl. Phys. 62, R15 (1987).

5) Herzer, G., Proc. Conf. on Soft Magnetic Materials 7, ed. by European Physical Society (Cardiff: Wolfson Centre for Magnetics Technology), 1985, p. 355.
6) Vazquez, M., Hernando, A. and Nielsen, O.V., J. Magn. Magn. Mat. 61, 390 (1986). Hernando, A., Vazquez, M., Madurga, V. and Kronmüller, H., J. Magn. Magn. Mat. 37, 161 (1983).
7) Warlimont, H. and Hilzinger, H.R., Proc. Conf. on Rapidly Quenched Metals, Sendai 1981, p. 1167.
8) Madurga, V., Barandiaran, J.M., Vazquez, M., Nielsen, O.V. and Hernando, A., J. Appl. Phys. 61, 3228 (1987).
9) Cochrane, R.W., Harris, R. and Plischke, M., J. Non-Cryst. Solids 15, 239 (1974).
10) Elsässer, C., Fähnle, M., Brandt, E.H. and Böhm, M.C., to be published in J. Phys. F.
11) Szymczak, H. and Zuberek, R., J. Phys. F12, 1841 (1982).
12) Szymczak, H., J. Magn. Magn. Mat. 67, 227 (1987).
13) Szymczak, H. and Lachowicz, H.K., IEEE Trans. Magn. MAG-24, 1747 (1988).
14) du Tremolet de Lacheisserie, E., J. Magn. Magn. Mat. 67, 102 (1987).
15) Furthmüller, J., Fähnle, M. and Herzer, G., J. Phys. F16, L255 (1986).
16) Furthmüller, J., Fähnle, M. and Herzer, G., J. Magn. Magn. Mat. 69, 79 (1987); 69, 89 (1987).
17) Fähnle, M. and Furthmüller, J., J. Magn. Magn. Mat. 72, 6 (1988).
18) Fähnle, M., Furthmüller, J. and Herzer, G., to be published in J. Physique.
19) Pawellek, R., Furthmüller, J. and Fähnle, M., to be published in J. Magn. Magn. Mat.
20) O'Handley, R.C. and Grant, N.J., Proc. Conf. on Rapidly Quenched Metals, Eds. Steeb, S. and Warlimont, H. (Amsterdam: Elsevier) 1985, p. 1125.
21) Suzuki, Y. and Egami, T., J. Magn. Magn. Mat. 31-34, 1549 (1983).
22) Tsuya, N. and Arai, K.I., J. Magn. Magn. Mat. 31-34, 1594 (1983).
23) Kurzyk, J., J. Magn. Magn. Mat. 73, 84 (1988).
24) Egami, T., Maeda, K. and Vitek, V., Phil. Mag. A41, 883 (1980).
25) Egami, T. and Srolovitz, D., J. Phys. F12, 2141 (1982).
26) Vitek, V. and Egami, T., phys. stat. sol. (b) 144, 145 (1987).
27) Kröner, E., "Kontinuumstheorie der Versetzungen und Eigenspannungen", Springer, Berlin-Göttingen-Heidelberg, 1958.
28) Fähnle, M. and Egami, T., J. Appl. Phys. 53, 231 (1982).
29) Hehl, F. and Kröner, E., Z. Naturforschung 20, 336 (1965).
30) Aero, E.L. and Kuvshinskii, E.V., Sov.-Phys.-Solid State 2, 1272 (1960).
31) Kröner, E. and Koch, H., SM Archives 1, 184 (1976).
32) Kröner, E., in: "Modelling Small Deformations of Polycrystals", Eds. Gittus, J. and Zarke, J. (Elsevier, Barking, Essex, England, 1986) p. 229.
33) Suzuki, Y., Haimovich, J. and Egami, T., Phys. Rev. B35, 2162 (1987).
34) Furthmüller, J., diploma work, University of Stuttgart, 1986.

STRESS DEPENDENCE OF MAGNETOSTRICTION IN METALLIC GLASSES, A CONSEQUENCE OF THE COEXISTENCE OF TWO AMORPHOUS PHASES?

A. Hernando

Departamento de Física de Materiales. Universidad Complutense. 28040-Madrid. Spain. Laboratorio de Magnetismo. RENFE. U.C.M.

Abstract

It is the aim of this work to show how the coexistence of two amorphous phases with different saturation magnetostriction constant to each other leads to a stress dependence of the macroscopic magnetostriction, λ, as that observed in nearly zero magnetostrictive metallic glasses. The dependence of λ on the magnetic field which has been experimentally detected can also been naturally explained under this assumption. It is concluded that the stress dependence of λ is an evidence of the multiphase character of the metallic glasses.

Introduction

O'Handley reported the possibility of transforming reversibly from one fairly well defined local atomic order to another in certain metallic glasses (1,2). More recently Riveiro (3,4) has observed, by measuring thermal dependence of resistivity, marthensitic like transformation for a wide range of compositions.

The dependence of the induced magnetic anisotropy on the pressure of the sputtering gas, in $Co_{1-x}Ti_x$ compounds, has been explained by Suran et al (5) as a consequence of progressive change of the local structure. They assumed the structural short range order, SRO, to be built up of clusters with icosahedral, trigonal and octahedral symmetry. Structural transformation can take place at certain temperature, T_c, and would consist in a small distorsion of a fraction of the icosahedral clusters. The volume fraction of each phase should be dependent of the fictive temperature of the glass. The stable phase at low temperature exhibit higher magnetic anisotropy than that corresponding at $T > T_c$. Hence it is suggested that the low temperature phase contains a significant amount of low symmetry clusters as the trigonal ones. The pressure of the sputtering gas and the cooling rate are relevant parameters for samples obtained by sputtering or melt spinning respectively.

At this point is worth noting the experiments performed by Madurga et al (6) related to the influence of the cooling rate on the magnetostriction of $Co_{80}Nb_{14}B_6$ composition, in which Corb et al (7) observed a structural transformation as a function of the temperature. Room temperature values of the magnetostriction constant depends on the quenching rate and range from 4×10^{-7} up to -1×10^{-6} as the wheel speed increases. Zero magnetostriction alloys were obtained at about 34 ms^{-1}. A possible interpretation of this behavior was outlined in ref.(6). By using a similar argument to that given later by Suran et al (5) we considered two phases with magnetostriction constant $\lambda_1 > 0$

and $\lambda_2 < 0$ and volume fractions V_1 and V_2 respectively. The macroscopic λ, which could be roughly written as $\lambda_1 V_1 + \lambda_2 V_2$, depends on the speed of the wheel through, V_1 and V_2.

Along this paper it will be shown that the stress dependence of the magnetostriction strongly supports the existence of different structural units with well defined SRO. The coexistence of two phases with different sign of magnetostriction constant leads to a dependence of the macroscopic magnetostriction with the applied stress as that observed in Co-rich alloys, when λ is measured by using magnetoelastic effects.

II) General remarks

a) Summary of experiments and interpretations

It has been recently shown that the saturation magnetostriction λ of low magnetostrictive metallic glasses decreases with the tensile stress, σ_{ap}, applied along the ribbon axis (8,9,10). Typical decreasing rates, $\frac{d\lambda}{d\sigma_{ap}}$, are centered at 10^{-10} MPa^{-1}. This effect has been discovered using magnetoelastic indirect methods which allow us to measure very low values of λ (11).

For metallic glasses with near-zero but positive λ a change of sign of λ occurs at a certain value of σ_{ap}, since the slope of the linear dependence of λ on σ_{ap} is always negative. The change of sign of λ can be seen directly by observing the effect produced by the tensile stress on the hysteresis loop. When an a.c. field, constant in amplitude, is applied along the ribbon axis the action of an applied stress produces a positive increment of the maximum magnetization but as the stress rises the increment decreases. At a certain stress there is no change of magnetization and for further values of the stress the increment of magnetization becomes negative so indicating that the sign of λ, positive at low values of σ_{ap}, has changed as σ_{ap} was increased toward negative values.

The stress dependence of λ is a rather astonishing effect which suggests two question. It is an intrinsic property or a consequence of the indirect methods used for measuring λ?. If is an intrinsic property which is its origin?

Criticisms of the magnetoelastic methods more frequently used have been reported (12, 13, 14). However the effect has been observed independently of the magnetoelastic experimental method from which λ is measured.

The attempts developed until now to explain the stress dependence of λ are mainly based on two ideas. Szymczak has proposed that the stress produces an anisotropic readistribution of atomic bonds giving rise to a magnetic anisotropy with the easy axis perpendicular to the ribbon axis (15). Furthmuller et al. have reported a complet micromagnetic theory in which the contribution to λ of the strain due to rigid rotation of the structural units is included (16).

A new idea is proposed here to account for the stress dependence of λ. Let us show how the coexistence of two amorphous phases, randomly mixed, with isotropic magnetostriction constant λ_1 and λ_2 and volume fraction V_1 and V_2 respectively becomes the more simple explanation of the observed behavior.

b) Evolution of the hysteresis loop, of a sample composed of two phases, with the applied stress.

Consider a ribbon with saturation magnetostriction λ, subjected to a longitudinal field, H_z, high enough to overcome the knee of the magnetization curve. Let us call $M(H_z)$ the longitudinal magnetization induced by H_z. It is well known that under the action of a tensile stress σ_{ap}, applied along the ribbon axis, $M(H_z)$ varies. We are interested in finding the derivative of magnetization respect to the applied stress, $(\frac{dM(H_z)}{d\sigma_{ap}})$ as a function of the applied stress. When the magnetization lies everywhere very nearly along the ribbon axis, as is the case above the knee of the magnetization curve, the effect of the applied stress can be depicted by means of an effective field $H_\sigma = \frac{3\lambda\sigma_{ap}}{\mu_0 M_s}$ which also acts along the ribbon axis. M_s being the saturation magnetization. Therefore when σ_{ap} is restricted to the range for which $H_z + H_\sigma$ remains above the value at which the knee of the magnetization curve occurs it can be written:

$$\frac{dM(H_z \sigma_{ap})}{d\sigma_{ap}} = \chi(H_z + H_\sigma) \frac{3\lambda}{\mu_0 M_s} \qquad |1|$$

where χ is the differential susceptibility.

In the region above the knee of the magnetization curve χ is a positive function which decreases as H increases. Hence $\chi' = \frac{d\chi}{dH}$ is a negative function everywhere with absolute value decreasing toward zero at a rate given roughly by $\frac{1}{H}\chi$. These general properties of χ allow us to distinguish two well defined different behavior of $(\frac{dM}{d\sigma_{ap}})$ according to the sign of λ. For positive λ, $(\frac{dM}{d\sigma_{ap}})$ is positive everywhere since χ and $\frac{3\lambda}{\mu_0 M}$ are positive. $(\frac{dM}{d\sigma_{ap}})$ decreases as σ_{ap} increases since H_σ is positive For negative λ, $(\frac{dM}{d\sigma_{ap}})$ is negative everywhere as a consequence of the negative sign of $\frac{3\lambda}{\mu_0 M_s}$ and the positive sign of χ in eq.$|1|$. The absolute value of $\frac{dM}{d\sigma_{ap}}$ increases as σ_{ap} increases since H_σ is negative and χ increases as H decreases.

Having considered these differences in the behavior of $\frac{dM}{d\sigma_{ap}}$ for samples with positive and negative λ one can predict roughly the expected behavior for a sample composed of two phases with similar saturation magnetization and magnetization curve but different sign of λ to each other. Independently of the value taken by $\frac{dM}{d\sigma_{ap}}$ at $\sigma_{ap} = 0$ it will be verified that $\frac{dM}{d\sigma_{ap}}$ evolves toward more negative values as σ_{ap} increases. This is in fact a characteristic of samples with negative λ, the difference however is that under some conditions $\frac{dM}{d\sigma_{ap}}$ changes its sign with σ_{ap} in a sample composed of two phases, and this change of sign cannot take place for a single phase system.

Let us assume both phases to have the same saturation magnetization and magnetization curve. This assumption is taken for the sake of clarity and does not affect the generality of the discussion. Let λ_1, λ_2, V_1 and V_2 the saturation magnetostriction and volume fraction of phases 1 and 2 respectively. Consider $\lambda_1 V_1 + \lambda_2 V_2$ to be near zero but positive, with $\lambda_1 > 0$ and $\lambda_2 < 0$.

Expanding $\chi_d(H+H_\sigma)$ in powers of σ_{ap} eq.|1| leads:

$$\frac{dM}{d\sigma_{ap}} = (\frac{dM}{d\sigma_{ap}}) V_1 + (\frac{dM}{d\sigma_{ap}}) V_2 = \frac{3\chi_d(H_z)}{\mu_0 M_s}(\lambda_1 V_1 + \lambda_2 V_2) +$$

$$+ \frac{9\chi_d'(H_\sigma)(\lambda_1^2 V_1 + \lambda_2^2 V_2)\sigma_{ap}}{\mu_0^2 M_s^2} \qquad |2|$$

For $\sigma_{ap} = 0$, $\frac{dM}{d\sigma}$ is positive since χ_d is positive everywhere and it was assumed $\lambda_1 V_1 + \lambda_2 V_2$ to be positive. The second term is strictly negative since χ_d' is negative everywhere and $\lambda_1^2 V_1 + \lambda_2^2 V_2$ is strictly positive. Hence there should be a critical value of σ_{ap} for which $\frac{dM}{d\sigma_{ap}}$ changes its sign from positive to negative

$$\sigma_{ap}^c = \frac{\mu_0 M_s \chi_d(H_z)}{3\chi_d'(H_z)} \frac{(\lambda_1 V_1 + \lambda_2 V_2)}{(\lambda_1^2 V_1 + \lambda_2^2 V_2)} \qquad |3|$$

Note that σ_{ap}^c increases as H_z does since $\frac{\chi_d(H_z)}{\chi_d'(H_\sigma)}$ is roughly proportional to H_z in the region of approach to saturation. This is a very important remark because it points out that the range of σ_{ap} for which $\frac{dM}{d\sigma_{ap}}$ is positive depends on the strength of the applied field. This phenomenon has been experimentally observed and reported (17). Previous theories developed to explain the stress dependence of λ and based on either anisotropic bond distribution or rotation of structural units produced by σ_{ap} cannot account so easily for this phenomenon.

In order to emphasise the simple idea summarized in eq.|2| we have neglected elastic interactions between the phases. In this first approximation it has been pointed out that the coexistence of two phases with opposite sign of λ is a reasonable point of view for analysing the characteristics of the metallic glasses with low magnetostriction. Under this assumption the more simple rules of magnetoelasticity and magnetization curve lead to a dependence of $\frac{dM}{d\sigma_{ap}}$ with σ_{ap} as that observed in the experiments. Moreover it is predicted that the range of σ_{ap} for which $\frac{dM}{d\sigma_{ap}}$ is positive increases with H_z in those samples which exhibits change of sign of $\frac{dM}{d\sigma_{ap}}$. Before proceeding with the analysis of the magnetoelastic methods used for measuring λ in the framework of a two phases model, some considerations about the elastic interaction between the phases will be outlined. As it will be seen the basic idea has been indicated in eqs.|1| and |2|.

c) Magnetoelastic energy of a sample composed of two phases and subjected to a tensile stress.

When a tensile stress, σ_{ap}, is applied along the ribbon axis of an amorphous ribbon an easy axis of magnetization is induced via magnetoelastic coupling. The easy axis is parallel or perpendicular to the ribbon axis for λ positive or negative respectively. The anisotropy constant takes the value $\frac{3}{2}\sigma_{ap}$. For the case of a sample composed of two phases randomly mixed with isotropic saturation magnetostriction λ_1 and λ_2 respectively, the average anisotropy constant can be written as:

$$\frac{3}{2}\lambda\sigma_{ap} = \frac{3}{2}\lambda_1\sigma_1 V_1 + \frac{3}{2}\lambda_2\sigma_2 V_2 \qquad |4|$$

where σ_1 and σ_2 are the average value of the stress acting on each phase along the ribbon axis.

With generality σ_1 and σ_2 can be expressed as:

$$\sigma_1 = \sigma_{ap} + \sigma_{12} \quad \text{and} \quad \sigma_2 = \sigma_{ap} + \sigma_{21} \qquad |5|$$

σ_{12} and σ_{21} are the average stresses mutually exerted between the phases, along the ribbon axis, which must hold the Albenga's relationship

$$\sigma_{12}V_1 + \sigma_{21}V_2 = 0 \qquad |6|$$

Two sources contributing to σ_{12} and σ_{21} can be distinguished. They have an elastic component proportional to σ_{ap} which arises from the possible difference in the elastic constants of both phases.

To keep constant strain in both phases under the action of σ_{ap}, some internal stresses are developed. They are expressed as:

$$\sigma_{12}^{el} = (\frac{Y_2-Y_1}{Y^*}) V_2\sigma_{ap} \quad \text{and} \quad \sigma_{21}^{el} = (\frac{Y_1-Y_2}{Y^*}) V_1 \sigma_{ap} \qquad |7|$$

where Y_1 and Y_2 are the Young's moduli of phases 1 and 2 respectively and Y^* the average value $Y_1 V_1 + Y_2 V_2$.

A second source of internal stresses is the magnetostrictive character of the phases. According to the relative orientation of the magnetization at each phase there should be mutual stresses to counterbalance the different magnetostrictive strain at each phase. A simple calculation yields

$$\sigma_{12}^{me} = \frac{Y_1 Y_2}{Y^*} V_2 \{\lambda_2 <\cos^2\phi>_2 - \lambda_1 <\cos^2\phi>_1\}$$

and

$$\sigma_{21}^{me} = \frac{Y_1 Y_2}{Y^*} V_1 \{\lambda_1 <\cos^2\phi>_1 - \lambda_2 <\cos^2\phi>_2\} \qquad |8|$$

where $<\cos^2\phi>_{2,1}$ are average values in phases 2 and 1 respectively and ϕ is the angle between the magnetization, at each domain, and the ribbon axis.

The elastic component given by $|11|$ being proportional to σ_{ap} does not introduce any physical significance different to a modification of λ_1 and λ_2 values with could be rewritten as:

$$\lambda_1^* = \lambda_1 (1 + \frac{Y_2-Y_1}{Y^*} V_2) \quad \text{and} \quad \lambda_2^* = \lambda_2 (1 + \frac{Y_1-Y_2}{Y^*} V_1) \qquad |9|$$

This modification will be disregarded.

The magnetoelastic components shown in eq. $|12|$ are dependent on σ_{ap}, although not linearly, and also depends on the applied field. Note that independently of the sign of λ, H tends always to approach $<\cos^2\phi>$ to 1. On the other hand σ_{ap} tends to approach $<\cos^2\phi>$ to 1 when is positive or to 0 when λ is negative.

The expression of the magnetoelastic energy given by $|4|$ becomes:

$$F_{me} = \frac{3}{2} \lambda \sigma_{ap} = \frac{3}{2} (\lambda_1 V_1 + \lambda_2 V_2) \sigma_{ap} + $$
$$+ \frac{3}{2} \frac{Y_1-Y_2}{Y^*} V_1 V_2 (\lambda_1-\lambda_2) \{\lambda_2 <\cos^2\phi>_2 - \lambda_1 <\cos^2\phi>_1\} \qquad |10|$$

Hence the magnetostriction, when defined as $\frac{2}{3} \frac{dF_{me}}{\sigma_{ap}}$, becomes:

$$\lambda = (\lambda_1 V_1 + \lambda_2 V_2) + \frac{3}{2} \frac{Y_1 Y_2}{Y^*} V_1 V_2 (\lambda_1-\lambda_2)$$
$$\cdot \{\lambda_2 \frac{<\cos^2\phi>_2}{\sigma_{ap}} - \lambda_1 \frac{<\cos^2\phi>_1}{\sigma_{ap}}\} \qquad |11|$$

III) SAMR method for a two phases model

The method developed by Narita (18) labelled small angle magnetization rotation, SAMR, has been widely applied to measure λ in metallic glasses (9, 19, 20) The ribbon is saturated under the action of an axial field H_z and subjected to a tensile stress, σ_{ap}, applied along the ribbon axis. A small transverse ac field, H_y, of frequency ω, produces a rotation of the magnetization around the ribbon axis. A pick-up coil wound around the ribbon detects a voltage of frequency 2ω which is proportional to the square of the angle, ϕ, rotated by the magnetization, ϕ is related to H_y, H_z and σ_{ap} through the relation:

$$\phi^2 = (\frac{\mu_0 M_s H_y}{\mu_0 M_s H_z + 3\lambda\sigma_{ap}})^2 \qquad |12|$$

The induced voltage can be written as

$$V_{2\omega} = \frac{1}{2} Ns\omega\mu_0 M_s \phi^2 \qquad |13|$$

where N is the number of turns of the pick-up coil and s is the cross section of the ribbon.

The method consists in drawing H_z-σ_{ap} diagrams for which $V_{2\omega}$ is constant when H_y is kept constant. λ can be obtained as:

$$\lambda = -\frac{\mu_0 M_s}{3} |\frac{d H_z}{d\sigma_{ap}}|_{V_{2\omega}, H_y} \qquad |14|$$

Experimental values of λ obtained by using this method in low magnetostrictive alloys have been found to verify |21|

$$\lambda = \lambda_{\sigma_{ap}=0} - (\frac{d\lambda}{d\sigma_{ap}})_{\sigma_{ap}=0} \sigma_{ap} \qquad |15|$$

$(\frac{d\lambda}{d\sigma_{ap}})$ being a positive value ranging from 1 to 6×10^{-10} MPa^{-1}

Let us analyse this method for a sample composed of two magnetostrictive phases. First we must find an expression for $V_{2\omega}$. Calling ϕ_1 and ϕ_2 the angle rotated by the magnetization at phase 1 and 2 respectively, $V_{2\omega}$ can be written, according to $|12|$, as

$$V_{2\omega} = \frac{1}{2} Ns\omega\mu_0 (M_1 V_1 \phi_1^2 + M_2 V_2 \phi_2^2) \qquad |16|$$

Where we have considered the identity existing between the cross section fraction and the volume fraction for each phase.

Putting $x_1 = \dfrac{3\lambda_1 \sigma_1}{\mu_0 M_s H_z}$ and $x_2 = \dfrac{3\lambda_2 \sigma_2}{\mu_0 M_s H_z}$, eq. $|12|$ yields

$$M_1 V_1 \phi_1^2 + M_2 V_2 \phi_2^2 = 2\left(\frac{H_y}{H_z}\right)^2 \left(\frac{M_1 V_1}{(1+x_1)^2} + \frac{M_2 V_2}{(1+x_2)^2}\right) \qquad |17|$$

Calculating the increment of $M_1 V_1 \phi_1^2 + M_2 V_2 \phi_2^2$ produced by $\Delta\sigma_{ap}$ and ΔH_z it is found, for $V_{2\omega}$ = = constant

$$|18|$$

$$-\frac{\mu_0}{3}\left(\frac{\Delta H}{\Delta\sigma_{ap}}\right)_{V_{2\omega}, H_y} = \frac{(\lambda_1 V_1 \frac{\partial\sigma_1}{\partial\sigma_{ap}})(1+x_2)^3 + (\lambda_2 V_2 \frac{\partial\sigma_2}{\partial\sigma_{ap}})(1+x_1)^3}{V_1 (M_1 + \frac{3\lambda_1}{\mu_0}\frac{\partial\sigma_1}{\partial H_z})(1+x_2)^3 + V_2 (M_2 + \frac{3\lambda_2}{\mu_0}\frac{\partial\sigma_2}{\partial H_z})(1+x_1)^3}$$

a) $\sigma_1 = \sigma_2 = \sigma_{ap}$.

When the elastic interactions are disregarded, $\sigma_1 = \sigma_2 = \sigma_{ap}$ then $\dfrac{\partial\sigma_1}{\partial\sigma_{ap}} = \dfrac{\partial\sigma_2}{\partial\sigma_{ap}} = 1$ and $\dfrac{\partial\sigma_1}{H_z} = \dfrac{\partial\sigma_2}{H_z} = 0$.

Expanding the right hand side term in eq. $|16|$ in powers of x_1 and x_2 it is obtained

$$|19|$$

$$-\frac{\mu_0}{3}<M>\left(\frac{\Delta H}{\Delta\sigma_{ap}}\right)_{V_{2\omega}, H_y} = \lambda_1 V_1 + \lambda_2 V_2 + \frac{9V_1 V_2}{\mu_0 <M> H_z}\left|2\lambda_1\lambda_2 - \frac{\lambda_1^2 M_2}{M_1} - \frac{\lambda_2^2 M_1}{M_2}\right|\sigma_{ap}$$

where $<M> = M_1V_1 + M_2V_2$.

The term proportional to σ_{ap} at the right hand side would become important when the sign of λ_1 and λ_2 are opposite to each other. For this case the first term $\lambda_1V_2 + \lambda_2V_2 = <\lambda>$ might be small whereas the second remains strictly negative and its absolute value increases with σ_{ap}. In metallic glasses with capability of undergoing very high stresses in the elastic range, the second term is expected to be observable for those composition with $<\lambda>$ small. When λ_1 and λ_2 differ in sign the experimental magnetostriction value obtained by applying SAMR method, given by eq. |19| can be written as

$$\lambda^{exp} = <\lambda> - A\sigma_{ap} \qquad |20|$$

where $A = \dfrac{9V_1V_2}{\mu_0 M H_z} \{-2 \lambda_1 |\lambda_2| - \dfrac{\lambda_1^2 M_2}{M_1} - \dfrac{\lambda_2^2 M_1}{M_2}\}$. For the particular case in which $M_1 = M_2 = M$, A becomes:

$$A = \dfrac{9V_1V_2}{\mu_0 M H_z} (\lambda_1 - \lambda_2)^2 \qquad |21|$$

Expression of λ^{exp} described by |20| accounts for the main characteristics of the behavior observed in low magnetostrictive metallic glasses. The physical origin of the term $-A\sigma_{ap}$ is the same we have seen before when analysing $\dfrac{\Delta M}{\Delta \sigma_{ap}}$ as a function of σ_{ap}. Note that in the phase with positive λ, ϕ^2 tends toward zero with decreasing rate as σ_{ap} rises, whereas ϕ^2 increases with increasing rate as σ_{ap} increases in the phase with negative λ. This difference, in the variation rate, break up the initial counterbalance of $\lambda_1V_1 + \lambda_2V_2$, at $\sigma_{ap} = 0$, as σ_{ap} increases.

From eq. |20| it comes out that for positive but near zero $<\lambda>$ there would be a critical σ_{ap} for which λ^{exp} is zero. The critical stress is predicted to increase linearly with H_z.

b) Effect of the elastic interactions

The true magnetoelastic energy stored at phases 1 and 2 is given by $\frac{3}{2}\lambda_1\sigma_1$ and $\frac{3}{2}\lambda_2\sigma_2$ respectively where σ_1 and σ_2 take the values expressed by eq. $|5|$. Taking into account the values of σ_{12} and σ_{21} given by eq. $|9|$ and substituting their derivatives respect to H_z and σ_{ap} in eq. $|16|$, a rather tedious expansion up to the first power of σ_{ap} yields:

$$\lambda^{exp} = <\lambda> - \frac{3V_1V_2}{\mu_0 M H_z}(\lambda_1-\lambda_2)^2 \cdot$$

$$\cdot \{3\sigma_{ap} + \frac{YH_y^2}{H_z^2}(\lambda_1 V_2 + \lambda_2 V_1)\} + f(\lambda^4)\,\sigma_{ap} \qquad |22|$$

where we have considered $M_1 = M_2 = M$ and $Y_1 = Y_2 = Y$ for the sake of clarity. $f(\lambda^4)$ is given by:

$$f(\lambda^4) = \frac{27\,YH_y^2 V_1 V_2}{(\mu_0 M_z)^2 H_z^4}(\lambda_1-\lambda_2)^2(\lambda_1^2 V_2 + \lambda_2^2 V_1) \qquad |23|$$

If this term is disregarded, what becomes reasonable after considering that its value is $\frac{3\,H_y^2}{\mu_0 M_z H_z^3}Y(\lambda_1^2 V_2 + \lambda_2^2 V_1)$ times the first term proportional to σ_{ap} given by eq. $|21|$, the effect of the interactions is restricted to introduce an effective stress given by:

$$\sigma_{ap}^{ef} = \frac{Y\,H_y^2}{3\,H_z^2}(\lambda_1 V_2 + \lambda_2 V_1) \qquad |24|$$

The physical meaning of the new terms can be understood from eq. $|11|$ which supplies the value of the magnetostriction when considered as the derivative of the magnetoelastic energy. For the case of ϕ_1 and ϕ_2 given by eq. $|12|$, eq. $|11|$ yields

$$\lambda = \lambda_1 V_1 + \lambda_2 V_2 - \frac{3Y_1 V_1 V_2 (\lambda_1-\lambda_2)^2 H_y^2}{\mu_0\,M\,H_z^3}\{1 - \frac{3(\lambda_1+\lambda_2)}{\mu_0 M H_z}\sigma_{ap} + \ldots\} \qquad |25|$$

For the single phase case the variation of H_z does not affect the magnetoelastic energy. However when the sample is composed of two phases with different λ values the variation of H_z produces a disturbance of the magnetoelastic energy since changes of H_z alter the values of σ_{12} and σ_{21}. This phenomenon which is reflected by the second term of eq. |25| also affects the experimental λ value obtained by SAMR method as illustrated in eq. |22|.

IV) Exchange iteractions:

Hitherto the exchange interactions between the magnetic moments of different phases have been disregarded. Therefore we have considered the angles rotated by the magnetization at each phase to be not correlated. This approximation holds when the diameter of the clusters with constant magnetostriction is larger than the exchange length $L_k = (\frac{C}{k})^{1/2}$ where C is the exchange constant and k the anisotropy constant. Taking $C = 4 \times 10^{-11}$ Jm^{-1} the exchange length is observed to range between 0.8 and 0.2 μm when k varies from 60 to 1000 Jm^{-3}. If the diameter of the clusters is lower than L_k the magnetic moments are coupled through the exchange interactions.

If we think the amorphous structure to be built up with a few of fairly well defined local atomic structures, the diameter expected for clusters with well defined magnetostriction is of the order of 10 Å, which is hundred times smaller than the exchange length. Hence the exchange interaction must be taken into account.

The effect produced by the exchange interaction can be envisaged as decreasing the difference between the angles rotated by the magnetization at each phase. For the case of the SAMR analysed previously the angle ϕ_1, and ϕ_2 would change, by effect of exchange, to ϕ_1' and ϕ_2' respectively which verify:

$$\phi_1' - \phi_2' < \phi_1 - \phi_2 \qquad |26|$$

Let us consider the average size δ of the structural units with a single well defined magnetostriction value as a parameter. When δ is much larger than L_k, ϕ_1' and ϕ_2' are uncorrelated and their values at the limit condition $L_k/\delta \to 0$, converge to

$$\phi_1' = \phi_1 = \frac{(\mu_0 M_s H_y)}{(\mu_0 M_s H_z + 3\lambda_1 \sigma_1)} \qquad |27|$$

and

$$\phi_2' = \phi_2 = \left(\frac{(\mu_0 M_s H_y)}{\mu_0 M_s H_z + 3\lambda_2 \sigma_2}\right) \qquad |28|$$

In the opposite limit, $\delta/L_k \to 0$, ϕ_1' and ϕ_2' tend toward a common value, ϕ, given by

$$\phi_1' = \phi_2' = \phi = \frac{\mu_0 M_s H_y}{\mu_0 M_s H_z + 3(\lambda_1 \sigma_1 V_1 + \lambda_2 \sigma_2 V_2)} \qquad |29|$$

This value can be easily derived by minimizing the total energy

$$F = (\frac{3}{2} \lambda_2 \sigma_2 \sin^2 (\phi - \phi_2) - \mu_0 M_s H_z \cos \phi_2 - \mu_0 M_s H_z \sin \phi_2) V_2 +$$
$$+ (\frac{3}{2} \lambda_1 \sigma_1 \sin^2 \phi_1 - \mu_0 M_s H_z \cos \phi_1 - \mu_0 M_s H_y \sin \phi_1) V_1$$

with the condition $\phi_1 = \phi_2$.

It can be concluded that the experimental value of λ given by eq. $|22|$ is that expected when $L_k/\delta \to 0$, however when the size of the clusters become much smaller than L_k the value of λ obtained by using SAMR method tends to

$$\lambda = \lambda_1 \sigma_1 V_1 + \lambda_2 \sigma_2 V_2 \qquad |30|$$

which according to eq. $|8|$ can be written as

$$\lambda^{exp} = <\lambda> - \frac{3}{2} Y V_1 V_2 (\lambda_1 - \lambda_2)^2 \frac{\partial \cos^2\phi}{\partial \sigma_{ap}} \qquad |31|$$

Eq. $|31|$ indicates a stress dependence of λ quite different to the one described by eq. $|22|$. By considering $\cos^2\phi \simeq -2\phi^2$ and taking into account the value of ϕ given by eq. $|29|$ one find that λ^{exp} depends on the stress through a fourth power of λ. This term is negligible when compared with the stress dependence of λ^{exp} given by eq. $|22|$.

The same influence of the exchange interactions can be envisaged when λ^{exp} is measured by using the initial susceptibility method. For applying this method the presence of an hemogeneous anisotropy with easy axis perpendicular to the ribbon axis is required. Let k be the anisotropy constant, then the initial susceptibility is related to the applied stress as

$$\chi = \frac{\mu_0 M^2}{2k - 3\lambda \sigma_{ap}} \qquad |32|$$

From the stress derivative of the inverse of χ, λ can be obtained through:

$$\lambda = -\frac{1}{3} \frac{d}{d\sigma_{ap}} (\frac{1}{\chi}) \qquad |33|$$

Consider now the sample to be composed of two phases with positive λ_1 and negative λ_2 magnetostriction constants respectively. Under the action of a small longitudinal field H_z, the magnetization would rotate from the transverse direction toward the longitudinal direction an angle ϕ_1 and ϕ_2 at each phase.

ϕ_1 and ϕ_2 can be obtained by minimizing the total energy,

$$F = \{-\mu_0 M_1 H_z \cos\phi_1 + (k + \frac{3}{2}\lambda_1\sigma_1) \sin^2\phi_1\} V_1 +$$
$$+ \{-\mu_0 M_2 H_z \cos\phi_2 + (k + \frac{3}{2}\lambda_2\sigma_2) \sin^2\phi_2\} V_2 \qquad |34|$$

When ϕ_1 and ϕ_2 are not correlated, i.e. $L_k/\delta \to 0$, the total susceptibility becomes

$$\chi = \chi_1 + \chi_2 = \frac{\mu_0 M_1^2}{2k - 3\lambda_1\sigma_1} V_1 + \frac{\mu_0 M_2^2}{2k - 3\lambda_2\sigma_2} V_2 \qquad |35|$$

Obviously $\frac{1}{\chi}$ should not be, in general, a straight line. Without calculating any expressions some qualitative considerations can be outlined. At low stresses χ_1 rises at a rate faster than the decreasing rate of χ_2. When $\lambda_1\sigma_1$ approaches to $2k$, χ_1 takes an almost constant value which increases slowly. However χ_2 proceeds decreasing according to the same law.

On the other limit, $\delta/L_k \to 0$, ϕ_1 and ϕ_2 are equals. By minimizing F given by $|34|$ with such condition, it is obtained,

$$\chi = \frac{(M_1 V_1 + M_2 V_2)}{2k - (3\lambda_1\sigma_1 V_1 + 3\lambda_2\sigma_2 V_2)} \mu_0 <M> \qquad |36|$$

where $<M> = M_1 V_1 + M_2 V_2$.

The inverse of χ is a straight line weakly affected by the fourth order magnetoelastic effects.

V) Conclusions

A few ideas about the expected dependence of the magnetostriction with the applied stress in a sample composed of two phases with different magnetostriction have been outlined.

Considerations have been carried out under two opposite limiting condition. First the effect of exchange between phases has been neglected. For this case λ^{exp} is expected to decrease with the applied stress. This dependence is essentially due to the different rate of variation of the susceptibility of each phase with the applied stress. Note that for most of magnetoelastic methods the susceptibility along the longitudinal direction is determined. Therefore when the susceptibility at each phase is not correlated to each other the average value depends on the stress if they do not change with the stress at the same rate. This approximation is expected to hold when the size of the structural units with well defined magnetostriction is larger than the exchange length. Second, the effect of the exchange interaction between phases has been supposed to be stronger than anisotropy effects. For this case the experimental λ is the average value $\lambda_1 V_1 + \lambda_2 V_2$. It is predicted however a field and stress dependence of λ through the elastic coupling between phases. This effect which also appears in the first approximation described above is negligible when compared with the stress dependence which arise from different evolution of susceptibilities at each phase.

Accordingly to the models usually proposed (5,7) the amorphous structure is composed of three type of clusters with size smaller than 10 Å. Since the magnetic anisotropy is closely linked to the local symmetry of the magnetic atoms it seems reasonable to assume the presence of three local anisotropy constants. Considering the magnetostriction as the strain derivative of the anisotropy it seems likely that there should also be three different magnetostrictive behavior. As the exchange length is much larger than 10 Å the magnetostriction would behave macroscopically as

$$\lambda = n_i \lambda_i V_i + n_0 \lambda_0 V_0 + n_t \lambda_t V_t$$

where n_i, n_0, n_t, λ_i, λ_0, λ_t, V_i, V_0 and V_t are the number per unit volume, magnetostriction and volume of icosahedral, octahedral and trigonal clusters respectively.

The observation of phase transformation suggests the possible coexistence of different phases. Each phase can be defined by a set of numbers n_i, n_0 and n_t. According to the experimental data (1,2,5,7) the phase stable at low temperature is richer in low symmetry trigonal prisms, therefore it can exhibit a different average λ than the phase stable at high temperature.

When phases with different λ are present the expected experimental λ depends on the size of the single phase regions. If this size is of the order of 10 μm the effect of the exchange interaction should be restricted to dimensions one hundred times smaller. For this case the expected behavior of λ agrees quite well with the experimental results if the magnetostriction of each phase is close to zero but with opposite sign.

A general theory will be now developed. The analysis of the thermal, compositional and structural relaxation dependence of λ in this framework should be carried out.

It is finally proposed that the coexistence of amorphous phases must be taken into account as a possible explanation of the stress dependence of λ in nearly zero magnetostriction amorphous alloys.

Acknowledgment

The author is indebted to Profs. H.Scymczak and H. Lachowicz for continuous stimulus to think on the stress dependence of λ. This work has been supported by the Spanish CAICYT project PA-84-0365 and by the Comité Conjunto Hispano-Norteamericano project CCA8411006.

References:

|1| O'Handley, R.C., McHenry, M.E., Eberhart, M.E., Johnson, K.H., Grant, N.J., in Amorphous Materials, edited by A.Jaffee, Acta Met., 1985.
|2| O'Handley, R.C., in M.R.S., Vol. 58, edited by B.G.Giessen, D.E.Polk and A.I.Taub., 1986, p. 141.
|3| Riveiro, J.M., Phys.Rev.B, 37, 13, 7731 (1988).
|4| Riveiro, J.M. and Pareja, R., Phys.Rev.B, 34, 2020 (1986)
|5| Suran, G., Dunadgela, K. and Machizaud, F., Phys.Rev.Lett. 57, 24, 3109, 1986.
|6| Madurga, V., Barandiarán, J.M., Vázquez, M., Nielsen, O.V. and Hernando, A., J.Appl.Phys., 61, 3228, 1987.
|7| Corb, B.W., O'Handley, R.C., Megusar, J. and Grant, N.J., Phys.Rev.Lett., 51, 1386, 1983.
|8| Herzer, G., in Proc. of Soft Magnetic Materials 7, Blackpool, edited by Wolfson Center (Publisher, Cardiff, England), 1987.
|9| Barandiarán, J.M., Hernando, A., Madurga, W., Nielsen, O.V., Vázquez, M. and Vázquez López, M., Phys.Rev. B, 35, 10, 5066, 1987.
|10| Siemko, A., Lachowicz, H.K., Lisowski, B., 3rd. Intern. Conf. on Phys. of Magn.Mater. 1986.
|11| Lachowicz, H.Z. and Szymczak, H., J.Magn.Magn.Mat., 41, 327, 1984.
|12| Hernando, A., Vázquez, M., Madurga, V., Ascasibar, E. and Liniers, M., J.Magn.Magn.Mat. 61, 39, 1986.
|13| Siemko, A. and Lachowicz, H.K., J.Magn.Magn.Mat., 66, 31, 1987.
|14| Santos, A.D., Severino, A.M. and Missel, F.P., J.Magn. Magn.Mat. 60, 153, 1986.
|15| Szymczak, H., J.Magn.Magn.Mat. 67, 227, 1987.
|16| Furthmuller, J., Fahnle, M. and Herzer, G. J.Magn.Magn. Mat., 69, 69 and 80, 1987.
|17| Hernando, A., Vázquez, M., Barandiarán, J.M. and van Hattum, W.J., ICM, Paris, 1988.
|18| Narita, J., Yanasaki, J. and Fukunaga, H., IEEE Trans. on Magn. MAG-16, 433, 1980.

MAGNETIC AND NMR SPIN ECHO STUDIES IN SOME Mn CONTAINING METALLIC GLASSES

R. Krishnan, H. Lassri and P. Rougier
Laboratoire de Magnétisme, C.N.R.S., F-92195 Meudon Cedex
K. Le Dang and P. Veillet
Institut d'Electronique Fondamentale, Bât. 220, F-91405 Orsay cedex.

ABSTRACT

We describe magnetization and NMR spin echo studies on some metallic glasses based on (Fe+Ni) and Co containing upto 12 at % of Mn. The study is however focussed on alloys with very small Mn concentration. The presence of Ni is needed to obtain high moment of 3.3 µB for Mn. For certain critical Mn concentration (0.2 to 0.5) both Fe and Ni moments increase and that of the latter by about 100 %. Mössbauer studies on a few selected samples are reported. When Fe alone is present, the Mn moment is small and that of Fe also is seen to decrease. Crystallization leads to phases unknown in the phase diagram. Co based alloys also show an increase in moment for 0.2 % of Mn. Spin wave excitations results are also discussed.

INTRODUCTION

The literature on transition metal-metalloid alloys is quite abundant (1). Alloys with the substitution various 3d-metals have been studied. However investigations on amorphous alloys containing Mn addition are small in number. Senno et al (2) have studied Mn additions to Co-based alloys whereas Obi et al (3) described the effect of metalloid content on the properties of Co-Mn B alloys. Ramanan proposed Co-Fe-Mn-B-Si alloys for high frequency applications (4). In all the above studies rather high concentration of Mn has been used. The interesting aspect of Mn addition is the fact that the crystallization temperature T_x is increased accompanied by a reduction in the Curie temperature T_c. This is particularly attractive for applications because these alloys could be annealed at $T > T_c$ without the risk of any accidental crystallization. Besides this point these alloys also offer rich possibilities to study certain magnetic phenomenon such as spin glass in Mn rich alloys (5). Thus far the magnetic moment of Mn atoms was thought to the less than $2\mu_B$ (3). However, we showed for the first time that in Co-Fe based alloys containing 4 at % Mn, Mn moment was as high as 3.5 μ_B (6). We had since carried out an extensive study of the effect on the magnetic properties, of the addition of rather small amounths of Mn. For instance in Ni-Fe-B-Si alloys, when Ni/Fe ≈ 1, a small addition of 0.5 at % Mn increases the magnetization by 19 % (7). We explored this type of alloys with various Ni-Fe ratios containing 0 to about 6 at % of Mn. We have carried out magnetization and NMR spin echo studies in a routine fashion and Mössbauer studies in certain cases in order to obtain as complete a picture as possible.

In the present work we wish to describe these studies and try to synthesize the results. Though this work is mainly concerned on Ni-Fe based alloys we would draw examples also from Co based ones wherever it appears interesting to do so. Low temperature data are discussed. A few studies have been made on a crystallized sample and also are reported.

EXPERIMENTAL DETAILS

Amorphous $Fe_{80-y-x}Ni_yMn_xB_{12}Si_8$ alloys with $0 < y < 50$ and $0 < x < 6$ all expressed in at % were prepared by simple melt spinning technique. The purity of B, Si and Mn was better than 99.9 % and that of Fe and Ni was 99.999 %. The ribbon samples were about 3 mm wide and about 30 to 40 microns thick. The amorphous structure was verified by X-ray diffraction. The exact chemical composition of the samples was determined by electron probe micro analysis. The magnetization was measured in the range 4.2 to 300 K with applied fields upto 15 T at Service National des Champs Intenses at Grenoble. Vibration sample magnetometer was used to measure for field upto 1.7 T and also in the high temperature range. The NMR measurements were carried out at 4.2 K with a frequency variable spin-echo apparatus. The rf field of the exciting coil was parallel to the ribbon plane. Mössbauer measurments were made in the conventional way. The spectral analysis was done using the Window method.

RESULTS AND DISCUSSIONS

Let us first discuss the results of magnetization measurments for all the samples in general and then NMR data. Finally let us analyze the results on one particular series of alloys obtained from different techniques of measurements.

Magnetic satruation was obtained for applied fields (in the ribbon plane) in the range 2-3 T and data discussed here are for saturation. Let us note that extrapolation to H = 0 would give pratically the same results. It is in order, to note here, that the high field susceptibility for all the samples studied here was small and on the order of 10^{-5}emu/Oe which indicates that there are no antiferromagnetic interactions present.

Fig. 1 shows at 4.2 K the Mn concentration dependence of magnetization (σ in emu.g^{-1}) for alloys with different Ni concentration. It is seen that for alloys with small Ni content (y), σ decreases with increasing Mn (x) content while for y = 30 and 35, the decrease in σ is very small. However, for y = 40, σ actually shows a strong increase for Mn(x) = 0.5 at %. We have prepared this particular alloy several times and found that the result is perfectly reproducible. In fact for x = 5.5, σ has not decreased from its initial value. For y = 45 a rather sharp increase in σ is observed but for a smaller concentration of Mn namely 0.2. Again for higher Ni concentration of 50 at %, σ remains practically the same for the range of x studied. It is noteworthy that for Ni(y) = 40 and 45, σ increases by the same amount even at 290 K. It can be generally said that for small Ni content, σ decreases strongly with the addition of Mn indicating that μ_{Mn} should be smaller than μ_{Fe}. For 30 < y < 50 (except for 40 and 45) the decrease in σ is practically negligible so $\mu_{Mn} \approx \mu_{Fe}$. So in order to get information on the magnetic state of Mn, let us analyze the NMR data. The ^{55}Mn spectra for three typical alloys at 4.2 K are shown in Fig.2. Though the spectra are not corrected for the frequency dependence of the signal intensity they clearly bring out the effect of Ni addition on the magnetic state of Mn. In alloys with relatively high Ni concentration Mn spectra is centered at a high frequency and have narrow line width. But on the contrary, for Ni = 0, the spectrum is very broad with a pronounced tail in the low frequency side besides being centered at a much lower frequency.

Now in what follows, let us concentrate on the alloys with Ni = 40 at %. Fig. 3 shows ^{55}Mn spectrum at 4.2 K for $Fe_{39.5}Ni_{40}Mn_{0.5}B_{12}Si_8$ alloy with a saturation field of 0.2 T applied in the ribbon plane. The skin depth δ at 400 MHZ is about 7 µm which is smaller than the ribbon thickness which is typically 40 µm. Therefore, the signal intensity should be corrected by a factor $f^{3/2}$ rather than f^2 where f is the resonance frequency. The spectrum centered at 380 MHZ corresponds to an unusually strong hyperfine field value of 362 kOe. It may be recalled that the values for Mn impurities in crystalline Fe and Ni are 228 and

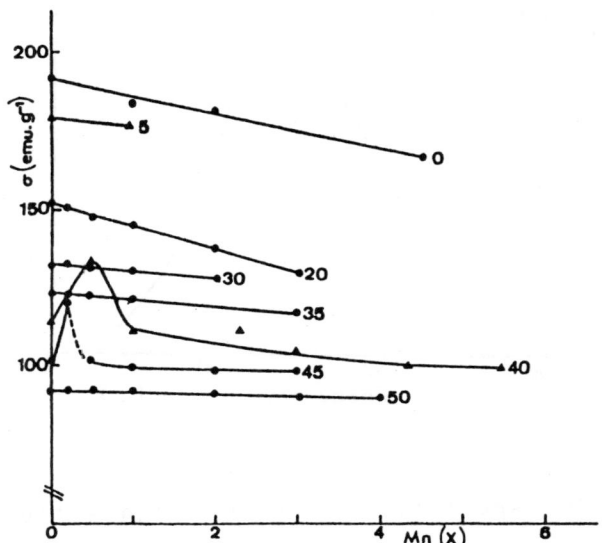

Fig. 1 Magnetization as a function Mn content at 4 K for different alloys. The numbers indicate the Ni content.

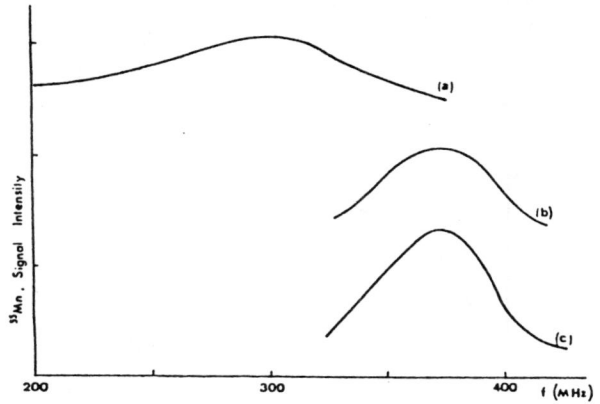

Fig. 2 ^{55}Mn spectra at 4 K for a) $Fe_{79.5}Mn_{0.5}$, b) $Fe_{34}Ni_{45}Mn_1$ and c) $Fe_{29}Ni_{50}Mn_1$ alloys. The metalloids are not indicated above.

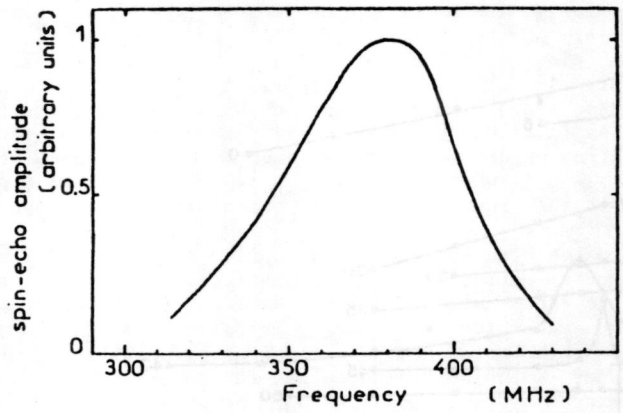

Fig. 3 ^{55}Mn spectrum at 4 K for $Fe_{39.5}Ni_{40}Mn_{0.5}B_{12}Si_8$

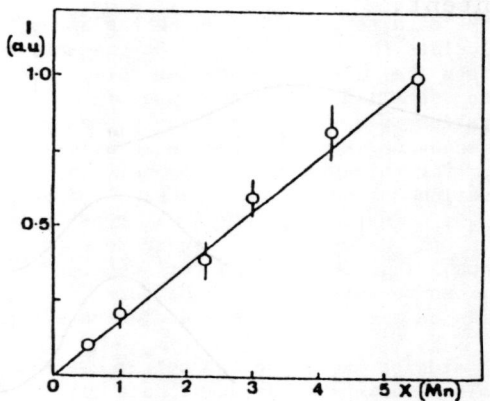

Fig. 4   Normalized ^{55}Mn signal intensity as a function of Mn content in $Fe_{39.5}Ni_{40}Mn_{0.5}$

328 kOe respectively. The Mn hyperfine field H_n (negative) arises from core polarization which is proportional to the Mn moment μ_{Mn} and from the conduction electron polarization due to the host magnetization $\bar{\mu}$. It is customary to write the semiempirical relation $H_n = a\mu_{Mn} + b\bar{\mu}$. Since these alloys are rich in Ni one could assume the a and b values to be the same as in Ni_3Mn, namely $a = -100$ kOe/μ_B and $b = -30$ kOe/μ_B (8). Knowing $\bar{\mu}$ from magnetization measurements, μ_{Mn} was calculated. For instance for $x = 0.5$ where μ peaks, μ_{Mn} is found to be the 3.3 μ_B. This value is close to that found by us in Co based amorphous alloys (6).

We studied the ^{55}Mn spectra in an applied field of 1.2 T when the frequency shows a decrease. This demonstrates that the Mn moment is parallel to the host magnetization. Furthermore, the Mn spectra are essentially the same for other Mn concentrations except that the low frequency tail gets reinforced with increasing Mn content. Thus, the average resonance frequency is decreased to 340 MHZ for $x = 5.5$. This may be understood by a decrease of μ_{Mn} and $\bar{\mu}$ in accordance with the magnetization data.

In the same alloys in order to check that all Mn atoms are coupled ferromagnetically we studied and compared the integrated Mn signal intensity of the different samples. As shown in Fig. 4 it is seen that the signal intensity per unit area is fairly proportional to the Mn concentration. This means that most of the Mn atoms are in a high spin state which constitutes an interesting behaviour of these amorphous alloys.

From the NMR data on μ_{Mn} and knowing the alloy moment $\bar{\mu}$, it is now possible to calculate for each value of Mn content, the magnetic moment contribution from Fe and Ni together $\mu_{(Fe+Ni)}$. Since μ_{Mn} remains practically the same (in fact a decrease of 10 % for $x = 5.5$) for the whole concentration range, the peak in the magnetization of the alloy for $x = 0.5$ cannot be explained unless we allow for a sudden increase in μ_{Fe} and μ_{Ni} for this particular composition. In the same manner, a similar situation should be present also in the alloys $Fe_{35-x}Ni_{45}Mn_xB_{12}Si_8$ (series B) where a peak in the magnetization is found to occur for $x = 0.2$. In order to know about the individual moments μ_{Fe}, μ_{Ni} it suffices to get some information on μ_{Fe} and by difference μ_{Ni} can be obtained from the total $\mu_{(Fe+Ni)}$. We carried out Mössbauer studies on these two alloys and which are described below.

Mössbauer studies were carried out at 290 K on the two series of alloys (9). Table I shows some relevent data. It is very interesting to note that in both the series for the Mn content where alloy moment peaks, the hyperfine field also shows an increase. It is reasonable to assume that this increase reflects a corresponding increase in μ_{Fe}. Assuming that the relative increase is the same at 4.2 K, we can now proceed to calculate μ_{Ni} in the two series of alloys and as a function of Mn content. As mentioned earlier, the alloy moment increases by about 20 % at the critical Mn concentration whereas increase in μ_{Fe} is only in the range 5-10 %. Therefore it turns out that μ_{Ni} increases

Table I : Mössbauer data on two series of alloys at 290K

	Ni at %	Mn at %	Hpeak kOe	Hav kOe
A	40	0	241	224
	40	0.5	253	234
	40	1.0	236	216
B	45	0	239	235
	45	0.2	263	247
	45	0.5	243	231
	45	1.0	234	220

Table II : Calculation of μ_{Fe} and μ_{Ni} at 4.2 K in two series of alloys. (See text for further explanation).

	X(Mn)	μ	μFe	μNi	μMn
A	0	0.994	2.18	0.31	--
	0.5	1.19	2.29	0.67	3.3
	1.0	0.99	2.10	0.35	3.3
B	0	0.9	2.18	0.30	--
	0.2	1.06	2.29	0.57	3.3
	0.5	0.89	2.14	0.30	3.3
	1.0	0.88	2.04	0.34	3.3

considerably. Table II shows the reuslts of the calculation. For $x = 0$, we have taken $\mu_{Fe} = 2.18\ \mu_B$ which explains our data well and μ_{Mn} has been fixed at $3.3\ \mu_B$ and the small decrease (≈ 1.8 % for $x = 1.0$) has been neglected. For each value of x, μ_{Fe} was adjusted to reckon the change in Fe average hyperfine field. It is interesting to note that μ_{Ni} increases by a large amount. It is however clear that one should not attach an importance to the numbers but appreciate the relative variations in μ_{Fe} and μ_{Ni}. For instance, the very high μ_{Ni} in both the alloys is noteworthy. We are led to conclude that the presence of Mn is very small amounts has led to an exaltation in the moments of Fe and Ni. Such an example in the crystalline alloys in our opinion, is not known.

Our study shows that the beneficial effect of Mn appears only with a certain amount of Ni present and that in alloys based on Fe alone, the magnetic moment of Mn is very small. It is not yet clear why the magnetization maximum is obtained only for a very narrow range of compositions. Does this mean that Mn needs two different transition metals to attain its high moment. Well some more information could be obtained by investigating Co-based alloy namely $(Co_{0.975}Fe_{0.025})_{74}Mn_8B_{12}Si_6$, where x had a maximum value of 8 (11). The Mn moments were still coupled ferromagnetically to that of the host. We can concluded that Mn still does not have another Mn as nearest neighbour. Again in these alloys a small quantity of Fe was present so we could not answer whether two transition metal atoms are needed. We then prepared and studied $Co_{80-x}Mn_xB_{12}Si_6$ alloys. Here again we noted a small peak in the magnetization (about 5 %) for $x = 0.2$. The Mn moment was on the order of $3.3 - 3.4\ \mu_B$. So it is clear that Mn attains a high moment in the presence of Co as in the case of Ni. Of course, alloys based on only Ni have to be prepared to bring additional support. As Ni moment vanishes easily alloys with high Ni content such as $Ni_{90}B_{10}$ need to be prepared and with Mn additions and studied. This study is being planned for the future. Our conclusion is coherent with the fact that Mn has high moment also in crystalline Ni_3Mn. However, it is not possible to elucidate the reason why the alloy magnetization peaks for very small Mn concentrations. Is there a specific local structure which leads to such moments for Ni ? More work on structure might hopefully throw some light on this question.

It was interesting to see if the amorphous state was essential for Mn to acquire such high moment. So we carried out some studies on crystallized samples. Here are briefly our reuslts. The alloy $Fe_{34.5}Ni_{40}Mn_{5.5}B_{12}Si_8$ after crystallization shows two ^{55}Mn resonance spectra centered at 320 and 413 MHz corresponding to Mn impurities present in two different compounds. We studied the effect of an external field of 1 T. The spectrum centered at 413 MHZ showed a negative shift in accordance with the gyromagnetic ratio of ^{55}Mn indicating that the compound in question is a normal ferromagnet. On the contrary, the other spectrum did not show any shift indicating that this is a canted antiferromagnet. In an attempt to identify the crystalline phases in which Mn impurities observed, we prepared and investigated three alloys namely Ni_3Fe, NiFe and $Ni_{10}Fe_{90}$ containing

1 wt % of Mn. The Mn spectra for the above three alloys were centered at 345, 376 and 245 MHZ respectively, a result which is totally different. Hence, we came to the conclusion that phases that appear on cyrstallization are metastable and cannot be easily produced by conventional method. We could not calculate the magnetization of the crystallized sample for the want of information on the magnetic fraction. However T_c was found to be as high as 672 K. Let us now recall our studies on $Co_{76.1}Fe_{1.9}Mn_4B_{12}Si_6$ after crystallization (6). In this case one has both Co and Mn spectra. As regards Co spectra, the one present near 220 MHZ corresponds to pure Co clusters and there are two more in the range 50-150 MHZ with a shoulder at 80 MHZ and a peak at 115 MHZ which could be ascribed to Co_2B and Co_3B. However Mn spectrum now narrows and shifts from 365 (in the amorphous state) to 373 MHz which is close to the resonance frequency of Mn impurity in f.c.c. Co. So we are led to conclude that Mn is present as an impurity in Co_3B. Application of an external field produces a negative shift of the Mn spectrum indicating that Mn moment is aligned ferromagnetically in the crystallized material. So to sum up, crystallization studies clarified only part of the problem and raised other questions to be answered particularly in Fe-Ni based alloys.

Let us now discuss the spin wave excitations in some of these alloys. As is well known, the temperature dependence of the magnetization in almost all the ferromagnetic amorphous alloys follows the Bloch's law, namely $\sigma(T) = \sigma(0) [1 - BT^{3/2} - CT^{5/2}]$. This behaviour is verified also by the alloys studied here (10). As a typical example Fig. 5 shows the $T^{3/2}$ dependence of σ in $Fe_{40-x}Ni_{40}Mn_xB_{12}Si_8$ alloys with x = 0, 0.5 and 5.5. The small deviation from the linear behaviour for T > 120 K arises from the C term which is two orders of magnitude smaller than B. B increases from 15.7×10^{-6} to 42.8×10^{-6} as x increases from 0 to 5.5. For all these samples C remains at $2.5 \cdot 10^{-8}$. In the treatment of simple spin-wave theory the coefficient B in the equation of Bloch's law is related to the spin-wave stiffness coefficient D by the equation.

$$B = 0.0587[g\mu_B/M(0)](k_B/D)^{3/2}$$

where the symbols have the usual meaning. The g factor in all these alloys varied in the range 1.95 to 2.1 and we took g = 2 for the calculation. We had shown in the case of amorphous Ni-Co-P films that D values thus calculated agrees very well with that obtained from standing spin-wave spectra (12). Fig. 6 shows the Mn concentration dependence of D for various Ni contents. For Mn = 0, the increase in D with an increase in Ni content is note worthy. D decreases with increasing Mn content which is in accordance with a decrease in T_c. As a typical result we show in Fig. 7 the variation of D with T_c for the series $Fe_{40-x}Ni_{40}Mn_xSi_8$. The linear dependence and the straight line passing through the origin could be noted. The datum point denoted by the letter A corresponds to the composition with x = 0, that is without Mn. The fact that it lies above the line is not due to error but it simply reflects the fact that D/T_c ratio is modified when Mn is added. This fact was verified in other series also. Since D/T_c characterizes

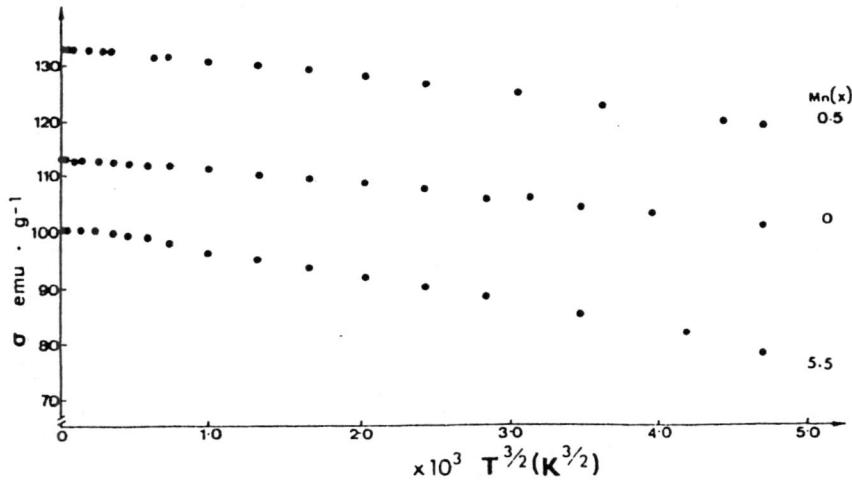

Fig. 5 $T^{3/2}$ dependence of σ in $Fe_{40}Ni_{40}$ alloy with different Mn content

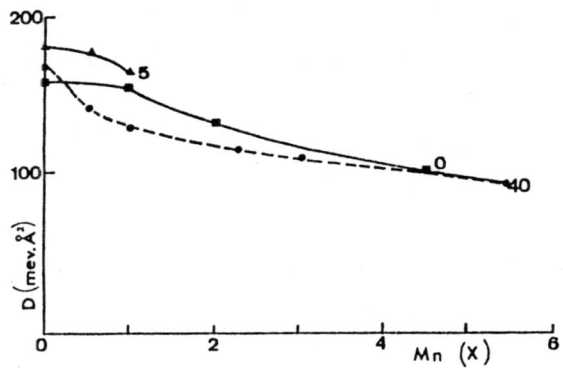

Fig. 6 Spin wave stiffness coefficient D vs Mn content in three different alloys. The numbers indicate the Ni content

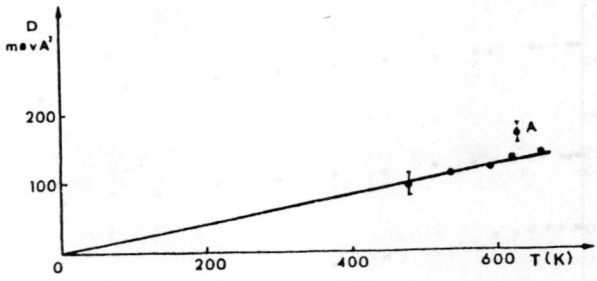

Fig. 7 D as a function of T_c in $Fe_{40-x}Ni_{40}Mn_x$ system of alloys.

simply reflects the fact that D/T_c ratio is modified when Mn is added. This fact was verified in other series also. Since D/T_c characterizes the range of exchange interactions our results indicate that they are modified by the presence of Mn.

In conclusion, we have shown that in metallic glasses in the presence of Ni and Co atoms, Mn posses a high moment on the order of 3.3 μ_B. For very small Mn concentration which seems to be critical, the Fe and Ni moments increase and particularly that of Ni by almost 100 %. Such a result has not been reported so far to our knowledge. However the beneficial action of Mn is lost if only Fe atoms were to be present. The study of ^{55}Mn nuclear spectrums indicates that upon crystallization new metastable phases are formed which have no equivalent in the usual alloy systems prepared by conventional methods. It is possible that structural relaxation studies on these alloys could help us to understand better the basic mechanisms.

This work is dedicated to the memory of Prof. R. Pauthenet with whom one of us (R.K.) had several interesting discussions. EPMA analyses by Mrs. Rommeluere is gratefully acknowledged.

REFERENCES

1. F.E. Luborsky in Ferromagnetic Materials Vol.1,
   P. 451, Ed. E.P. Wohlfarth, North Holland, Holland 1960.

2. H. Senno, H. Sakakima and E. Hirota
   Sup. to Sci Rep. RITU A 276 1980.

3. Y. Obi, H. Morita and H. Fujimori
   I.E.E.E. Trans. Mag. MAG-16 1132 (1981).

4. V.R.V. Ramanam
   J. Appl. Phys. 53 7822 (1982).

5. M.A. Manheiwar, S.H. Bhagat and H.S. Chen
   J. Appl. Phys. 53 7737 (1982).

6. R. Krishnan, K. Le Dang, P. Veillet and V.R.V. Ramanam
   J. Appl. Phys. 57 1394 (1985).

7. R. Krishnan, P. Rougier, K. Le Dang and P. Veillet
   Proc. ISPMM, Sendaï (world Scientific, Singapore p.387 1987).

8. Y. Kitoaka and K. Asayama
   J. Phys. Soc. Jap. 40 1521 (1976).

9. Anuradha Lagu, S.N. Shingri, A.K. Nigam, Girish Chandra,
   Shiva Prasad and R. Krishnan
   To be published in J. Hyperfine Int. (1988).

10. R. Krishnan, K. Le Dang and P. Veillet
    J. Appl. Phys. 63 2992 (1988).

11. R. Krishnan, K. Le Dang, V.R.V. Ramanan and P. Veillet
    J. Mag. Mag. Mat. 54-57, 263 (1986).

12. R. Krishnan, Shiva Prasad, J. Sztern, C. Battarel
    and R. Morille
    Phys. rev. B21, 1246 (1980).

NEW RESULTS IN THE APPLICATION OF SOFT MAGNETIC METALLIC GLASSES

G. Konczos and T. Tarnóczi

Central Research Institute for Physics,
Hungarian Academy of Sciences
1525 Budapest POB 49, Hungary

ABSTRACT

Recent progress in the application of soft magnetic metallic glasses is reviewed on the basis of papers published since 1986. Special emphasis is paid to the applications in distribution transformers, in electronic components, and in sensors. The developments in related fields (manufacturing, heat treatments and characterization) are also summarized.

1. INTRODUCTION

The study of soft magnetic amorphous alloys is interesting both from a fundamental point of view /1/ and with regard to the possibility of practical use /2/. From time to time, the progress that has been made in applications has been reviewed in invited lectures at conferences /3-8/ and in chapters of monographs.

Based on the scope of the present conference the trends of practical applications are discussed in terms of materials science: this means that the recent progress in the development of materials, technologies and devices is summarized. Only the latest results are reviewed; for the background and earlier results we refer to the literature.

2. MATERIALS

In the last one or two years, no metallic glasses of essentially new composition were developed for soft magnetic applications. Merely the properties of the well-known three families of amorphous magnets:
  i.) iron based alloys with high saturation induction,
 ii.) iron-nickel based alloys with medium saturation induction and higher permeability,
iii.) cobalt based alloys with nearly zero magnetostriction and extremly high permeability
were improved.

The purpose of these works was to improve the magnetic properties, first of all the reduction of disaccommodation and loss, moreover the increase of permeability or decrease of coercive force was also important. Beside these, the main problem is the stability of the properties.

The above mentioned aims could be reached by the addition of different elements to the basic alloy /9-11/ or by different kinds of annealing causing structural relaxation /12-15/ or surface crystallization, too /16/.

3. PREPARATION

Nowadays the majority of soft magnetic amorphous alloys are manufactured in the form of ribbons by planar flow casting, which is one of the versions of the single roller rapid quenching technologies. Besides the improvement of this method, intensive research and development activity can be observed in the field of producing rapidly solidified wires and powders.

For many years the development of planar flow casting has been motivated by the demand for the wider ribbons which are used in distribution transformers. The preparation of metallic glass ribbons up to 150-250 mm width has been realized on an industrial scale by a number of firms (mainly in the USA, Japan and FRG). A charge mass of about 150 kg is typical, and casting equipment of up to 500 kg is under development.

Infrared measurements have given useful information on heat and mass flow and have helped towards a quantitative understanding of the complex phenomena of ribbon formation /17,18/. Hargitai and co-workers have studied the correlations and interdependences among the process parameters and magnetic properties by multiple regression analysis in a metallic glass with nearly zero magnetostriction /19,20/. The interactions of metallic melts and substrates are systematically studied by several authors /21,22/. Östlund and West examined the influence of wheel surface roughness on the microstructure of Fe-Ni-B-Si ribbons /21/. Sato and co-workers have investigated the variation of magnetic properties along the ribbon length /23/ and have made observations on the influence of thickness on the magnetic properties /24/.

Metallic glass ribbons can also be produced by the twin-roller technique /25/. This method enables both the ribbon thickness and surface roughness to be much better controlled than by planar flow casting.

In addition to ribbons, quenched wires may also be of importance for soft magnetic applications /26/, especially for some type of sensors. The Japanese firm UNITIKA Ltd (Uji, Japan) is now capable of producing amorphous wires as long as several kilometers, in a diameter of 80-100 μm.

Increasing interest is being shown in amorphous soft magnetic powders and in their consolidation. Amorphous powders can be produced by a variety of methods. The high-pressure gas atomization process has successfully been applied by Inoue et al. for the production of powders from iron-, cobalt- and palladium- based metallic glasses /27/. Fe-Co-B amorphous powders have been prepared by chemical reduction in aqueous solutions /28,29/. Promising results are described by Minakawa et al. in the warm consolidation of cobalt-based amorphous powders with resins for high frequency applications /30/. Omuro and co-workers describe details on warm consolidation of amorphous Ni-Fe-P-B-Al powders /31/. The preparation of ferromagnetics amorphous powders is now in the development stage but it seems likely that these will find industrial applications in the future.

4. HEAT TREATMENTS

The technically important soft magnetic properties (coercivity, permeability, loss) of metallic glasses are strongly dependent on induced uniaxial anisotropy and magnetoelastic energy.

These factors can be modified by different kinds of annealing, one of the most important being stress relief annealing. Internal stresses originate from rapid quenching itself or by further handling (e.g. winding). The stress relief treatments are carried out below the crystallization temperature. A harmful effect is that the material may become brittle which means a problem for device manufacture and may lead to the devices having a limited life-time.

Various methods are available for avoiding embrittlement. Taub /32/ suggested to perform the stress relief annealing at higher temperatures but for a shorter time under dynamic conditions. In this respect the recent developments in flash annealing (sometimes called "current annealing") seems to be promising. Polish scientists played a pioneering role in the development of this new, non-conventional heat treatment method (see e.g. Jagielinski /33/, Zaluska and Matyja /34/). The effect of rapid heating on the magnetic and thermal properties including crystallization behaviour have been studied systematically by different authors /35-37/. From a practical point of view the work by Yavary and co-workers is especially interesting /38/. Amorphous Fe-Si-B tapes have been rapidly annealed for a short time (typically seconds) at about 800 K by Joule heating. It is found that this type of annealing in air can result in nearly the same effect as obtained by two hours of conventional annealing in a furnace under inert gas or vacuum.

The excellent magnetic properties of cobalt based non-magnetostrictive alloys can further be improved by annealing in a magnetic field. Kohmoto et al. /13/ and Sawa et al. /39/ studied the influence of heat treatment on the high frequency properties of Co-based glasses in a rotating magnetic field and in magnetic field transverse to the ribbon axis, respectively.

The surface crystallization of metallic glasses has intensively been studied /40/. It was found that in several iron based alloys precipitation of a very low volume fraction of α-Fe leads to ferromagnetic domain refinement and thereby to the reduction of anomalous eddy current losses /16/.

Several papers deal with the influence of different surface treatments, e.g. isolating coatings /41/ or oxidations /42/.

5. RELAXATION, AGEING

For the practical application of magnetic materials, one of the most important questions is the stability of magnetic properties. In the case of metallic glasses, beside the magnetic properties ductility is also very important. As it is well known, sometimes amorphous magnetic materials become brittle which makes impossible their application in the majority of cases. That is why fabrication of amorphous magnetic materials with stable and suitable properties is a very delicate problem.

For stabilizing the properties, the only way is the ageing. Due to practical reasons, ageing is generally performed at elevated temperature (to accelerate the processes) and so beside stress relief, some structural relaxation also takes place.

Recently, a very interesting comparision could be made in connection with accelerated ageing behaviour of METGLAS^R 2605 S2 amorphous ribbons (METGLAS^R is the registered trademark of ALLIED-SIGNAL, Inc. for amorphous alloys of metals). A French group /43/ found that ageing of this material in silicon oil at 200 ^{o}C for 2000 hours gives perfectly stable properties if there is no mechanical stress during annealing, while a Japanese group /44/ obtained the optimal annealing procedure of isothermal heat treatment at 400 ^{o}C for 20 minutes. This annealing leads to a complete stress relief and, at the same time, structural relaxation is not developed to such an extent which could cause any significant embrittlement.

In amorphous powders, there is a possibility to recover ductility of a previously embrittled material. This procedure consists of further pulverisation associated with adiabatic heating and it re-introduces excess free volume annealed out before /45/.

6. METHODS OF QUALITY CONTROL

The demand for practical applications gave an impetus to improvement of magnetic measurements. Only some of them are shortly mentioned here.

The applications of non-magnetostrictive metallic glasses needed the determination of very low magnetostriction coefficients. As a consequence, Narita's method was introduced in many laboratories.

The measurement of power losses cannot be realized easily by the standard Epstein frame method. Therefore the interpretation /46/ and determination of different components of power losses became actual, even at relative high frequencies /47,48/. Sasaki et al. studied the power losses in Fe-B-Si metallic glasses under various stresses at 50 Hz frequency /49/.

Rabinkin /50/ compared three different methods:
a) thermogravimetry with magnetic field (force balance)
b) differential scanning calorimetry (DSC)
c) magnetic susceptibility ($\chi(T)$)
for determining the Curie temperature T_c of the following METGLAS^R alloys: 2605S-2, 2605SC, 2605S-3A, 2705M, 2826MB and 2605CO.

It was found that in the case of five compositions all three methods gave the same result if an appropriate point of registered curve was chosen for establishing the Curie temperature. These points were: a) extrapolated temperature to zero force, b) place of the minimum, c) end of the "tail" on $\chi(T)$ curve for the above methods, respectively.

In the case of 2605CO, DSC was not able to detect any magnetic transformation. The other two methods gave the same result using standard extrapolation procedure due to the very high T_c value.

For a basic understanding of soft magnetic properties the knowledge of domain structure is necessary /51,52/. Recently one can observe an essential improvement in methods used for domain structure observation e.g. by Kerr technique /53/ and by scanning electron microscopy /54/. Saenz et al. developed a new method of observing magnetic domain structure based on the idea of measuring magnetic forces with an atomic force microscope (AFM) /55/.

7. DEVICES

The most important fields of application of soft magnetic amorphous alloys have been established in the last decade.
These are as follows:
- power electrical devices (mainly distribution transformers)
- inductive components used in electronics

- sensors.

Many review papers have been published in the last years on the soft magnetic applications of metallic glasses /56-60/.

7.1 Distribution transformers

One of the most important fields of soft magnetic applications is that of distribution transformers. Energy saving is achieved here by the lower non-load losses.

Iron-based amorphous ribbons - very frequently METGLAS^R 2605 S-2 alloys - are used in this type of transformer. Various technical problems arise by substituting the polycrystalline Fe-Si sheets for metallic glasses. These problems are due to the high magnetostriction coefficient, the mechanical hardness and the embrittlement after stress relief annealing /61/. As the amorphous cores are more "sensitive" than the traditional ones, unusual transformer design and manufacture is necessary. In several countries manufacturers have managed to solve this problem.

In the USA one thousand 25 kVA single phase amorphous core distribution transformers have been built by the General Electric Company for use under normal operating conditions; the installation of a production line has been completed by Westinghouse Electric Corporation. The Electric Power Research Institute (EPRI) and the Empire State Electric Energy Research Corporation (ESEER), in cooperation with manufacturers, organized a standard test program. After nearly two years of continuous operation no significant change in power loss was found /62/. Recently Whitley gave detailed information on the tests /63/. It has been confirmed that the use of METGLAS^R 2605 S-2 ribbons allows an essential reduction in no-load losses. As a new development a three phase stacked core transformer has been described by Nathasingh and Liebermann /64/. The transformer core is made of POWERCORE^R consolidated amorphous alloy strips (POWERCORE^R is the registered trademark of ALLIED-SIGNAL, Inc., for consolidated strips of amorphous ribbons).

Systematic studies have been reported by research workers of the Institut de Recherche d' Hydro-Québec (IREQ), Québec, Canada /65/. On the basis of comparative performance evaluation of different types of amorphous distribution transformers a new design was suggested. The experimental 25 kVA single phase, dry transformer contains about 76 kg amorphous metal; the core loss is about 18.5 W at 1.3 T /66/.

7.2 Inductive components for electronics

The useful soft magnetic properties of metallic glasses (low coercive field, high remanence ratio, low magnetostriction) are exploited in various inductive components. By appropriate choice of composition, manufacture and heat treatment, different types of hysteresis curves can be obtained. The most important aspects of material selection are discussed by Warlimont /60/, Boll and Hilzinger /57/.

In many cases metallic glasses are competitive with the traditional soft magnetic materials (ferrites, permalloys and others), especially in high frequency applications up to 100 kHz. A typical field of application is in switched-mode power supplies where amorphous cores can be used in various types of components, e.g. in inverter transformers, saturable core reactors, current compensated chokes and spike killers. Amorphous soft magnetic alloys very often substitute the existing (traditional) materials. The present situation is characterized more by the improvement of material properties than by the introduction of new technical solutions. Recent developments have led to a further decrease of losses at high frequencies and to the frequency range being extended beyond 100 kHz.

The firm VACUUMSCHMELZE GmbH (Hanau, FRG) achieved an improvement in the high frequency behaviour of cobalt based alloys by modifying the ribbon composition and reducing its thickness. In the new type of VITROVAC^R 6510 ribbons the saturation induction (B_s) and the Curie temperature are higher than in the earlier VITROVAC^R types (VITROVAC^R is the registered trademark of VACUUMSCHMELZE GmbH for amorphous alloys). The reduction in power losses is advantageous in switched-mode power supplies. Yamauchi and co-workers tailored the magnetic properties of a cobalt-rich amorphous alloy to meet the requirements of saturable cores. They investigated the effects of induced magnetic anisotropy on the remanence ratio, the core loss and on ageing. On the basis of this study a new type of material - HITACHI type ACO-5SH - has been elaborated for saturable cores /67/.

As an example of new design the so called "cloth inductor" is mentioned. It consists of a network of amorphous and conductive fibers and it makes possible the miniaturization of some inductive components /68/.

Further applications of soft magnetic metallic glasses (magnetic heads, magnetic shields, high harmonics generators) are cited in the references /56,57,59,60/. Amorphous alloys are commercially available for these applications in the form of ribbons or wires or are available as ready-made components.

7.3 Sensors

Currently, interest is rapidly increasing towards different types of sensors from the aspect of mechatronics, robotics and other branches of industry /69/. High sensitivity, independence on environment (temperature, humidity), robustness, small size: these are only some of the strict requirements against these devices. The special magnetostrictive behaviour (either very low or very high λ_s), the high tensile stress simultaneously with good soft magnetic properties and the corrosion resistance combine to make the metallic glasses possible candidates for sensor materials. Several types of amorphous sensors are already commercially available but the majority of them are now under development. The principle and the practical use of amorphous sensors are summarized by Mohri /70/. Here, only the latest results are mentioned.

The amorphous ferromagnetic sensors can be divided into two main categories: magnetometers using zero-magnetostrictive (cobalt-rich) alloys, and stress sensors containing high magnetostrictive (iron-based) alloys /59/.

7.3.1 Magnetometers. The bias field dependence of reversible permeability is often utilized in magnetic field sensors. There is a wide possibility to use amorphous materials either as toroidal cores or in a "rod" like arrangement to detect magnetic field, torque, replacement or temperature.

A new type of torque sensor has been described by Mohri and co-workers: one pair or twin pairs of amorphous star-shaped cores are combined with one pair of multi-pole ring magnets. This arrangement can detect a twisted angle as small as $0.002°$ for static shafts and $0.01°$ for rotating shafts /71/.

A similar amorphous star-shaped core has been developed for an accurate mechanocardiogram sensor; a magnet displacement from 0.02 μm to 15 mm can be detected by this device /72/

7.3.2 Stress sensors. Stress sensors utilize the stress dependence of magnetic properties in high magnetostrictive materials. The amorphous alloys can essentially surpass the crystalline ones because they have a higher elastic limit.

Wun-Fogle et al. measured the change of permeability under bending /73/. In such experiments good contact is necessary between the sensitive amorphous ribbon and the substrate. Recently highly viscous liquids have been used to solve this problem /74/. Mermelstein and co-workers elaborated a new type of magneto-elastic sensor for detecting low-frequency magnetic fields. The principle of operation is based on the field-dependent coupling between the magnetization and strain modes /75/.

Some of the high magnetostrictive amorphous ribbons or wires exhibit large Barkhausen discontinuities. Such materials can be used as bistable magnetic switching transducers in high resolution rotary encoders and high harmonics generators.

Bistable hysteresis loops have been produced by Mälmhäll et al. in cold-drawn amorphous Fe-Si-B wires (with positive magnetostriction) after flash current annealing under applied axial tension /76/. The same authors investigated the role of magnetostrictive anisotropy in the case of Co-Si-B wires with negative magnetostriction, too /77/.

Another type of stress sensor utilizes the ultrasonic propagation effect. Squire and Gibbs describe a new method of measuring low magnetic fields based on changes in the speed of acoustic waves in an amorphous ribbon due to a magnetic field-induced change in the elastic modulus /78/. Murakami and co-workers have developed an amorphous magnetostrictive delayline cordless digitizer /79/. The behaviour of a half-wave ultrasonic transducer has been investigated by Kaczkowski /80/.

CONCLUSIONS

More than ten years ago a number of papers predicted the rapid spreading of soft magnetic amorphous alloys. It is now clear that these opinions were too optimistic: market penetration has taken longer. The commercialization of a new material is a complex process in which not only the technical characteristics of the materials but other factors also have to be taken into account. These factors include the need for new manufacturing and design, the improvement of competitive materials as well as economic considerations. The cited papers demonstrate that soft magnetic metallic glasses offer a wide range of possibilities for newer and newer practical applications.

REFERENCES

/1/ O'Handley, R.C., J. Appl. Phys. $\underline{62}$, R15 (1987).

/2/ Froes, F.H., Carbonara, R., J. Metals $\underline{40/2/}$, 20 (1987).

/3/ Matyja, H., Zielinsky, P.G. (eds.)., Amorphous Metals (Proc. Summer School, Wilga, Poland, 1985), World Scientific Publ. Co., Singapore, 1985

/4/ Haasen, P., Jaffee, R.I. (eds.), Amorphous Metals and Semiconductors (Proc. International Workshop, Coronado, USA, 1985), Pergamon Press, Oxford, New York, Beijing, 1986.

/5/ Bhatnagar, A.K. (ed.), Metallic and Semiconducting Glasses (Proc. International Conf., Hyderabad, India 1986) in Key Engineering Materials $\underline{13-15}$, (1987).

/6/ Hernando, A., Madurga, V., Sánchez-Trujillo, M.C., Vázquez, M., (eds.)., Magnetic Properties of Amorphous Metals (Proc. Symp., Benalmádena, Spain, 1987), North-Holland, Amsterdam, Oxford, New York, Tokyo, 1987.

/7/ Rapidly Quenched Metals (Proc. Sixth International Conf., Montréal, Canada, 1987), in: Mater. Sci. Engr. $\underline{97-99}$, (1988).

/8/ Soft Magnetic Materials (Proc. Eight International Conf., Badgastein, Austria, 1987) to be publ. in Physica Scripta (1988).

/9/ Barrue, R., Bigot, J., Faugierres, J.C., Perron, J.C., Railland, J.F., Robert, J.,Schwartz, F., Physica Scripta $\underline{37}$, 356 (1988).

/10/ Grössinger, R., Sassik, H., Wezulek, R., Tarnóczi, T.,Paper no.3P K-7, presented at the International Conference on Magnetism, Paris, France, July 1988, to be publ. in J. de Physique Coll.

/11/ Schwartz, F., Bigot, J., Mater Sci. Engr. 99, 39 (1988).

/12/ Buttino, G., Cecchetti, a., Poppi, M., Zini, G., Physica Scripta 35, 721 (1987).

/13/ Kohmoto, O., Fujishima, H. Shibata, K., Mater. Sci. Engr. 99, 53 (1988).

/14/ Guo, H.-Q., Kronmüller, H., Moser, N., ibid. 97, 519 (1988).

/15/ Komatsu, T., Seiwa, A., Matusita, K.,J. Mater. Sci. 23 687 (1988).

/16/ Hilzinger, H.R., Herzer, G., Mater. Sci. Engr. 99, 101 (1988).

/17/ Stephani. G., Mühlbach, H., Fiedler, H., Richter, G., ibid. 98, 29 (1988).

/18/ Vogt, E., Frommeyer, G., Z. Metallkde 78, 262 (1987).

/19/ Nagy, I., Hargitai, C., Kopasz C., Key Engineering Materials 13-15, 837 (1987).

/20/ Hargitai, C., Kopasz, C., Németh, S., Albert, B., Mater. Sci. Engr. 99, 81 (1988).

/21/ Östlund, A., West, R., Internat. J. Rapid Solid. 3, 177 (1988).

/22/ Adler, R.P.I., Hsu, S.C., J. Mater. Sci. 23, 25 (1988).

/23/ Sato, J., Fujine, T., Miyazaki, T., J. Magn. Magn. Mater. 71, 255 (1988).

/24/ Sato, T., Otake, H., Miyazaki,T., ibid. 71, 263 (1988).

/25/ Shibuya, K., Kogiku, F., Yukumoto, M., Miyake, S., Ozawa, M., Kan, T., Mater. Sci. Engr. 98, 25 (1988).

/26/ Liu, J., Arnberg, L., Bäckström, N., Savage, S., ibid. 98, 21 (1988).

/27/ Inoue, A., Masumoto, T., Ekimoto, T., Furukawa, S., Kuroda, Y., Chen, H.S., Metall. Trans. 19A, 235 (1988).

/28/ Dragieva, I., Buchov, D., Mehandjiev, D., Slacheva, M., J. Magn. Magn. Mater. 72, 109 (1988).

/29/ Corrias, A., Ennas, G., Licheri, G., Marongiu, G., Musinu, A., Paschina, G., Piccaluga, G., Pinna, G., J. Mater. Sci. Lett. 7, 407 (1988).

/30/ Minakawa, S., Masumoto, T., IEEE Trans. Magn. MAG-23, 3245 (1987).

/31/ Omuro, K., Mura, H. Isa, S., Ikuta, K., Mater. Sci. Engr. 98, 399 (1988).

/32/ Taub, A.J., IEEE Trans. Magn. MAG-20, 564 (1984).

/33/ Jagielinski, T., ibid. MAG-19, 1925 (1983).

/34/ Zaluska, A., Matyja, H., Mater. Sci. Engr. 89, L11 (1987).

/35/ Barandiarán, J.M., Hernando, A., Nielsen, O.V., IEEE Trans. Magn. MAG-22, 1864 (1986).

/36/ González, J., Vázquez, M., Barandiarán, J.M., Madurga, V., Hernando, A., J. Magn. Magn. Mater. 68, 151 (1987).

/37/ Lovas, A., Potocky, L., Cziráky, Á., Kisdi-Koszó, É., Zsoldos, É., Pogány, L., Novák, L., Z. für Phys. Chem. Neue Folge 157, 371 (1988).

/38/ Yavari, A.R., Barrue, R., Harmelin, M., Perron, J.C., J. Magn. Magn. Mater. 69, 43 (1987).

/39/ Sawa, T., Hashimoto, S., Inomata, K., IEEE Trans. Magn. MAG-23, 3509 (1987).

/40/ Köster, V., Mater. Sci. Engr. 97, 233 (1988).

/41/ Okazaki, Y., Kanno, H., Kousaka, S., Sakuma, E., IEEE Trans. Magn. MAG-23, 3515 (1987).

/42/ Huang, G.X. Shi, S.Y., Xu, Q.Z. Yang, G.B., ibid. MAG-23, 3596 (1987).

/43/ Kedous-Lebouc, A., Brissoneau, P., Paper no.3.22, to be publ. in Ref. /8/

/44/ Sasaki, T., Hosokawa, T., Takada, S., Paper no.3.15, to be publ. in Ref. /8/

/45/ Small, C.J., Davies, H.A., Mater. Sci. Engr. 97, 457 (1988).

/46/ Bertotti, G., IEEE Trans. Magn. MAG-24, 621 (1988).

/47/ Yamaguchi, M., Murakami, K., ibid. MAG-22, 958 (1986).

/48/ Nakajima, S., Yoshizawa, Y., Yamauchi, K., Matsumoto, Y., ibid. MAG-23, 3272 (1987).

/49/ Sasaki, T., Shimomura, E., Yamada, K., ibid. MAG-23, 3587 (1987).

/50/ Rabinkin, A., ibid. MAG-23, 3874 (1987).

/51/ Zhao-Hua, L., IEEE Trans.Magn. MAG-23, 2990 (1987).

/52/ Smith, R.H., Jones, G.A., Lord, D.G., ibid. 24, 1868 (1988).

/53/ Rawe, W., Schäffer, R., Hubert, A., J. Magn. Magn. Mater. **66**, 31 (1987).

/54/ Záveta, K., Jurek, K., Duhaj, P., Czech. J. Phys. **B37**, 42 (1987).

/55/ Sáenz, J.J., Garcia, N., Gruütter, P., Meyer, E., Heinzelmann, H., Wiesendanger, R., Rosenthaler, L., Hibder, H.R., Güntherodt, H.J., in Ref. 6 p.83

/56/ Lachowicz, H.K., in Ref.3 pp. 313-339

/57/ Boll, R., Hilzinger, H.R., Elektronik 99 (Oct. 1987).

/58/ Ramanan, V.R.V., Smith, C.H., Fish, G.E., in Ref. 4 pp. 849-861

/59/ Konczos, G., Kisdi-Koszó, É., Lovas, A., Report KFKI-1987-54/E pp.1-23, to be publ. in Ref.8

/60/ Warlimont, H., Mater. Sci. Engr. **99**, 1 (1988).

/61/ Davies, L.A., Ramanan, V.R.V., in Ref. 4. pp. 733-748

/62/ Bailey, D.J., Lowdermilk, L.A., Lee, A.C., J. Magn. Magn. Mater. **54-57**, 1618 (1986).

/63/ Whitley, D.W., IEEE Trans. on Power Delivery **PWRD-2**, 827 (1987).

/64/ Nathasing, D.M., Liebermann, H.H., ibid. **PWRD-2**, 843 (1987).

/65/ Alexandrov, N., Schulz, R., Roberge, R., ibid. **PWRD-2**, 420 (1987).

/66/ Schulz, R. Chretien, N., Alexandrov, N., Aubin, J., Roberge, R., Mater. Sci. Engr. **99**, 19 (1988).

/67/ Yamauchi, K., Yoshizawa, Y., Nakajima, S., Mater. Sci. Engr. **99**, 95 (1988).

/68/ Matsuki, H., Miyazawa, H., Murakami, K., Yamamoto, T., J. Appl. Phys. 63, 3394 (1988).

/69/ Jones, B.E., J. Phys. E. 20, 1113 (1987).

/70/ Mohri, K., in Ref. 6 pp.360-366

/71/ Mohri, K., Mukai, Y., Yasuda, K., Takayama, K., IEEE Trans. Magn. MAG-23, 2191 (1987).

/72/ Mohri, K. Jinnouchi, T., Kawano, K., IEEE Trans. Magn. Magn. MAG-29, 2212 (1987).

/73/ Wun-Fogle, M., Savage, H.T., Clark, A.E., Sensors aand Actuators 12, 323 (1987).

/74/ Mermelstein, M.D., Alskin, C., Dandridge, A., Electronics Letter 23, 280 (1987).

/75/ Mermelstein, M.D., Dandridge, A., Appl. Phys. Lett. 51, 545 (1987).

/76/ Malmhäll, R., Mohri, K., Humprey, F.B., Manabe, T., Kawamura, H., Yamasaki, I., in Ref. 6 pp.89-91

/77/ Yamasaki, J., Humphrey, F.B., Mohri, K., Kawamura, H., Takamure, H., Mälmhäll, R., J. Appl. Phys. 63, 3949 (1988).

/78/ Squire, P.T., Gibbs, M.R.I. IEEE Trans. Magn. 24, 1755 (1988).

/79/ Murakami, A., Hosaka, K., Fukushima, M., Maeda, M., Tsuya, N., ibid. 24, 1758 (1988).

/80/ Kaczkowski, Z., ibid. 24 1990 (1988).

APPLICATION OF MAGNETICALLY SOFT AMORPHOUS ALLOYS IN ELECTRONIC AND ELECTRICAL DEVICES

Ying-Shan Yang*

Central Iron and Steel Research Institute, Beijing, China

ABSTRACT

The unique characteristics of magnetically soft amorphous alloys provide a favourable condition to make the electronic and electrical devices. Many examples such as switching power supply, pulse transformer, 400Hz transformer, middle-frequency transformer, magnetic shielding, magnetic recording head, the earth leakage circuit breaker etc are presented.

1. INTRODUCTION

Amorphous alloys produced by melt-spun technology exhibit excellent soft magnetic properties. The magnetostriction of Co-based amorphous alloys tends to zero. There is not apparent magneto-crystalline anisotropy in the as-quenching samples. Comparing with the conventional soft magnetic alloys such as Si-Fe steel, Fe-Ni alloys and ferrites, the amorphous alloys (Fe-based, Co-based and FeNi-based) have many unique characteristics: very low coercive force Hc, very high DC-permeability μ_- and AC-permeability $\mu_\sim$; extremely low magnetic losses, especially in the 50Hz-1MHz region; relatively high resistivity, hardness and chemical corrosion resistance.

Amorphous alloys could be used as soft magnetic materials,

---

* on visit: PROCESSI PREPARATIVI MATERIALI AMORFI, TIB-CHIA, ENEA, C.R.E. CASACCIA, S.P. Anguillarese, 301, ROMA A.D., ITALY

ductile brazing ribbon and corrosion-resisting materials. The great quantity of amorphous alloys is applied as soft magnetic materials.

Although amorphous alloys have many advantages, when they are used for manufacturing the distribution transformer, some advantages could be turned into faults. For example, the thin thickness is a advantage for the high frequency application but for the 50-60Hz distribution transformer it is a obvious fault, the packing coefficient could be decreased. The competitive object of amorphous alloys is Si-Fe steel. In the respects of price, magnetic flux density Bs, tape thickness and annealed toughness, Si-Fe steel possesses obvious superiority. Just a important advantage, extremely low iron loss, belongs to amorphous alloys. So in the market, there will be a long period competition between amorphous alloy core transformer and Si-Fe core transformer. Here, the application of amorphous alloy on distribution transformer will not be described in detail.

The electronic and electrical devices were designed in accordance with the magnetic properties of amorphous alloys. Some typical parameters of amorphous alloys and crystalline alloys for contrast are shown in Table 1.

Application of magnetic amorphous alloys usually may be divided into two groups: those for transfering the energy(current) and those for transfering the information(voltage). For the former, the high Bs materials such as Fe-based amorphous alloys are chosen. But when the working frequency is relatively high and transfered power is not very large, the Co-based amorphous alloys are usually chosen. For the transfering of information, the high AC-permeability and low Hc are necessary. So Co-based amorphous alloys are usually chosen. Many application examples of electronic and electrical devices are presented. They are switching power supply, pulse transformer, 400Hz transformer, middle-frequency transformer, magnetic shielding, magnetic

Table 1 The soft magnetic properties in amorphous alloys (Fe-, Co- and FeNi-based)
(Contrast with crystalline alloys)

Trade-mark	Composition	Bs(G)	Br/Bm	Hc(Oe)	$\mu_m(\times 10^4)$	$\mu \sim$	$\lambda_s \times 10^6$	Iron Loss(W/Kg)	Country
METGLAS 2605S2	$Fe_{78}B_{13}Si_9$	15600	0.83	0.03	30	$\mu_1=13000$ 60Hz,100G	27	P10/60=0.10 P14/60=0.22	USA
FJ-301Z	$Fe_{79}B_{15.5}Si_{3.5}C_2$	16100	0.92	0.03	31	$\mu_1=45000$ 60Hz,1T		P10/60=0.11 P10/1K=4.7	China
3.2Si-Fe(Crys.)		20000	0.72	0.1	4		10	P13/60=1.5	
METGLAS 2826	$Fe_{40}Ni_{40}P_{14}B_6$	7800	0.83	0.006	88-110		11	P4/1K=0.8	USA
Vitrovac 4040Z	$Fe_{40}Ni_{40}(MoSiB)_{20}$	8000	0.80	0.013	50		8		FRG
FJ-111	$Fe_{40}Ni_{40}P_{12}B_8$	9000	0.85	0.04	50			P4/10K=33.6	China
Amomet B	$Fe_5Co_{70}Si_{15}B_{10}$	8400		0.002	120		±0.1		Japan
Vitrovac Z	$Fe_4Co_{66}(MoSiB)_{30}$	5500	0.80	0.004	100		0.5	P2/20K=10	FRG
FJ-103	$Fe_4Co_{64}V_2B_{22}Si_8$	6940	0.95	0.003	165	$|\mu|=35900$ 20KHz,0.5T		P5/20K=33	China
1J79(Crys.)	$Fe-Ni_{79}-Mo_4$	7500		0.015	20			P2/100K=32.5	China
Sendust	$Fe-Si_{9.5}-Al_{5.6}$	10000		0.05		$\mu_1=3\times10^4$ 1KHz,H=3mOe			
Mn-Zn Ferrite		4000		0.02		$\mu_1=2\times10^4$ 1KHz,H=3mOe	14		

recording head, the earth leakage circuit breaker etc. Most of the examples are from China.

2. SWITCHING POWER SUPPLY

With the miniaturization of electronic elements (such as transistor), the contradiction of size and weight between the transistors and 50-60Hz power transformer is becoming increasingly acute. Because of the improvement of the high frequency high power transistor, the 50-60 Hz power transformer is abandoned in switching power supply. But the high frequency transformer is still necessary. We know that the weight and the size of transformer are decreased with the increasing of working frequency. Now the woring frequencies are 10KHz-1MHz. Since the working frequency is very high, the ferrite conventially is used. The Bs and permeability of amorphous alloys are better than that of ferrites. A contrast between the ferrite and amorphous alloy on 10KHz switching power supply from USA is shown in Table 2.

Table 2   10KHz switching power supply

	Total volume (cm^3)	Total weight (g)	Working point (G)	Core loss (100°C)	Exciting power (100°C)
Amorphous alloy	200	520	5500	5.4W	7.6VA
Ferrite	300	1000	2500	6.3W	72.7VA

A 5KVA switching power supply was made in China. The specification of this power supply is shown in Table 3.

It was a switching power supply with very high power. A same size grain-oriented silicon iron steel (thickness 0.03mm) core was made for the same 5KVA switching power supply. Under the same test condition, the Si-Fe core worked for 11 minutes, the core temperature rised from 28°C to 100°C and continued rising up; the amorphous alloy core worked for 90 minutes, the core temperature kept at 100°C steadyly. In fact, the Si-Fe core could not be used. The comparison of switch-

Table 3   Specification of the 5KVA switching power supply

Input voltage	380V/50Hz, 3 phases
D.C. output voltage	40KV
D.C. output current	130mA
Working frequency	10KHz
Instability of voltage	0.1%
Efficiency	86%

ing power supply between the 5KVA made in China and the 1KVA made in USA is shown in Table 4.

Table 4   The comparison of switching power supply

	Material	Freq. (KHz)	Core weight (g)	Working point (B)	Output (VA)	Effic.	T. rise (°C)	Packing coefficient
USA	2605 S-3	20	118	4200	1000	0.81	70	0.7(Toroid)
CN	FJ-301Z	10	950	3500	5000	0.89	70 (load) 36 (unload)	0.61(Track type)

CN--China; Effic.--Efficiency; T.--Temperature; Freq.--Frequency

The performance mode of switching power supply is divided into "single direction" and "double direction". For the latter, the high remanence Br is necessary. The 5KVA supply above was a double directions mode. For the former, the electronic circuit is simple than that of double directions mode and the low Br is necessary. The low Br can be obtained by the transverse magnetic field annealing. Many switching power supplies are worked at the single direction mode(SDM). A SDM switching power supply(5V,30A,f=20KHz) with amorphous core was made in China. Before using the amorphous alloy to make core, the conventional material was Fe-Ni crystalline alloy (1J67h), it was the best alloy for this kind of device. Because the losses of Fe-based amorphous alloy (FJ-301HA) is obviously lower than that of 1J67h, for the same output power (5V,30A), the amorphous alloy core could be smaller and lighter. The losses of the two kinds of alloys is shown in Table 5. The comparison of the core and transformer is shown in Table 6.

Table 5   The losses(W/Kg) of FJ-301HA amorphous alloy and 1J67h(cry.)

	P5/20K	P7/20K	P8/20K	P9/20K	P10/20K
FJ-301HA	72	162	220	290	389
1J67h	131	233	294	359	434

Table 6   The comparison of the core and transformer (5V,30A,20KHz)

	core			transformer	
	volume (cm^3)	weight (g)	output/weight (VA/g)	volume (cm^3)	weight (g)
FJ-301HA	12.6 (ϕ25/40X12.5)	49.5	3.04	75.2	157.8
1J67h	19.6 (ϕ40/50X15)	66.5	2.26	146.5	293.5
AM/1J67h	64.1%	74.3%	134%	51.3%	57.2%

Table 6   (continued)

	temperature rise of core	temperature rise of coil
FJ-301HA	42.5°C	33.5°C
1J67h	67.5°C	57.5°C

3. PULSE TRANSFORMER

Pulse transformer is used for transfering pulse power. The pulse waveform is shown in Fig.1. The ideal waveform is a square shape. The deform is demonstrated by rise time t_r and top drop Ud. We assume that the waveform before transform is a ideal square wave. After transform the waveform changes to the shape as shown in Fig.1. Of course, the more less of the t_r and Ud the better.

$$t_r = K_t \left(\frac{U_p \tau_p}{\Delta B} \right)^2 \qquad (1)$$

$$Ud = K_d \frac{\tau_p}{\mu_p} \qquad (2)$$

ΔB—the change of B; μ_p—pulse permeability. It is very clear, the material which possesses high ΔB and μ_p is the best. Amorphous alloy possesses the good characters. The comparison of the pulse characters between the amorphous and crystalline alloys is shown in Table 7.

A pulse transformer with Fe-based amorphous alloy core was made in China.

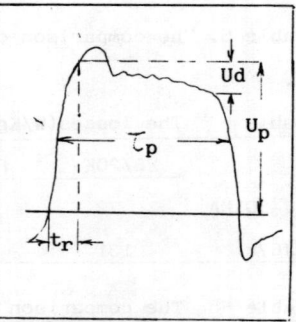

Fig.1 Pulse waveform

Table 7 The comparison of pulse characters (τ_p=3µsec, f=300Hz)

		pulse	characters	
Fe-based amorphous alloy	ΔB	3000	6000	9600
	μ_p	6580	4720	2300
Si-Fe steel(0.03mm)	ΔB	3000	6300	9600
	μ_p	1200	1320	1280
Permalloy (1J67h)	ΔB	3000	6600	9100
	μ_p	2140	2230	1290

It transformed a τ=0.25µsec square pulse wave, its t_r=0.035µs and Ud< 2%. It was much better than that with the crystalline alloys such as Si-Fe and Fe-Ni alloys.

There are two kinds of magnetizing manner for the pulse transformer. One is single direction magnetizing, there is not bias magnetic field supplied by the circuit. The low Br is necessary. The above pulse transformer belonged to this. Another is double direction magnetizing, there is a bias magnetic field supplied by the circuit. The material is magnetized from -Bm to +Bm. Certainly the ΔB and transfer power are larger than that of single direction manner.

A double direction manner pulse transformer with a track type core was made in China. The pulse permeability is the function of core form, τ_p and ΔB. The double direction pulse transformer needs the high Br material. The Fe-based FJ-301ZA amorphous alloy was used for this manner. The typical magnetic parameters were shown in Table 8. For the track type core, the best pulse permeability ($\tau_p=3\mu s$) were 8460(B=1.5T) and 6040(B=1.8T). The parameters(keeping the same rise time t_r) of pulse transformers with amorphous track core and Si-Fe track core were shown in Table 9.

Table 8 The typical magnetic parameters of FJ-301ZA amorphous alloy (contrast with Si-Fe steel)

Core form	DC magnetic properties				
	$B_{10}(T)$	$Br(T)$	Br/B_{10}	$Hc(Oe)$	$\mu_m(10^4)$
FJ-301ZA Toroid	1.57	1.47	0.936	0.078	15.5
FJ-301ZA Track type	1.50	1.12	0.75	0.09	4.1

B_{10}--Induction at H=10 Oe; μ_m--maximal permeability.

Table 8 (continued)

Core form	μ_p ($\tau_p=3\mu s$)				μ_p ($\tau_p=10\mu s$)	
	B=1T	1.5T	1.8T	2.0T	B=1.5T	1.8T
FJ-301ZA Toroid	11000	10000		8900		
FJ-301ZA Track type		5000	4000		9710	7300
Si-Fe(0.05mm) Track type		2630	2570			

Table 9 The comparison of the pulse transformer(amorphous & Si-Fe)

	t_r (μs)	Ud (%)	Temperature rise (°C)
Amorphous alloy FJ-301ZA	1.0	1.4	21.5
Si-Fe steel (0.05mm)	1.0	3.5	42.0

4. 400Hz TRANSFORMER

400Hz transformer was used extensively in the electronic and aero fields. The conventional core was made of grain-oriented Si-Fe steel(ordinarily 0.2mm thickness). Since the thickness and the resistivity(50$\mu\Omega$-cm) of Si-Fe steel are relatively thicker and lower, the temperature rise at 400Hz is relatively higher. The more the power of transformer the more the temperature rise. Sometime for decreasing the temperature rise, the working induction value B was decreased. For the 400Hz Si-Fe steel transformer, when the power is

100VA, the B is chosen at 1.5T; 400VA, it is chosen at 1.2T; more than 1KVA, it is chosen at 0.825T.

The resistivity(125μΩ-cm) and tape thickness(35μm) of Fe-based amorphous alloy are higher and thiner than that of Si-Fe steel. Therefore it is more suitable to use for the 400Hz application. Since the Bs of amorphous alloys is less than that of Si-Fe steel, its working point could not be chosen at 1.5T. For the low power(100VA) 400Hz transformer, its weight with amorphous alloy core instead of Si-Fe steel can not be reduced obviously. When the power is more than 1KVA, the working point of Si-Fe steel has to be chosen at 0.825T, however the amorphous alloy can work at 1.2T. Therefore all of the advantages in amorphous alloy in respects of weight, volume, exciting power and temperature rise are demonstrated evidently.

A 3KVA 400Hz transformer was made in China. The Fe-based amorphous(Fe-B-Si) alloy(FJ-302Z) was used in the transformer. The core form was track type. The weight of core was 3443g. The efficiency of transformer was 98.5%. The comparison of DC magnetic properties and iron losses between the toroid core and track type core is shown in Table 10 and Table 11.

Table 10 The DC magnetic properties of toroid core & track type core

	$B_{10}(T)$	$Br(T)$	$Hc(Oe)$	$\mu_m(10^4)$
Toroid core	1.53	1.25	0.051	19
Track type core	1.53	1.33	0.15	6

Table 11 The iron losses of toroid core and track type core(f=400Hz)

	Bm (T)	0.5	0.6	0.7	0.8	0.9	1.0	1.1	1.2	1.3
Toroid	Loss(W/Kg)							1.07	1.59	
Toroid	Exciting power (VA/Kg)							1.36	2.17	
Track	Loss(W/Kg)	0.35	0.61	0.83	0.95	1.37	1.64	2.01	2.47	2.86
Track	Exciting power (VA/Kg)	0.80	1.12	1.34	1.69	2.02	2.56	3.17	4.1	5.50

The comparison of 400Hz transformers with amorphous alloy core and Si-Fe steel(0.2mm) core is shown in Table 12.

Table 12 The comparison of 400Hz transformer

	Work point (T)	Output (W)	Core W. (Kg)	Exci. power (VA)	Iron loss (W)	Copper weight (Kg)	Cu loss (W)	C.T. rise (°C)	W.T. rise (°C)	Volume (cm^3)
Amorphous	1.2	3000	3.44	7.1	13	2.75	35	24	30	2520
Si-Fe	0.74	3000	6.38	125	40.5	3.1	37.5	50	48	3018
AM/Si-Fe	1.34	1	0.54	0.14	0.32	0.81	0.93	0.48	0.62	0.83

Core W.--Core weight; Exci.power--Exciting power; Cu loss--Copper loss C.T. rise--Core temperature rise; W.T. rise--Wire temperature rise.

5. MIDDLE-FREQUENCY TRANSFORMER

The sort of middle-frequency transformer worked at 1-10KHz requires advanced material with low magnetic loss.

A 8KHz transformer as a water quenching transformer was used to crankshaft quenching hardening machine. Since the transformer must be rotated around the crankshaft, its shape ought to be flat and weight to be light.

The conventional middle-frequency transformer is made of Si-Fe sheet. The core weight of a 500KVA transformer was about 45Kg. For cooling the Si-Fe sheets, some copper plates and copper water cases were required. Therefore the weight of transformer was raised to 85Kg. It was too heavy to use to crankshaft quenching hardening machine.

A 500KVA,8KHz transformer with amorphous alloy core was made for this purpose in China. For comparison, a 500KVA, 8KHz transformer with ferrite core (core weight and copper weight were the same) was made.

The parameters of 8KHz transformers with amorphous core and ferrite core are shown in Table 13.

It is obvious that the amorphous alloy transformer is better than that of ferrite.

We have to point out that the pressing of a fixing plate against

Table 13 The comparison between amorphous core and ferrite core(f=8KHz)

Material	Core weight (Kg)	Density (g/cm^3)	Copper weight (Kg)	Work point (T)	Turn ratio	Iron loss (W)	Exciting current (A)	Remarks
Ferrite	10	4.8	10	0.409	8:1	500 750	63 74	Initial After working 10 minutes
Amorphous alloy	10	7.3	10	0.547 0.608 0.684	10:1 9:1 8:1	200 250 350 350	0.25 0.375 0.5 0.5	No change Heating to 90°C.

the amorphous core effects obviously the loss and exciting current. The results are shown in Table 14.

Table 14 The effect of the pressing against amorphous core in the middle-frequency transformer.

The degree of pressing	Turn ratio	Iron loss (W)	Exciting current (A)
Light	10:1 9:1 8:1	237.5 300 400	0.25 0.375 0.625
Normal (~6Kg/cm^2)	10:1 9:1 8:1	262.5 375 500	0.25 0.5 1

Conclusion: 1. The 8KHz 500KVA amorphous core transformer have been manufactured. The average (from 5 sets) iron loss (8:1) was 665W and the average exciting current (8:1) was 1.2A. It was better than normal values--1000W and 66.7A.

2. The five 8KHZ 500KVA amorphous core transformer were the first batch transformer in China or in the world as we know.

6. MAGNETIC SHIELDING

One of the earliest applications of soft magnetic amorphous alloys was in the magnetic shielding.[1] The as-quenched amorphous ribbons were woven into a fabric and the magnetic properties of these sheets at 60Hz were compared with those of crystalline permalloy tape. While the shielding ratio are comparable, amorphous alloy fabric offer a greater flexibility and resistance to mechanical strain. The shielding result is shown in Table 15.

Table 15 The shielding result of the amorphous ribbons fabric barrel

Condition		Barrel size: diameter 5.1cm height 40.6cm				Barrel size: diameter 10.2cm height 40.6cm			
		layers' number	2	4	10	layers' number	2	4	10
Peak of external field	2 Oe	Shielding ratio	35	50	100		29	45	60
	4 Oe		50	85	150		6	70	110
	8 Oe		9	110	280		2.5	9	180

Usually the shielding case made of crystalline permalloy has to be annealed. Some time the shielding case is very large, so that the furnace must be as large as it. It is not convenient. The Co-based amorphous alloy is used in as-quenched state. A 75X75X75 cm^3 double layers shielding box winded crossly by Co-based amorphous tape (20-25mm width) was made in China.

The shielding ratio is:

$$S = Ho/Hi \qquad (3)$$

Ho--External field density;

Hi--Internal field density.

The experimental result was S=10081nT/293nT=34.4.

A conventional permalloy shielding barrel was used in a Ampex product--video camera tube. A Co-based amorphous alloy shielding barrel was made in China instead of permalloy. The weight of the amorphous barrel was only 1/3 to the permalloy's.

7.   MAGNETIC RECORDING HEAD

Because Co-based amorphous alloys possess high initial permeability, high resistivity and high hardness, people certainly thought to use it to make magnetic recording head. Especially for the metal recording tape, an excellent wear resistance is necessary. The magnetostriction of Co-based amorphous alloy is near-zero. So there is very little effect on the properties resulting from stresses. Because it possesses high AC permeability at high frequency, its audio-frequency response curve is very even and wide. The frequency compensation circuit is not necessary in the recorder.

The ferrite and FeSiAl alloy are also used to make magnetic head. The properties of several head materials are shown in Table 16.

Table 16   Properties of several head materials

	FJ-101*	Permalloy	Ferrite	FeSiAl
Hc (mOe)	4	7	100	30
μ_i (H=1mOe)	12×10^4	5×10^4	4000	3×10^4
μ_m	55×10^4	25×10^4	--	10×10^4
Hv	965	120	650	480
Resistivity (Ω-cm)	126×10^{-6}	55×10^{-6}	5×10^{-3}	85×10^{-6}
Bs (T)	0.7	0.78	0.4	1.1

* A Chinese trade-mark, CoFeVSiB alloy.

Several audio-recording magnetic heads were made in China. The third harmonic distortion of amorphous was less 50% than that of Fe-Ni-Nb hard-permalloy. The wear resistance of the amorphous magnetic audio-head was 4 times than that of permalloy audio-head. The amorphous head almost had not phase delay over the total frequency range as compared to permalloy head. The noise from the amorphous head was about 5 db lower than that of permalloy head.

Several 1600 bit/inch digital heads for computer magnetic tape recording were also made in China. The standard value of resolving power is 50%, the amorphous head reached 70%. The wear resistance of amorphous digital head was 5-6 times than that of Fe-Ni-Nb hard-permalloy. A comparison of digital heads for computer magnetic tape recording is shown in Table 17.

8. THE EARTH LEAKAGE CIRCUIT BREAKER

The principle working sketch is shown in Fig.2. When the earth leakage current occurs, the current passing though the core loses its balance. Then a votage is induced in the coil, the breaker cut off the

Table 17 Comparison of digital heads with amorphous and FeSiAl core

	Amor. head	FeSiAl head
Output frequency character 14K/315Hz	+17 db	+15 db
Recording frequency character 14K/1K	-10 db	-12 db
High freq. max. output voltage level (10KHz)	-1 db	-2 db
Bias current (μA)	300	600
Noise (ref. on amor. head)	0 db	+3 db

circuit.

Fe-Ni based and Co-based amorphous alloys were researched for the ELCB core in China. The breaking sensitivity is 30-100mA, the rated current is 6-60A.

So far, the main products made of amorphous alloy are earth leakage circuit breaker core in China. About 30 ton amorphous alloy tape per year is produced for it. In China, the ELCB core made of amorphous alloy was much cheaper than permalloy core.

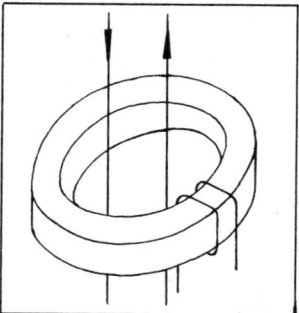

Fig.2. The principle working sketch.

9. MAGNETIC SEPARATION

The excellent magnetic properties of amorphous alloy, combined with thier strength and corrosion resistance, make amorphous alloy fibers become filter media for high-gradient magnetic separation[2].

When the soft magnetic alloy fiber is magnetized, a high magnetic field gradient is induced at the edge of fiber. By the high magnetic field gradient (HMFG) the ferromagnetic or weak-magnetic material powder could be extracted at the edge of fiber. The amorphous fiber should possesses the properties of noncorrosion and high magnetization.

$Fe_{75}Cr_5P_{13}C_7$ amorphous alloy fiber was made in China 1981. The width was 0.2-0.5 mm, thickness about 30μm.

The FeCrPC amorphous fiber as a magnetic medium was used in the high extraction magnetic filtration of Kaolin clay. The medium can decrease the Fe_2O_3 content in Kaolin clay from 2.01% (crude ore slurry) to 0.70%.

Another kind of FeCrPB amorphous alloy fiber was made in China 1984. Using it, the high votage transformer insulation oil could be cleaned. The insulation property can be improved from working voltage 48KV to 63KV, medium loss tgδ (90°C) was decreased from 0.5 to 0.25.

The amorphous alloy fiber not only to be used in magnetic extraction but also in electric deposition of gold and silver as the deposit pole.

10. THE OTHERS

Magnetic amplifier is used usually in the circuit of switching power supply. Two kinds of Co-based amorphous toroidal cores (Di=6mm, Do=12mm, h=5mm and 3mm) were made in Japan. The powers of the switching power supply were 24V, 2A and 5V, 10A respectively. Its work frequency was 200KHz.

A 180KHz magnetic amplifier used in switching power supply circuit was made in China 1982.

A pick-up element of electromagnetic sensor was made of amorphous alloy ribbon. The pick-up element was buried under the road for detecting the traffic. Amorphous alloy pick-up element had some advantages: high sensitivity, small volume and low cost (1/10 cost to pick-up element of ultrasonic). Now it is used to control automaticly the traffic lights in Shanghai China.

REFERENCE

1. L.T. Mendelsohn et al, IEEE Transactions on Magnetics, MAG-12 924 (1976)
2. Ying-Shan Yang et al, Digest of Intermag Conf., Boston, P21-1(1980)

COERCIVITY AND MAGNETIZATION PROCESSES IN NdFeB - MAGNETS

J. Schneider, A. Handstein, D. Eckert, and K.-H. Müller
Zentralinstitut für Festkörperphysik und Werkstofforschung
der AdW der DDR, Helmholtzstr. 20, DDR-8027 Dresden
GDR

ABSTRACT

The effects of substitution on the coercive force, the thermal stability, and other magnetic parameters are briefly summarized. Furthermore the nature of the magnetization processes in different sintered and melt-spun NdFeB-based magnets is analyzed. Finally model calculations for the temperature dependence of the coercive force and the hysteresis loops are discussed.

1. INTRODUCTION

The recent development of permanent magnetic materials based on the ternary compound $Nd_2Fe_{14}B$ has attracted considerable scientific and technological interest. NdFeB permanent magnets exhibit excellent hard magnetic properties at room temperature. However, the low Curie temperature T_c and the rapid decrease of the intrinsic coercive force $_JH_c$ on heating restrict the application of NdFeB magnets to below 100 °C. Different alloy modifications have been tried to overcome this situation. Desirable are high values of $(BH)_{max}$ ($(BH)_{max} = B_r^2 / 4\mu_o$), a low temperature coefficient of B_r as well as low irreversible losses.

The temperature coefficient of the remanent induction B_r may be improved by partially substitution of Fe by Co or Nd by heavy rare earth metals, which raise T_c and compensate the rapid decrease of the Nd moment with increasing temperature, respectively. Unfortunately, both of these approaches influence unfavorable the other magnetic parameters, e.g. $_JH_c$. The irreversible losses may be even simply reduced by increasing $_JH_c$. In this way the "knee" in the demagnetization curve appears at higher fields. Again appropriate substitutions like Dy and Al, which increase $_JH_c$, cause unfavorable effects with respect to the other characteristics of the permanent magnets.

Analyzing the behaviour of the coercive force in different NdFeB-based magnets, one should therefore also discuss the behaviour of the other magnetic quantities. In any case further efforts for a better understanding of the hardening mechanism, of the physical parameters affecting the temperature coefficient of $_JH_c$, and of the dependence of these parameters on the chemical composition and microstructure as well as technological conditions in preparing magnets will be essential to reach better qualities of NdFeB-based magnets.

2. EXPERIMENTAL FACTS
2.1 Modifications of $_JH_c$ and Other Properties by Substitution

A high magnetocrystalline anisotropy like in Co-R alloys is important to find a high coercivity in FeRB magnets. The highest values in $Fe_{14}R_2B$ compounds have been found for Tb and Dy [1]. Substitution of Nd by Dy in NdFeB alloys gives a remarkable increase of $_JH_c$, Fig. 1. Unfortunately the remanence B_r decreases progressively with the Dy-content [3]. The influence of a substitution of Dy for Nd on the temperature coefficient of B_r (TC(B_r)) and on the irreversible losses for $B/\mu_0H = -2$ is illustrated in Fig. 2 and 3, respectively. One finds a decrease of the irreversible losses, which is mainly related to the increase of $_JH_c$, as well as a small decrease of TC(B_r) (see also [4]). The anisotropy field H_A in $Nd_{2-x}Tb_xFe_{14}B$ shows a considerable increase for $x \lesssim 0.2$, while the decrease of the magnetic saturation polarization J_s is relatively small [6].

A substitution of Co for Fe in NdFeB results in an increase of the Curie temperature T_c. Fig. 4 shows the changes of T_c by substitutions of Fe by Co and other elements. A considerable increase one finds only for Co. The increase of T_c results in a decrease of $TC(B_r)$ in Nd(FeCo)B magnets. J_s shows a small maximum at $x \approx 2$ in $Nd_2Fe_{14-x}Co_xB$ [9]. However, $_jH_c$ decreases drastically in Nd(FeCo)B magnets [10]. This may partly attributed to the decrease of the anisotropy field H_A with increasing Co, Fig. 5. Addition of Co in NdDyFeB alloys improves $TC(B_r)$, but it is not benefical for $TC(_jH_c)$ [12].

Al was found to increase $_jH_c$, as shown in Fig. 6, despite the fact that H_A decreases [13 to 15]. T_c, B_r, and $(BH)_{max}$ were found to decrease with the Al-content. For an Al-content less then about 1 wt% magnets with reasonable magnetic properties may be made. Reducing the B-content, the hard magnetic properties seem to be deteriorated already at lower Al-contents [13]. Substitution of Al for Fe in NdFeB magnets causes an increase of $TC(B_r)$ as well as of the temperature coefficient of $_jH_c$ (see [17]). Similar variations of T_c, B_r, $_jH_c$, and $(BH)_{max}$ upon substituting Al for Fe like in NdFeB have been observed for Co- and Dy-containing NdFeB [10, 18]. A combination of Co and Dy seems to be more effective to improve the temperature coefficients of $_jH_c$ and B_r than the addition of Co and Al [12]. When Al is added to (NdDy)FeB or (NdDy)(FeCo)B alloys much less Dy is required to find high values of $_jH_c$ [10, 19]. However, the values of $TC(B_r)$ and $TC(_jH_c)$ were found to increase with rising Al-content in $(Nd_{0.88}Dy_{0.12})(Fe_{80-x}Co_{0.12}B_{0.08}Al_x)_{5.5}$ [18].

Beside the effects of addition of Dy, Co, and Al to NdFeB also the effects of substitution of other elements like Nb, Ti, Zr, Ga, or Mo on the coercivity and the thermal stability have been investigated. It was shown that small amounts of Nb, Ti, Zr can increase the coercive force, Fig. 7. Nb gives in comparison to Ti, Zr only a small reduction of J_s and the largest increase in $_jH_c$ without any remarkable change of $TC(B_r)$ and $TC(_jH_c)$ [20]. Small addition of Ga increases remarkable $_jH_c$ and improves the thermal stability in $Nd(Fe_{0.72}Co_{0.2}B_{0.08})_{5.6}$ [21]. With respect to the addition of Mo to NdFeB-based magnets an increase of $_jH_c$ for $Nd(FeCo)B$ [22] and an improvement of $TC(B_r)$, $TC(_jH_c)$ for $(NdDy)(FeCoAl)B$ [10] are reported.

2.2 Coercivity and Microstructure

The coercivity in the NdFeB-based magnets like in other R-TM magnets depends not only on the magnetocrystalline anisotropy and the saturation magnetization of the hard magnetic phase but also, as well known, on metallurgical parameters like grain size distribution of the main phase, type and distribution of additional phases, grain boundary phases, and defect structure. The microstructure is influenced by the chemical composition, the alloy preparation as well as by the technological conditions during the various steps of making magnets.

For instance $_jH_c$ depends critically on the quenching rate in melt spun magnets, by which the average size of the grains is changed [3]. Growth of the grains is influenced by diffusion and rearrangement of atoms. The appropriate final microstructure of the optimal quenched ribbons evolves through a solid state reaction as opposed to a solidification process. Similar properties like for optimal quenched ribbons were obtained for overquenched ribbons upon annealing above the crystallization temperature, where crystallites nucleate and grow [24].

As it is well known, a thermal treatment improves considerably the coercive force in sintered magnets. The mechanism of enhancement of $_jH_c$ during post sintering processes is still not clear [25]. It is suggested that atomic disorder at the grain boundary or high distortion in the surface layer of the grains or even small-scale roughness at the surface of the grains of the main phase may be responsible for this effects.

Furthermore, the grain size, $_JH_c$ as well as B_r of the sintered magnets depend on the composition and initial particle size, Fig. 8. The initial particle size as well as the oxygen content and the amount of oxygen containing phases, which appear in the intergrain region between the grains of the main phase [26, 27], depend on the milling time. Beside these additional phases three phases occur in the standard alloy $Nd_{15}Fe_{77}B_8$ [28, 29]:
- $Nd_2Fe_{14}B$, responsible for the hard magnetic properties,
- Nd-rich phases, which enables a liquid phase sintering,
- $NdFe_4B_4$, which is nonmagnetic in the interesting temperature range.

The Nd-rich phases differ in their Nd-content. The Nd-contents are about 85 at% and 75 at%, respectively [30]. The ideal microstructure of sintered "fine-particle" magnets would consist of grains of the main phase with no precipitations or lattice defects and ideal smooth and clean surfaces, isolated by a thin nonmagnetic phase. Any non-ferromagnetic inclusions as well as additional ferromagnetic phases in the surface region of the grains or in the intergrain regions should be avoided. They cause a decrease of B_r and $_JH_c$. According to the phase diagram of NdFeB the appearence of the $NdFe_4B_4$ phase may be suppressed by reducing the B-content [28]. In this way "two phase magnets" of the composition $Nd_{18.5}Fe_{75}B_{6.5}$ [27] and $Nd_{16.7}Dy_{1.8}Fe_{75}B_{6.5}$ [31] were obtained, which exhibit remarkable improved values of $TC(_JH_c)$ compared to their corresponding "three phase" magnets.

With respect to the substitution of Nd by Dy, Tb or of Fe by Co, Al as well as the addition of small amounts of elements such as Nb, Zr, Ti, Ga, Mo, or Si the knowledge of the changes in the microstructure and microcomposition are important to understand the changes in the magnetic properties. It was found that Dy partitions preferentially into the $Nd_2Fe_{14}B$ phase. There is no segregation of Dy to the interface [32]. The observed increase of $_JH_c$ with Dy-content is larger than the expected increase due to the increased values of the anisotropy field H_A (see Fig. 1). This is attributed to a less excessive grain growth in Dy-containing magnets [2].

The decrease of the coercivity in Co-substituted NdFeB is attributed to the formation of soft magnetic Nd(Fe, Co) phases [33]. This may also explain the larger decrease of $_JH_c$ compared with those of H_A. For Al-containing NdFeB magnets beside on partition in the $Nd_2Fe_{14}B$ phase small regions rich in Al appear especially nearby the Nd-rich regions [13, 34]. The composition of a new intermetallic phase, found at higher Al-contents, was given as $Fe_{64}Nd_{23}Al_{13}$ [35]. The increase of $_JH_c$ may be due to a better grain surface quality.

There are also attempts to find a precipitation structure in the hard magnetic phase, which may result in a "pinning dominated" magnetization reversal behaviour. Precipitations in the main phase were found for Nb- and Zr-containing NdFeB magnets [36 to 38]. The present observed form of these precipitations, however, does not give rise to pinning fields, which would overcome the nucleation fields.

3. MAGNETIZATION PROCESSES, HARDENING MECHANISM

Generally magnetization reversal may occur by coherent or incoherent rotation of magnetization or by nucleation of regions with reverse magnetization, their growing and movement of the resulting expanded walls (domain processes). One may distinguish between the following critical fields for magnetization reversal:

H_n — nucleation field,
H_G — critical field for expansion of nucleous,
H_{pB} — critical field for pushing a wall throgh a phase boundary in a two phase system or through a precipitation at the grain boundary,
H_p — pinning field, which inhibit the displacement of extended walls within a given phase (bulk wall pinning),
H_{pS} — pinning of extended walls in nonperfect surface regions within the grains,
H_{pD} — critical field for disappearing of residual domains or fragments of walls.

Both types of magnetization reversal and their relevant critical fields are influenced by the microstructure and microcomposition. Considering one important parameter, the grain size, the known NdFeB-based magnets belong like other RE-magnets [39] approximately to one of the following idealized types:
- single phase aligned grains with $D \gg D_c$ (D_c — single domain particle diameter),
- aligned grains with $D \gg D_c$ containing a precipitation structure,
- single phase aligned or unaligned grains with $D \ll D_c$.

The critical size concept deals with energy differences connected with the occurence or disappearing of domain structure within the grains in the demagnetized state. However, the magnetization processes and the coercivity depend on energy gradients (displacement fields for walls or fragments of walls, nucleation, ...). Thus, if H_n is large, even grains with $D > D_c$ may behave like single-domain particles.

Because of different magnetic structures in the demagnetized and magnetically saturated state investigations of the magnetization behaviour in both cases may much improve our understanding of the magnetization processes and the relevance of the different critical fields. From the magnetization behaviour one distinguish between a "nucleation controlled" and "pinning controlled" coercivity. In pinning type magnets precipitations give rise to high values of H_p, which determines the magnetization processes, while in nucleation type magnets the grains have a high degree of perfection ($H_p \approx 0$). The nucleation controlled behaviour includes nucleation of reverse domains (H_n, H_G) as well as their pinning at the surface region of the grains (H_{pB}). With these remarks in mind we will now analyse the experimental results.

3.1 Virgin Permeability

In the demagnetized state in sintered as well as in melt-spun magnets nearly all grains were multi-domain [40 to 44]. The domain patterns within individual grains of sintered magnets differ mostly in detail from each other, which may indicate an exchange decoupling between the grains. In the melt-spun magnets as well as die-upset magnets the domain walls passes through many grains and are pinned at the boundaries of the outer grains [24].

Field induced motion of domain walls within the grains is very easy in sintered magnets, as can be directly seen by domain observations. This is also confirmed by the high values of the initial susceptibility $\mu_o dJ_m/dH_m$ (see Fig. 9). The virgin permeability is much higher for the thermal demagnetized state than for the ac-field demagnetized or even the direct field demagnetized state. Much lower values of the virgin permeability are observed for isotropic and anisotropic melt-spun magnets (compare Fig. 10 and 11). This indicates that domain walls are generally more strongly pinned in melt-spun magnets.

3.2 Dependence of $_JH_c$, B_r and of the Magnetic Polarization J_m on the Magnetizing Field H_m

The experimental dependence of the coercive field determined from the minor loops obtained after thermal demagnetization and applying a field with maximum value of H_m is very often used to discuss the coercive mechanism [28, 31, 48, 49]. However, it seems more convenient to analyse $_JH_c$, B_r and J_m as function of H_m to get a deeper understanding of the magnetization and demagnetization processes and the differences between them (see also [46]).

Fig. 9 shows the behaviour for a high-coercive (NdDyTb)FeB sintered magnet. Similar curves were observed for NdFeB, (NdDy)FeB, and (NdDy)(FeCo)B magnets [31, 48]. Already for low H_m the remanent induction B_r^{th} and J_m increase drastically, while $_JH_c$ is very low. The much slower increase of B_r^{th} compared with those of J_m at somewhat higher H_m indicates remarkable reversible magnetization processes. $_JH_c$ at low H_m characterizes the pinning forces H_p within the grains. A linear relation between H_p and $1/d$ (d - average diameter of the grains) was found, which proves the relevance of the equation [50)]

$$H_p \approx 3\gamma / J_s \cdot d \qquad (1)$$

(γ - wall energy). This relation gives the coercivity for the displacement of domain walls in soft magnetic materials. J_m reaches maximum values for values of H_m, where $_JH_c$ is small, while $_JH_c$ reaches maximum values $_JH_c^{sat}$ (region of approach to saturation) for $H_m^{sat} < {_JH_c^{sat}}$. For $H_m > H_m^{sat}$ only a small increase of J_m, B_r^{th} and $_JH_c$ is observed. This behaviour suggests different magnetization processes for magnetizing and demagnetizing NdFeB magnets. The dependence of $_JH_c$, J_m vs. H_m is equivalent with $H_{pD} > H_n^{eff} > H_p$, which characterizes a "nucleation controlled" coercivity (H_n^{eff}: H_n, H_G, H_{pB}) [31)].

A quite different behaviour of $_JH_c$, B_r^{th} and J_m versus H_m is observed for melt-spun magnets, Fig. 10. The obtained curves are similar to those of precipitation hardened Sm(Co, Fe, Cu, Zr)$_{7.6}$ [52)]. B_r^{th} and J_m increase more or less in the same way with H_m, suggesting essentially irreversible magnetization processes, and $_JH_c$ reaches maximum values for $H_m^{sat} \gtreqless {}_JH_c^{sat}$ ("pinning controlled" coercivity). The continuously increase of $_JH_c$ with H_m indicates a broad distribution of the pinning fields. An analogous behaviour like in melt-spun isotropic magnets is also observed for die-upset anisotropic magnets, Fig. 11. Due to the isotropic distribution of grain orientations (angular dependence of $_JH_c$) one obtains a stronger increase of $_JH_c$ in the region of approach to saturation for isotropic magnets compared to anisotropic magnets.

3.3 Minor Loops After Saturation

The minor hysteresis loops measured after saturation and subsequent application of a demagnetizing field with maximum values of H_m behave quite different for various NdFeB-based magnets [48, 53, 54]. The resulting loops obtained by reducing the applied demagnetizing field H_m to zero and increasing it again to H_m (recoil loops) show practically no hysteresis. As an example the minor loops for a sintered NdFeB magnet are given in Fig. 12. The behaviour of the minor loops may be analysed using the quantity:

$$\Delta J_{rev}^d = J^d(H_m) - J_r^d , \qquad (2)$$

where $J^d(H_m)$ denotes the magnetic polarization at H_m and J_r^d denotes the direct field demagnetization remanence. ΔJ_{rev}^d gives also a measure for the appearance of reversible magnetization processes upon reducing H_m to zero [45]. It is found that ΔJ_{rev}^d dependes critically on the technological parameters during the fabrication of the magnets. Thus $\Delta J_{rev}^d \approx 0$ as well as large values of ΔJ_{rev} are observed for magnets of a (NdDy)FeB alloy, which differ in their value of $_JH_c$ [48]. For the high $_JH_c$-(NdDy)FeB magnets, where $\Delta J_{rev}^d \approx 0$, the recoil loops turned out to be contracted to straight lines. The value of ΔJ_{rev}^d as well as the type of the recoil loops (straight or curved lines) are also influenced by the appearance of a dip in the demagnetization curve. In any case finite ΔJ_{rev}^d indicate soft magnetic regions, which contain easy movable domain walls. Such a complex behaviour of the minor loops and recoil loops cannot be described within a model for magnetization reversal with only one critical field. It has been shown [53, 55] that a double well potential for the pinning at the surface of the grains can account for the observed effects.

3.4 Thermal Remagnetization

In a magnet with "nucleation controlled" coercivity most of the grains will fully reverse once nucleation occurs. If a grain really transforms from the fully magnetized state to the fully magnetized state in the opposite direction, it depends on the ratio of H_n^{eff} to H_{pD}. H_{pD} is mainly related to internal stray fields (see [56]). Thus in a nucleation-type magnet mostly single-domain grains appear in the direct field demagnetized state, whereas multi-domain grains appear for a pinning-type magnet. The former state has a high internal energy. Therefore partial remagnetization is observed upon heating. This is related to the decrease of H_n^{eff} with increasing temperature. This phenomena, which was observed for NdFeB-based magnets [40] is characteristic for nucleation-type magnets with high values of H_n^{eff} (e.g. $(H_n^{eff} - H_{pD}) \longrightarrow 0$). It is nearly absent for pinnig-type magnets.

4. TEMPERATURE DEPENDENCE OF $_JH_c$

Measurements of the dependence of $_JH_c$ on the angle θ of the applied field with respect to the easy axis of the magnet or on the temperature is, in principle, also a useful method to investigate the coercivity mechanism. $_JH_c(T)$ in NdFeB-based magnets has been already widely studied [14, 25, 31, 57, 58]. Mostly a formula of the type

$$_JH_c = c\, H_n^{eff} - N_{eff}\, J_s\, /\, \mu_o \qquad (3)$$

has been used to analyse the measured data. The nucleation

field H_n may be expressed in the simplest form as $2K_1/(J_s/u_o^{-1})$ or $(2K_1+4K_2)/(J_s/u_o^{-1})$, if the second order anisotropy constant is taken into account. c describes the magnetic decoupling of the grains. If $H_n=2K_1/(J_s/u_o^{-1})$, the grains would be perfectly isolated for c=1. The second term in (3) gives the effective demagnetizing field taking into account demagnetizing fields due to surface poles and local internal demagnetizing fields, which may be as high as about $2J_s/u_o^{-1}$ in the case of neighbouring reversed grains with large surface irregularities. The effective internal demagnetization factor N_{eff} depends sensitively on the real mesoscopic material structure. More accurate calculations using statistical methods give [56]

$$1/3 \leq N_{eff} \leq 1/3 + 0.3 \phi(1 - \delta_B^2/d^2)^{-1} , \quad (4)$$

where $\phi(x)^{-1}$ is the Gaussian probability distribution function and δ_B the domain wall width. As a typical example one gets $N_{eff} = 1.7$ for $\delta_B=5$ nm and d=5 μm. The upper bound in (4) is nearly realized, if the nonmagnetic phases are mainly localized in large spheric grains. The lower bound corresponds to the case that the grains of the main phase are surrounded by the nonmagnetic phase (ideal shell structure).

Plotting $(_JH_c +J_s)$ versus H_A, the parameters c and N_{eff} may be calculated from the slope of the curve and the intercept, respectively. Typical experimental curves for sintered magnets are given in Fig. 13. The deviations of the experimental determined curves from a linear relationship between $(_JH_c + J_s)$ and H_A at higher temperatures demonstrate the growing importance of the internal stray fields at increasing temperature. Different nucleation and pinning theories yield different relations between H_n and K, J_s and other parameters, which characterize the domain wall and the type of defect. Using the expression

$$c H_n = (2 K_1 / u_o^{-1} J_s) (\delta_B / \pi r_o) \quad (5)$$

in equ. (3), which describes the nucleation field in a

rather thin precipitation at the grain boundary, characteristic values for the half width of the precipitation r_o and N_{eff} in various NdFeB-based magnets were determined in [31, 60]. Surprisingly the same values for r_o were obtained for sintered and melt-spun materials.

In the case of "nucleation controlled" coercivity it is difficult to distiguish between reverse domain nucleation H_n and domain-wall unpinning from localized defects near the grain boundary H_{pB}. In both cases the same type of demagnetization curves (major loop) and virgin curves appear. Also the temperature dependence of $_JH_c$ can be hardly used to this purpose. For instance for H_{pB} in the case of an extended phase boundary one finds within the same model as for H_n (see (5)):

$$c\ H_{pB} = \sqrt{2}\ (K_1\ \delta_B) / (\mu_o^{-1}\ J_s\ \pi\ r_o) . \quad (6)$$

5. MODEL CALCULATIONS FOR MAJOR AND MINOR LOOPS

Hysteresis loops of "fine particle" magnets may be calculated in the simplest way by regarding an assembly of noninteracting particles. One difficulty of this classical Stoner-Wohlfarth model is attributed to the fact that the "particle interaction" is not taken into account. The degree to which magnetic interaction among particles influence the magnetization processes may be estimated by regarding the temperature dependence of $_JH_c$ or by studying the reversal of isolated individual grains in comparison to bulk magnets. The Wohlfarth relationship

$$J_r^d(H_m') = J_r(\infty)\ \ \ 2\ J_r^{th,ac}(H_m) \quad (7)$$

may be also used for this purpose. $J_r^{th,ac}$ gives the static remanence after thermal or ac-field demagnetization, respectively. Equ. (7) holds for an assembly of noninteracting uniaxial single domain particles or multidomain particles with identical magnetization processes starting from the demagnetized or magnetically saturated state.

Fig. 14 shows J_r^d versus $J_r^{th,ac}$ for melt-spun and sintered magnets. One observes no differences in the behaviour of J_r^d versus J_r^{th} and J_r^d versus J_r^{ac} for melt-spun magnets ("pinning controlled" magnetization processes), whereas large differences appear for sintered magnets, which are caused by quite different magnetization processes upon magnetizing or demagnetizing the sintered magnets (see also [45,46]). Only for J_r^d versus J_r^{ac} a resonable aggreement with the relation (7) is found. The analysis of the remanence relationship indicates that model calculations for an assembly of noninteracting particles may be used for a description of the direct field demagnetization behaviour of sintered NdFeB-based magnets. It was shown in [53, 55], that the experimentally observed types of major- and minor loops upon demagnetizing the magnetically saturated state in various sintered NdFeB-based magnets can be well described within such a model with given distributions of H_n, H_{pD} and H_{pS}.

6. REFERENCES

1) Livingston, J.D., Report 85CRD081, May 1985.
2) Ma, B.M. and Krause, R.F., Proc. 5th Internat. Symp. on Magnetic Anisotropy and Coercivity in RE-TM Alloys, Bad Soden, 1987, p. 141.
3) Ramesh, R., Thomas, G., and Ma, B.M., Mater. Res. Soc. Symp. Proc., Vol. 96, Pittsburg, 1987, p. 203.
4) Edeling, M. and Herget, C., 9th Internat. Workshop on Re-Magnets and Their applications, Bad Soden, 1987, paper WP8.2.
5) Dauermagnete auf der Basis NdFeB-Vacodym, VAC-Vacuumschmelze GmbH, Hanau, 1987.
6) Buschow, K.H.J., as 3), p. 1.
7) Rodewald, W. and Fernengel, W., IEEE Trans. on Magnetics MAG-24, 1638 (1988).
8) Pedziwiatr, A.T. et al., IEEE Trans. on Magnetics MAG-23, 1795 (1987).

9) Sagawa, M. et al., Jap. J. Appl. Phys. 26, 785 (1987).
10) Ma, B.M., see 3), p. 143.
11) Grössinger, R., see 2), p. 15.
12) Wei, Li et al., see 4), p. 503.
13) Handstein, A. et al., see 4), p. 601.
14) Grössinger, R., et al., IEEE Trans. on Magnetics MAG-24 (1988).
15) Ma, B.M. and Willman, C.J., see 3), p. 133.
16) Maocai, Zhang et al., Proc. 8th Internat. Workshop on RE-Magnets and Their Applications, Dayton, 1985, p. 541.
17) Rodewald, W., see 4), p. 609.
18) Xiao, Y. et al., see 3), p. 155.
19) Kim, A.S., J. Appl. Phys. 1988 (Proc. MMM-Conf., 1987).
20) Xiao, Y., see 4), p. 467.
21) Endoh, M., Tokunaga, M., and Harada, H., IEEE Trans. on Magn. 1988 (Proc. INTERMAG 1987).
22) Shen, X. et al., J. Appl. Phys. 61, 3433 (1987).
23) Croat, J.J. et al., J. Appl. Phys. 55, 2078 (1984).
24) Mishra, R.K., see 3), p. 83; J. Magn. Magn. Mater. 54-57, 450 (1986).
25) Sagawa, M. et al., see 2), p. 229.
26) Ramesh, R., Chen, J.K., and Thomas, G., J. Appl. Phys. 61, 2993 (1987).
27) Schneider, G., see 2), p. 347.
28) Handstein, A. et al., Mater. Letters 3, 200 (1985).
29) Fidler, J., IEEE Trans. on Magn. MAG-21, 1955 (1985).
30) Fischer, W., ZFW Dresden, private communication.
31) Durst, K.D., Kronmüller, and Schneider, G., see 2), p. 209.
32) Ramesh, R., Thomas, G., ans Ma, B.M., see 3), p. 203.
33) Yamamoto, H. et al., IEEE Trans. on Magn. MAG-23 (1987)
34) Chen, J.K. and Thomas, G., see 3), p. 221.
35) Schrey, P., IEEE Trans. on Magn. MAG-22, 931 (1986).
36) Parker, S.F.H., Grundy, P.J., and Fidler, J., J. Magn. Magn. Mater. 66, 74 (1987).
37) Schrey, P., J. Magn. Magn. Mater. (1988).

38) Pollard, P.J., to be published.
39) Livingston, J.D., Technical Information Series (General Electric Company), No. 86, CRD 159 (1986).
40) Livingston, J.D., J. Appl. Phys. $\underline{57}$, 4137 (1985).
41) Sagawa, M. et al., see 16). p. 587.
42) Mishra, R.K., Chen, J.W., and Thomas, G., J. Appl. Phys. $\underline{59}$, 2244 (1986).
43) Pastushenkov, J., Durst, K.D., and Kronmüller, H., phys. stat. sol. (a) $\underline{104}$, 487 (1987).
44) Handstein, A. et al., this conference.
45) Heinecke, U., Handstein, A., and Schneider, J., J. Magn. Magn. Mater. $\underline{53}$, 236 (1985).
46) Handstein, A. et al., Proc. ICM 1988.
47) Pinkerton, F.E. and Van Wingerden, D.J., J. Appl. Phys. $\underline{60}$, 3685 (1986).
48) Schneider, J. et al., Proc. 3rd Internat. Conf. on Physics of Magnetic Materials, Szczyrk, Poland, 1986, p. 225.
49) Pinkerton, F.E., see 3), p. 65.
50) Fernengel, W., see 2), p. 259.
51) Becker, J.J., IEEE Trans. on Magn. $\underline{MAG-12}$, 965 (1976).
52) Kronmüller, H. et al., IEEE Trans. on Magn. $\underline{MAG-20}$, 1569 (1984).
53) Heinecke, U. et al., see 2), p. 237.
54) Fernengel, W. and Adler, E., see 2), p. 247.
55) Heinecke, U. et al., this conference.
56) Müller, K.-H., this conference.
57) Schneider, J. et al., Mater. Letters $\underline{3}$, 401 (1985).
58) Grössinger, R. et al., see 4), p. 593.
59) Sagawa, M. et al., see 3), p. 161.
60) Durst, K.D. and Kronmüller, H., J. Magn. Mater. Mater. $\underline{68}$, 63 (1987).
61) Kronmüller, H., Proc. 3rd Internat. Symp. on Magnetic Anisotropy and Coercivity in RE-TM Alloys, Baden, 1982, p. 555.

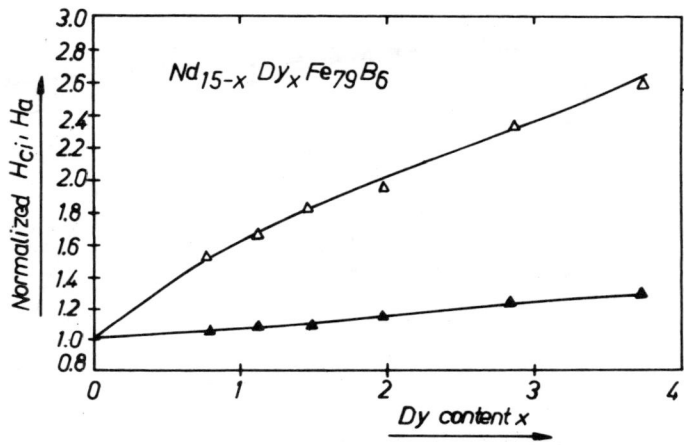

Fig. 1 $_JH_C$ and H_A vs. Dy-content in $Nd_{15-x}Dy_xFe_{79}B_6$ [2]
($\triangle$ - $H_{ci}/H_{ci}(x=0)$; ▲ - $H_A/H_A(x=0)$).

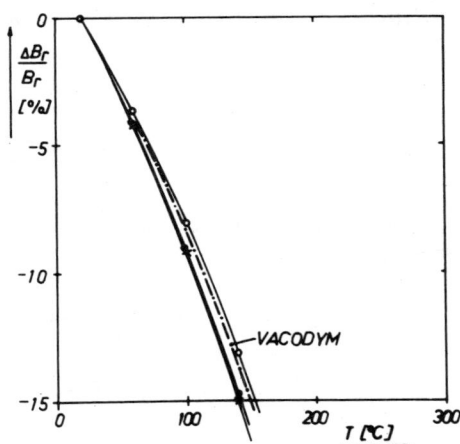

Fig. 2 $\Delta B_r/B_r$ vs. T for sintered magnets:
○ $(Nd_{.83}Dy_{.17})_{14}Fe_{79}B_7$, × $(Nd_{.92}Dy_{.08})_{17}Fe_{75}Al_1B_7$
● $Nd_{16}Fe_{77}B_7$, —·—·— Vacodym (after [5]).

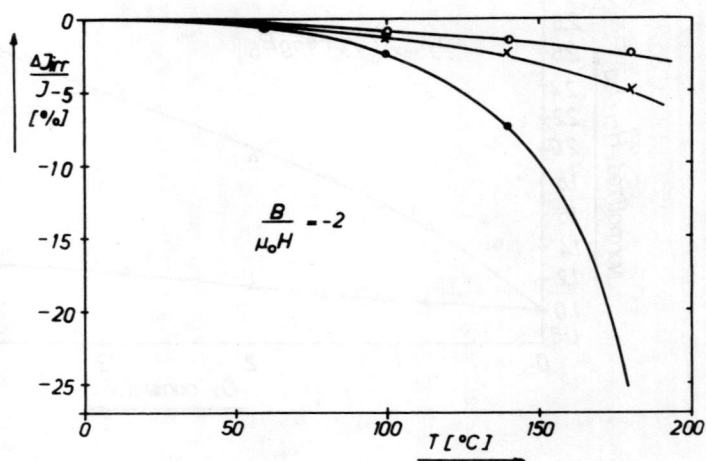

Fig. 3 $\Delta J_{irr}/J$ vs. temperature for sintered magnets (symbols and samples as in Fig.2; o $_JH_c$(RT)= 1.8 MA/m; x $_JH_c$(RT)=1.4 MA/m; ● $_JH_c$=0.8 MA/m).

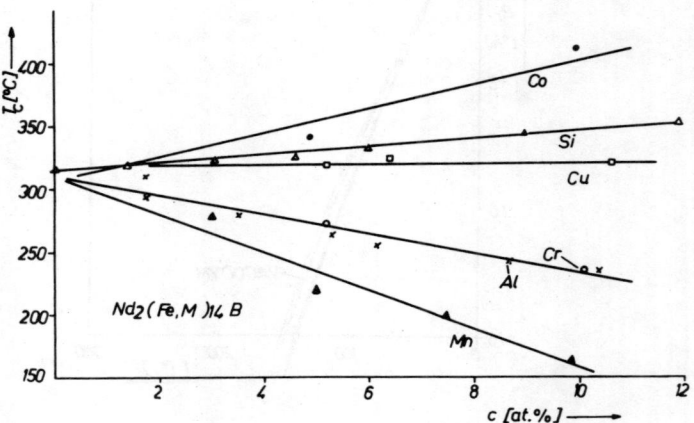

Fig. 4 T_c vs. concentration of M in $Nd_2(FeM)_{14}B$ (after [7, 8]).

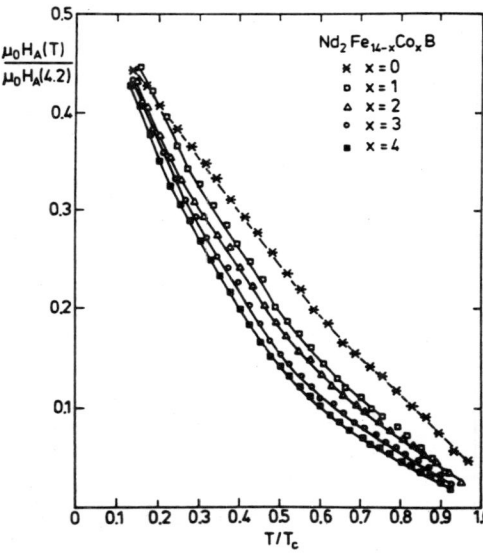

Fig. 5
Reduced anisotropy field H_A vs. temperature T for $Nd_2Fe_{14-x}Co_xB$ [11].

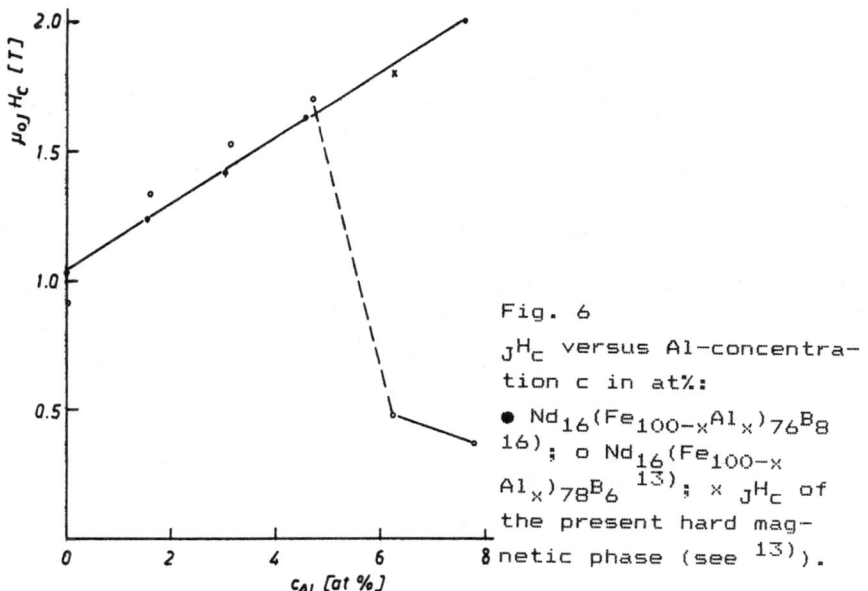

Fig. 6
$_JH_C$ versus Al-concentration c in at%:
● $Nd_{16}(Fe_{100-x}Al_x)_{76}B_8$ [16]; ○ $Nd_{16}(Fe_{100-x}Al_x)_{78}B_6$ [13]; × $_JH_C$ of the present hard magnetic phase (see [13]).

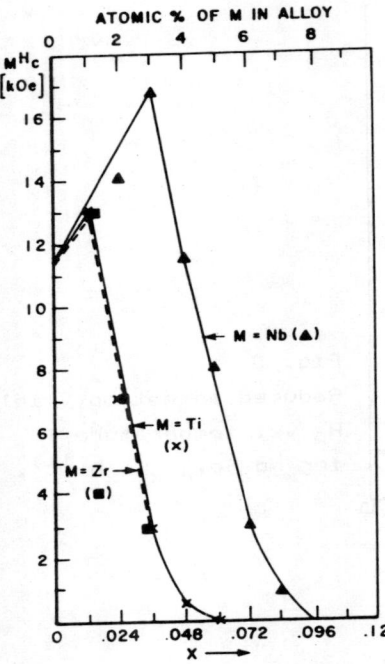

Fig. 7
$_JH_C$ vs. content of M = Nb, Ti, and Zr in sintered $(Nd_{.88}Dy_{.12})(Fe_{.88-x}Co_{.12}B_{.08}M_x)_{5.5}$ [20].

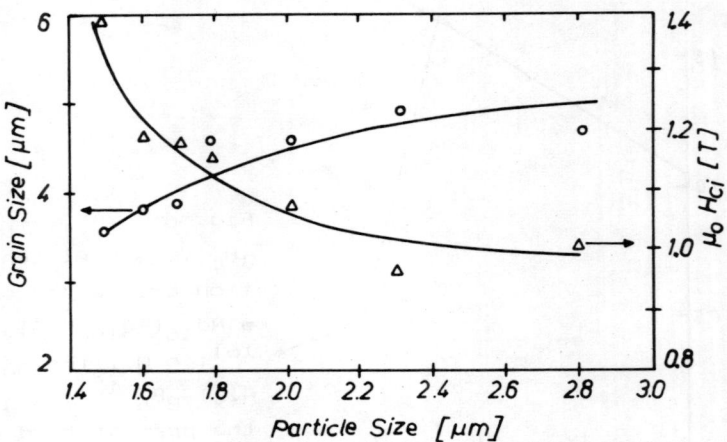

Fig. 8 Variation of grain size and coercive force in dependence on the initial particle size [2].

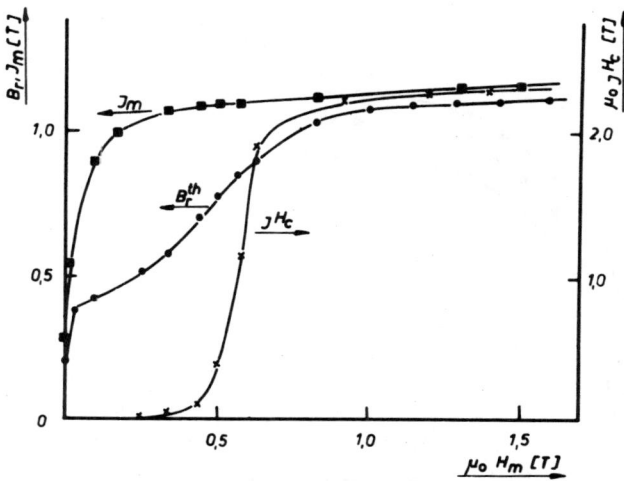

Fig. 9 $\mu_0 J H_c$, J_m and B_r^{th} vs. $\mu_0 H_m$ for sintered $(NdDyTb)_{15}Fe_{78}B_7$.

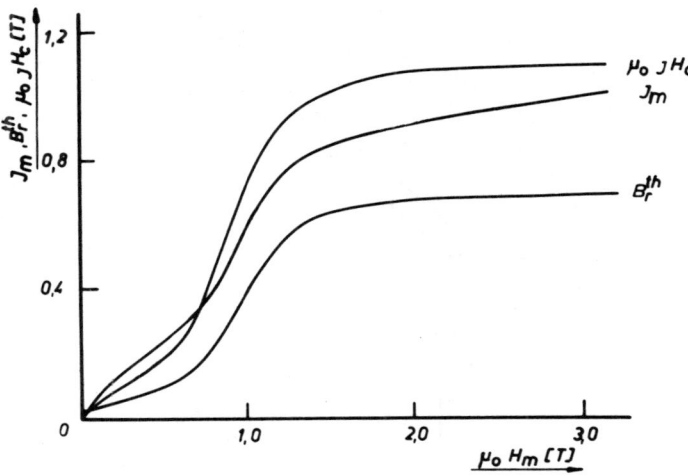

Fig. 10 $\mu_0 J H_c$, J_m, and B_r^{th} vs. $\mu_0 H_m$ for melt-spun $Nd_{15}Fe_{77}B_8$.

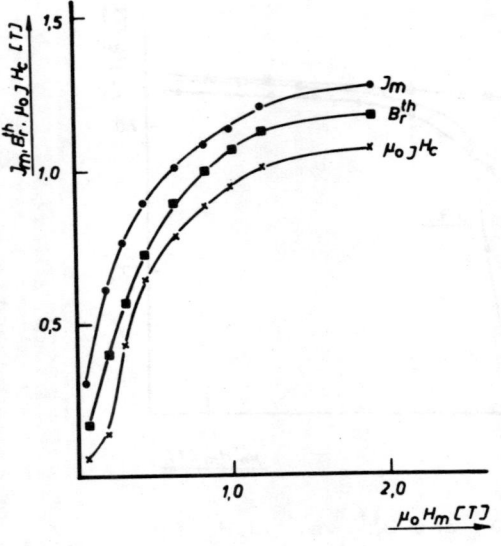

Fig. 11
$\mu_0{}_JH_c$, J_m, and B_r^{th} vs. $\mu_0 H_m$ for a die-upset magnet (data taken from [47]).

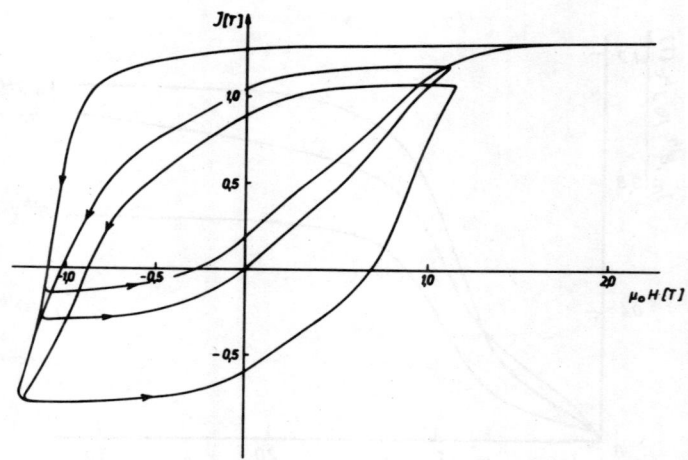

Fig. 12 Demagnetization curve and minor loops for a sintered $Nd_{18}Fe_{75}B_7$ magnet.

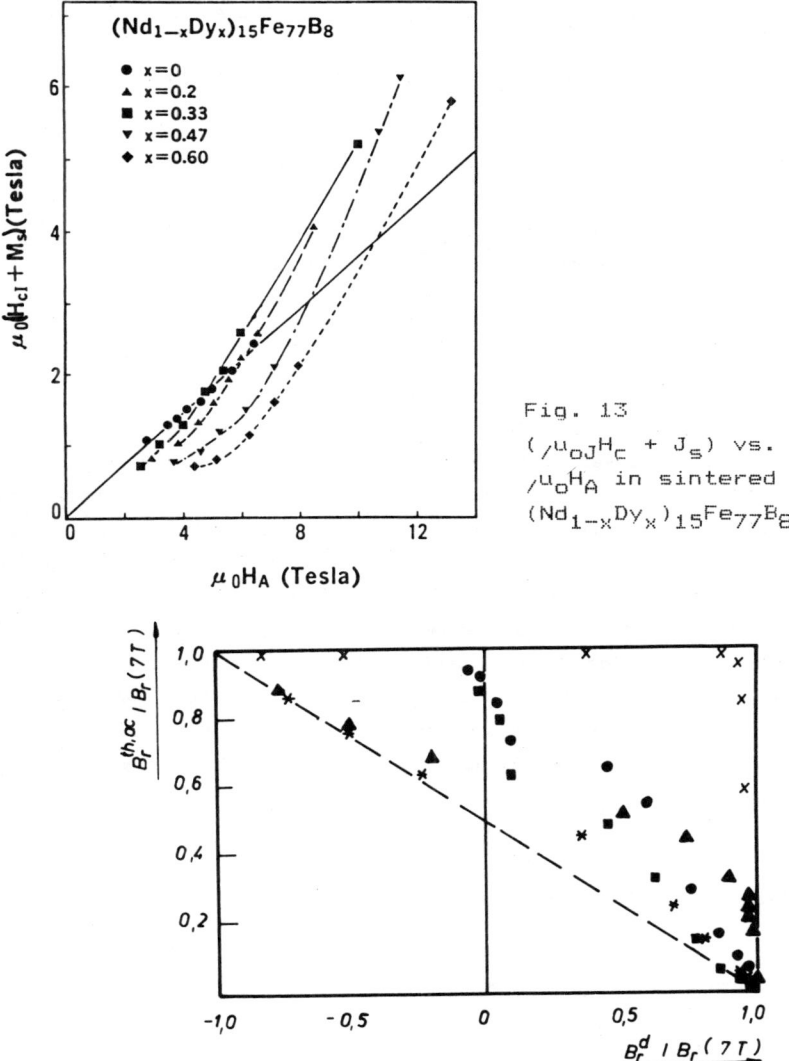

Fig. 13 ($\mu_0 J^H_c + J_s$) vs. $\mu_0 H_A$ in sintered $(Nd_{1-x}Dy_x)_{15}Fe_{77}B_8$.

Fig. 14 $B_r^{th,ac}$ in dependence on B_r^d;
melt-spun $Nd_{14}Fe_{81}B_5$ (● B_r^{th}, ■ B_r^{ac}),
melt-spun $Nd_{15}Fe_{77}B_8$ (▲ B_r^{th}),
sintered $(NdDyTb)_{15}Fe_{78}B_7$ (× B_r^{th}, ∗ B_r^{ac}).

EXCHANGE AND CRYSTAL - FIELD INTERACTIONS
IN Ho_2Co_{17} AND $Ho_2Fe_{14}B$:
TWO EXAMPLES OF R - T INTERMETALLICS

J.J.M.Franse and R.J.Radwański[*]

Natuurkundig Laboratorium der Universiteit van Amsterdam,
Valckenierstraat 65, 1018 Amsterdam, The Netherlands.

ABSTRACT

The magnetic properties of the trivalent Ho ion in Ho_2Co_{17} and $Ho_2Fe_{14}B$ are analysed in terms of crystal- and molecular-field interactions. A full set of CF parameters of the two non-equivalent rare-earth ions in the hexagonal Ho_2Co_{17} structure has been evaluated reconciling Mössbauer, inelastic-neutron-scattering, specific-heat and high-magnetic-field results. A tentative set of crystal-field parameters is presented for the tetragonal $Ho_2Fe_{14}B$ compound with the orthorhombic local symmetry of the rare-earth ion. The magnetic tilted structures, with the cone with respect to the c axis of 85° for Ho_2Co_{17} and 24° for $Ho_2Fe_{14}B$, observed in these compounds result solely from crystal-field interactions. Exchange interactions between the 3d and 4f spins in these compounds are compared and discussed.

[*] On leave from the Solid State Physics Dept., University of Mining and Metallurgy, 30-059 Kraków, Poland.

1. Introduction.

A detailed description of the magnetic behaviour of rare-earth - 3d (Fe,Co) intermetallics is a long-standing problem which solution is hampered by the large number of parameters that is involved. The basic principles behind the magnetic properties in these compounds are quite well understood: the Curie temperature of the 3d-rich compounds is mainly governed by the 3d-3d exchange interactions, whereas an extraordinary large magnetocrystalline anisotropy is a special feature of many of rare-earth ions. The 3d and rare-earth moment are coupled by exchange interactions that are usually described in a molecular-field type of model. The interaction between the rare-earth spins are assumed to be weak compared to the interactions mentioned before.

Moreover, a really detailed description of the magnetisation of these intermetallic compounds has to take into account the different inequivalent sites of the rare-earth and 3d ions in the unit cell. In the 2:17 and 2:14:B series, there are two inequivalent sites for the RE ion and 4 or 6 for the 3d ions, respectively. Each rare-earth site has its own set of crystal-field parameters that determine the energy level scheme of the R^{3+} ion and each 3d site has its own magnetic moment and local magnetocrystalline anisotropy energy. Mössbauer, NMR and neutron-diffraction experiments have improved our knowledge about the distribution of the 3d moment over the different sites considerably [1-7]. The experimentally observed 3d moment distribution over the different sites can be reproduced by band-structure calculations rather well [8]. The theoretical understanding of the anisotropy contributions of the different 3d sites is still poor. In principle, this situation is easier for the rare earth ions. Magnetization studies [9-11] in combination with the Mössbauer experiments [12,13] on rare-earth nuclei provided some insight in the contributions to the magnetocrystalline anisotropy energy arising from the different rare-earth sites. Magnetic moment and magnetocrystalline anisotropy of rare-earth ions are determined by the energy level scheme and assuming that the ground-state multiplet is given by Hund's rules, the magnetic behaviour can be described in terms of crystal- and molecular-field interactions.

In the hexagonal and tetragonal R-T intermetallics, the number of relevant CEF parameters is 4 or 9. Assuming the 3d-4f exchange interactions to be isotropic, the total number of parameters that represent the magnetic behaviour of the trivalent rare-earth ion in the R_2T_{17} and $R_2Fe_{14}B$ compounds is at least 9 or 19, respectively. In this situation it is helpful if some parameters, or some relations between them, can be established by questionless independent methods.

In previous publications we pointed out that the 3d-4f exchange interaction parameter can be deduced accurately from exchange-driven transitions that are observed in the magnetisation curves of some ferrimagnetic R-T intermetallic compounds for external magnetic field applied along the easy direction of magnetisation [10,14,15]. These transition fields are typically in the field range 20 and 100 T. At present they can be experimentally studied.

In several compounds of the above mentioned series, the zero field, zero temperature magnetic structures turns out to be tilted. It means that the easy direction of magnetisation is not along any of the main symmetry direction of the structure. This tilted structure imposes some specific relations between the crystal-field coefficients of different order and is again helpful in reducing the number of free parameters. Still, the number of free parameters is too large to deduce an unique set of crystal-field coefficients from an extended study of one particular compound by, for instance, inelastic-neutron, Mössbauer, NMR or high-field magnetisation studies at low temperatures. In fact, one has to consider the whole series of compounds of one particular composition, the R_2T_{17} compounds, for instance. By putting the additional requirement that within such a series the exchange and crystal-field parameters for the different compounds in the isostructural series are related in a consistent way, one can be confident concerning the results deduced from such a highly correlated analysis. The set of crystal-field and exchange parameters derived in this way provides an energy-level scheme for the trivalent rare-earth ions that can be further verified in specific-heat, magnetisation and magnetic anisotropy studies at elevated temperatures. A review of magnetisation curves shows that a magnetic tilted structure is a special feature of many holmium compounds at low temperatures [16-19]. The easy direction of magnetisation does not lie along one of the main crystallographic directions. It is noted in magnetisation measurements by a non-zero value of the spontaneous magnetic moment along the "hard" axis.

The aim of this contribution is to discuss the origin of the tilted magnetic structure in the hexagonal Ho_2Co_{17} and the tetragonal $Ho_2Fe_{14}B$ compounds on the basis of results obtained from inelastic neutron scattering, Mössbauer, specific heat and magnetisation data available at present. Special attention is devoted to the analysis of high-field magnetisation curves. A tentative set of crystal-field and exchange parameters is provided for both compounds. The presented sets of crystal-field coefficients and the value of the exchange field B_{ex} do not only represent the trivalent holmium ion but are also relevant for other compounds of 2:17 and 2:14:B isostructural series.

2. Theoretical outline.

The high-field magnetic behaviour of these 3d-4f compounds has been analyzed by taking into account the contribution originating from rare-earth and 3d ions. The contribution arising from the rare-earth ions is considered to be describable by a Hamiltonian including the crystal-field and isotropic exchange interactions, the latter being represented by a molecular field B_{mol}^R. As the spin-orbit interactions are much stronger than the crystal- and molecular-field interactions, the ground-state multiplet as given by Hund's rules (for Ho: $J = 8$) is considered only. Therefore, the rare-earth Hamiltonian in the presence of an external field H is considered to be of the form:

$$H_R = H_{CF} + g \mu_B J (B_{mol}^R + B_o) \tag{1}$$

where $B_o = \mu_o H$.

The molecular field experienced by the rare-earth moment is composed from two contributions originating from the 3d and 4f surrondings

$$B_{mol}^R = \mu_o n_{RT} M_T + \mu_o n_{RR} M_R \tag{2}$$

where M_T and M_R are the magnetisations of the 3d and 4f sublattices, respectively. n_{RT} and n_{RR} are the inter- and intra-sublattice molecular-field coefficients. It is generally accepted that the first term strongly dominates the second term, that is usually neglected. The molecular field experienced by the rare-earth magnetic moment, $B_{RT}^R = \mu_o n_{RT} M_T$, is related to the exchange field, experienced by the rare-earth spin moment B_{ex}^R, by the relation

$$B_{ex}^R = \frac{g_R}{2 (g_R - 1)} B_{RT}^R. \tag{3}$$

This relation can be obtained by comparing the 3d-4f interaction written in two forms;

$$g_R \mu_B J_R B_{RT}^R = 2 \mu_B S_R B_{ex}^R \tag{4}$$

with taking into account that for a rare-earth ion the following relation holds; $S_R = (g_R - 1) J_R$. Since the 3d-4f interactions are generally accepted to be interactions between the spins we expect the exchange field to be the fundamental parameter rather than the molecular field. Moreover, it is reasonable to expect the exchange field to be rather constant or to show a smooth behaviour in an iso-structural series. In case of a constant value of B_{ex}^R the value of the molecular field B_{RT}^R will change considerably over the series following the factor of $2(g_R - 1)/g_R$ [20].

The crystal-field Hamiltonian for the rare-earth ion is taken in the form relevant to the point symmetry of the crystallographic site:

$$H_{CF} = B_2^0 O_2^0 + B_4^0 O_4^0 + B_6^0 O_6^0 + B_6^6 O_6^6 \quad (5)$$

in the hexagonal Th_2Ni_{17}-type of structure for Ho_2Co_{17} and

$$H_{CF} = B_2^0 O_2^0 + B_4^0 O_4^0 + B_4^4 O_4^4 + B_6^0 O_6^0 + B_6^4 O_6^4 \quad (6)$$

$$+ B_2^2 O_2^2 + B_4^2 O_4^2 + B_4^2 O_4^2 + B_6^6 O_6^6$$

in the tetragonal $Nd_2Fe_{14}B$-type of structure for $Ho_2Fe_{14}B$. The last four terms appear in eq. 6 as the local symmetry of the rare-earth ion is lower than tetragonal, i.e. orthorhombic. B_n^m are the CF parameters associated with the Stevens operators O_n^m.

The 3d ions are treated phenomenologically by taking in first approximation the values for the magnetic moment and for the anisotropy energy of the 3d sublattice equal to those observed in an isostructural compound with a non-magnetic rare-earth partner, usually yttrium. In the local-site picture of the 3d anisotropy in this class of compounds [21,22] these values should be regarded as averages over different crystallographic sites of the 3d ions, four in this case. The 3d moments are coupled together by exchange interactions between the 3d spins, that are regarded to be sufficiently strong, into one sublattice with resultant values for the magnetic moment and for the anisotropy energy. The Curie temperatures for Y_2Co_{17} and $Y_2Fe_{14}B$ lead to values for the 3d-3d exchange coupling parameter of $J_{Co-Co}/k_B = +115\,K$ and $J_{Fe-Fe}/k_B = +34\,K$, respectively. Both values are much larger than the value of about 7 K that is found for the exchange interaction parameter between the 3d and the heavy rare-earth spins [10,20,23]. A quantitative treatment of the influence of the magnetic rare-earth partner on the magnetic parameters of the 3d sublattice is still hampered by the limited experimental accuracy, mainly connected with the uncertainty of the exact composition of the compound.

3. Results and discussion.

In fig. 1 we show the high-field magnetisation curves for the compound Ho_2Co_{17} along the different crystallographic directions. The magnetisation curves along the b-axis and a-axis show transitions that are due to a breaking of the collinear ferrimagnetic structure.

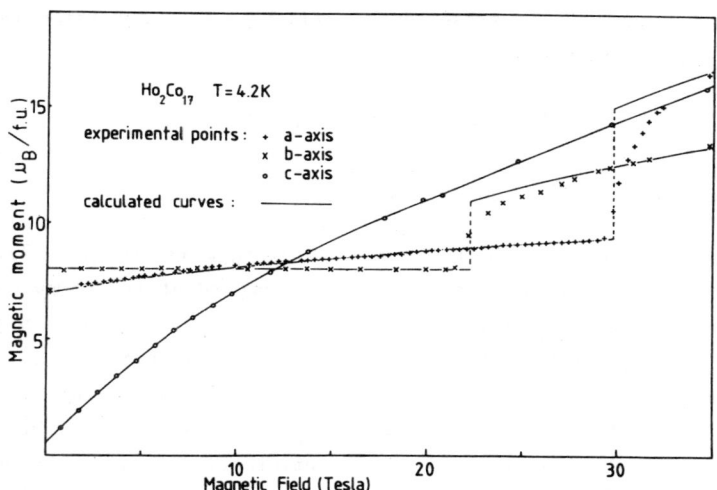

Fig. 1. The high-field magnetisation curves of single-crystalline Ho_2Co_{17} at 4.2 K. Experimental data taken after ref. 10. The lines represent the theoretical fits.

In the case that a large number of CF parameters is present, the analysis is not easy. The existence of two different crystallographic sites for the rare-earth ions makes the problem even more puzzling. For that reason we start with a classical approach for the description of the magnetisation curves that, nevertheless, provides good insight into the high-field-magnetisation process [8,24,25]. In this approach the free energy of a 3d-4f system is considered as:

$$E_{R-T} = E_a^{R,i} + E_a^{T} - \mu_o n_{RT} M_T \cdot M_R^i - M_T \cdot B_o - M_R \cdot B_o \qquad (7)$$

where i refers to different sites of the rare earth ions.

The anisotropy energy is expressed by anisotropy coefficients associated with the Legendre functions $P_n^m(\cos\theta)$:

$$E_a(\theta,\alpha) = \sum_{n,m} \kappa_n^m P_n^m(\cos\theta) \cos m\alpha \qquad (8)$$

where θ and α are the polar and azimuthal angles of the magnetisation with respect to the crystallographic axes. A list of the functions $P_n^m(\cos\theta)$ employed in our analysis is presented in ref. 26. The expression (8) for the hexagonal symmetry takes the form:

$$E_a(\theta,\alpha) = \kappa_2^o P_2^o(\cos\theta) + \kappa_4^o P_4^o(\cos\theta) \qquad (9)$$
$$+ \kappa_6^o P_6^o(\cos\theta) + \kappa_6^6 P_6^6(\cos\theta) \cos 6\alpha$$

A standard analysis for searching the minimum of E_{R-T} as given by (7), with as fitting value the resultant magnetisation composed of the components of the two sublattices magnetisations along the field direction, provides pairs of the angles θ and α for the sublattice magnetisations.

In order to understand the transitions that take place in easy-plane ferrimagnets we introduce a simplified picture in which the rare-earth anisotropy within the hexagonal plane is infinitely strong and with zero in-plane anisotropy for the 3d sublattice. The appropriate expression for the free energy in this case is the sum of the intersublattice interaction $\mu_o n_{RT} M_R \cdot M_T$ and the Zeeman magnetostatic energy $-B_o \cdot (M_R + M_T)$.

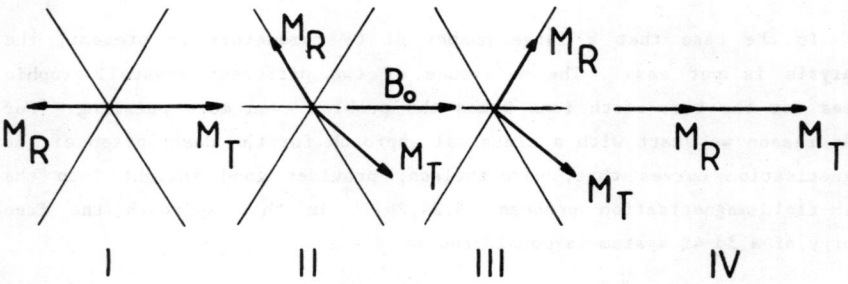

Fig. 2. Different configurations of the rare-earth and 3d magnetisations of a ferrimagnetic easy-plane hexagonal compound realised with an increasing value of the external magnetic field, B_o.

The direction of the rare-earth moment is completely determined by the in-plane anisotropy. Taking at the transition field the value of B_o $(M_T + M_T)$ equal to zero, we deduce from the transition field of 22.0 T, as observed for Ho_2Co_{17}, a value of 46.5 for the parameter n_{Ho-Co}. In a full analysis, in which the finite anisotropy of the rare-earth sublattice is taken into account, the resulting value for n_{Ho-Co} is 47.8 [26]. Due to the six-fold symmetry of the in-plane anisotropy, the non-collinear structure after the transition requires a rather small canting angle of approximately $20°$. As a consequence, small amounts of exchange energy are lost at the transition and they can occur at relatively low fields. For easy-plane ferrimagnetic compounds, three transitions should occur along the easy axis, whereas along the hard axis in the plane, two transitions are expected [15]. By increasing the external field along one of the easy axes in the basal plane different configurations as shown in fig. 2 can be achieved.

The values for the anisotropy coefficients κ_n^m of the holmium sublattice derived from the magnetisation curve measured for external fields applied along the c-axis, the "hard-axis" of magnetisation, are collected in table 1.

Table 1. Magnetocrystalline anisotropy coefficients κ_n^m for the two Ho sublattices in Ho_2Co_{17} assigned to the crystallographic sites, b and d, as derived from high-magnetic-field studies at 4.2 K. The density amounts to $9.04 \cdot 10^3 kg/m^3$.

κ_n^m	b		d	
	K/Ho-ion	J/kg	K/Ho-ion	J/kg
κ_2^0	+70.0	+440	+24.4	+152
κ_4^0	+13.0	+81	-18.0	-112
κ_6^0	-13.2	-82	-20.5	-128
κ_6^6	+15.2	+95	+15.2	+95

The angular dependence of the magnetocrystalline anisotropy energy of the holmium ion is shown in fig. 3 for the two crystallographic sites, b and d. The assignment to the sites b and d is made on the basis of a point-charge-model result that indicates a strong positive value for B_2^0 for the site b [27]. One can see from fig. 2 that there is a broad plateau of the MCA energy of the Ho ion in Ho_2Co_{17} in the interval 75 - 90° with a slight minimum at 85°. Such a shape of the anisotropy energy is produced by a mutual cancellation of the leading second- and higher-order contributions

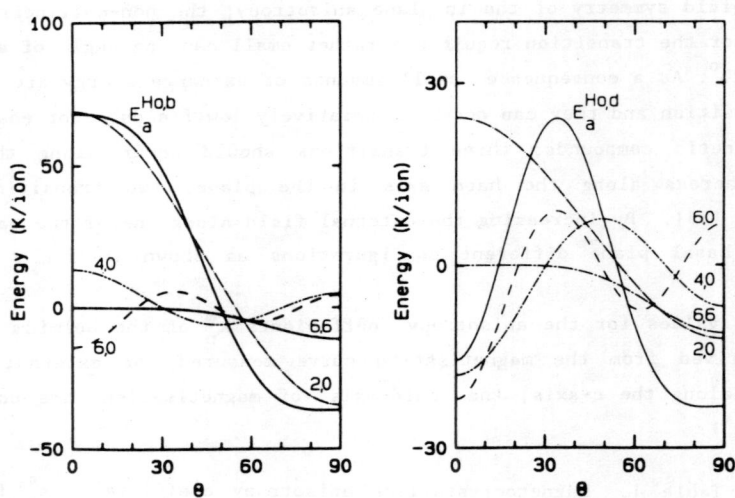

Fig. 3. The angular dependence of the MCA energy of the holmium ion at the two crystallographic sites, b and d, in Ho_2Co_{17}. The different multipole contribution $\kappa_n^m P_n^m(\cos\theta)$ are shown; values for (n,m) are indicated.

to the MCA energy. The second-order contribution strongly dominates the other ones in case of site b. One should note that the anisotropy coefficients of the cobalt sublattice, $\kappa_2^0 = +6.0\,\text{K/f.u.}$ and $\kappa_4^0 = -0.09\,\text{K/f.u.}$ are one order of magnitude smaller than those of the holmium sublattices. The anisotropy of the cobalt sublattice has been derived by a magnetic torque studies on single-crystalline Y_2Co_{17} [28].

As indicated before, a value for n_{Ho-Co} of 47.8 for the intersublattice molecular field coefficient has been obtained. It results in a value for the molecular field experienced by the holmium moment, $B_{RT}^{Ho} = \mu_o\, n_{RT}\, M_T$ of 63.5 T, in perfect agreement with the result reported previously by Franse et al. [10] from a much simpler analysis. Note that for Ho_2Co_{17}, M_{Co} is equal to $28.00\,\mu_B/\text{f.u.}$ which value corresponds to $1.06\,\text{MA/m}$. In the large exchange limit the anisotropy coefficients κ_n^m are directly related to the crystal-field parameters B_n^m (for the relations see, for instance, ref. 26). The values for the CF parameters are collected in table 2.

Table 2. Crystal-field parameters B_n^m and coefficients A_n^m for the trivalent Ho ion at the two crystallographic sites, b and d, in Ho_2Co_{17} as derived from high-magnetic-field studies at 4.2 K. a_o is the Bohr radius.

B_n^m	b K/Ho-ion	d K/Ho-ion	A_n^m	b $K\,a_o^{-n}$	d $K\,a_o^{-n}$
B_2^0	+0.58	+0.20	A_2^0	-348	-121
B_4^0	$+6.0\cdot 10^{-4}$	$-8.2\cdot 10^{-4}$	A_4^0	-13.1	+17.8
B_6^0	$-13.5\cdot 10^{-6}$	$-18.1\cdot 10^{-6}$	A_6^0	+1.9	+2.6
B_6^6	$+137\cdot 10^{-6}$	$+137\cdot 10^{-6}$	A_6^6	-19.7	-19.7

In table 2, CEF coefficients $A_n^m = B_n^m/\theta_n \langle r_{4f}^n \rangle$, where θ_n is the Stevens factor of order n and where values of $\langle r_{4f}^n \rangle$ are taken after ref. 29, are included because these coefficients better characterize the crystal field. Identical values for the parameter B_6^6 at both Ho sites in Ho_2Co_{17} are in agreement with point-charge calculations [30].

The energy-level scheme of the Ho^{3+} ion under the combined action of the crystal and molecular fields is presented in fig. 4. The internal molecular field lifts completely the degeneracy of the energy levels. The difference between the ground-state energies for the internal field acting along the angles $\theta = 85°$ and $0°$ is a measure for the MCA energy at helium temperature. It is negative for both sites and amounts to -120 K and -10 K for the sites b and d, respectively. The ground state is characterized by a value of the magnetic moment of $9.91\mu_B$ at both sites. Despite the quite different values for the CEF parameters at the sites b and d, the energy separations to the three lowest excited levels are practically the same for the two sites (table 3). These energy separations are in good agreement with results obtained from inelastic neutron scattering studies by Clausen [31] and Clausen and Lebech [32]. It is worth noting that recently we have succeeded in analysing the specific heat measurements of Ho_2Co_{17} with the crystal-field contribution calculated with these sets of CEF parameters

[33]. The values reported here for the parameters B_2^0 are consistent with results obtained by Mössbauer spectroscopy for the isostructural compound Tm_2Co_{17} [34]. According to these Mössbauer studies the ratio of the parameters B_2^0 at the two sites amounts to 3.0, whereas in case of Ho_2Co_{17} our analysis results in a value of 2.9.

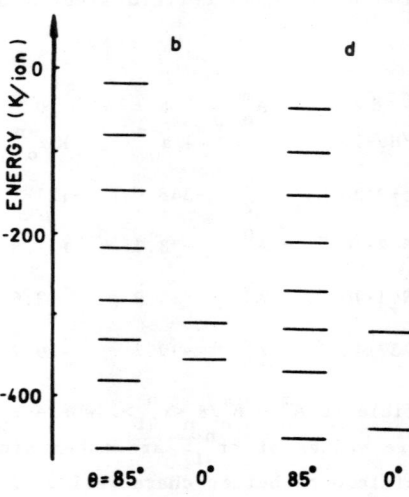

Fig. 4. The energy-level scheme of the trivalent holmium ion in Ho_2Co_{17} at the two sites, b and d, under the combined action of the crystal field and the molecular field acting along the polar angle θ. The angle θ is indicated. Only eight lowest levels for $\theta = 85°$ and two for $\theta = 0°$ are shown. The level scheme for $\theta = 90°$ is quite similar to that for $\theta = 85°$.

Table 3. The energy separations Δ_n (in units of K) to the five lowest excited states for the trivalent Ho ion in Ho_2Co_{17} at the two crystallographic sites, b and d, as derived from the set of CF parameters presented in table 1. The separations obtained from inelastic-neutron-scattering experiments in ref. 32 are included.

Δ_n	b	d	Exp [32]
Δ_1	81.7	82.1	82(1)
Δ_2	134.9	135.7	138(3)
Δ_3	179.0	179.4	182(3)
Δ_4	245.0	240.0	-

A similar analysis of high-field magnetisation curves has been performed for $Ho_2Fe_{14}B$. The magnetisation curves are presented in fig. 5.

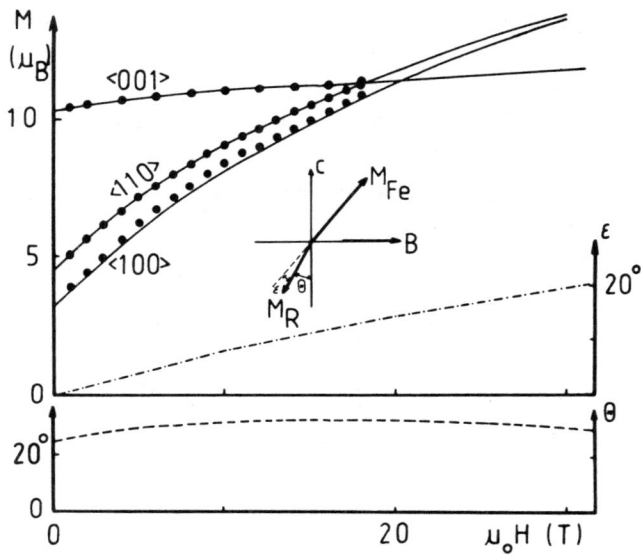

Fig. 5. High-field magnetisation curves of $Ho_2Fe_{14}B$ at 4.2 K [25] with the experimental data taken after ref. 11. The variation of the canting angle ε with the increase of the external magnetic field applied along the direction <110> is shown. The rotation of the holmium-sublattice magnetisation described by the angle θ is also shown.

The molecular field, in this case, can not directly be deduced from a first-order moment-reorientation transition like for Ho_2Co_{17}. From the other hand, there are several ways to arrive at an approximate for this field value on the basis of a microscopic model for the exchange interactions. Assuming a nearest neighbour exchange between the 4f and 3d spins only, we estimate a value of about 80 T for the molecular field experienced by the holmium moment in $Ho_2Fe_{14}B$ [23]. The fitting of the magnetisation curve measured up to 18 T provides a value of 86 T [25] in good agreement with ref. 23. The relatively large value for the molecular field assures that the molecular-field effect largely dominates the crystal-field effect with the consequence that the magnetic moment at 4.2 K is very close to that expected from Hund's rules. A fitting procedure of all magnetisation curves shown in fig. 5 enables us to evaluate the holmium contribution to the anisotropy energy in $Ho_2Fe_{14}B$ and to decompose it in the different terms. Since the Mössbauer analysis indicates the B_2^0 values for the two different sites in $Gd_2Fe_{14}B$ are almost the same, we restrict ourselves to one set of crystal-field parameters only.

The angular dependence of the holmium anisotropy energy is presented in fig.6. There is a minimum in the vicinity of $24°$ showing that this direction is preferred by the holmium magnetic moment.

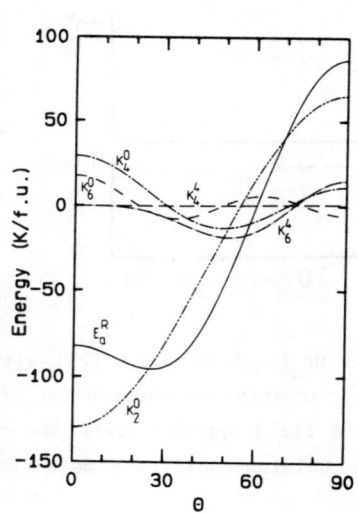 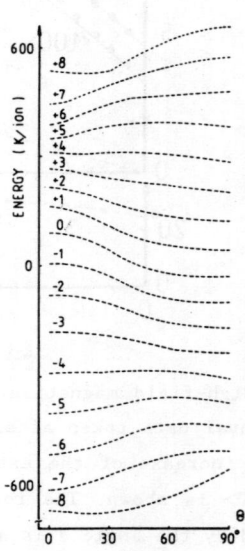

Fig. 6 (on left). The angular dependence of the anisotropy energy of the holmium moment in $Ho_2Fe_{14}B$ together with the different contributions.

Fig. 7 (on right). The energy level scheme of the trivalent holmium ion under the combined action of the crystal field and the molecular field acting along the polar angle θ, in $Ho_2Fe_{14}B$. The angle θ is indicated. The levels are labeled with the largest component of $|J_z>$ to the eigenfunctions.

The energy-level scheme of the trivalent holmium ion calculated with the crystal- and molecular-field Hamiltonian (1) as a function of the direction of the internal (molecular) field arising from the iron sublattice, also shows a minimum at 24° (fig. 7). By inspection of figs 6 and 7, one notices that the angular dependence of the ground-state level follows the angular dependence of the total anisotropy energy of the holmium sublattice. The sets of the anisotropy coefficients as well as crystal-field parameters are presented in table 4.

Order		κ_n^m		B_n^m	A_n^m
		K/f.u.	J/kg	K/Ho-ion	$K\,a_o^{-n}$
2	0	-130	-963	-0.56	+334
4	0	+30	+222	$+6.6\cdot 10^{-4}$	-14.4
6	0	+18	+133	$+6.3\cdot 10^{-6}$	-0.9
4	4	-1	-7	$-1.8\cdot 10^{-4}$	+3.9
6	4	+14	+103	$+83\cdot 10^{-6}$	-11.9
2	0,f	-	-489	-0.57	+340
2	0,g	-	-474	-0.55	+330

Table 4. Magnetocrystalline anisotropy coefficients κ_n^m of the holmium sublattice in $Ho_2Fe_{14}B$ as derived from high-field magnetization experiments together with CEF parameters for the trivalent Ho ion. The density amounts to $8.12\cdot 10^3 kg/m^3$. In the bottom part the sub-division of the second-order CF term according to the Mössbauer result of Bogé et al. [12] is included.

The anisotropy energy is frequently expressed as:

$$E_a = K_1 \sin^2\theta + K_2 \sin^4\theta + K_3 \sin^6\theta + K_4 \sin^6\theta \cos 6\alpha, \tag{10}$$

where the parameters K_i are known as the anisotropy constants. The anisotropy coefficients κ_n^m are related to the anisotropy constants K_i (see for instance ref. 26). We would like to point out that in the two-sublattice model the expression (10) is used separately for the rare-earth and the 3d sublattice anisotropy, whereas in the standard analyses, based on the Sucksmith-Thompson plot, the anisotropy constants $\{K_i\}$ are attributed to the whole magnetic system. In this standard analysis the free energy of a magnetic system is considered to be of the form:

$$E = E_a - \mathbf{M}_s \cdot \mathbf{B}_o \tag{11}$$

where the value of the resultant magnetisation M_s ($= M_T - M_R$) is considered to be constant during the magnetisation process. Such an approach leads, especially for ferrimagnetic systems, to anisotropy constants that are only effective parameters [35]. These constants are a complex function of the inter-sublattice interactions and of the intrinsic local anisotropies of the different sublattices. A similar conclusion has been also drawn in ref. 36 and 37 in an analysis of $Nd_2Fe_{14}B$. These constants, however, are still useful for a brief characterisation of magnetic materials with respect to technical purposes.

The effective anisotropy energy are much smaller than the intrinsic anisotropy energy. The values for κ_n^m of the holmium sublattice in $Ho_2Fe_{14}B$ from table 4 lead to $K_1^{Ho} = -1085$ J/kg, $K_2^{Ho} = +4150$ J/kg and $K_3 = -1940$ J/kg, values that are nearly one order of magnitude larger than those derived from the standard analysis. The effective anisotropy constants have been reported in ref. 17 to be $K_1 = -160$ J/kg and $K_2 = +566$ J/kg or in ref. 25 $K_1 = -204$ Jkg, $K_2 = +661$ J/kg and $K_3 = -218$ J/kg.

The value for the molecular field, experienced by the holmium moment and arising from the 3d sublattice, is considerably larger in $Ho_2Fe_{14}B$ compared to Ho_2Co_{17}. This is attributed to the larger iron moment. The ratio of the molecular fields follows perfectly the ratio of the 3d magnetic moment in these compounds: $2.23\,\mu_B$ and $1.65\,\mu_B$, respectively. A similar behaviour has been found also in the $Ho_2(Co-Fe)_{17}$ series [38] indicating that the coupling parameter between the 3d and 4f spins is hardly sensitive to the type of the 3d spin, at least in case of Fe and Co.

4. Conclusions

The analysis of magnetisation curves within a classical approach based on a two-sublattices model provides a realistic description of the high-field magnetisation process. It permits to take explicitly into account the field-induced non-collinearity of a ferrimagnetic system as well as to determine the rare-earth contribution to the magnetocrystalline anisotropy energy of 3d-4f intermetallic systems. A description of the magnetocrystalline anisotropy in terms of coefficients associated to the Legendre functions has been found to be very useful for the decomposition of the rare-earth contribution into different components originating from different crystal-field parameters. In a single-ion concept of the anisotropy energy the coefficients κ_n^m should follow the Stevens factors and should be roughly related in isostructural compounds to the values for $\theta_n \langle r_{4f}^n \rangle J_n$. Values for this product are collected in ref. 26. Values of J_n are connected with values of the Stevens operators $O_n^m(J_z)$ for the ground state with $J_z = J$.

It is worth to mention that due to a non-zero value of the 3d MCA anisotropy and a finite value of the intersublattice interactions in both cases the 3d-sublattice magnetisation is not exactly antiparallel to the holmium sublattice magnetisation. In case of Ho_2Co_{17}, the magnetic structure of the two rare-earth and the 3d sublattice magnetisations is planar in the b-c plane, whereas in case of $Ho_2Fe_{14}B$ off-diagonal terms mainly of the second order, produce a non-planar ferrimagnetic fan structure [25]. This spatial fan structure has been deduced to be limited to a few degrees. Quite recently such fan structures have been confirmed in polarized-neutron experiments by Fruchart et al. [39].

The conclusion of this paper is that the magnetic properties of the intermetallic compound Ho_2Co_{17} as studied by Mössbauer, inelastic-neutron-scattering, specific-heat and high-magnetic-fields experiments can be well understood on the basis of isotropic exchange interactions between the 3d and 4f spins and crystal-field interactions. The tilted magnetic structure observed in hexagonal Ho_2Co_{17} and tetragonal $Ho_2Fe_{14}B$ at zero external field entirely result from crystal-field interactions. The effect of higher-order crystal-field parameters is largest for Ho and Nd ions and is one of the reasons that these compounds are studied so extensively.

Values derived for the exchange field experienced by the holmium ion of 159 T in Ho_2Co_{17} and of 215 T in $Ho_2Fe_{14}B$ indicate that the coupling parameter J_{RT} between the 3d and 4f spins is hardly sensitive neither to the type of the 3d spin, at least in case of Fe and Co, nor to the crystallographic structure. Using the relation (4) from ref. 20, the values for the exchange field lead to $J_{Ho-Co}(Ho_2Co_{17}) = -6.72$ K and to $J_{Ho-Fe}(Ho_2Fe_{14}B) = -7.2$ K.

The present approach, based on the crystal- and molecular-field Hamiltonian, is applicable to many rare-earth intermetallic compounds e.g. RCo_5, R_2Co_7, RCo_3, RCo_2, $RFe_{11}Ti$. It can be also quite useful for compounds where strong exchange interactions are absent and one deals solely with crystal field interactions.

Acknowledgements. This work was partly supported by the Commission of the European Communities within the EURAM project and by the Dutch Foundation for Fundamental Research of Matter (FOM).

References

[1] H.H.A.Smit, R.C.Thiel and K.H.J.Buschow, Physica B145 (1987) 329; J.Phys.F: Met.Phys. 18 (1987) 295.
[2] F.Grandjean, G.J.Long, D.E.Tharp, O.A.Pringle and W.J.James, ICM-1988, 3P H-15.
[3] C.Kapusta, Z.Kąkol, H.Figiel and R.J.Radwaśki, J.Magn.Magn.Mat. 59 (1986) 169.
[4] K.Erdmann, M.Rosenberg and K.H.J.Buschow, ICM-88, 3P H-16.
[5] M.Wójcik, E.Jędryka, M.Rosenberg, S.Hirosawa and M.Sagawa, ICM-88, 3P H-18.
[6] D.Givord, H.S.Li and F.Tasset, J.Appl.Phys. 57 (1985) 4100.
[7] W.B.Yelon and J.F.Herbst, J.Appl.Phys. 59 (1986) 93.
[8] Zong-Quan Gu and W.Y.Ching, Phys.Rev. B36 (1987) 8530.
[9] M.Yamada, Y.Yamaguchi, H.Kato, H.Yamamoto, Y.Nakagawa, S.Hirosawa and M.Sagawa, Solid State Comm. 56 (1985) 663.
[10] J.J.M.Franse, F.R. de Boer, P.H.Frings, R.Gersdorf, A.Menovsky, F.A.Muller, R.J.Radwański and S.Sinnema, Phys.Rev. B31 (1985) 4347;
[11] D.Givord, H.S.Li, J.M.Cadogan, J.M.D.Coey, J.P.Gavigan, O.Yamada, H.Maruyama, M.Sagawa and S.Hirosawa, J.Appl.Phys. 63 (1988) 3713.
[12] M.Bogé, J.M.D.Coey, G.Czjzek, D.Givord, C.Jeandey H.S.Li and J.L.Oddou, J.Phys.F: Met.Phys. 16 (1986) L67-72.
[13] H.R.Rechenberg, J.P.Sanchez, P.L'Héritier and R.Fruchart, Phys.Rev. B36 (1987) 1865.
[14] R.J.Radwański, J.J.M.Franse and S.Sinnema, J.Magn.Magn.Mat. 51 (1985) 175.
[15] R.J.Radwański, J.J.M.Franse and S.Sinnema, J.Phys.F: Metal Phys. 15 (1985) 969.
[16] H.Yamauchi, M.Yamada, Y.Yamauchi, H.Yamamoto, S.Hirosawa and M.Sagawa, J.Magn.Magn.Mat. 54-57 (1986) 575.
[17] A.Fujita, H.Onodera, H.Yamauchi, M.Yamada, H.Yamamoto, J.Magn.Magn.Mat. 69 (1987) 267.

[18] J.J.Rhyne, in: Magnetic Properties of Rare Earth Metals, ed. R.J.Elliott (Plenum Press, London, 1972) ch.4.
[19] M.Slaski, J.Leciejewicz and A.Szytuła, J.Magn.Magn.Mat. **39** (1983) 268.
[20] R.J.Radwański, Phys.Stat.Sol.(b) **137** (1986) 487.
[21] Z.Kąkol and H.Figiel, Phys. Stat.Sol.(b) **138** (1986) 151;
[22] N.P.Thuy and J.J.M.Franse, J.Magn.Magn.Mat. **54-57** (1986) 315; J.J.M.Franse, N.P.Thuy and N.M.Hong, J.Magn.Magn.Mat. **72** (1988) 361.
[23] R.J.Radwański, Z.Phys.B-Condensed Matter **65** (1986) 65.
[24].R.J.Radwański, Physica B **142** (1986) 57.
[25] R.J.Radwański and J.J.M.Franse, J.Magn.Magn.Mat. **74** (1988) 43;
[26] R.J.Radwański and J.J.M.Franse, Physica B **154** (1988).
[27] J.E.Greedan and V.U.Rao, J.Solid State Chem. **6** (1973) 387.
[28] B.Matthaei, J.J.M.Franse, S.Sinnema and R.J.Radwański, J.Physique (Paris), ICM-88, 3P B-14.
[29] A.J.Freeman and J.P.Desclaux, J.Magn.Magn.Mat. **12** (1979) 11.
[30] R.J.Radwański, J.Phys.F: Metal Phys. **17** (1987) 267.
[31] K.N.Clausen, Thesis, Riso-R-426 (1981).
[32] K.N.Clausen and B.Lebech, J.Phys.C: Solid State Phys. **15** (1982) 5095.
[33] R.J.Radwański, J.J.M.Franse, J.C.P.Klaasse and S.Sinnema, ICM-88. J.Physique (1988), 3P B-13.
[34] P.C.M.Gubbens, A.M. van der Kraan, J.J. van Loef and K.H.J.Buschow, J.Magn.Magn.Mat. **67** (1987) 255.
[35] R.J.Radwański, J.J.M.Franse and S.Sinnema, J.Magn.Magn.Mat. **70** (1987) 313.
[36] J.M.Cadogan, J.P.Gavigan, D.Givord and H.S.Li, J.Phys.F: Met.Phys. **18** (1988) 779.
[37] J.P.Gavigan, Thesis, University of Dublin, 1988.
[38] S.Sinnema, J.J.M.Franse, R.J.Radwański, A.Menovsky and F.R. de Boer, J.Phys: Metal Phys. **17** (1987) 233.
[39] Fruchart et al. (1988) to be published

SPIN GLASS AND INVAR PROPERTIES OF IRON-RICH AMORPHOUS ALLOYS

K. Fukamichi, T. Goto*, H. Komatsu and H. Wakabayashi**

Department of Materials Science, Faculty of Engineering, Tohoku University, Sendai 980, Japan, * The Institute for Solid State Physics, The University of Tokyo, Tokyo 106, Japan, ** Department of Applied Physics, Faculty of Science , Tokyo 152, Japan

ABSTRACT

Several kinds of Fe-rich amorphous alloys were prepared by sputtering and melt-quenching in order to investigate the spin glass and Invar properties because these properties in the disordered materials are important from the view point of fundamental research.

With decreasing solute concentration, that is, coming close to pure Fe, the Curie temperature drastically decreases and eventually no long-range order is established, resulting in the spin glass state. In the magnetic phase diagram, the re-entrant spin glass behavior is observed above the triple point composition. In association with the spin glass state, the high-field susceptibility is extremely large. On the other hand, no significant anomaly in the thermal expansion and in the electrical resistivity curves is confirmed at the spin freezing temperature.

In connection with the Invar effect, a remarkably large pressure effect on the Curie temperature and a large compressibility are obtained. From various experimental results, it is concluded that the magnetic state of the present Fe-rich amorphous alloys is inhomogeneous.

INTRODUCTION

Magnetic properties of a number of amorphous alloys have been investigated from both physical and practical viewpoints. Fe-rich amorphous alloys exhibit various peculiar properties in contrast with Co- and Ni-rich amorphous alloys[1]. Physical properties of Fe and its alloys have been discussed extensively. Recently, it has been pointed out that YFe_2 amorphous alloy shows a spin glass behavior and this alloy has been called the concentrated spin glass[2]. Other composition amorphous alloys of Fe-Y also show the spin glass behavior accompanying no long-range magnetic order[3]. More recently, re-entrant spin glass behavior has been demonstrated in Fe-Zr amorphous alloys obtained by melt quenching[4,5]. It has been confirmed that several kinds of Fe-metal alloy systems also show similar behavior[6]. Therefore, the spin glass behavior is considered to be common in Fe-rich amorphous alloys.

The magnetic properties of Fe and its alloys are markedly affected by crystalline structure. Especially, γ-Fe and fcc alloys are interesting materials in connection with the antiferromagnetic interactions and the Invar properties. It is considered that the amorphous structure is similar to fcc because its coordination number is very close to 12[7]. This is the main reason why Fe-rich amorphous alloys exhibit the Invar properties, e.g., the anomalous thermal expansion, large pressure effect on the Curie temperature in analogy with Fe-Ni crystalline fcc Invar alloys. In the case of Fe-rich crystalline Invar alloys, their concentration range is very close to the martensitic phase transition. On the other hand, Fe-rich amorphous alloys have no such transformation. Furthermore, the long-range antiferromagnetic order is not established in the disordered materials such as amorphous alloys because of the spin frustration[8]. From these reasons, the latter alloys are good candidates for the study of the spin glass and Invar properties.

Since many physical properties are interrelated with one another, we can see many anomalous behaviors in Fe-rich amorphous alloys. In the present paper, the following various experimental results are discussed and compared with the existing data on Fe-rich amorphous and crystalline alloys with respect to the spin glass and Invar properties. The discussion is mainly concentrated on the data on Fe-metal amorphous alloys, because the magnetic and Invar properties of Fe-metalloid amorphous alloys were already reviewed[1].

(A) The Curie Temperature of Fe-ET Alloy Systems
(B) Ac Susceptibility and Magnetic Cooling Effect
(C) Magnetic Phase Diagram

(D) Magnetic Moment and High-field Susceptibility
(E) Low Temperature Specific Heat
(F) Thermal Expansion and the ΔE Effect
(G) Pressure Effect on the Curie and Spin Freezing Temperatures
(H) Electrical Resistivity Minimum and Compressibility

EXPERIMENTAL

Alloy targets with about 50 mm in diameter were made by arc-melting in an argon atmosphere. Almost all samples were prepared by high-rate dc sputtering on a Cu substrate. The argon pressure during sputtering was 40m Torr and the target voltage was 1.0 kV. In order to make thick samples about 0.2 mm the sputtering was carried out for about 3 days because such a thick sample is necessary for the various following measurements[9]. The Cu substrate was dissolved away in a solvent of H_2O(1000 cc) + CrO_3(500 g) + H_2SO_4(27 cc). Some Fe-Zr and Fe-Hf amorphous alloys in the vicinity of the eutectic composition were prepared by melt-quenching in an argon atmosphere. Their amorphous state was examined by X-ray diffraction.

The magnetizations were measured with a vibrating sample magnetometer up to 145 kOe by using a water-cooled magnet and also by an induction method up to 60 kOe by using a superconducting magnet. Very high field measurements up to 300 kOe were made by an induction method using a wire wound pulse magnet. The measurements of magnetic cooling effect and ac susceptibility were carried out by an induction method and by a mutual induction method, respectively, using about 100 mg samples.

The temperature dependence of electrical resistivity was measured by a conventional four terminal method. The thermal expansion was measured with a differential transformer type dilatometer at a heating rate of 2.5 K/min. The temperature dependence of Young's modulus was measured at zero and external magnetic fields by an electrostatic driving method.

Pressure was applied to the specimen in a Teflon pressure cell filled with a silicon oil by using a piston-cylinder type device and the shift of the Curie temperature was determined from the shift of initial permeability[10].

RESULTS AND DISCUSSION

(A) The Curie Temperature of Fe-ET Alloy Systems

The concentration dependence of the Curie temperature of several kinds of Fe-ET(ET:early transition metal) amorphous alloys is presented in Fig. 1. With decreasing solute concentration, the Curie temperature increases until about 20 %, but the drastic decrease occurs in the low concentration range in the similar manner as that of Fe-rich fcc alloys having the Invar properties. Such a drastic decrease suggests that the magnetic long-range order is not established in an amorphous pure Fe as predicted from the band calculations. The density of states of four kinds of pure Fe were obtained by using a LCAO method[11]. The result of amorphous Fe is very similar to that of γ-Fe and liquid Fe except for the fine structures, being very different from that of α-Fe. Furthermore, the self-consistent spin-polarized energy-band calculation for γ-Fe does not satisfy the well-known Stoner criterion of the appearance of ferromagnetism[12]. The band calculation by using a LMTO method indicates that amorphous pure Fe does not become ferromagnetic in the similar manner as γ-Fe[13].

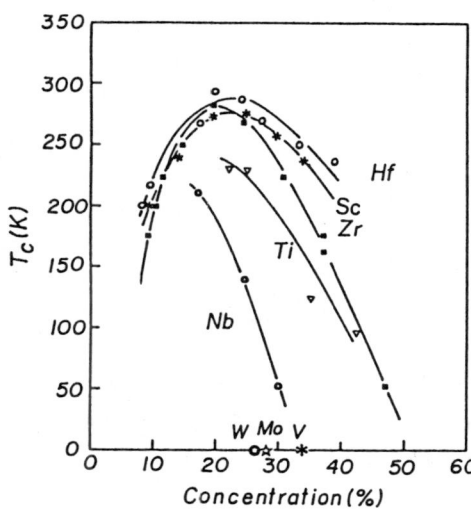

Fig.1. Concentration dependence of the Curie temperature of Fe-ET amorphous alloys. Note that Fe-W, V and Mo alloys are not ferromagnetic.

It has been pointed out that the magnetic properties of Fe-rich fcc alloys are drastically affected by the environment such as the coordination number and the atomic distance. Figures 2 (a) and (b) show the histogram of ratio of coordination number for fcc alloys containing 15 and 7% solute element with the same atomic size. In the case of Fe-rich alloys, a high coordination number state becomes magnetically unstable. A drastic increase in the high coordination number is observed in the low concentration alloy. The structure analysis of γ-Fe by X-ray shows that the atomic distance is very close to 2.54 A which is a critical value[14], competing the ferromagnetic and antiferromagnetic interactions according to the Bethe-Slater curve. Moreover, in the amorphous state, the long-range antiferromagnetic order is not established due to the spin frustrations[8] as mentioned in the preceding section.

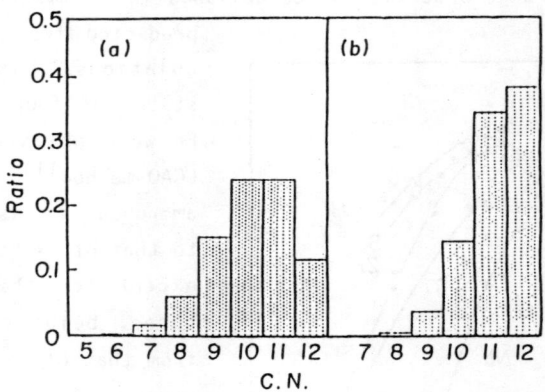

Figs.2(a) and(b). Histogram of ratio of coordination number for fcc alloys containing 15 and 7 % of solute element with the same atomic size.

(B) Ac Susceptibility and Magnetic Cooling Effect

In such circumstances mentioned above, it is expected that the spin glass state brings about in such Fe-rich amorphous alloys. Figures 3(a) and (b) show the temperature dependence of ac susceptibility of Fe-La[15], Fe-Lu[16] and Fe-Ce[17] amorphous alloys. Ce ion in the trivalent state is magnetic, but it becomes nonmagnetic with tetravalent in Fe-Ce alloys because they do not show a large coercive force due to the random axial anisotropy in the amorphous state[17]. Therefore, Fe-Ce alloy system is treated in analogy with Fe-Zr, Fe-La and so on. In the very low concentration range, the characteristic cusp of the spin glass is observed as shown in Fig.3(a). This is the transition from the spin glass to the paramagnetic state with increasing temperature. On the other hand, as shown in Fig.3(b), the ac susceptibility increases sharply at the Curie temperature to a maximum value and gradually decreases and then drops off abruptly as the temperature decreases in the higher concentration[15-17]. This behavior of the low field susceptibility is typical of re-entrant spin glasses and mictomagnets. Therefore, these

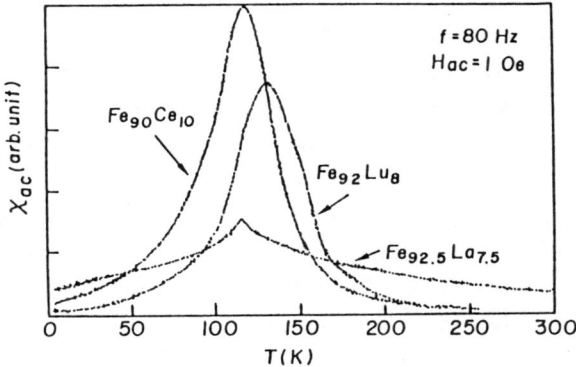

Fig.3(a). Temperature dependence of ac susceptibility of some Fe-rich low concentration amorphous alloys.

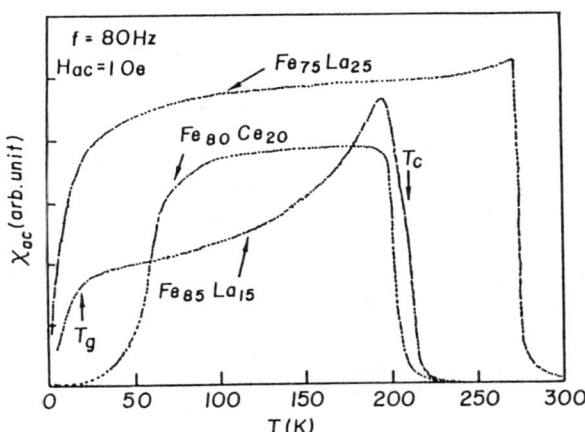

Fig.3(b). Temperature dependence of ac susceptibility of some Fe-rich high concentration amorphous alloys.

alloys exhibit the transition from the paramagnetic to the ferromagnetic and subsequently to the spin glass state(re-entrant transition) with decreasing temperature. Similar behaviors have been observed in other Fe-rich amorphous alloy systems such as Fe-Zr[4] and Fe-Hf[18].

Figure 4 shows the magnetic cooling effect of $Fe_{92.5}La_{7.5}$ amorphous alloy. The two branches merge into one curve at the freezing temperature T_g and the magnetization is reversible above T_g for H > 50 Oe, as usual with other spin glass systems. However, the magnetization M in the lower external magnetic field H is not reversible even near room temperature. The value of M/H for various fields does not become unique even well above T_g. The temperature dependence of inverse susceptibility does not show a linear for any value of H. These results indicate that the high temperature phase is not simply paramagnetic but contains some ferromagnetic clusters[15]. The torque measurement of $Fe_{91.4}Zr_{8.6}$ amorphous alloy has revealed that this alloy exhibits an induced unidirectional anisotropy constant K_{ud} associated with the antiferromagnetic interaction. The temperature dependence of K_{ud} for this alloy is presented in Fig.5[19]. As seen from the figure, the value

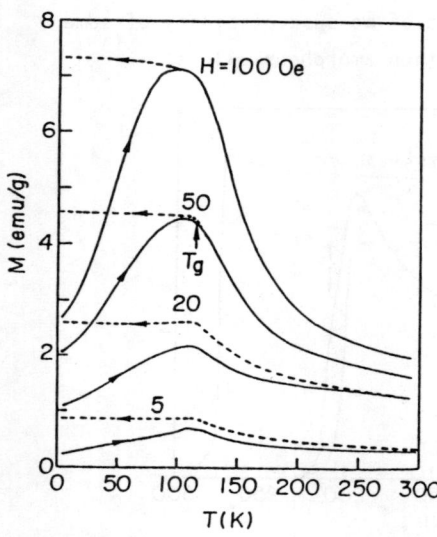

Fig.4. Magnetic cooling effect of $Fe_{92.5}La_{7.5}$ amorphous alloy.

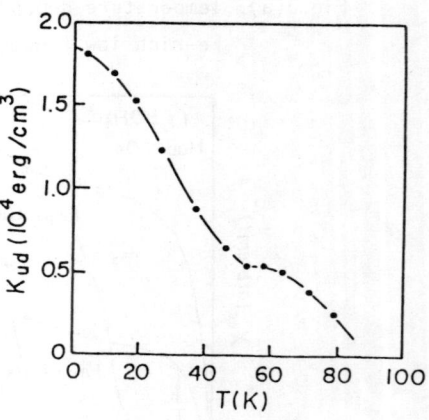

Fig.5. Temperature dependence of the induced unidirectional anisotropy of $Fe_{91.4}Zr_{8.6}$ amorphous alloy.

decreases with increasing temperature and becomes almost zero in the vicinity of 90 K, corresponding to the re-entrant spin glass transition temperature.

Fe-Zr and Fe-Hf amorphous alloys with low concentrations show the thermomagnetic history and the asymmetric hysteresis loop after magnetic field cooling [4,18]. The change of time dependence of magnetization for Fe-Hf amorphous alloys is given by the following well-known expression;

$$\Delta \sigma = \sigma_o + S \ln t \qquad (1)$$

where σ_o and S are the parameters depending on both temperature and the change of field. This relation is kept below 40 K and $10 < t < 1000$ sec[18].

(C) Magnetic Phase Diagram

Figure 6 shows the magnetic phase diagram of several kinds of Fe-rich amorphous alloy systems. In general, the alloys with a smaller magnetic moment exhibit a lower Curie temperature and a higher spin freezing temperature. That is, the ferromagnetic region becomes narrower as the Curie temperature decreases and eventually the Curie temperature collapses. It should be noted that the spin freezing temperature is

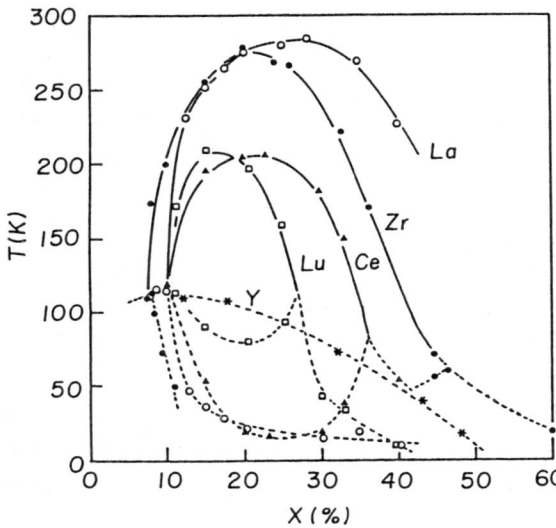

Fig.6. Magnetic phase diagrams of several kinds of Fe-rich amorphous alloys.

relatively higher than that of other magnetically dilute amorphous alloys[20]. A linear extrapolation of three points of the spin freezing temperature below the triple point to zero % La gives about 110 K. It is considered that pure amorphous Fe is not ferromagnetic but becomes a spin glass with the freezing temperature of about 110 K. Similar result has been obtained in Fe-Zr amorphous alloy system[5]. It is interesting to note that the magnitude order of local magnetic moment of Fe is $ZrFe_2$ > $LuFe_2$ > YFe_2[21], being the same tendency of the collapse of the Curie temperature.

The domain structure of $Fe_{90}Zr_{10}$ re-entrant spin glass amorphous film has been observed with the Lorentz electron microscope and found the existence of very large domains and domain walls with quite unusual shapes, that is, closed and somewhat concentric loops with many branches stretching from the small loops towards the larger ones, fine magnetization structures or ripples inside the domains themselves[22]. The comparison of the domain structure with that of the amorphous alloys below the triple point in the magnetic phase diagram is very important in order to understand in more detail the spin glass state of Fe-rich amorphous alloys.

The spin glass behavior is found in Fe-Ni Invar alloys at very low temperatures in the limited concentration range close to the phase boundary of the martensitic transformation[23]. Therefore, it is considered that the spin glass behavior is common in Fe-rich amorphous and crystalline alloys. But in the case of the latter alloys, the martensitic transformation prevents a precise discussion. On the other hand, much lower concentration alloys are obtained in the amorphous alloys.

(D) Magnetic Moment and High-field Susceptibility

Figures 7(a) and (b) show the magnetization curves at 4.2 K for several Fe-Zr[24], Fe-Hf[18], Fe-Sc[25], Fe-Ce[17] and Fe-La amorphous alloys[16]. In Fig.6(a), in the dilute concentration range, the curves are not easily saturated, suggesting the existence of the antiferromagnetic interactions. The magnetic moment is not determined easily in such a dilute concentration range, therefore, the Mossbauer experiment and high field magnetization measurements are necessary. The total Fe moment determined from the Mossbauer spectra of Fe-Zr amorphous alloys in the composition range from 8 to 11 %Zr remarkably increases at a certain temperature with decreasing temperature[26,27]. This increase begins at the freezing temperature determined from the ac susceptibility in the weak dc magnetic field. It would result from the transverse spin freezing, and the spin freezing temperature line given in Fig.6 may cor-

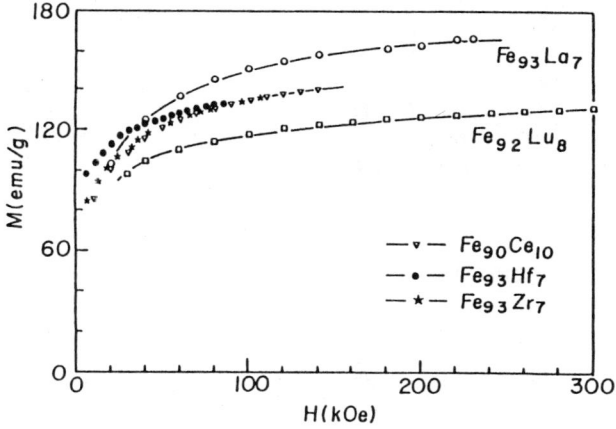

Fig.7(a). Magnetization curves of several kinds of Fe-rich amorphous alloys with low solute concentration.

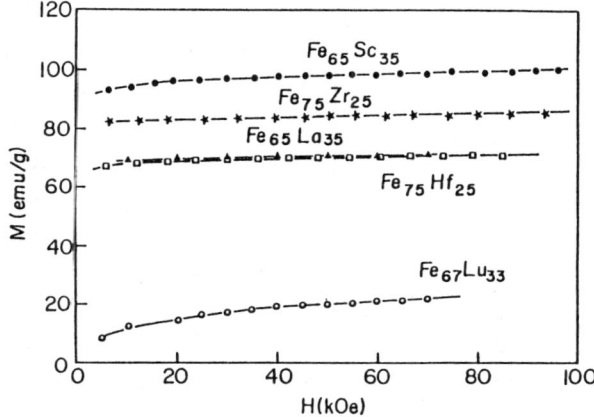

Fig.7(b). Magnetization curves of several kinds of Fe-rich amorphous alloys with high solute concentration.

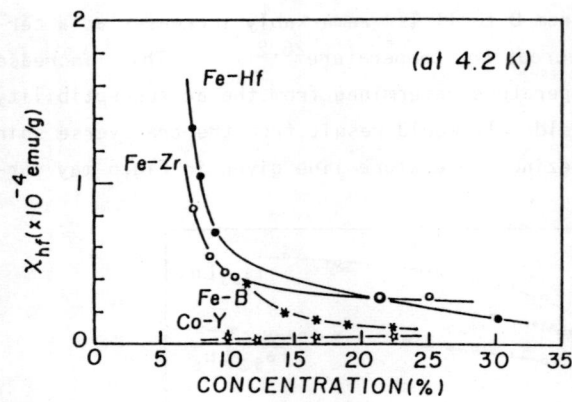

Fig.8. Concentration dependence of the high-field susceptibility of some amorphous alloys.

respond to the so-called G-T line[28]. On the other hand, it should be emphasized that $Fe_{93}Zr_7$ amorphous alloy which has no re-entrant spin glass behavior does not show such an increase in the temperature dependence curve[26].

By taking into account the antiferromagnetic interactions, a new law of approach to saturation magnetization is given by[24]

$$M = M_s(1 - a/H^2) + \chi_{hf}H \qquad (2)$$

where M_s is the saturation magnetization, a the constant and χ_{hf} the high field susceptibility. In the higher concentration range of solute element, the magnetization changes linearly in the high field range, but its slope(high-field susceptibility χ_{hf}) is much larger than that of conventional ferromagnetic alloys. The concentration dependence of χ_{hf} determined from Eq.(2) is shown in Fig.8[1,6]. The values of χ_{hf} of both Fe-Ni crystalline Invar alloys and Fe-B amorphous alloys are 4×10^{-5} emu/g at most, being smaller by one order of magnitude than those of the present alloys in the low concentration range of the solute element.

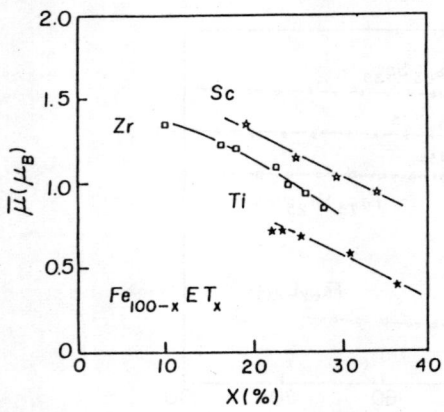

Fig.9. Concentration dependence of the average magnetic of Fe-rich alloys.

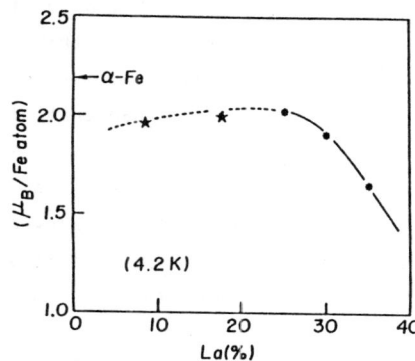

Fig.10. Concentration dependence of the magnetic moment per Fe atom of Fe-La.

Concentration dependence of the average magnetic moment of Fe-Zr, Fe-Ti and Fe-Sc amorphous alloys is shown in Fig.9[25]. The band calculations of Fe_3ET predicts that the magnitude of average magnetic moment depends on the size and valence of solute element, that is, the larger the element size and the smaller the valence, the larger the magnetic moment[12]. Therefore, the order of magnitude of average magnetic moment becomes as $Fe_3Sc > Fe_3Zr > Fe_3Ti$. The experimental results are consistent with the calculations as seen from the figure. Other various data also show the consistent tendency[12]. In the low concentration ranges, the magnetic moment is not easily obtained from the magnetization curves because of lack of saturation. Figure 10 shows the concentration dependence of the magnetic moment per Fe atom of Fe-La amorphous alloy system[8]. The values below 20 % are obtained from the Mossbauer spectra and the magnitude of magnetic moment hardly depends on the composition, being about 2 μ_B which is smaller than the value of α-Fe. Similar results are obtained in Fe-Zr amorphous alloys in the composition range from 7 to 12 %Zr[26].

(E) Low Temperature Specific Heat

Low temperature specific heat measurement was carried out in the temperature rang from 1.5 to 6 K by employing a semi-differential calorimeter[29]. The apparent electronic specific heat coefficient γ of Fe-Zr amorphous alloys is large in the wide range of composition as shown in Fig.11[30,31]. Especially, the value of γ becomes very large in the dilute concentration range. These extremely large values should be interpreted as being composed not only of the electronic specific heat coefficient, which is proportional to the density of states at the

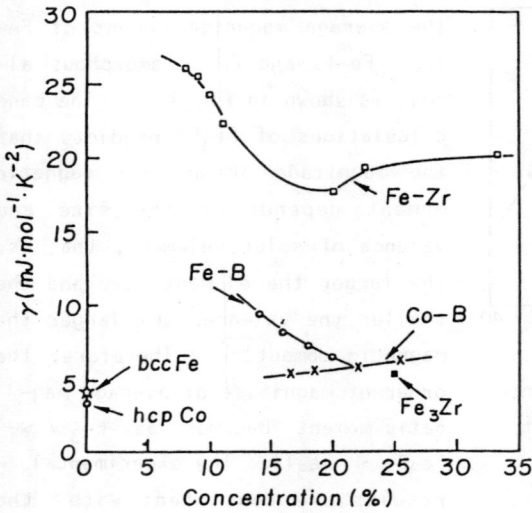

Fig.11. The γ values of several kinds of amorphous and crystalline materials.

Fermi level, but also of the magnetic contribution. These values are much larger than that of pure bcc Fe as seen from the same figure. Furthermore, the observed value of $Fe_{75}Zr_{25}$ in the amorphous state is four times that estimated from the band calculation by assuming Cu_3Au-type Fe_3Zr compound. It should be noted that this calculated value is fairly close to the calculated value for amorphous pure Fe[13]. On the other hand, the values of γ for Fe-B amorphous alloy system are not so large as shown in the same figure and comparable to the value obtained by the band calculation[32]. Fe-B amorphous alloys are weak ferromagnetic but Co-B are strong ferromagnetic, and then the γ values of the latter alloys are smaller than those of the former ones as shown in the same figure[33].

It is known that a large linear term contribute to the specific heat in the spin glass state[34]. From these results, it is considered that such very large values of Fe-Zr amorphous alloys are not due to the weak ferromagnetic but due to the spin glass state. Therefore, the careful measurements of magnetic properties in the wide range of composition below 4.2 K are very important in this alloy system because the specific heat data was obtained between 1.5 and 6 K.

(F) Thermal Expansion and the ΔE Effect

The Invar properties are associated with the magnetic instabilities in Fe-rich crystalline fcc alloys and accompany a large thermal expansion anomaly, significantly large pressure effect on the Curie temperature, large high-field susceptibility, remarkably large compressibility and so on[1]. Such peculiar phenomena have been found in Fe-B amorphous alloys. In this alloy system, the concentration dependences of the Curie temperature and the magnetic moment are similar to those of Fe-Ni crystalline Invar alloys. By comparison of many magnetic properties between Fe-B amorphous alloys and Fe-Ni crystalline alloys, it has been shown that almost all properties are very similar and concluded that the Invar properties are observed not only Fe-rich crystalline fcc alloys but also Fe-rich amorphous alloys[1].

Figures 12, 13 and 14 show the thermal expansion curves of Fe-Zr, Fe-Hf and other Fe-rich amorphous alloys, respectively[9]. We can see the following many interesting phenomena in these curves. a) a large thermal expansion anomaly occurs even far above the Curie temperature. b) in the low concentration alloys, a distinct shrinkage is observed even above the Curie temperature. c) there is no additional anomaly at

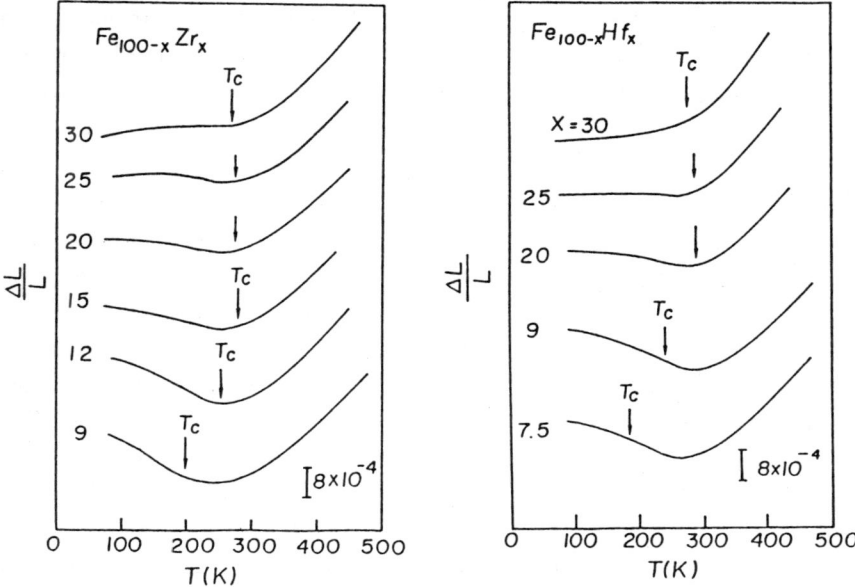

Fig.12. Thermal expansion curves of Fe-Zr amorphous alloys.

Fig.13. Thermal expansion curves of Fe-Hf amorphous alloys.

the spin freezing temperature. d) Fe-Y amorphous alloys have no long range magnetic order but show a similar large anomaly.

The Invar effects in crystalline alloys have been interpreted by the spin fluctuation theory developed by Moriya and Usami[35]. According to this model, in the case of weak ferromagnets, the spontaneous volume magnetostriction ω_s is given by

$$\omega_s = M_o(\partial\omega/\partial H) / \chi_{hf}(\eta - 1) \qquad (3)$$

$$\text{with } \eta = <M_{loc}^2> / M_o^2 \qquad (4)$$

where M_o is the uniform magnetization per atom at 0 K, M_{loc} the local amplitude of the spin density times the atomic volume and $<>$ is the average. A large thermal expansion anomaly or a large spontaneous volume magnetostriction comes from the strong temperature dependence of $<M_{loc}^2>$. Experimentally, the determination of M_{loc} is not so easy, but the thermal expansion measurement gives a important information[36]. In order to discuss the contribution from the spin fluctuation, the thermal expansion should be measured up to well beyond the Curie temperature. However, the relaxation of the amorphous state and the crystallization prevent the precise measurement. Then, this contribution is not discussed exactly in the present alloy systems. The existence of inhomogeneity or the magnetic clusters has been confirmed by

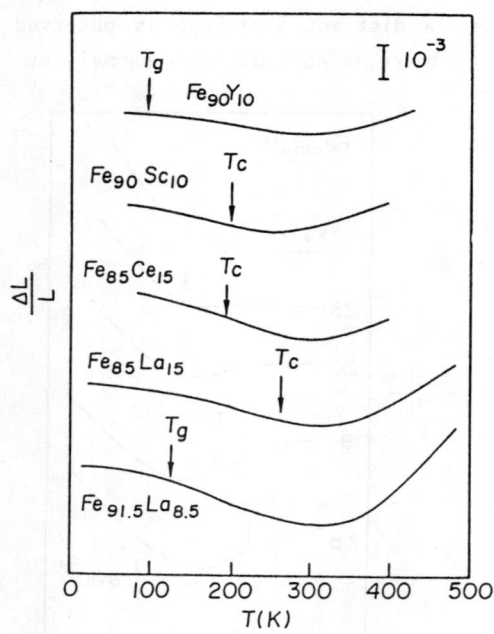

Fig.14. Thermal expansion curves of several kinds of Fe-rich amorphous alloys.

the small angle neutron scattering[37] and by the magnetic measurement in the vicinity of the Curie temperature[24,38]. Therefore, such an anomalous thermal expansion is also closely correlated with the magnetic clusters existing above the Curie temperature. However, the inhomogeneity is not the intrinsic condition for the occurrence of the Invar properties because Fe-Pt ordered fcc alloys also show a distinct anomalous thermal expansion[39].

The ΔE effect is obtained by measuring elastic constant in zero and applied magnetic fields and defined by

$$\Delta E/E_0 = (E_s - E_0) / E_0 \qquad (5)$$

where E_s and E_0 are Young's moduli at the saturated and zero magnetic fields, respectively. Magnetic domains and the spontaneous volume magnetostriction mainly contribute to the ΔE effect[40];

$$\Delta E = \Delta E_\lambda + \Delta E_\omega + \Delta E_A \qquad (6)$$

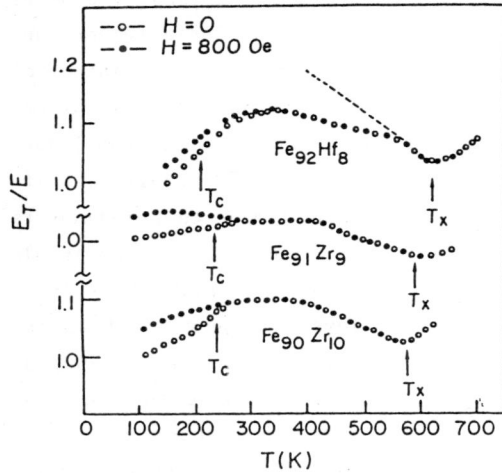

Fig.15. Relative change of Young's modulus of Fe-Zr and Fe-Hf amorphous alloys.

where the first, second and third terms are associated with the linear magnetostriction, forced volume magnetostriction and the spontaneous volume magnetostriction, respectively. Therefore, the anomalous thermal expansion is responsible for such an anomalous temperature dependence of Young's modulus. In the case of non-Invar type ferromagnetic materials, Young's modulus generally decreases with increasing temperature in the magnetic field as shown by the dotted line because the linear magnetostriction term is eliminated. However, Young's modulus in a magnetic field for Fe-Zr[41] and Fe-Hf amorphous alloys shows a softening even well beyond the Curie temperature as shown in Fig.15. Similar anomaly has been observed in other Fe-metal amorphous alloys. This behavior is also attributed to the existence of magnetic clusters having a large spontaneous volume magnetostriction. Another change at T_x is not concerned with the magnetic contribution but also with the crystallization. At low temperatures, no distinct anomaly is observed.

(G) Pressure Effect on the Curie and Spin Freezing Temperatures

One of the characteristic properties of Invar alloys is the remarkable pressure effect on the magnetic properties, being associated with the large spontaneous volume magnetostriction. The pressure effect on the Curie temperature of the crystalline Invar alloys is remarkable ranging from about -2 to -6 K/kbar. Since Fe-rich amorphous alloys show a large thermal expansion anomaly as mentioned above, the pronounced pressure effect on the Curie temperature is expected. Figure 16 shows the permeability vs. hydrostatic pressure curves of $Fe_{77}Zr_{23}$ amorphous alloy. The Curie temperature was defined as the point of of intersection of two dotted lines extrapolated from linear portions. The pressure shift of the Curie temperature is significant, being comparable to that of crystalline Invar alloys. Figure 17 shows the concentration dependence of the pressure effect on the Curie temperature of Fe-Zr, Fe-Hf and Fe-Sc amorphous alloys, together with that of Fe-Ni crystalline alloys for comparison. Conventionally the effect in the weak fer-

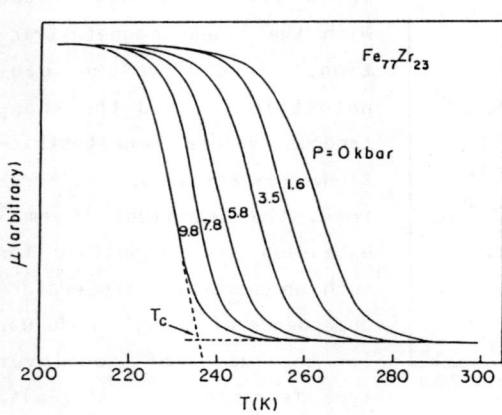

Fig.16. Temperature dependence of the permeability as a function of pressure for $Fe_{77}Zr_{23}$ amorphous alloy.

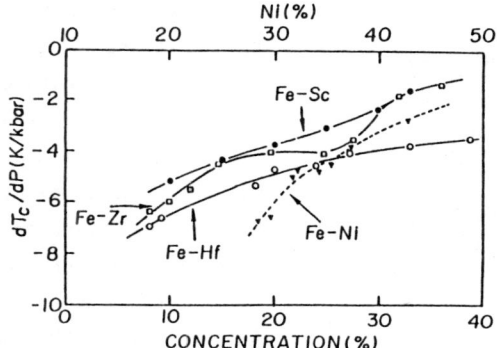

Fig.17. Concentration dependence of the pressure effect on the Curie temperature of Fe-rich amorphous alloys, together with that of Fe-Ni crystalline alloys.

romagnets is given by the following expression;

$$dT_c/dP = -\alpha/T_c \quad (7)$$

where α is the positive parameter and is assumed to be constant though the same alloy system. However, the data on Fe-rich amorphous alloys do not obey this expression but follow the expression given by[42,43]

$$dT_c/dP = +aT_c - bT_c^2 \quad (8)$$

$$dT_c/dP = -cT_c + dT_c^2 \quad (9)$$

The expressions of Eqs.(8) and (9) have been obtained from experimental data on many amorphous alloys and from the Landau-Ginzburg model, respectively. Dividing by T_c, they are rewritten as follows;

$$(1/T_c)(dT_c/dP) = \pm A \mp BT_c \quad (10)$$

Following the expression(10), the data on three amorphous alloy systems are presented in Fig.18. Each result exhibits a linear relationship changing its slope at a certain concentration, corresponding to the maximum point in the concentration dependence of the Curie temperature as seen from Fig.1. The high concentration alloys follows Eq(9), showing the inhomogeneous state. It has been confirmed from many magnetic data that the low concentration alloys are also magnetically inhomogeneous, and the pressure effect on the Curie temperature obeys the empirical law of Eq.(8). Therefore, Fe-rich amorphous alloys have a

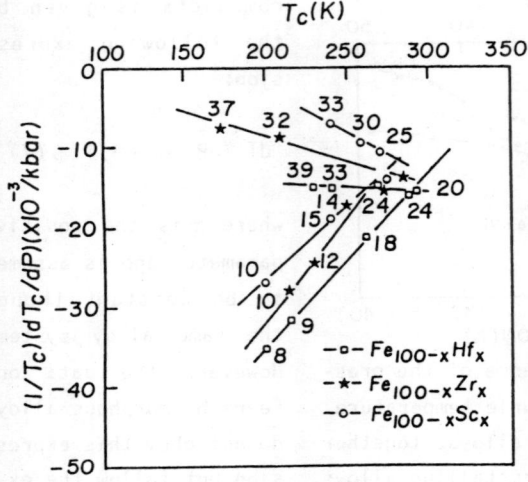

Fig. 18. The Curie temperature vs. its pressure coefficient of Fe-Zr, Fe-Hf and Fe-Sc amorphous alloys.

different tendency of the pressure effect on the Curie temperature in the inhomogeneous state, drawing the line at the maximum point of the Curie temperature in the concentration curve. In contrast to the Curie temperature, the spin freezing temperature increases initially with increasing hydrostatic pressure. Such a typical example is shown in Fig. 19 for $Fe_{87.5}La_{12.5}$, $Fe_{92.5}La_{7.5}$ and $Fe_{85}Ce_{15}$ amorphous alloys[44].

The spin freezing temperature T_g may be given by[45]

$$T_g^2 = z/2 \, [\, c_A \xi_{AA}^2 + c_B \xi_{BB}^2 \\ + \{(c_A \xi_{AA}^2 - c_B \xi_{BB}^2)^2 + 4c_A c_B \xi_{AB}^4\}^{1/2} \,] \qquad (11)$$

where z is the coordination number, c_A and c_B the compositions, and ξ_{AA} and ξ_{BB}, the ferromagnetic and antiferromagnetic couplings, respectively. By applying pressure, the numbers and compositions associated with the different couplings would be remarkably changed. It should be noted that the pressure dependence of the Curie and spin freezing temperatures is similar with the concentration dependence curves. That is, the application of the hydrostatic pressure gives the same result with that of the decrease of solute concentration as seen from Fig. 6.

The magnetic properties of Fe are strongly affected by the atomic distance and also the coordination number. It is interesting to note

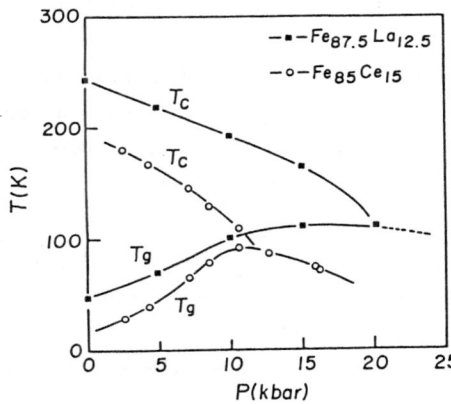

Fig.19. Pressure effect on T_c and T_g of $Fe_{87.5}La_{12.5}$ and $Fe_{85}Ce_{15}$ amorphous alloys.

that γ-Fe films oriented parallel to {111} and {110} planes are ferromagnetic and {100} films are antiferromagnetic, reflecting the atomic distance of Fe[46]. Detailed experiments on anomalous lattice contraction and magnetism of γ-Fe precipitates in Cu have been recently carried out[47].

In some Fe-metal amorphous alloys, the atomic distance is about 2.54 A which is the critiamorphous competing the ferromagnetic and antiferromagnetic interactions. In order to obtain the atomic distance, X-ray diffraction measurements were carried out θ-2θ reflection mode using Mo-K_α radiation with a curved quartz monochrometer. The range of scattering vector in the measurements was from 1.2 to 15.7 A^{-1}. Figure 20 shows the Fe-Fe and Fe-La distances of Fe-La amorphous alloys[48]. The

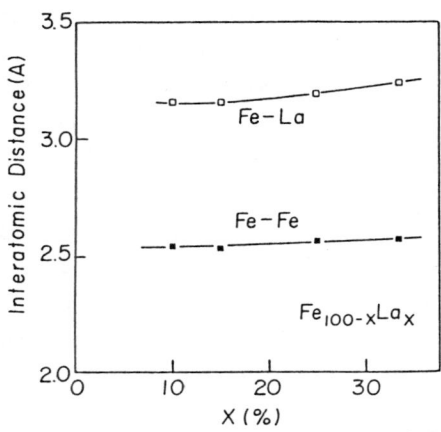

Fig.20. Concentration dependence of Fe-Fe and Fe-La distances in the amorphous state.

Fe-Fe distance is slightly larger than 2.5 A, and it hardly depends on the La concentration. It should be noted that the Fe-Fe distance in RFe_2(R:rare earth metal) amorphous alloys becomes smaller than 2.5 A for heavy rare earth metal alloys[49]. The plot of the Curie temperature vs. de Gennes factor for Fe-heavy rare earth metal alloy systems indicates that the Fe-Fe exchange interaction is almost zero[49]. This behavior is explained by taking into account the critical atomic distance mentioned above.

Another explanation of the spin glass state has been made by Kakehashi[50]. According to his itinerant electron model, even a positive exchange interaction between Fe-Fe can bring about a spin glass state due to the anomalous magnetic coupling between Fe local moments. There are a large fluctuation of Fe local moments and a nonlinear coupling between Fe local moments in the Fe fcc alloys, that is, the amplitude of thermal average of Fe local moment becomes larger below the critical coordination number and shows the ferromagnetic coupling, but it does smaller above the critical number and shows the antiferromagnetic coupling.

(H) Electrical Resistivity Minimum and Compressibility

Figures 21 and 22 show the relative change of electrical resistivity of Fe-Zr, Fe-Hf, and other several kinds of Fe-rich amorphous alloys[9]. A distinct minimum is observed in the temperature dependence curves. It is worth remarking that this minimum does not occur at the Curie temperature, being always higher by several tens of degree. Its interval becomes wider with increasing content of solute element. The magnetoresistance of Fe-Zr amorphous alloys is positive[52] and thus inconsistent with the resistance minimum being associated with a Konko-like anomaly. No clear anomalous temperature dependence is observed at the freezing temperature. Note that $Fe_{80}Y_{20}$ is a spin glass with no magnetic long-range or-

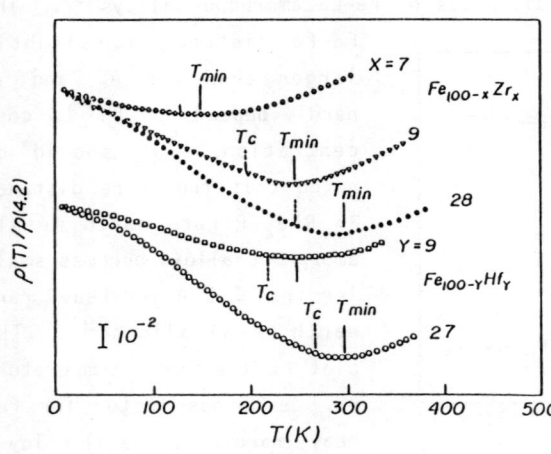

Fig.21. Temperature dependence of the electrical resistivity of Fe-Zr and Fe-Hf amorphous alloys.

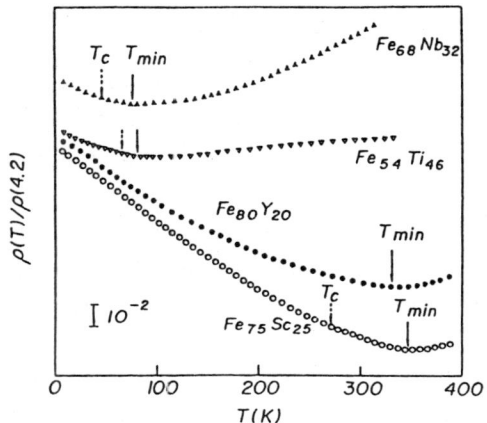

Fig. 22. Temperature dependence of the electrical resistivity of Fe-rich amorphous alloys.

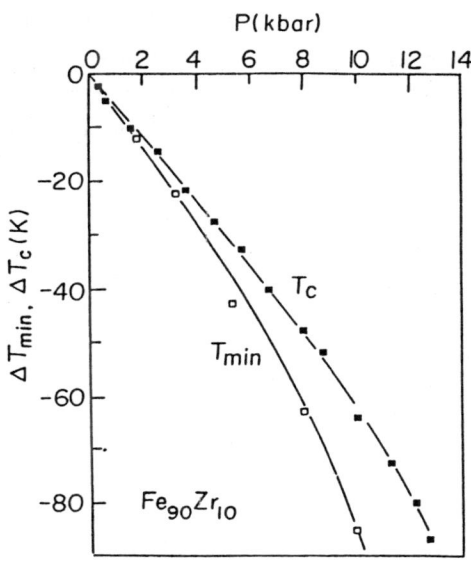

Fig. 23. Pressure effect on T_{min} and T_c of $Fe_{90}Zr_{10}$ amorphous alloy.

rder, but it also exhibits a minimum around 330 K, although the previous report did not confirm[53].

Figure 23 shows the shift of the minimum point T_{min} and the Curie temperature T_c as a function of hydrostatic pressure for $Fe_{90}Zr_{10}$ amorphous alloy. The minimum point is shifted to lower temperature ranges by applying hydrostatic pressure. The initial slopes of these curves exhibit almost the same value and the feature of pressure dependence is very similar, being convex upward. Therefore, the T_{min} point may be closely concerned with T_c. Generally speaking, the compressibility κ of Invar alloys is significantly large due to the large magnetoelastic effects. The relation between the pressure coefficient of electrical resistance and the compressibility is given by

$$(1/R)dR/dP \simeq 4\kappa/3 \quad (12)$$

Therefore, the remarkable decrease of the electrical

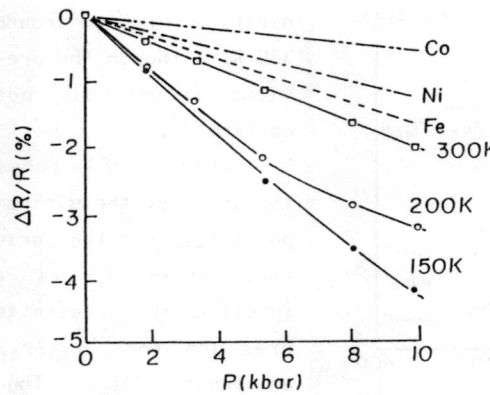

Fig.24. Relative change in the resistance of $Fe_{90}Zr_{10}$ amorphous alloy, Fe, Co and Ni.

resistance is expected in the present alloy. In order to make this point clear, the decrease at 300, 200 and 150 K is shown as a function of pressure is plotted in Fig.24, together with that of pure crystalline Fe, Co and Ni measured at room temperature[54]. The relative change at 300 K of $Fe_{90}Zr_{10}$ amorphous alloy is almost the same as that of crystalline Fe, but the value at 150 K is about twice the value at 300 K. Note that the slope at 200 K is very similar with that at 300 K, but it becomes almost the same as that at 300 K at higher pressures. Then, it is considered that magnetic transition from ferromagnetic to paramagnetic state takes places around 6 kbar with increasing hydrostatic pressure. Such behaviors are closely correlated with the large compressibility of $Fe_{90}Zr_{10}$ amorphous alloy in the ferromagnetic state.

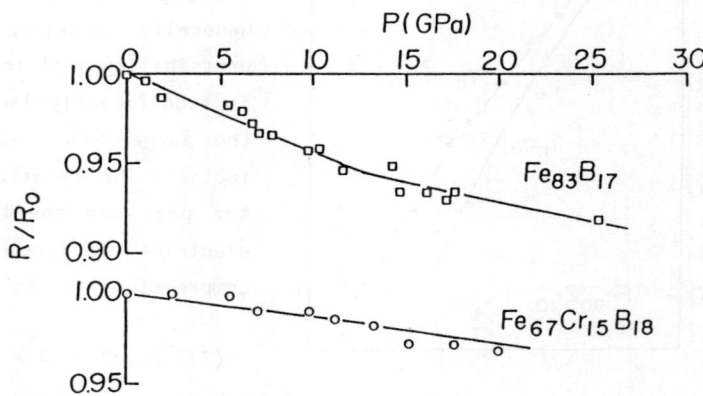

Fig.25. Relative change in the interatomic distance of $Fe_{83}B_{17}$ and $Fe_{67}Cr_{15}B_{18}$ amorphous alloys.

Figure 25 shows the interatomic distance, normalized to that at zero pressure, of $Fe_{83}B_{17}$ and $Fe_{67}Cr_{15}B_{18}$ amorphous alloys[55]. The former alloy exhibits a very excellent Invar property in a wide temperature range and the latter is paramagnetic at room temperature. The high pressure was applied at room temperature by using a diamond-anvil cell. The compressibility estimated from the initial slope for $Fe_{83}B_{17}$ and $Fe_{67}Cr_{15}B_{18}$ amorphous alloys is 4.0 and 1.4×10^{-3} GPa, respectively. According to Kamarad et al, T_c of the former decreases with increasing pressure at an initial rate of 25 K/GPa[56]. If the pressure increases to 12 GPa, a linear extrapolation places T_c at 300 K. This suggests that the pressure induced transition from the ferromagnetic to paramagnetic state occurs. It should be noted that both alloys show almost the same compressibility above this pressure.

SUMMARY

Various kinds of Fe-metal amorphous alloys were prepared by melt quenching and sputtering in order to investigate the spin glass and Invar properties. The coexistence of the spin glass and Invar properties has been confirmed and the main results are summarized as follows:

1) The Curie temperature of almost all Fe-ET amorphous alloys is lower than room temperature and a drastic decrease occurs with decreasing solute element in the dilute concentration ranges.
2) Since the critical atomic distance is 2.54 Å, the competition between ferromagnetic and antiferromagnetic interactions easily occurs, resulting in the spin glass state due to the spin frustration. The triple point is located around 10 % in many alloys, and the re-entrant spin glass behavior is observed above this concentration. The direct transition from a paramagnetic to a ferromagnetic state brings about below this composition.
3) Because of the transverse freezing, the remarkable increase in the magnetic moment determined from the Mossbauer spectra is observed with decreasing temperature in the re-entrant spin glass composition range.
4) In the magnetic phase diagram, the higher the Curie temperature,

the lower the spin glass freezing temperature. Amorphous pure Fe is considered to be a spin glass. The spin freezing temperature is estimated to be 110 k and the magnetic moment is close to 2 μ_B.

5) Thermal expansion anomaly is distinct in the wide temperature ranges, and the shrinkage is observed even above the Curie temperature. Similar anomaly is confirmed in Fe-Y amorphous alloys which have no long-range magnetic order. There is no marked anomaly at spin freezing temperature.

6) There are no obvious anomalies in both electrical resistivity and elastic property at the spin freezing temperature.

7) Anomalous temperature dependence of Young's modulus is caused by the large spontaneous volume magnetostriction and by the magnetic clusters.

8) Pressure effect on the Curie temperature and the spin freezing temperature is remarkable, reflecting the environment effect on the magnetic properties

9) The Curie temperature vs. its pressure coefficient plots of many alloys and other magnetic data indicate that Fe-metal amorphous alloys are the inhomogeneous alloy systems.

10) Fe-rich amorphous Invar alloys show a large compressibility, being about three times that of a paramagnetic amorphous alloy.

Acknowledgments

The authors would like to thank Prof. Y. Nakagawa, Drs. H. Hiroyoshi and N. Saito of Tohoku University, Prof. F. E. Fujita and Dr. S. Nasu of Osaka University, Prof. U. Mizutani of Nagoya University, Prof. M. Matsuura of Miyagi National College of Technology, Prof. H. Iwasaki of National Laboratory for High Energy Physics, Prof. N. Mori and Dr. C. Murayama of the University of Tokyo, Prof. A. Itoh and Dr. S. Morimoto, who participated in much of the study discussed in the present paper, for many illuminating discussions.

The sputtering samples and amorphous ribbons were prepared, respectively, at the laboratories of Prof. H. Fujimori and Prof. T. Masumoto of Institute for Materials Research, Tohoku University. The high-field up to 145 kOe was produced by using the water-cooled magnet of the High Field Laboratory for Superconducting Materials (HFLSM) of Tohoku University.

The present work was supported in part by Grad-in-Aid for Scientific Research Project, No. 63460190, from the Ministry of Education, Japan.

REFERENCES

1) Fukamichi, K., "Amorphous Metallic Alloys", Butterworths Co. Ltd.,317-340(1983).
2) Forester, D. W., Koon, N. C., Schelleng, J. H. and Rhyne, J. J., J. Appl. Phys., 50, 7336(1979).
3) Coey, J. M.D., Givord, D., Lenard, A. and Robouillat, J. P., J. Phs. F11, 2707(1981).
4) Hiroyoshi, H. and Fukamichi, K., J. Appl. Phys. 53, 2226(1982).
5) Saito, N., Hiroyoshi, H., Fukamichi, K. and Nakagawa, Y., J. Phys. F16, 911(1986).
6) Fukamichi, K., Komatsu, H., Goto, T., Wakabayashi, H. and Matsuura, M., Proc. Int. Conf. on Advanced Materials(MRS Meeting, Tokyo, 1988), in press.
7) Fukunaga, K., Misawa, M., Fukamichi, K., Masumoto, T. and Suzuki, K., Proc. Int. Conf. on RQM, Vol.2, 325(1978).
8) Kaneyoshi, T., "Glassy Metals", CRC Press, 31-63(1983)
9) Fukamichi, K. and Hiroyoshi, H., Sci. Rep. RITU, A32, 154(1985).
10) Shirakawa, K., Fukamichi, K., Kaneko, T. and Masumoto, T., Physica, 119B+C, 192(1983).
11) Bose, S. K., Ballentine, L. E. and Hammerberg, J. E., J. Phys. F13, 2089(1983).
12) Malozemoff, A. P., Williams, A.R., Terakura, K., Mruzzi, V. L. and Fukamichi, K., J. Magn. Magn. Mater., 31-34, 1(1983).
13) Fujiwara, T., J. Non-Cryst. Solids, 61-62, 192(1983).
14) Kummerle, W. and Gradmann, U., Solid State Commun. 24, 33(1977).
15) Wakabayashi, H., Fukamichi, K., Komatsu, H., Goto, T., Sakakibara, T. and Kuroda, K., to be submitted.

16) Kuroda, K., Goto, T., Wakabayashi, H., Morimoto, S. and Itoh, A., to be submitted.
17) Fukamichi, K., Komatsu, H., Goto, T. and Wakabayashi, H., Physica, 149B, 276(1988).
18) Hiroyoshi, H., Noguchi, K., Fukamichi, K. and Nakagawa, Y., J. Phys. Soc. Jpn. 54, 3554(1985).
19) Morita, H., Hiroyoshi, H. and Fukamichi, K., J. Phys. F16, 507(1986).
20) Rammel, R. and Souletie, J., "Magnetism of Metals and Alloys", North-Holland Co. 379-486(1982).
21) Yamada, H. and Simizu, M., J. Phys. F16, 1039(1986).
22) Senoussi, S., Hadjoudj, S., Jouret, P., Bilotte, J. and Fourmeaux, R., J. Appl. Phys. 63, 4086(1988).
23) Miyazaki, T., Ando, Y. and Takahashi, M., J. Magn. Magn. Mater. 60, 219(1986).
24) Hiroyoshi, H., Fukamichi, K., Hoshi, A. and Nakagawa, Y., Proc. Int. on High Field Magnetism, 113(1983).
25) Fukamichi, K., Hiroyoshi, H., Shirakawa, K., Masumoto, T. and Kaneko, T., IEEE Trans. Magn. MAG-23, 424(1986).
26) Nasu, S., Kitagawa, H., Fujita, F. E. and Fukamichi, K., to be submitted.
27) Saito, N., Onodera, H., Fukamichi, K. and Nakagawa, Y., to be submitted.
28) Gabay, M. and Toulouse, G., Phys. Rev. Lett. 47, 201(1981).
29) Massalski, T.B., Mizutani, U., Hartwig, K.T. and Hopper, R. W., Proc. 3rd Int. Conf. on RQM, Vol.2, 81(1978).
30) Mizutani, U., Matsuura, M. and Fukamichi, K., J. Phys. F14, 731(1984).
31) Matsuura, M., Mizutani, U. and Fukamichi, K., Proc. 5th Int. Conf. on RQM, 1019(1984).
32) Fujiwara, T., J. Phys. F12, 611(1982).
33) Matsuura, M. and Mizutani, U., J. Magn. Magn. Mater. 31-34, 1481(1983).
34) Marshall, W., Phys. Rev. 118, 1519(1960).
35) Moriya, T. and Usami, K., Solid State Commun. 34, 95(1980).
36) Nakamura, Y., J. Magn. Magn. Mater. 31-34, 829(1983).

37) Ishida, A., Thesis of Tohoku University.
38) Yamamoto, Y., Onodera, H., Hosoyama, K., Masumoto, T. and Yamauchi, H., J. Magn. Magn. Mater. 31-34, 1579(1983).
39) Nakamura, Y., Sumiyama, K. and Shiga, M., J. Magn. Magn. Mater. 12, 127(1979).
40) Hausch, G. and Warlimont, H., Z. Metallke. 64, 152(1973).
41) Fukamichi, K., Kikuchi, M., Masumoto, T., J. Non-Cryst. Solids, 61-62, 961(1984).
42) Wagner, D. and Wohlfarth, E. P., J. Phys. F11, 2417(1981).
43) Schneider, J., Arnold, Z., Kamarad, J. and Handstein, A., Phys. Stat. Sol. (a)64, K133(1981).
44) Goto, T., Murayama, C., Mori, N., Wakabayashi, H., Fukamichi, K. and Komatsu, H., ICM(Paris, 1988), to be published.
45) Kakehashi, K., J. Phys. Soc. Jpn. 50, 3177(1981).
46) Keune, W., Halbauer, R., Gonser, U., Lauer, J. and Willamson, D.L., J. Appl. Phys. 48, 2976(1977).
47) Tsunoda, Y. and Imada, S. and Kunitomi, N., J. Phys. F18, 1421(1988).
48) Matsuura, M., Fukunaga, T., Fukamichi, K. and Suzuki, K., Solid State Commun. 66, 333(1988).
49) Matsuura, M., Fukunaga, T., Fukamichi, K., Satoh, Y. and Suzuki, K., Z. Phys. Chem. 157, 85(1988).
50) Fukamichi, K., Satoh, Y. and Komatsu, H., IEEE Trans. Mag. MAG-23, 2548(1987).
51) Kakehashi, Y., Phys. Rev. B38, 474(1988).
52) Dahlberg, D., Rao, K. V. and Fukamichi, K., J. Appl. Phys. 55. 1942(1984).
53) Cochrane, R. W., Strom-Olsen, J., Williams, G., Lienard, A. and Rebouillat, J. P., J. Appl. Phys. 49, 1677(1978).
54) Shirakawa, K., Fukamichi, K., Kaneko, T. and Masumoto, T., Phys. Lett. 97A, 213(1983).
55) Tomizuka, A., Iwasaki, H., Fukamichi, K. and Kikegawa, T., J. Phys. F14, 1507(1984).
56) Kamarad, J., Arnold, Z., Schneider, J. and Krupicka, S., J. Magn. Magn. Mater., 15-18, 1409(1980)

NON-EXPONENTIAL RELAXATION IN SPIN GLASSES AND OTHER DISORDERED SYSTEMS

IA. Campbell, J.M. Flesselles[*], R. Botet and R. Jullien

Physique des Solides, Université Paris-Sud,
91405 Orsay, France

[*]Laboratoire des Solides Irradiés, Ecole Polytechnique
91128 Palaiseau, France

ABSTRACT

In spin glasses, as in all other glassy systems, relaxation is strongly non-exponential both below the freezing temperature T_g and for a range of temperature above T_g. We propose that the freezing process in a spin glass or glassy material can be identified with a percolation transition sequence in configuration space. The non-exponential relaxation is then a direct consequence of the "sparse" geometry of the restricted part of phase space which the system is allowed to sample when close to or below freezing. We expect that near T_g relaxation should tend to the stretched exponential form $\exp -(t/\tau)^\beta$ with $\beta = 1/3$, and that there should be a distinct upper transition temperature above which relaxation becomes exponential.

As is well known, spin glasses are magnetic systems (which can be metallic, insulating or semi-conducting) where there exist both strong positional disorder of the local moments and competing interactions. As a result, when the sample is cooled to low temperature, the system freezes into one of many ground states in which the overall moment is zero but where each individual spin has a fixed orientation. The ground state degeneracy is very high for mean field models[1] and we will assume this is the case for realistic systems[2].

One of the best known characteristics of a spin glass is the fact that the relaxation in response to a small change of the external field is highly non-exponential. This non-exponential type of response is in fact observed in *all* materials showing glassy ordering -- polymers, dieletrics, superconductors, proteins and so on [3]. This universality suggests that linking the glassy ordering and the relaxation behaviour there is a connection which is universal and which does not depend on the detailed physical properties of the various materials.

We will briefly present a scenario for the glass transition which can provide this link ; various aspects of the model have been discussed elsewhere [4-7].

We will first concentrate on Ising spin glasses (ISG) which are probably the simplest examples of "complex" systems. For any Ising system of N spins the 2^N possible configurations can be mapped onto the 2^N vectices of a hypercube of dimension N. At any instant, the system is in a given configuration, i.e. the "system point" occupies a definite site in the hypercube configuration space. We will impose relaxation by successive single spin flip events as in ref. 8 . A single spin flip is equivalent to the system point making a jump from one site to a near neighbour site on the hypercube.

For a given ISG sample with fixed interactions between the spins, each configuration will have an energy. At high T all configurations are thermodynamically accessible, and as the temperature is lowered the higher energy states will progressively become forbidden until at very low T only the low energy ground states will be allowed. These ground states will be distributed more or less at random over the hypercube and will be isolated from each other by regions of high energy states or "infinite barriers" in phase space. The sequence as T is lowered is like a percolation process[9]. High T is like high concentration p in a conventional percolation problem, with all the sites joined together in one large cluster, and low T is like low p with many isolated small clusters. At high T the system carries out a random walk between all the permitted sites, which are all linked. As T is lowered, the permitted set of sites becomes sparser. At T = 0 the system will end up in one particular ground state. When the system is cooled a second time it will generally end up in a <u>different</u> ground state.

To get from one limit to the other, it seems clear that we must go through a percolation type of transition at some critical T. Percolation transitions are usually discussed for Euclidean flat spaces of infinite extent, but the hypercube is a closed sphere-like space (even though it has 2^N sites, which is a huge number as soon as N is large). However the hypercube type of percolation problem has been studied by graph theory[10] in the case of a hypercube where the sites are occupied at random with probability p (like the problem of percolation with sites on a lattice occupied at random with probability p). In the hypercube case there are <u>two</u> critical concentrations. For p > 1/2 all the sites form one single cluster. For p < 1/N all sites are in "small" clusters localized in the hypercube. In between, for 1/N < p < 1/2, there exists a giant cluster (like a percolating cluster) together with small clusters. We conjecture that for the real ISG systems, the set of configurations with a maximum energy E(T) will follow the same geometrical sequence as T is reduced. At high T all the states will be linked together. At intermediate T there will be a giant cluster together with small clusters and the system point will be constrained to lie on the giant cluster of sites. Finally at low T where there are only small clusters, the system will have to choose to live on one or other of the small clusters. At high T the system is ergodic because all states are accessible from one another ; in the intermediate region it becomes non ergodic because the giant cluster of states coexists with small clusters which have the same energy but which are inacessible (because of "infinite barriers"). However symmetry in the Edwards-Anderson sense [11] is still not not broken in this intermediate regime because $q_{EA} = 0$ for the giant cluster. We denote the upper transition from ergodic to non-ergodic by T_{gr}. In the lowest temperature region where there are only small clusters, symmetry is broken because for each small cluster q_{EA} is not zero. We identify the transition to the low temperature symmetry broken regime with T_g.

Now as far as the relaxation is concerned, arguments and simulations on random walks in the different geometrical situations[4,5,6] lead to the conclusion that above T_{gr} relaxation will be exponential, while between T_{gr} and T_g we can expect non-exponential relaxation tending to the stretched exponential form $\exp-(t/\tau)^\beta$ with $\beta = 1/3$ as T tends to T_g. Physically, the non-exponential relaxation is related to the complexity of the available phase space, the permitted cluster of states becoming progressively more sparse (or fractal with low dimension) as T drops, with numerous dead ends. A random walk on this complicated geometry leads to a complicated relaxation with a spectrum of effective relaxation times, wheras a random walk for a simple compact phase space geometry will always lead to an exponential decay.

What evidence is there that this description is in any way relevant to real situations ? We will focus on the prediction of stretched exponential decay with exponent 1/3 as T tends to T_g from above. Large scale ISG simulations in 3d [8] lead to a relaxation which was phenomenologically parametrized in terms of an stretched exponential ; the simulation results show β tending to very close to 1/3 at T_g. Less extensive simulations on an ISG in 2d [12] also seem compatible with $\beta = 1/3$ as T tends to zero, which is the accepted value for Tg in a 2d ISG. Turning to real materials, results on relaxation close to T_g in Cu Mn 5 at %[13], in Cd $In_{0.3}Cr_{1.7}S_4$ [14], in EuSrS[15], and in $FeMnTiO_3$ [16] spin glasses all indicate non-exponential relaxation for T just above T_g which has been parameterised as stretched exponential relaxation with β close to 1/3. There is thus strong evidence in favour of the suggestion that in the neighbourhood of T_g stretched exponential relaxation with exponent $\beta = 1/3$ is universal in spin glasses, whether Ising or Heisenberg, metallic or insulating.

In addition, the 3d and 2d ISG simulations[8,12] show a clear transition from non-exponential to exponential relaxation at some temperature well above T_g; in fact the exponential onset temperature is equal to the Griffiths temperature[17], i.e. the ordering temperature of the equivalent ferromagnet. Evidence for a transition to exponential relaxation at a temperature T_{gr} well above T_g in real spin glass materials is much less clear, because this requires very high frequency measurements. Hopefully neutron scattering should be able to test this prediction in the future.

Now in the introduction we mentioned that the relaxation in other glassy systems was very similar to that in spin glasses. To the extent that the scenario sketched out above can be accepted as being valid for ISG's, we can extend it in principle to other glassy materials. The configuration space of these systems is much less easy to visualize than is the case for the hypercube corresponding to the ISG, but we can accept the general idea that the configuration space will again be some sort of high dimensional closed space. For any glassy system, we must have the same high T and low T limits as for the ISG (a single cluster of linked states at very high T, and a multiplicity of isolated ground states at very low T). It seems reasonable to expect that as T is lowered the phase space of each of these systems will follow the same geometrical sequence as that of the ISG. Then we should again find an onset of non-exponential relaxation at an upper temperature T_{gr}, and $\exp-(t/\tau)^{1/3}$ relaxation at T_g. For a number of these systems there are practical experimental limitations, related principally to the frequency window necessary for the experiments, but at least in certain well studied cases it appears that relaxation with $\beta \sim 1/3$ at T_g is consistent with the experimental data, and that exponential relaxation does set in beyond some higher temperature[18]. Once again we interpret the non-exponential relaxation as being the signature of a non-compact phase space geometry intimately related to the break up of phase space into disconnected clusters at the glass transition.

We have not discussed a number of other problems associated with spin glasses and glassy systems - power low relaxation and aging below T_g, critical exponents at the transition, and even the "two level systems". Hopefully a phase space approach could be useful for understanding these phenomena as well as the relaxation above T_g which we have discussed.

In summary, we propose a scenario for the glass transition and non-exponential relaxation in terms of percolation transitions in configuration space. A great deal needs to be done concerning the mathematical and thermodynamical basis of the model, and obviously further and more stringent tests from simulations and experiments are very necessary. As things stand, we have made some non-trivial predictions which appear to be in good agreement with existing experimental results. If this behaviour is confirmed there are for reaching implications for the physics of disordered systems.

References

1/ K. Binder and A.P. Young, Rev. Mod. Phys. **58** 801 (1986)

2/ N. Sourlas, Europhys.Lett. **6** 561 (1988)

3/ J. Jäckle, Rep. Prog. Phys. **49** 171 (1986), L.C.E. Struik, Physical Aging in Amorphous Polymers and other materials (Elsevier, Amsterdam 1978),

A.K. Jonscher, Nature **267** 673 (1977), G. Richter, Ann. Phys. Lpz. **29** 605 (1937), Y.B. Kim, C.F. Hempstead and A.R. Strnad, Phys. Rev. Lett. **9** 306 (1963), K.A. Müller, M.Takashige and J.G. Bednorz, Phys.Rev.Lett. **58** 1143 (1987), A. Ansari, J. Berendzen, S.F. Bowne, H. Frauenfelder, I.E. Iben, T.B. Sauke, E. Shyamsunder and R.D. Young, Proc. Natl.Acad.Sci. **82** 5000 (1985)

4/ I.A. Campbell, J. Phys. Lett. **46** L1159 (1985), Phys. Rev. B **33** 3587 (1986), Phys. Rev.B **37** 9800 (1988)

5/ I.A. Campbell, J.M. Flesselles, R. Jullien and R. Botet, J.Phys.C **20** L47(1987), Phys.Rev. B **37** 3825 (1988), and to be published

6/ J.M. Flesselles and R. Botet, J. Phys. A to be published

7/ J.M. Flesselles, R. Botet, I.A. Campbell and R. Jullien, to be published

8/ A.T. Ogielski, Phys. Rev. B **32** 7384 (1986)

9/ D. Stauffer, Phys.Rep. **54** 1 (1979)

10/ P. Erdös and J. Spencer, Comp. Math. Appl. **5** 33 (1979)

M. Ajtai, J. Komlos and E. Szemerdi, Combinatorica **2** 1 (1982)

K. Weber, J. Inf.Proc.Cyber. **22** 601(1986)

11/ S.F. Edwards and P.W. Anderson, J.Phys. F **5** 965(1975)

12/ W.L. McMillan, Phys. Rev. B **28** 5216 (1983)

13/ F. Mezei and A.P. Murani, J.Mag.Mag.Mat. **14** 211 (1979)

14/ P. Refrigier, E. Vincent, M. Ocio, and J. Hammann, Jap.J.App.Phys.**26** Supp. 3, 783 (1987)

15/ N. Bontemps and R. Orbach, Phys. Rev. B **37** 4708 (1988)

16/ K. Gunnarson, P. Svedlindh, P. Nordblad, L. Lundgren, H. Aruga and A. Ito, Phys. Rev. Lett. **61** 754 (1988)

17/ R.B. Griffiths, Phys. Rev. Lett. **23** 17(1969), A.J. Bray, Phys. Rev. Lett. **60** 720 (1988)

18/ K.L. Ngai and U. Strom, Phys. Rev. B **27** 6031 (1983), C.A. Angell and L.M. Torell, J. Chem. Phys. **78** 937 (1983), M. Adam, M. Delsanti, J.P. Munch and D. Durand Phys. Rev. Lett. **61** 706 (1988)

SPIN GLASS APPROACH TO NEURAL NETWORK MODEL

S. Kobe

Technische Universität Dresden, Sektion Physik
Mommsenstr. 13, DDR-8027 Dresden, GDR

Abstract

A review is given on an analogy between the Ising spin glass system and a model of a formal neural network. Information can be stored by a suitable choice of synaptic couplings (interactions) in the network and can be retrieved by relaxation. The network formulation of an information system is an attempt to capture the behaviour of a nervous system for a development of a new type of computer with a high level of parallelism.

1. Introduction

Magnetic models play an important role not only in explanation of properties of solids, but they also contribute to the understanding of other complex systems. The most famous example is the exact solution of the two-dimensional ferromagnetic Ising model with nearest-neighbour interactions by Onsager [1], which allows to study the general principles of phase transitions. The reason is that magnetic models are relatively simple, although not trivial. Therefore the question is allowed, if it is possible to use the models known in solid state magnetism to decribe other complex systems in nature as well. One of the most fascinating and unsolved problem is that of the human brain. Its processing capabilities is enormous and includes motor control, information processes as e.g. pattern and speech recognition, association, reasoning etc. Although the complete understanding of all these working principles cannot be expected in the next decades (the problem is, to understand the brain functions with the help of our brain!), it might be possible, that the study of some aspects prove to be helpful as a first step on this way.

Restricting the problem to the investigation of principles of the brain as an information storage system, fundamental differences to a present computer are being found: The storage and the retrieval of information is a collective effect of the neural network, neither special processors and programs, nor addresses and marks are needed. The memory is distributed on the whole network, the access to information is associative.

In the following Hopfield's model [2] of a formal neural network in analogy with an Ising spin glass model is introduced. It is shown that such a model contents some ingredients of an information storage system with the behaviour wanted. The aim is to implement such new basic models in computers with brain-like capabilities.

2. Model

Hopfield's idea [2] was to analyze the "experimental" situation of a neural network and to simplify it in such a way that the remaining model approaches an Ising spin glass. He believed that the relevant ingredients of a nervous system were being kept. This analogy is given in table 1. However, there is an important difference between both models: Whereas the interactions in the spin glass are fixed, the synaptic couplings can be altered in strengths during the process of storing a set of given information ("learning").

Using the analogy given in table 1 Hopfield [2] has introduced a model of a formal neural network. The states of the neurons "i" (i = 1 ... N) are represented by binary variables $S_i = \pm 1$ and the synaptic couplings by the interactions I_{ij} with different signs and strengths. Then the Ising energy of the system (whatever it means in the case of a neural network) is given by

$$E = -\sum_{i<j}^{1..N} I_{ij} S_i S_j \qquad . \qquad (1)$$

A given information (e.g. a pattern) can be coded as a set of binary variables $S_1 \ldots S_N$. Every set of variables (i.e. every state of the system) is characterized by a value of E according to (1). An information system must have the ability to simultanously store different sets of variables, e.g. p patterns. In Hopfield's model this demand is fulfilled by a suitable choice of the strengths of the synaptic couplings I_{ij} in (1): The I_{ij} have to be chosen in such a way

Table 1. Analogy between the models of a neural network (brain) und the Ising spin glass.

Neural network	Ising spin glass
$N = 10^{10\ldots12}$ neurons	$N = 10^{23}$ spins
assumption: every neuron can have two states: "active" (firing) or "quiescent"	spin states: up or down
every neuron interacts by axons/synapses with about 10^4 other neurons	every spin interacts with n neighbours (n = 4 ... N)
the influence of synapses can be "restraining" or "amplifying"	interaction is ferro- and antiferromagnetic
the state of a neuron depends on the sum of synaptic couplings with other neurons compared with a threshold value	the spin configuration in the ground state depends on the given interactions

that the p given patterns become stable low-energy states of the network. This process is called "learning". The questions, whether "learning" is possible and how it can be realized, are positively answered. Various "learning rules" are known in literature (see e.g. [3]), one of the most simple of them follows a idea of Hebb [4] with

$$I_{ij} = \sum_{\alpha=1}^{p} S_i^{(\alpha)} S_j^{(\alpha)} \qquad , \qquad (2)$$

where $S_i^{(\alpha)}$ is the ith binary variable of the αth pattern.

The situation after "learning" is shown in the energy landscape of fig. 1. The abscissa is a one-dimensional representation of the high-dimensional phase space of states and E is the energy corresponding to (1). Depending on the used "learning rule" and on the ratio p/N all minima or most of them belong to the stored patterns.

It is obvious, that every "noise" pattern has a higher energy than the "learned" one. Therefore, the input of a noise pattern and a simple relaxation procedure lead to a retrieval of the corresponding original pattern, provided that the energy of the input pattern lies within the attractor region of the corresponding minimum.

Fig. 2 demonstrates that frustration, i.e. the competition of positive and negative I_{ij}, is a necessary condition for this kind of storage system: The energy landscape of a non-frustrated (e.g. a ferromagnetic system with $I_{ij} = I > 0$) shows only two minima

belonging to two trivial "patterns", a black and a
white area corresponding to both ground states of a
ferromagnet with all spins up and all spins down,
respectively.

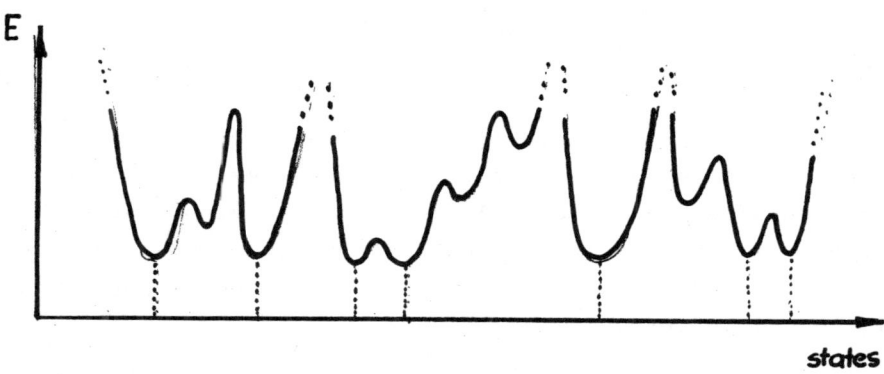

Fig. 1. Energy E versus states of the neural network.
Note, that the abscissa is a one-dimensional
representation of the high-dimensional phase space.
The minima corrsponds to the "learned" states.

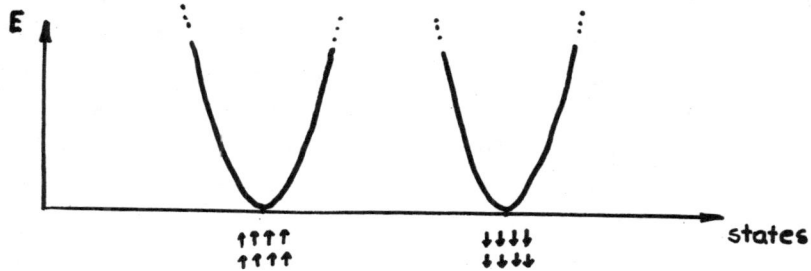

Fig. 2. In a system without frustration only two
trivial states can be stored, e.g. a black and a white
pattern.

3. Example

As a simple example the results of a simulation on a personal computer is shown in fig. 3. A system of N = 130 neurons/spins/pixels/nodes is used, each of them connected with all the others by synaptic couplings I_{ij}. The task is to store 21 different patterns in the network, e.g. the numbers "0" to "9" and the letters "A" to "K". Obviously, some of these characters are correlated, e.g. "E" and "F" or "O" and "D". In such case it is convenient to use an improved "learning rule" instead of the simple Hebb rule (2). The advantage of the projection rule (see [5,6] for more details) is, that all "learned" patterns become global minima of the network energy (1). Following Kanter and Sompolinsky [7] we set all diagonal terms I_{ii} equal to zero.

After the "learning" procedure all I_{ij} are fixed and thereby all 21 characters are stored corresponding to the right column of fig. 3 (only "E" and "F" are shown). Every row demonstrates the retrieval process by relaxation. The left picture in the row results, if 25 % of the neurons of a pattern are randomly flipped into the opposite state. It serves as input for the process. The second picture is obtained after 1 Monte-Carlo-step per neuron/spin. A Monto-Carlo-step means, that for a randomly chosen S_i it is proved, whether the condition $S_i h_i \geq 0$ is fulfilled, where $h_i = \Sigma I_{ij} S_j$ is the corrsponding local field. In the opposite case S_i is flipped.

The retrieval process is very fast; the stored pattern is found in two or three Monto-Carlo-steps per spin. No adresses are needed, because the content of the information plays the role of the address itself.

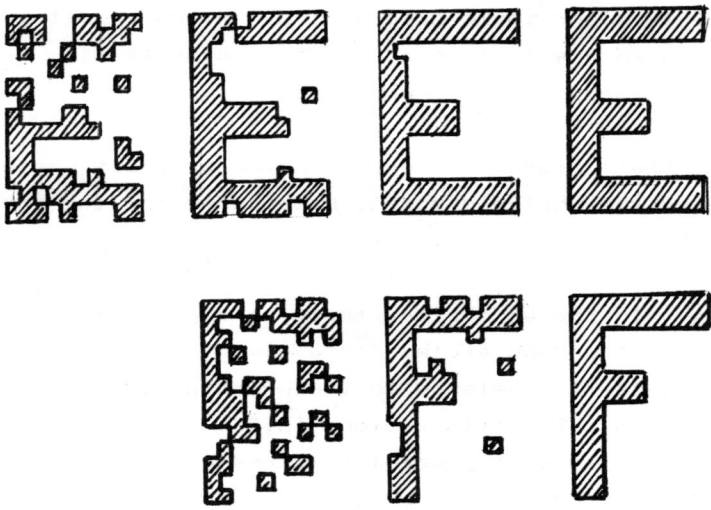

Fig. 3. 21 characters ("0" to "9" and "A" to "K") are stored in a Hopfield network with 130 neurons. Starting from a state with 25 % noise the pattern (e.g. "E" and "F") are retrieved after 2 or 3 relaxation steps.

4. Summary

The investigation of neural network models is a challenge for researchers in information processing, which had already become an established discipline in this field. For special tasks with a high degree of complexity the principles of neural networks are better than those of conventional computers and alternative to algorithmic programming. A neural network can act as an associative and content-addressable memory, it is insensitive to hardware-faults.

The interest in neural network models was prompted by the Hopfield model. Especially for the community of solid state physics and researchers into magnetism it is a good example for demonstrating the basic idea. In the spin glass theory the problem of finding the exact ground states for a given set of interaction constants I_{ij} belongs to the NP-complete problems of combinatorial optimization [8,9]. In the Hopfield model the inverse problem is stressed: The interaction constants have to be chosen in such a way that given states become ground states of the system.

However, the ability of the Hopfield model is limited with respect to the complexity of tasks and memory capacity (see e.g. [10,11]). Therefore the attention is shifted towards other networks, e.g. generalizations of earlier ideas such as the perceptron [12-15]. First successful attempts are done by Sejnowski and Rosenberg [16], who trained a perceptron-like network to learn "reading like a child". The input was a sliding window, seven letters long, and the desired output was a phonetic transcription of the central letter out of the seven. More surprising results in this field can be expected in the future.

Acknowledgement

I am very grateful to Andreas Schütte for his collaboration and stimulating discussions.

References

[1] L. Onsager, Phys. Rev. **65** (1944) 117.
[2] J.J. Hopfield, Proc. Natl. Acad. Sci. USA **79** (1982) 2554.
[3] S. Grossberg, Neural Networks **1** (1988) 17.
[4] D.O. Hebb, The Organisation of Behaviour, Wiley, New York, 1949.
[5] T. Kohonen, Associative Memory; A System-Theoretical Approach, Springer, Berlin, 1977.
[6] L. Personnaz, I. Guyon, G. Dreyfus, J. Physique Lett. **46** (1985) L359.
[7] I. Kanter, H. Sompolinsky, Phys. Rev. **A35** (1987) 380.
[8] S. Kobe, A. Hartwig, Computer Phys. Commun. **16** (1978) 1.
[9] A. Hartwig, F. Daske, S. Kobe, Computer Phys. Commun. **32** (1984) 133.
[10] A. Engel, H. Englisch, A. Schütte, subm. to J. New Generation Computer Systems.
[11] S. Kobe, A. Schütte, subm. to Acta Phys. Polon.
[12] F. Rosenblatt, Principles of Neurodynamics, Spartan Books, New York, 1962.
[13] M. Minsky, S. Papert, Perceptrons: An Introduction to Computational Geometry, MIT Press, Cambridge MA, 1969.
[14] D.E. Rumelhart, G.E. Hinton, R.J. Williams, in: D.E. Rumelhart, J.L. McClelland, Parallel Distributed Processing, MIT Press, Cambridge MA, 1986, p. 318.
[15] K.A. Benedict, J. Phys. A: Math. Gen. **21** (1988) 2643.
[16] T.J. Sejnowski, C.R. Rosenberg, Johns Hopkins Technical Report JHU/EECS-86/01 (1986)

ACOUSTIC EMISSION, DOMAIN WALLS AND HYSTERESIS IN VARIOUS FERRO AND FERRIMAGNETS.

M. GUYOT, T. MERCERON AND V. CAGAN, Laboratoire de Magnétisme, CNRS, 1, Place Aristide Briand 92195 Meudon Principal, France.

ABSTRACT

The state of the present knowledge of the AE phenomenon in magnetic materials is first reviewed. By using the data of the present work as well as those already published, it is demonstrated that the shape demagnetizing effect can greatly affect the AE profile along the hysteresis loop. From a series of YIG polycrystals it is shown that, for each sample, the AE activity is proportionnal to the hysteresis losses. This law is consistent with our own interpretation, which is that the DW creation/annihilation process generates AE, ie ultra sonic waves, which are then dissipated into heat in the sample. The classical interpretation - AE results from abrupt motion of non-180° DW - which implies a proportionality between AE and magnetostriction is questionned by the strong grain size dependence of AE in YIG polycrystals, which all have the same magnetostriction.

I - INTRODUCTION

The acoustic emission (AE) is a general phenomenon observed in various materials which involves bursts of stress waves generated internally during a dynamical process (1). In some cases a pure mechanical origin has been suggested (dislocation or crack propagation, interparticle movement, fracture, etc...) and various models have been proposed accordingly (1). The practical uses of AE is mainly non-destructive testing (NDT) : for example it is well known that the spontaneous AE activity increases significantly prior some catastrophic events : mines or large concrete constructions (bridges, nuclear power plants ...) are then under permanent AE control.

In the case of magnetic materials, Spanner (2) (1970) predicted the appearance of AE in correlation with the classical Barkhausen noise (3) ; the first experimental work on this basis has been done by Lord, who observed the variation of AE activity along the hysteresis loop on a Ni wire (4). The most detail studies have been done by Kusanagi and coworkers (5, 6) on steels and Ni, and by Ono and coworkers (7, 8) on Fe, Ni and on Fe-Si and Fe-Ni alloys. A theory based on these works has been proposed which attributes the AE activity to the release of the magnetoelastic energy associated with the abrupt motion of non-180° domain walls (DW). The motion of a 180° DW does not involve a priori, any change in the magnetoelastic energy. It must be pointed out that this theory underlies a relation between AE and the "defects" (precipitates, residual stress etc), which leads to a practical uses of AE in magnetism, as a NDT of defects. like in the case of non magnetic materials. We thus can understand the reason of the detailed work of Ono and Shibata entitled "Magnetomechanical acoustic emission for residual stress and prior strain determination" (7). The same NDT approach of AE seems to have guided the series of works by Buttle and coworkers about "Dislocation dependence of AE on iron" (9) or "The study of neutron irradiation damage on α-iron with magnetoacoustic and Barkhausen noise" (10). Still the same motivation for Jiles and co-workers when studying "The effects of grain size and carbon contents on the magnetic properties of decarburized steels" (11).

Such an approach of the AE - related to the defects - is also guided by what we could call a dominant idea : the hysteresis of the magnetization in such "soft" ferromagnetics metals or alloys - on which materials all the previous AE studies have been - is essentially related to the presence of defects and the domain wall motion is essentially controlled by its pinning at the various defects.

We are of the opinion that the hysteresis is better described if in addition to the classical DW pinning, one also takes into account the process of DW creation/annihilation. It must be stressed that in ferromagnetic materials, everybody agrees that there is no DW in a saturated sample. On the other hand in the demagnetizated state (coercive field). The DW are present and observable. Thus, there might be a field region where the DW are created and another region where they are annihilated. We have already shown that the DW creation/annihilation process is a dissipative process responsible for an important contribution to the hysteresis losses (12, 13). If at the time of this study we were not able to propose a physical mechanism to account for the magnetic energy conversion into heat, it seems, now, that the acoustic emission is a good candidate for such a mechanism. The main reason is the observation by Mohammad et al. (14) of AE bursts correlated with the collapse or creation of DW in ferroelectric. Then here is our goal : first, to show how AE in magnetic materials can be related to the process of DW creation/annihilation ; secondly, to discuss whether or not this explanation is more appropriated than the non-180° DW motion theory (6, 7).

II - EXPERIMENTAL

1- AE MEASUREMENTS.

In general the AE consists of bursts of ultrasonic waves of which the central frequency as well as the spectral repartition are not

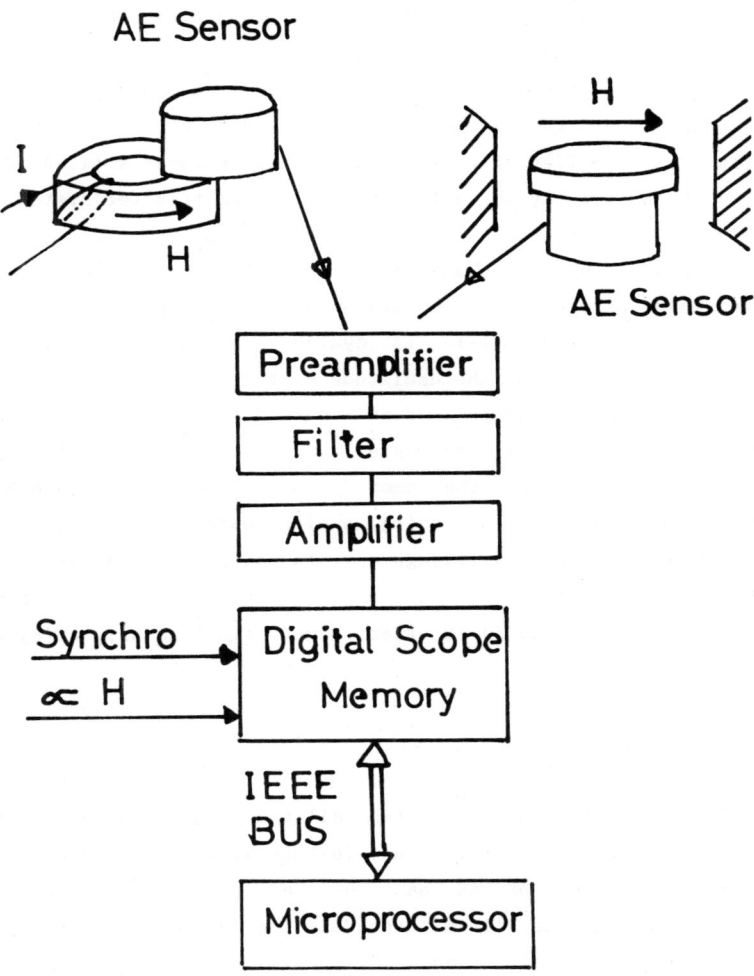

Fig. 1 - Schematic diagram of the set-up for AE measurements.

known. The general methods for AE detection use piezo sensors (which detect surface waves) ; according to the weakness of the signals, the system is generally of a resonance type. In our case we use a Bruel & Kjaer chain sets at a 200 kHz central frequency ; we will discuss later the possible consequence of such a limited bandwidth upon the understanding of the AE phenomenon. The scheme of our system is shown on fig. 1. It differs from the classical arrangements (1, 6, 7, 8, 9, 11, 17) essentially by the data acquisition and transfer system. We use a high sampling rate Digital Storage Scope (Gould - 400 mega sample/seconde) which includes a special built-in function for storage of fast events such as the AE bursts, acting as a peak detector. Then, when the magnetizing field is cycled along an hysteresis loop, we record the variation of the amplitude of the AE activity as a function of H. These data are then stored and treated using a micro-computer. As schematized on fig. 1 the magnetizing field is applied i) through a winding directly on the sample in the case of toroids (100 Hz-triangular) or ii) by using an electromagnet in the case of disk shaped samples (10^{-1} Hz-triangular). In addition to the AE signals we also record the field H and the flux changes $d\phi/dt$ (through an appropriate pick-up coil) in order to get simultaneously AE(H) as well as B(H).

2 - THE SAMPLES.

If we except our own works, all the previous studies were done exclusively on ferromagnetic metals (either single or polycrystals) (4, 11). The existence of AE has then been shown to exist in magnetic materials and to be field dependent on various other parameters (composition, residual or applied stress, inclusion, etc...) Although we have also observed AE along the hysteresis loop in conducting materials (15), we do prefer using the insulating ferrimagnets particularly because they show no complicated effect related to eddy currents, which depend upon the magnetizing field frequency. Another advantage is that we have a good knowledge of the magnetization process in such ferrimagnets, that we have previously used to establish the direct relation between the hysteresis losses and the process of DW creation-annihilation (12, 13) which process we want here to correlate with the AE phenomenon. Consequently in this paper we will present the results obtained on YIG single crystal and polycrystals (other compositions such as Ni and Ni-Zn ferrites, YIG:Mn will be presented in a forthcoming paper (16)). The YIG single crystal is a disk-shaped sample nearly 0.4 mm thick, cut parallel to the [110] plane ; the measurement field is applied along a [111] direction. This crystal is used mainly for qualitative observation of AE. On the other hand,for quantitative measurements, we have specially prepared a series of high purity and stoichiometric , toroid-shaped YIG samples. By changing the preparation parameters we have obtained polycrystalline samples with a mean grain size ranging from 2 to 30 μm.

III - QUALITATIVE OBSERVATIONS OF AE.

1 . YIG SINGLE CRYSTAL

On fig. 2, from the top to the bottom, are represented the time dependence of the applied field (triangular, 0.2 Hz), the flux changes (Barkhausen activity) and the AE activity respectively. The maximum amplitude of the magnetic field is high enough to approach the saturation of the sample, as seen from Fig. 3 top where the ascending branch of the hysteresis loop has been recorded. From these figures it is evident that there is no AE activity close to the saturation (AE signal is at the noise level of the system). There exist two pronounced but broad AE maxima, roughly located in the so called "knee" of the hysteresis loop and a relative minimum in the vicinity of the zero field (in fact close to the coercive field). This observation is compatible with the mechanism of DW creation/annihilation : starting from the maximum negative field we have first some magnetization rotations, which do not give AE ; then comes the field zone of DW creation (first peak of AE in negative field region) ; then the DW move through the sample without an appreciable DW surface variation, so they

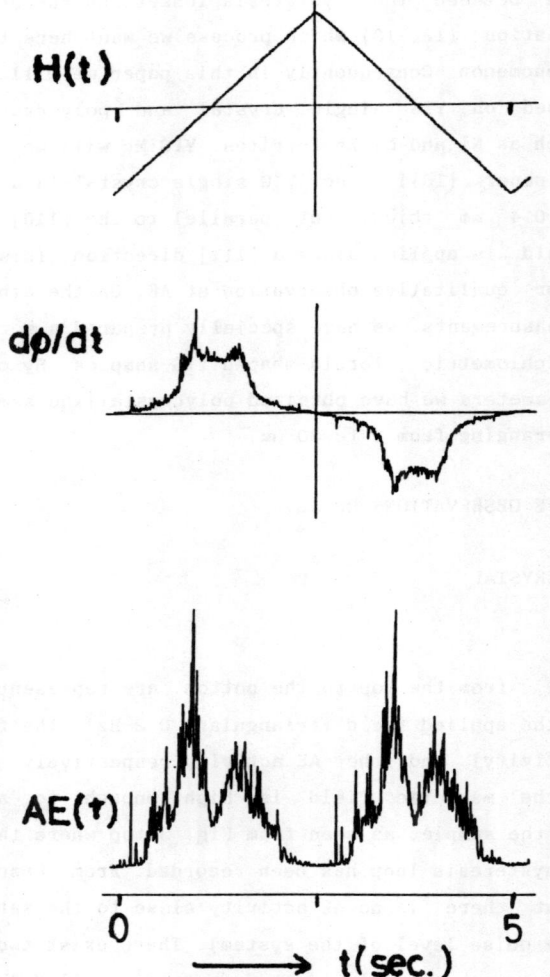

Fig. 2 - From the top to the bottom : applied magnetic field H, derivative of the flux dϕ/dt and acoustic emission versus time for a YIG single crystal.

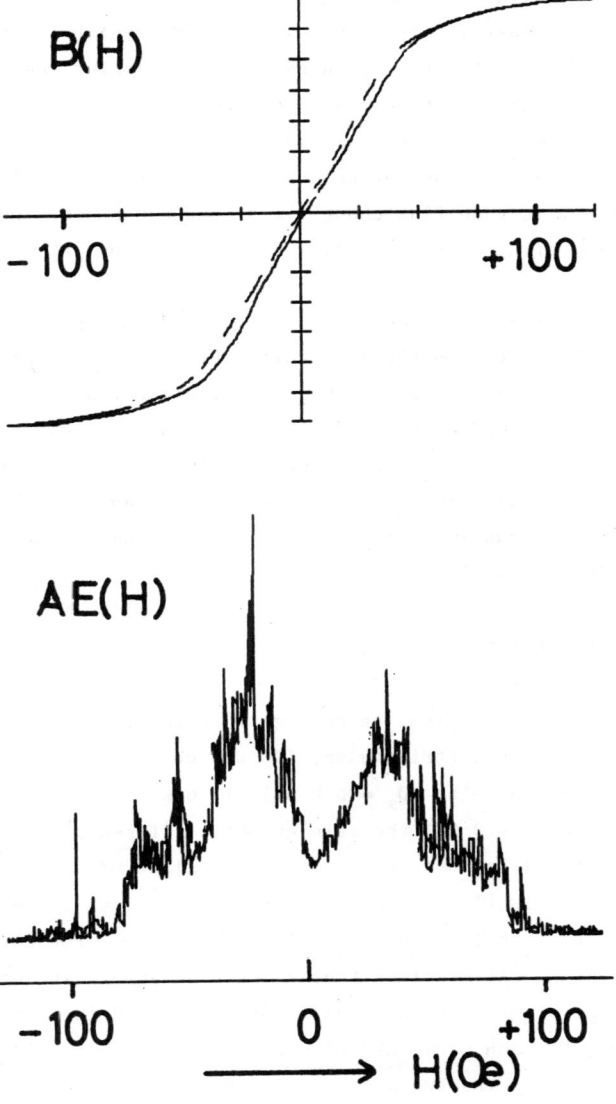

Fig. 3 - For a YIG single crystal, induction B (top) and AE activity (bottom) as a function of H.

give no AE (minimum of AE close to H_c) ; then the DW are annihiled (second AE maximum in the positive field region ; then, when approaching the saturation there is no longer DW, but only magnetization rotations : no AE is seen. The shape of the recorded flux changes (fig. 2 middle) shows some loose correlation between AE and BN : on the one hand the appearance or disappearance of a DW gives rise both to AE and to some flux changes ; on the other hand the DW motion in the vicinity of H_c gives only flux changes, without AE.

Our observations of AE on a YIG single crystal are in good qualitative agreement with those by Buttle et al. (9, 17) on a metallic single crystal (Incoloy 904). But it must be stressed that the interpretation is different : Buttle et al. attributes the AE to the release of elastic energy associated with the abrupt motion of a non-180° DW , while we speak of a DW creation/annihilation process. This will be discussed later, but let's now look at some results on polycrystalline materials.

2. POLYCRYSTALLINE YIG SAMPLE.

Fig. 4 shows a typical recording of AE (bottom) as a function of the magnetizing field (triangular, 100 Hz) obtained on a toroidal YIG sample (mean grain size D_m = 8.5 µm). If one looks at the ascending branche (- 3 Oe to + 3 Oe) the main evident difference with respect to the single crystal is that only one broad maximum of AE is seen around the coercive field (instead of two maximum away from H_c and a minimum at H_c for the single crystal). If we compare the corresponding hysteresis loops, it is clear that a shape demagnetizing effect tilts the loop of the single crystal, while for the toroïd the "knees" of the loop are close to each other and close to H_c. Then one could think that for the polycrystal the two broad maxima expected at the knees overlap each other and appear as a broad maximum. We will give a confirmation of the influence of the shape demagnetizing effect by looking at some published data and by doing a specific experiment on our sample.

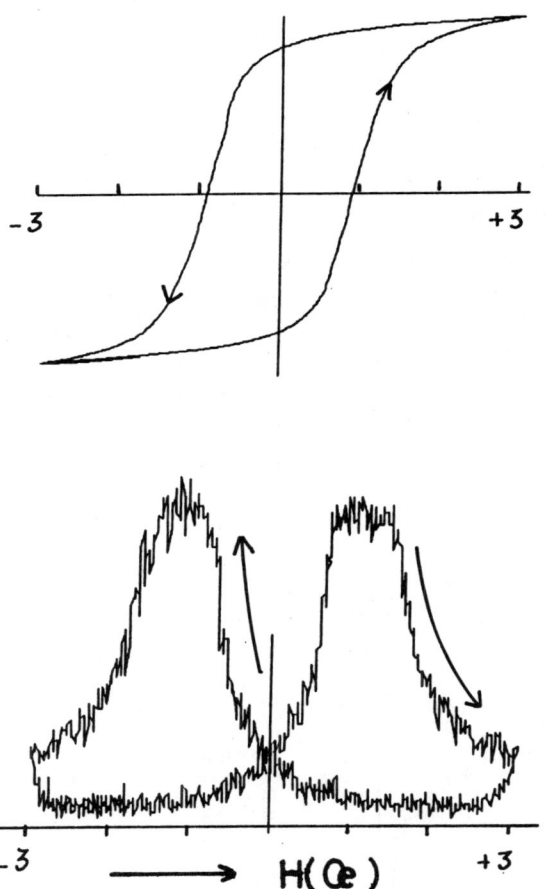

Fig. 4 - For a YIG polycrystal, induction (B) (top) and AE activity (bottom) as a function of H.

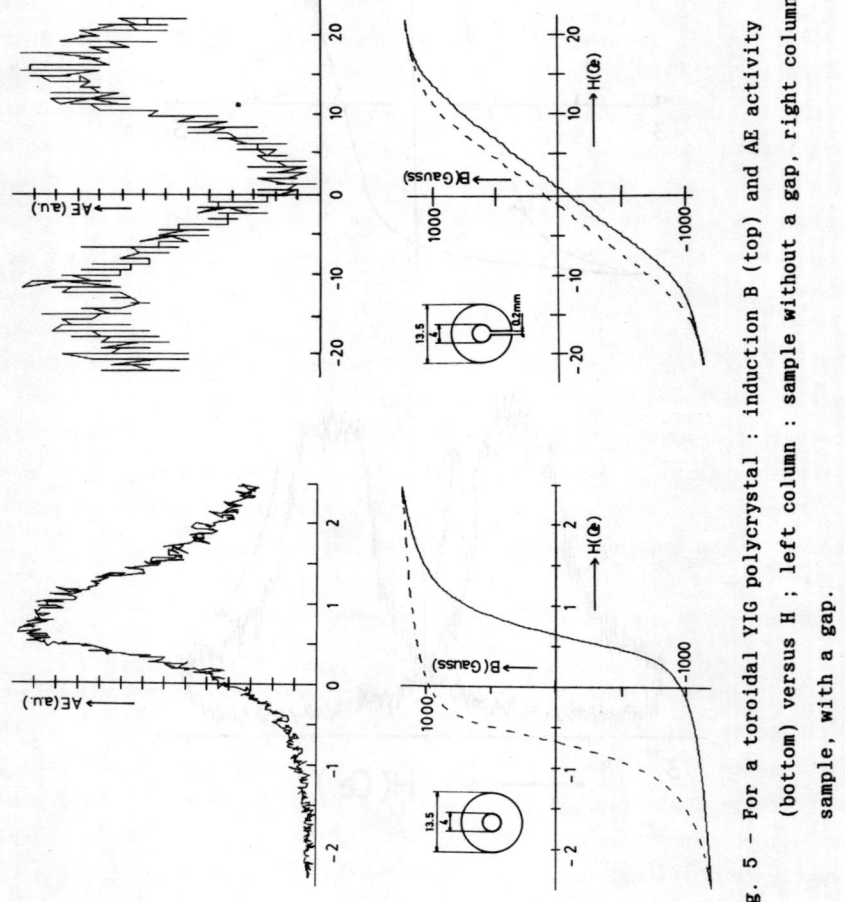

Fig. 5 - For a toroidal YIG polycrystal : induction B (top) and AE activity (bottom) versus H ; left column : sample without a gap, right column: sample, with a gap.

3. SHAPE DEMAGNETIZING EFFECT AND AE PROFILE.

May Man Kwan, Ono and Shibata (8) have published typical results of AE on Fe sheets (lenght 190 mm, width 15 mm, thickness 0.45 mm) which show around the coercive field some splitting of the AE profile in two close maxima. According to our explanation, the shape demagnetizing effect in such a geometry is rather small but large enough to produce the separation between the DW creation zone and the annihilation zone. This is more pronounced on the Buttle et al. AE profile obtained on a polycrystalline α-Iron rod (lenght 50 mm, diameter 5 mm) (10) : the demagnetizing effect is there very large and consequently the AE profile consists of two large maxima separated by a relative minimum.

To definitly confirm such a demagnetizing effect on the AE profile, we have performed the simple experiment shown on fig. 5. In a toroidal polycrystal we have created an artificial demagnetizing effet by making a small gap (0.2 mm thick). As a result, the broad maximum of AE shown on the virgin sample (left) is splitted in two maxima (right) located in the knee region of the tilted loop (top left).

4. SUMMARY OF QUALITATIVE AE OBSERVATIONS ALONG HYSTERESIS LOOPS.

In general, the AE profile along hysteresis loops of ferro or ferrimagnets consists of two assymetric broad maxima located on the knee regions of the hysteresis loop, separated by a relative minimum in the coercive field region. The distance between these two peaks depends upon the shape demagnetizating effect of the particular sample investigated, the extreme case being the toroid shaped sample where the absence of shape demagnetizing effect results in an overlaping of the two peaks into a broad maximum around H_c .

The qualitative origin of the AE could be satisfactory explained by both the theories : non-180° DW motion or DW creation/annihilation process. The next step will then discuss the quantitative results with

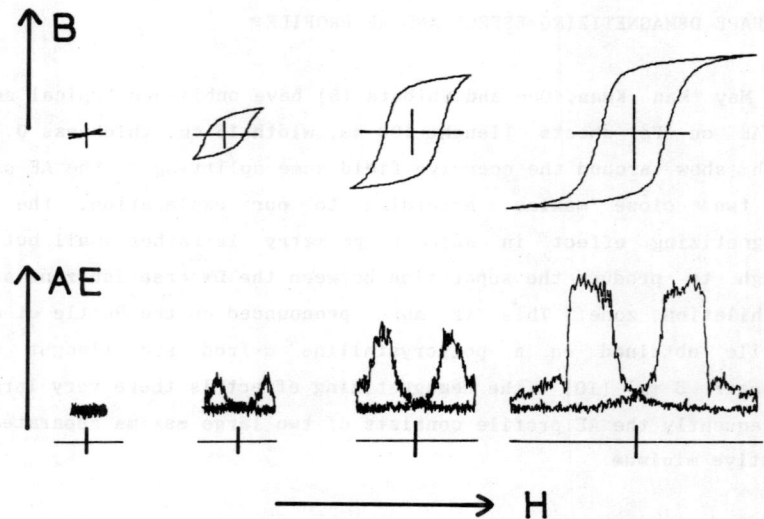

Fig. 6 - Hysteresis (top) and AE activity (bottom) for various maximum applied fields H_m recorded on a toroidal YIG polycrystal. From the left to the right : H_m = 0,5 - 1 - 2 and 4 Oe.

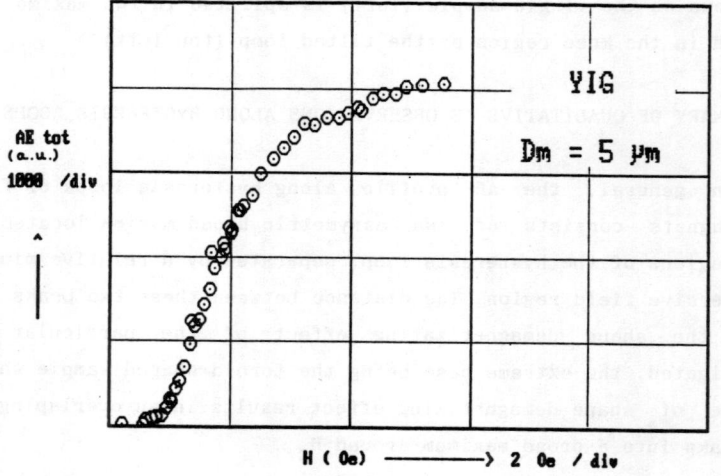

Fig. 7 - Total Acoustic Emission Versus maximum magnetizing field for a YIG polycrystal.

respect to each theory.

IV - QUANTITATIVE STUDY OF FIELD DEPENDENCE OF AE

1 - AE AND HYSTERESIS LOSSES

Fig. 6 gives some selected AE_{Hm} profiles and the corresponding magnetization loops measured on a toroidal YIG polycrystal ($D_m = 8.5$ μm) for various values of the maximum applied field H_m (triangular 100 Hz). It can be seen i) that for H_m values in the Rayleigh region, no AE occurs (the signal is at the noise level), ii) that the onset of AE is related to the onset of irreversible phenomena on the hysteresis loop, iii) that the total amount of AE increases with the loop surface, independently from the AE profile shape.

More quantitatively, using the facilities of our computer-aided system, we define :

$$AE_{tot.}^{Hm} = \int_0^T \{ AE^{Hm}(t) - A_0 \} dt$$

as the total AE activity per period (per loop) for a loop taken a the maximum field H_m (where A_0 is the noise level of the system).

Fig. 7 shows the variation of AE versus measured on a 5 μm grain size sample. The shape of the curve is qualitatively similar to those published by Kusanayi et al., who measured the RMS value of AE.

In fig. 8 we have plotted $AE_{Hm}(tot)$ as a function of the hysteresis losses measured for the same loop.

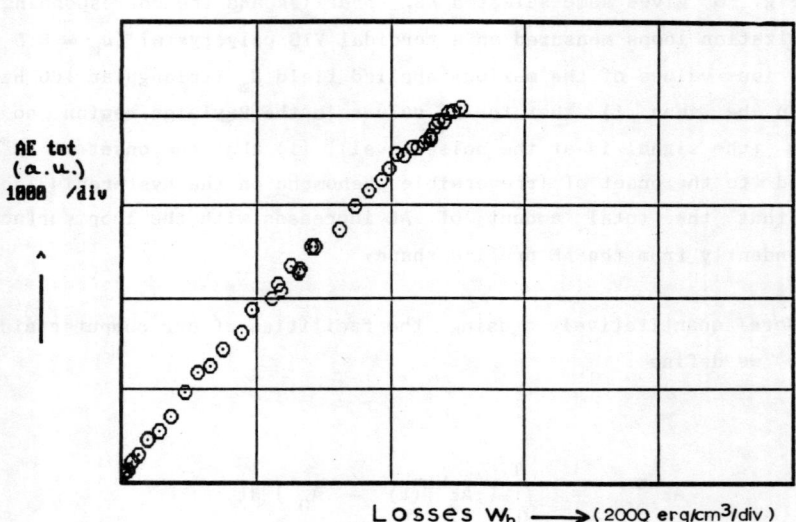

Fig. 8 - For a polycrystalline YIG sample ($D_m = 5$ μm), total cumulative acoustic emission versus hysteresis losses.

$$W_h^{Hm} = \oint H \, dM$$

Such a plotting shows a clear proportionality between the two phenomena, which could implie a common origin. We are of the opinion that the common source is the DW creation/annihilation process because we have shown earlier that the hysteresis losses are due to this process. Let's recall this demonstration.

2 - HYSTERESIS LOSSES AND DOMAIN WALL CREATION/ANNIHILATION

In order to explain the grain size dependence of the hysteresis loop and of related quantities in polycrystalline ferrimagnets, we have developed a model where the domain wall size is the fundamental parameter (18, 19, 20). This model has sucessfully described many of the effects of the grain size on the technical magnetization parameters. In order to fully explain the field dependence of the hysteresis losses we were led to improve the model by introducing the hypothesis that the DW surface variation is a dissipative process (12, 13).

Formally, for the mathematical treatment we postulate that any surface variation ΔS gives an energy loss $E = \gamma \Delta S$ where γ is the specific domain wall surface energy. Using our model, where an unique DW pinned at the grain boundary in a spherical grain (diameter D_m), allows us to relate ΔS to the relative magnetization $m_r = M/M_s$. In a first approximation we get the energy loss E for a loop versus the relative magnetization m_r :

$$E \simeq \frac{8}{D} \left(f.m_r + \frac{\gamma}{3} m_r^2 \right)$$

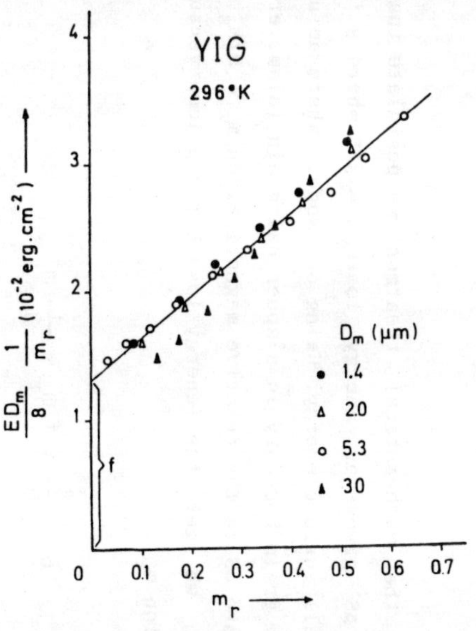

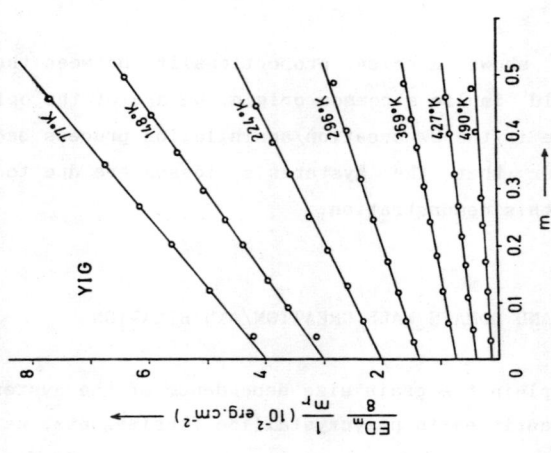

Fig. 9 – Hysteresis losses as a function of m_r for four YIG samples with different mean grain sizes measured at room temperature.

Fig. 10 – For one of the four YIG samples reported in fig. 9 hysteresis losses as a function of m_r at different temperatures.

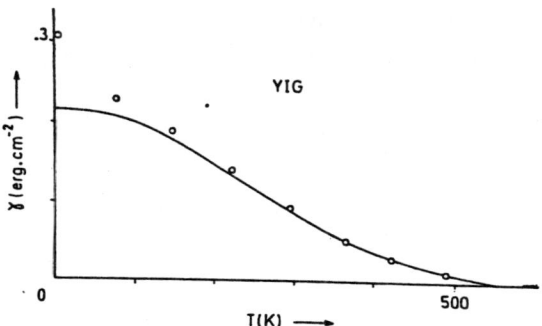

Fig. 11 - Domain wall energy per surface unit versus temrature : circle = experimental ; continuous line = theory $\gamma_0 = 2\sqrt{AK_1}$

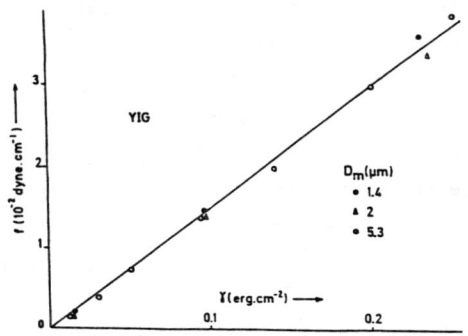

Fig. 12 - Pinning parameter f as a function of the domain wall energy γ.

The first term $f \cdot m_r$ corresponds to the energy lost by the pinning/unpinning mechanism of DW at the grain boundaries, while the term $(\gamma/3) \cdot m^2_r$, is the dissipative term due to the DW surface variation. In a more formal presentation the formula (1) could be written :

$$\frac{ED_m}{8} \cdot \frac{1}{m_r} \simeq f + \frac{\gamma}{3} m_r$$

The relation (2) has been experimentally checked, both as a function of the grain size and at various temperatures (fig. 9 and fig. 10 respectively, in the case of YIG samples).

This general law allowed us to deduce the domain wall energy γ (from the slope of the E(m) curves). As an example, fig. 11 top shows the temperature dependence of γ so determined (circle) compared to the theoretical value $\gamma_0 = 2\sqrt{AK_1}$ (continuous line), which as been calculated from the know values of A(T) and K (T). This remarkable agreement - without the use of any adjustable parameter - has been also observed on other ferromagnetic materials all along the ferrimagnetic state from 4K to the Curie temperature T_c (13, 21), which confirmed "in fine" the dissipative nature of the DW creation/annihilation process.

The other remarkable result was the linear relation between the pinning parameter f and the DW energy γ (Fig. 12). This implies that the pinning/unpinning process can be considered as a creation/annihilation process of micro-portions of a domain wall at the grain boundaries (or at the pinning points). Then the hysteresis losses can be totally attributed to the DW creation/annihilation process. The dissipative nature of the irreversible DW creation/annihilation process was also introduced independently by Brailsford (22) and by Haller &

Kramer (23). Nevertheless, if we except a recent publication by Sakaki and Matsuoka at Intermag 86 (24) , the impact of such an idea was nearly null among the magnetic community.

We believe that the lack of echo was mainly due to the fact that we did not propose at that time a physical mechanism for the conversion of the magnetic energy into heat. Presently, it appears that AE could be the missing link.

3 - EFFECT OF MICROSTRUCTURE ON AE

From our vision of the magnetization processes it turns out presently, that the DW creation/annihilation process generates AE, ie ultrasonic waves (through the magnetoelastic coupling) which are then dissipated in the materials into heat. This interpretation implies the proportionality between AE activity and hysteresis losses that we have already observed.

None among the previous studies reports such a relation, mainly because no one tried to relate the AE activity to the hysteresis losses. One more reason is that the other theory which relates AE to the non-180° DW motion, predicts a relation with the magnetostriction λ_s, via a certain distribution of non-180° domains. If we look at the most extensive study in that field, (7, 8) only a poor relation between AE and magnetostriction λ_s, is seen, although the authors try to improve there relation by selecting λ_{100} for their SiFe samples.

Now we will show that it exists a first order effect of the grain size on the observed AE, which invalides the relation with the magnetostriction

In fig. 13 we have plotted the results of AE $_{(Hm)}$ as a function of $W_{h(Hm)}$ for a series of 7 samples of YIG, with grain size ranging from 2 to 30 μm.

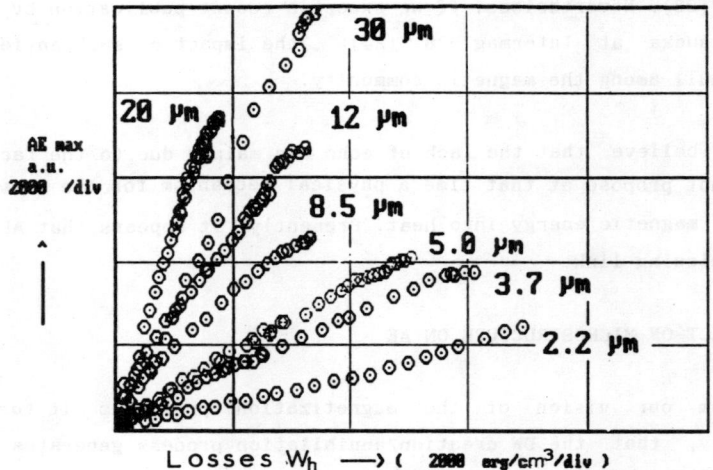

Fig. 13 – For each of the seven investigated YIG samples, the total cumulative acoustic emission AE_{Hm} versus hysteresis losses.

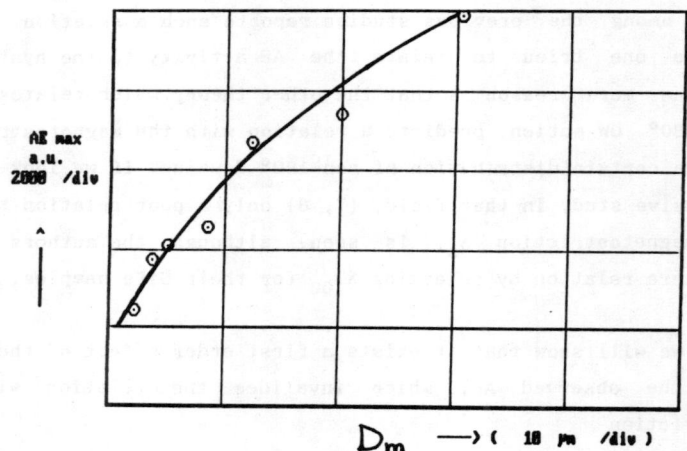

Fig. 14 – The total cumulative acoustic emission as a function of the mean grain size for polycrystalline YIG material.

For each sample we still have the proportionality between AE and W_h as shown before. But the important point is that, while the saturation value of W_h decreases with increasing grain size (which is consistent with our previous work (see eq. 2)) on the contrary the saturated value of AE increases with the grain size. This is more clearly seen in fig. 14. If we look to the literature, Jiles et al. (11) had already mentionned some tendancy of an increasse of AE for increassing grain size in a series of Steel samples. It follows that there is no an unique (or typical) value of AE activity for a given compound, while there is an unique value of λ_s.

V - DISCUSSION

1 - COMPARISON OF THE THEORIES OF AE IN MAGNETIC MATERIALS

Consequently the "non-180° DW motion" theory of AE generation seems disqualified for the following reasons :

- the predicted relation with magnetostriction is not well verified experimentally,

- the grain size effect is not taken into account.

Does that mean that our explanation - AE generation during the DW creation/annihilation process is fully justified ? The answer is : only partly.

Indeed, if our theory predicts a proportionality to the hysteresis losses, the grain size dependence might be different.

As the grain size increases, we must except a decrease of the total DW surface per volume unit in the sample, which implies lower hysteresis losses (as observed) and lower AE, which is not observed. Fig. 15 top shows a scheme of the predicted AE (Wh), compared to the observed one (fig. 15 bottom).

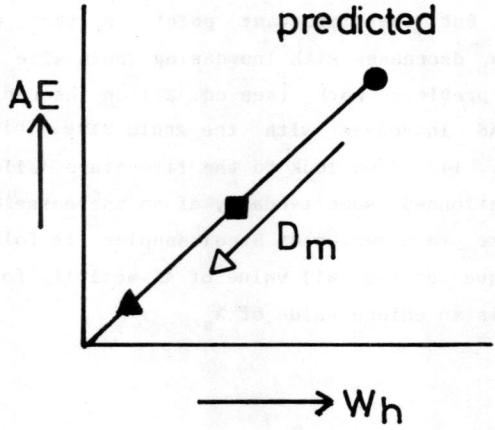

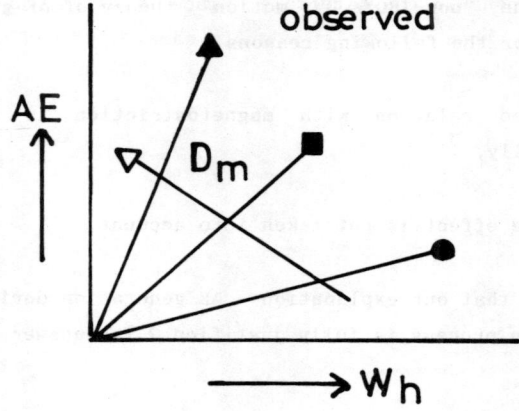

Fig. 15 – For YIG polycrystalline samples with different mean grain sizes, the total cumulative acoustic emission as a function of hysteresis losses ; top : predicted, bottom : observed.

2 - EXPLANATION OF THE GRAIN SIZE EFFECT ON AE.

- Sample auto-absorption

When a burst of AE is emitted somewhere in the sample, it travels through the sample and interacts with the matter. One can notice that when the grain size D_m decreases, the density of grain boundaries increases which could give a first contribution to the partial absorption of AE bursts.

Another source of local absorption of ultrasonic waves in magnetic materials is the DW themselves. Indeed, Le Craw and Comstok (25) have previously shown on a YIG single crystal sphere that 2,44 MHz ultrasonic waves are very well transmitted when the sphere is magnetically saturated (no DW) while they are drastically absorbed as soon as the DW appear in the materials. Since it exists more DW in a small grain sample than in a large grain sample, the DW absorption might result in a relative decrease of the observed AE in the small grain samples.

- Spectral repartition of AE bursts

We could also think that the observed grain size dependence of AE is some kind of experimental artefact, due to the way we look at the AE phenomenon. One must have in mind that nearly all the AE detection systems are of a resonance type (200 kHz in our case). Such systems have a very narrow bandwidth (see on fig. 16) and we have presently no idea of the spectral repartition of the AE bursts. Consequently, if for example, two samples with different grain size have the proposed AE spectral repartition, as in fig. 16, the observation at the sensor frequency is not representative of the actual ultrasonic waves energy involved in the processes.

It must be mentionned that Ono et al. (7)used a series of 6

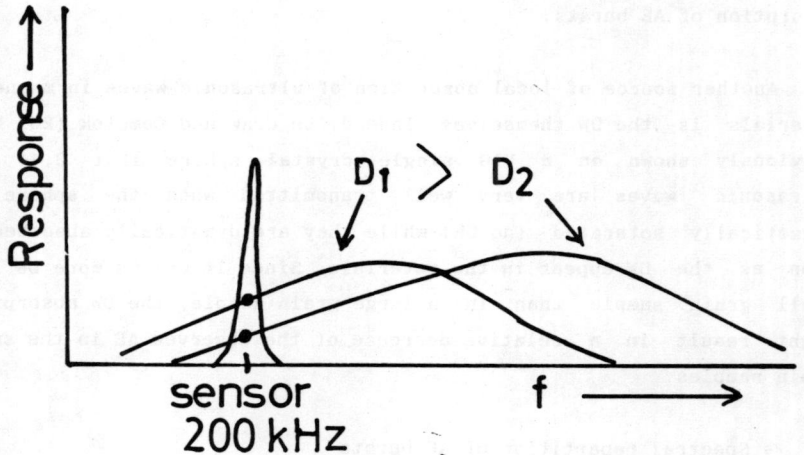

Fig. 16 – The narrow bandwidth of the AE sensor as a possible alteration of the grain size effect on the AE observation.

captors centered between 0,1 and 1,5 MHz ; however, although they have reported some frequency effects, they were not able to draw any spectral repartition, because of the lack of an absolute calibration of the captors.

We are of the opinion that detailed studies of the spectral repartition of AE should be carried out to clarify this point and that the observed frequency range must be extended as far as possible, particularly in the 10 - 100 MHz range where it is known that DW relaxes : the AE emission could also contribute to the DW damping, a phenomenon which is still not satisfactory explained.

REFERENCES

1 - A.E. LORD Jr., in Physical Acoustic, edited by MASON (Academic Press New York) (1975), Vol. XI, 289.
2 - J.G. SPANNER M.S., Thesis, Washington State University, Pullman.
3 - H. BARKHAUSEN, Physik Z. 20 (1919), 401.
4 - A.E. LORD Jr., R. USATSCHEW and M. ROBINSON, Letters in Applied and Engineering Sciences 2 (1974), 1-9.
5 - H. KUSANAGI, H. KIMURA and H. SASAKI, J. Appl. Phys. 50 (1979), 2965.
6 - H. KUSANAGI, H. KIMURA and H. SASAKI, in Fundamentals of Acoustic Emission, ed. by K. ONO, UCLA (1978) 309-334.
7 - K. ONO and M. SHIBATA, in Advances in Acoustic Emission, eds DUNEGAN and HARTMAN, DUNHART, KNOXWILLE, 1981, 154-174.
8 - MAY MAN KWAN, K. ONO and M. SHIBATA, J. of Acoustic Emission, 3 (1984), 144-156.

9 - D.J. BUTTLE, C.B. SCRUBY, J.P. JAKUBOVICS and G.A.D. BRIGGS, Phil. Mag. $\underline{55}$, (1987), 717-756.

10 - D.J. BUTTLE, E.A. LITTLE, C.B. SCRUBY, G.A.D. BRIGGS and J.P. JAKUBOVICS, Proc. R. Soc. London, A414 (1987), 221-236.

11 - R. RAJAN, D.C. JILES and P.K. RASTOGI, IEEE Trans. Mag. MAG-23 (1987), 1869.

12 - M. GUYOT and A. GLOBUS, Phys. Stat. Solidi (b) $\underline{59}$ (1973), 447.

13 - M. GUYOT, Thèse d'Etat, Paris (1975).

14 - J.J. MOHAMAD, L. ZAMMIT-MANGION, E.F. LAMBSON and G.A. SAUNDERS, J. Phys. Chem. Solids $\underline{43}$ (1982), 749.

15 - M. GUYOT, T. MERCERON and V. CAGAN, J. Appl. Phys. $\underline{63}$ (1988), 3955.

16 - M. GUYOT, T. MERCERON and V. CAGAN, to be published in Proc. Int. Conf. Ferrite 5 (ICF 5), BOMBAY, January 1989.

17 - D.J. BUTTLE, G.A.D. BRIGGS, J.P. JAKUBOVICS, E.A. LITTLE and C.B. SCRUBY, Phil. Trans. R. Soc. London, A 320 (1986) 363.

18 - A. GLOBUS, P. DUPLEX and M. GUYOT, IEEE Trans. Mag., MAG-7 (1971) 617.

19 - M. GUYOT and A. GLOBUS, AIP Proc. Conf. $\underline{5}$ (1971) 902.

20 - A. GLOBUS and M. GUYOT, Phys. Stat. SolL. (b) 52 (1972) 427.

21 - M. GUYOT and A. GLOBUS, J. Physique CI (1977) 157.

22 - F. BRAILSFORD, Proc. IEEE 117 (1970) 1052.

23 - T.R. HALLER and J.J. KRAMER, J. Appl. Phys. 41 (1970) 1034.

24 - Y. SAKAKI and T. MATSUOKA, IEEE Trans. Mag. MAG-22 (1986) 623.

25 - R.C. LE CRAW and R.L. COMSTOCK, in Physical Acoustics, vol. IIIB, ed. MASON (Acad. Press. New York) 127-199.

INSTABILITY OF ITINERANT ANTIFERROMAGNETISM IN MANGANESE LAVES PHASES

R. Ballou, J. Déportes, R. Lemaire, B. Ouladdiaf* and P. Rouault

Laboratoire Louis Néel, C.N.R.S., 166X
38042 Grenoble cedex
FRANCE

ABSTRACT

Experimental investigations of the RMn_2 Laves phases give evidence of new magnetic properties associated with the onset of itinerant antiferromagnetism. As in most intermetallics, compact atomic packings imply that two nearest neighbors of an atom can be nearest neighbors of each other. This topology induces frustration of negative exchange interactions. The Mn antiferromagnetism being close to its instability, amplitude change of the magnetic moment is allowed. Therefore, in addition to the classical effects of frustration characteristic of localized systems new effects are observed : cancellation of magnetic moments corresponding to cancellation of local band splitting ($ThMn_2$), reduction of magnetization density in frustrated zones, large amplitude fluctuations of the moments in the para- magnetic state which increase with temperature reflecting the thermal weakening of the frustrations (YMn_2).

1 - INTRODUCTION

The metallic radius of 3d transition metals is known to be smaller than the radial extension of the 3d wavefunctions which allows an overlap of orbits of neighboring atoms and in consequence 3d electron conductivity. In that sense the magnetism associated to these electrons in metals or alloys is an itinerant magnetism, that is of collective electron states. However, the 3d states are fairly localized and the overlap of orbits is sufficiently weak so that the 3d electrons are well described in a tight binding approximation. The atomic character is preserved with predominant intraatomic Coulomb interactions. These latter induce local spin polarisation stabilizing a short range

---

*present address : Physics Department, University of Guelma, Algeria.

exchange coupling ; only nearest neighbor exchange interactions can therefore be considered to a good approximation. They lead to a ferromagnetic coupling for nearly filled bands but can favour an antiferromagnetic coupling for half filled bands, in agreement with Hund's rule promoted by the large intraatomic Coulomb interactions [1].

The instabilities associated with the onset of such itinerant magnetism are of wide theoretical interest [2], but further improvements of the experimental investigations are still needed. The R-M intermetallics where M stands for a 3d-transition metal and R for a rare earth (or yttrium and thorium) are appropriate tools for such studies. The strong hybridization between the 3d and 5d electrons (or 4d and 6d electrons) plays to reduce the 3d magnetism at increasing concentration of the R atoms. On the other hand in such systems a large number of well-defined compounds exists, among which the critical concentration for the 3d magnetism collapse is reached. The Y-Co and Y-Ni systems, in which ferromagnetism is favoured, have allowed to get the first experimental evidences for collective electron metamagnetism and thermally induced magnetization [3-4]. When such instabilities compete against well-established 3d ferromagnetism large longitudinal spin fluctuations manifest themselves in finite range of temperatures and lead to anomalous thermal variation of the magnetic anisotropy in Y_2Co_7 [5].

The magnetic behaviours associated to the collapse of itinerant antiferromagnetism are much less known. In the R-Fe intermetallics, a competition between ferromagnetic and antiferromagnetic couplings results in complex properties which are difficult to handle. Moreover, an antiferromagnetic coupling leads already by itself to a large complexity in intermetallics even when only nearest neighbors interactions are considered. Indeed, the crystal structure of intermetallics reflects a compact atomic packing in which two nearest neighbors of a given atom are often nearest neighbors of each other. It becomes then impossible to have all pairs of nearest moments antiparallel and strong frustration of the antiferromagnetic exchange interactions must occur. As a result the magnetic ground state includes several close minima associated with a variety of complex magnetic structures. There always exist pairs of nearest moments for which the antiferromagnetic coupling is not achieved. The corresponding cost in energy $-2J_{ij}\ \vec{S}_i.\vec{S}_j$ is then minimized in several ways. Through the dependence of the exchange integral J_{ij} upon interatomic distances, crystallographic distortions occur below the Néel temperature. Other effects are related to the trigonometry associated with the $\vec{S}_i.\vec{S}_j$ scallar products. The number of nearest neighbor pairs with parallel moments are minimized and some atoms can even be uncoupled from their nearest neighbors (second kind of antiferromagnetic ordering in a f.c.c. lattice [6]). This trigonometrical minimization can also show itself through non-collinear structures corresponding to a homogeneous weakening of the frustration over all the nearest neighbor pairs (tetrahedral magnetic structure in a f.c.c. lattice [6] or triangular magnetic structure in a hexagonal lattice [7]). All these effects are common to localized magnetism as well as itinerant magnetism. An additional feature is to be expected in the case of an itinerant antiferromagnet close to its instability. The spin amplitude becomes an extra degree of freedom which allows an additional reduction of exchange energy in frustrated nearest neighbor pairs.

The most suitable series of intermetallics to study the instabilities of itinerant antiferromagnetism are the RMn$_2$ Laves phases thanks to their relatively simple crystallographic structures. As described in the following sections, these intermetallic compounds exhibit all the features that we have mentioned above.

2 - MAGNETIC INSTABILITY OF MANGANESE IN THE RMn$_2$ LAVES PHASES

For a light rare earth or for a heavy one, that is for large or small R atoms (R = Th, Pr, Nd or Er, Tm, Lu), the RMn$_2$ compounds crystallize in the hexagonal C14 Laves phase. For intermediate R atoms (R = Gd, Y, Tb, Dy) they crystallize in the f.c.c. C15 Laves phase. A dimorphism is observed for R = Sm or Ho. Both crystallographic structures consist of a compact packing of atoms of two different sizes. The smaller atom, that is Mn, lies on the tops of regular tetrahedrons. These tetrahedrons are stacked in chains along the hexagonal axis in the C14 structure and have a diamond stacking in the cubic C15 structure (figure 1). The different atomic sites in both structures reported in table 1 are labelled as in the International tables for X-ray crystallography.

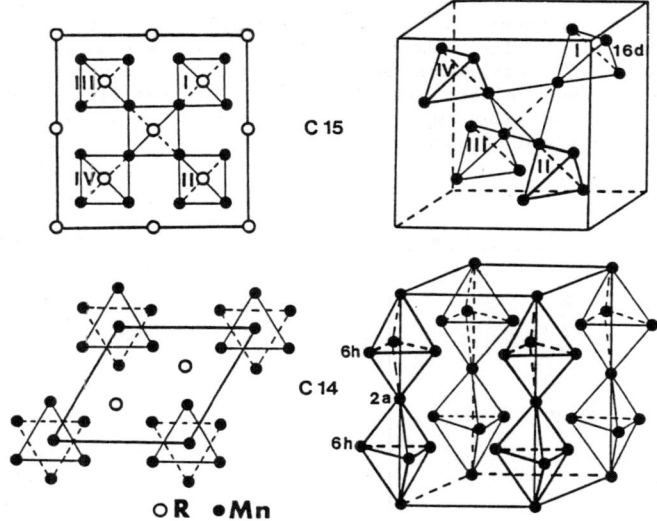

Figure 1 : Crystallographic structures of the f.c.c. C15 and the hexagonal C14 Laves phases.
a - projection in (001) planes
b - stackings of Mn tetrahedrons.

Table 1 : Atomic sites in the f.c.c. C15 and the hexagonal C14 Laves phase

Atom	C15 (Fd3m)		C14 (P6$_3$/mmc)	
R	8a	$\bar{4}$3m	4f 3m	z = 1/4
M	16d	$\bar{3}$m	2a $\bar{3}$m	
			6h mm	x = -1/6

A characteristic signature of the magnetic instability of manganese in these RMn$_2$ intermetallics is well-illustrated by the large magnetovolumic anomalies observed for some R atoms. In fact, such anomalies occur depending upon the Mn-Mn interatomic distances as shown by Wada et al [8] through thermal expansion and NMR measurements. There exists a critical distance (2.73 Å) which shares the compounds in three different subsets. With light rare earths, that is with Mn-Mn distance larger than the critical distance, the compounds exhibit an antiferromagnetic ordering accompanied with a steep volume increase of relative magnitude of 1 % at the Néel temperature. With heavy rare earths, that is with Mn-Mn distance smaller than the critical distance, the Mn atoms are non-magnetic. The compounds exhibit a ferromagnetic ordering characteristic of rare earth moments only. In the cubic C15 compounds, the magnetovolumic effect associated with the magnetic ordering goes through a maximum reaching up to 5 % in YMn$_2$ in which the critical distance is achieved (figure 2).

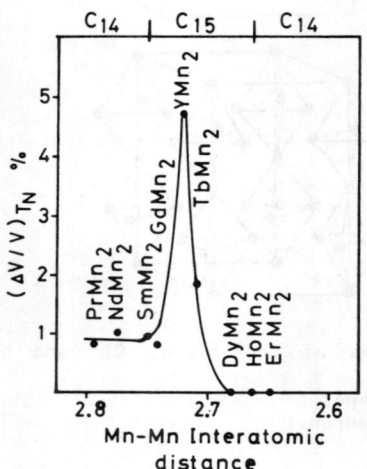

Figure 2 : Relative volume increase at the magnetic ordering temperature in the RMn$_2$ compounds. The Mn moment has collapsed with heavy rare earths.

These magnetovolumic effects reflect an increase of Mn moments in the antiferromagnetic ordered state associated with an increase of the splitting of the local 3d band. Electrons are transferred from the minority spin band into the majority spin band that is on less bonding orbitals [9]. The increase of the band splitting is significant, then subject to strong dependence on external parameters such as applied [10] or chemical [8] pressure. An applied field is not a relevant external parameter since the ordering is an antiferromagnetic one, but the exchange field H_{ex} associated with the Mn-Mn interactions has the appropriate magnetic symmetry. This staggered field when acting on the local band is at the origin of the first order transition under which the antiferromagnetic ordered state is stabilized. Indeed, it has been shown that first order transition occurs if the exchange energy is a sufficiently sensitive function of lattice volume [11]. In the RMn$_2$ Laves phase the exchange energy is very sensitive on the amplitude of the magnetization induced selfconsistently. The magnetization in each magnetic sublattice can be expanded as

$$m = m_o + x_1 H_{ex} + x_3 H_{ex}^3$$

leading in molecular field approximation to

$$T_N = T_{No} (1 + \alpha x_3 m^2).$$

When the temperature increases, the Mn magnetization decreasing, T_N decreases. If this decrease is fast enough, T_N can be reached when the magnetization is not yet cancelled. The antiferromagnetic order is then no more stable and the paramagnetic phase is reached through a first order transition. This transition occurs for large positive values of x_3, that is strong non linear dependence of the 3d magnetization which exchange field. Such a dependence is also at the origin of the collective electron metamagnetism in YCo$_2$ and of the first order magnetic transitions in the RCo$_2$ Laves phases [3]. Unlike in these latters, in the RMn$_2$ compounds the Mn moment does not cancel in the paramagnetic state but is stabilized at a lower value [12] ($m_o \sim 1.6\ \mu_B$). The existence of two different magnetization states of the manganese in the RMn$_2$ compounds has been recently predicted through band structure calculations [13].

On the other hand whereas the transition temperature in the RCo$_2$ compounds is highly sensitive on the spin of the R atoms due to the R-Co exchange interactions on the cobalt magnetization, in the RMn$_2$, the R-Mn exchange field has not the required symmetry to be efficient on the Mn sublattice leading to a weak sensitiveness of the Néel temperatures on the R species.

3 - DISTORTION AND MAGNETIC COUPLING IN NdMn$_2$

The thermal variation of the initial susceptibility as measured in the NdMn$_2$ is reported in figure 3. The large discontinuity observed at $T_N = 105$ K accounts for a first order transition to an antiferromagnetic state. In the paramagnetic state, the reciprocal susceptibility follows a law of the form :

$$\chi = \chi_b + C_{Nd^{3+}}/(T - \theta_p)$$

with $\chi_b = 13\ 10^6$ e.m.u./g and $\theta_p = 37$ K.

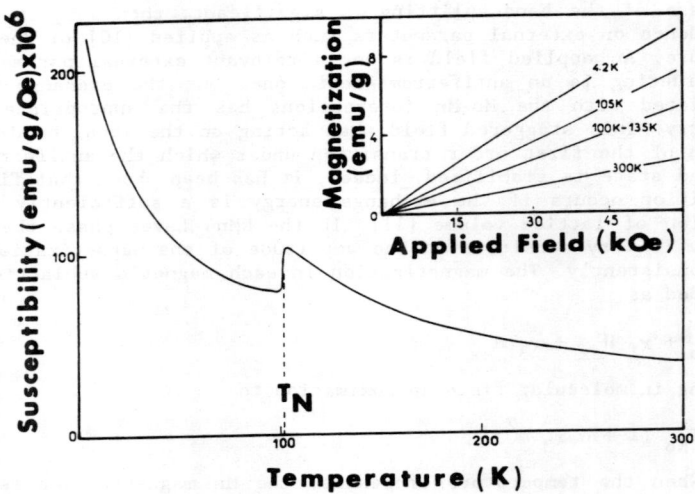

Figure 3 : NdMn$_2$: Thermal variation of the susceptibility. Inset : field variation of the magnetization at different temperatures.

The thermal variations of the lattice parameters have been measured by X-ray diffraction using the monochromatic radiation Kα_1 of chromium. Below T$_N$ a large monoclinic distortion is observed. This is put into evidence through the splitting of the (220) Bragg peak in three peaks (figure 4-A). The measured thermal variations are reported in figure 4-B. The relative volume discontinuity reaches 0.9 %. A

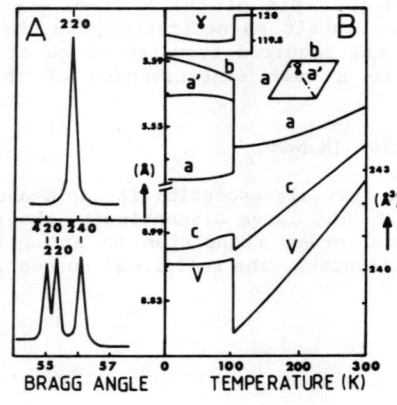

Figure 4 : NdMn$_2$:
A - signature of the monoclinic distortion : splitting of the (220) Bragg peak in three peaks.
B - thermal variations of the crystallographic parameters.

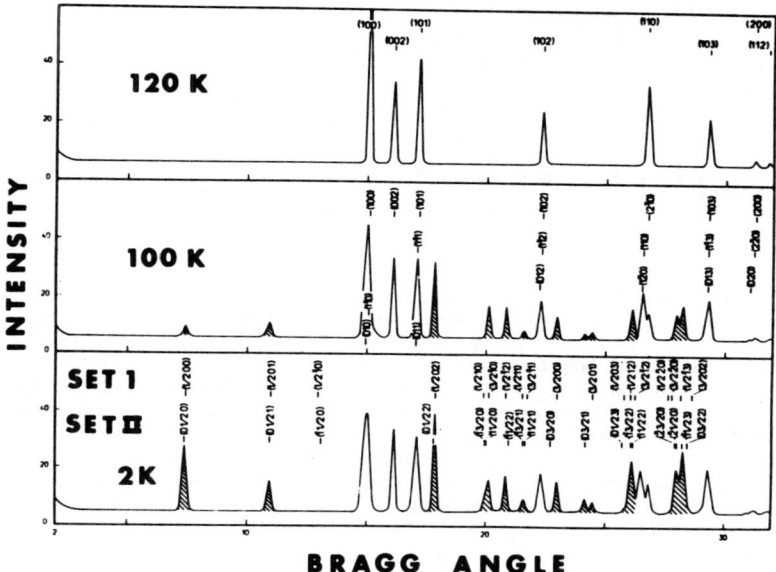

Figure 5 : NdMn$_2$: indexing of powder neutron diffraction patterns.

powder neutron diffraction study has also been performed. The obtained patterns are reported in figure 5. The pattern at 120 K is characteristic of the crystallographic structure in the paramagnetic phase. Thanks to the good knowledge of the monoclinic distortion all the extra peaks of magnetic origin observed below T_N could be indexed. The peaks can be classified in two sets of peaks of different thermal variations. The first set is indexed as (h/2, k, l) ; the intensities are almost temperature independent and essentially represent the ordering of the Mn moments. The intensities in the second set indexed as (h, k/2, l) decrease when the temperature increases, therefore are more characteristic of Nd magnetism. The existence of such two sets, i.e. two propagation vectors, indicates that the magnetic structure consists of two uncoupled sets of moments. The magnetic structure, as deduced from intensity fitting (reliability factor : 6 %), is reported in ref. 14. This structure, complex, also found in PrMn$_2$ [14], is schematized in figure 6. Due to the symmetry breaking associated with the monoclinic distortion, the 6h Mn atoms are shared out in three new sites 2e$_1$, 2e$_2$, 2e$_3$. At 100 K the magnetic structure is collinear with the moments along the [120] direction. The first set of moments propagating with [1/2 0 0] are those of the 2a, 2e$_1$ and 2e$_2$ Mn atoms. The second set propagating with [0 1/2 0] contains the moments of the

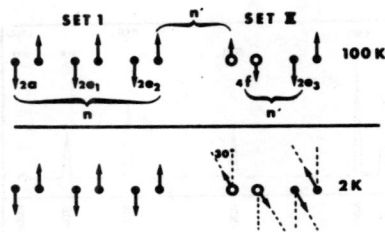

Figure 6 : NdMn$_2$: schematization of the magnetic structure.

2e$_3$ Mn atom and the Nd atom. At decreasing temperature a reorientation of the moments of the second set only occurs below 50 K. This reorientation results from crystal field effects on Nd^{3+} ions responsible for a local magnetic anisotropy favouring a [100] axis. It does not occur in PrMn$_2$ since in that case the crystal field effects on Pr^{3+} leads to a local magnetic anisotropy favouring a [120] axis. The analysis of the magnetic structure shows that the exchange field of Mn atoms of the first set cancels on the Mn atoms of the second set and conversely. Nevertheless these two sets of Mn moments are indirectly coupled through the Nd-Mn exchange. It should be noted that as in the R-M (M = Fe, Co, Ni) compounds, the 3d spin is coupled antiparallel to the 4f spin which leads to a parallel coupling of the moments of nearest Nd and Mn atoms. This coupling is weaker than the Mn-Mn coupling and even weaker than the anisotropies of Mn 2a and Nd atoms since the Mn magnetic moments of the first set do not rotate at all down to low temperature.

In conclusion, the magnetic structure of NdMn$_2$ carries most of the characteristics associated with frustration of negative exchange interactions induced by magnetic ordering. Large distortions occur which are generally associated with a strong dependence of the exchange integrals on interatomic distances. These distortions are here enhanced through the possible decrease of the amplitude of the Mn moment. On the other hand an uncoupling of pairs of nearest Mn moments is induced by the frustration which allows weaker magnetic interactions to manifest themselves in place of the predominant ones.

4 - TRIANGULAR CONFIGURATION AND CANCELLATION OF Mn MOMENT IN ThMn$_2$

As previously reported [15], ThMn$_2$ exhibits no spontaneous magnetization down to low temperature (inset of figure 7). However, ThMn$_2$ is not a Pauli paramagnet since careful measurements put into evidence a weak maximum of the susceptibility at T_N = 115 K (figure 7).

Unlike NdMn$_2$ and PrMn$_2$, the crystallographic structure of ThMn$_2$ remains hexagonal. Below T_N = 115 K, anomalous thermal expansion is observed for the a and c parameters [16]. The relative anomalous changes of a and c are opposite in sign and reach $\Delta a/a$ = 0.09 % and $\Delta c/c$ = -0.13 % at low temperature. No significant volume anomaly is observed at T_N.

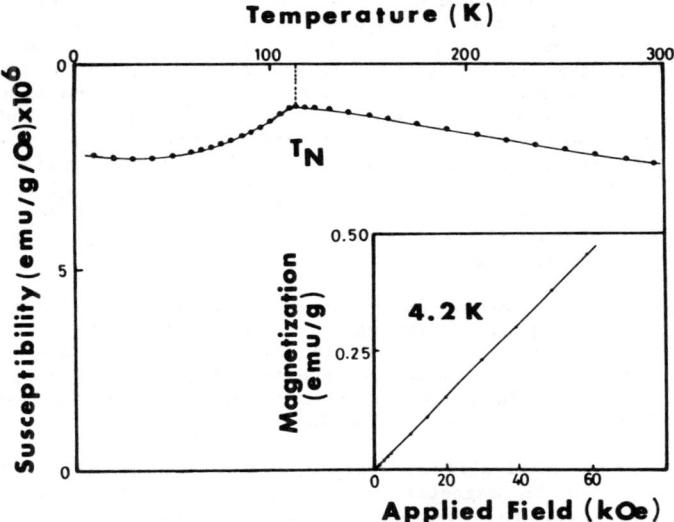

Figure 7 : ThMn$_2$: thermal variation of the susceptibility. Inset : field variation of the magnetization at 4.2 K.

The antiferromagnetic structure has been determined by powder neutron diffraction [16]. It is a triangular configuration of the moments of the 6h atoms coupled antiparallel (figure 8) : the moments of the atoms at z = 1/4 are parallel and antiparallel to those at z = 3/4. The propagation vector is [1/3 1/3 0]. The 2a Mn atoms are found non-magnetic ; the exchange fields acting on them cancel.

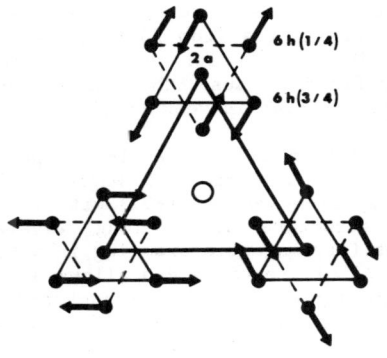

Figure 8 : ThMn$_2$: triangular configuration in the magnetic structure.

Thorium is tetravalent, loosing four of its electrons in the conduction band, and is non-magnetic. These four electrons through the 3d-6d hybridization tend to weaken the Mn 3d magnetism more than in RMn$_2$ with trivalent R. ThMn$_2$ is then characteristic of Mn-Mn interactions which are negative and predominant only between nearest neighbors. A strong frustration of these interactions is to be expected owing to the topology of the Mn packing in regular tetrahedrons. The magnetic structure reflects this frustration through the triangular configuration. A new effect associated with the itinerant character of the Mn magnetism is also found, that is the cancellation of the Mn moment on the 2a site. This is really a non-magnetic state and not a paramagnetic one which would have resulted from a simple exchange field cancellation. Indeed the susceptibility does not increase at decreasing temperatures as expected for local moments in paramagnetic states under applied field. Consequently the splitting of the local 2a band is cancelled. It may be due to large staggered exchange fields acting on this band. Such a fact allows to understand the unexpected parallel coupling of the moments of the 6h Mn atoms.

The anomalous thermal expansion of the lattice parameters have previously been explained through the dependence of the exchange integral on interatomic distances [16]. It should be noticed however, that amplitude changes of moments is also accompanied by large volume changes [9] and the shrinkage of the c parameter can be well-accounted by the collapse of the moment on the 2a site.

The lack of a significant volume anomaly at T_N results from the weakening of the Mn magnetism due to the tetravalent character of thorium. Therefore, the exchange field is no more strong enough to induce the transition to the higher magnetization state.

5 - FRUSTRATION AND INHOMOGENEITIES OF THE MAGNETIZATION DENSITY IN YMn_2

Thermal expansion and magnetization measurements [17] show that YMn_2 exhibits a first-order antiferromagnetic transition accompanied by a huge volume anomaly and a large thermal hysteresis. This hysteresis is dependent upon sample preparation. As studied by neutron diffraction the main characteristics of the hysteresis are the following : at decreasing temperatures, the compound remains paramagnetic down to a temperature T_1' which may reach 45 K. In this sample the coexistence of the paramagnetic state with the antiferromagnetic ordering is observed below T_1' down to T_1 - 18 K. At increasing temperature the two magnetic phases coexist in a wide range of temperatures from 25 K up to 110 K. This hysteresis shows itself also in the thermal dependence of the susceptibility. It is less well-defined because the change in the susceptibility between the two magnetic states is weak (6 %).

The powder neutron diffraction pattern obtained at low temperature in the antiferromagnetic state exhibits a series of magnetic peaks which can be indexed using the propagation vector $[\tau 01]$ with τ - 0.02 [18] (figure 9). The magnetic structure is a long wavelength propagation of the first kind of antiferromagnetic ordering in a f.c.c. lattice [6]. When Mn is replaced by Al, τ increases and allows a better insight of the long wavelength propagation. On the contrary, the replacement of Y by Tb cancels the splitting of the (110) magnetic peak and allows to study the magnetic configuration which is propagated (figure 9).

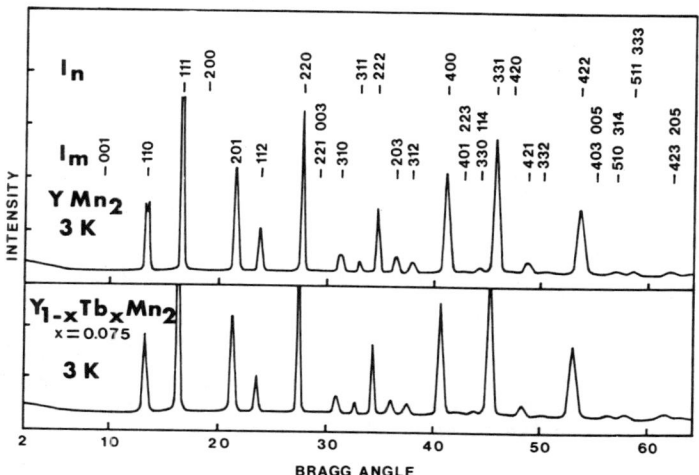

Figure 9 : YMn$_2$ and Y$_{1-x}$Tb$_x$Mn$_2$; x = 0.075 : indexing of powder neutron diffraction patterns.

Recent NMR measurements performed on YMn$_2$ and Y$_{0.9}$Tb$_{0.1}$Mn$_2$ have shown that the spin echo spectrum exhibits only one resonance line at 120 MHz which is sharper and more symmetric in Y$_{0.9}$Tb$_{0.1}$Mn$_2$ [18]. This indicates that the hyperfine field on Mn nucleous is almost independent on the propagation of the magnetic configuration. According to the value of the Mn magnetic moment (2.6 μ_B) the relatively low value of the hyperfine field (120 kOe) accounts for a significant orbital contribution to the Mn magnetism. If the resulting Mn anisotropy, also evidenced in NdMn$_2$, is considered the long wavelength propagation must have an antiphase character rather than reflecting an angular or amplitude modulation. The corresponding high order satellites are not observed in the neutron pattern for their intensities are too weak. Moreover, the third order satellites are still not well-resolved and overlap with the tail of the main peaks.

There are several features which enables to get an insight of the magnetic structure. The Mn atoms in YMn$_2$ are shared out on the tops of four tetrahedrons related to each other through the four f.c.c. translations t$_I$ (000), t$_{II}$ (0 1/2 1/2), t$_{III}$ (1/2 0 1/2) and t$_{IV}$ (1/2 1/2 0). The translation t$_I$ generates the tetrahedron I (figure 1) at the tops of which lie the Mn atoms 1, 2, 3 and 4 reported in table 2. For the magnetic structure two of these translations are anti-translations.

Table 2 : Positions of the Mn atoms 1, 2, 3 and 4 at the tops of tetrahedron I.

1	2	3	4
5/8, 5/8, 5/8	5/8, 7/8, 7/8	7/8, 5/8, 7/8	7/8, 7/8, 5/8

No intensity is observed for the magnetic peak of the (001) family. Furthermore, the exchange frustration has to be the weakest. Therefore, the arrangement of the magnetic moment in a tetrahedron is such that the four atoms are coupled in pairs of antiparallel moments not necessarily coupled to each other (figure 10).

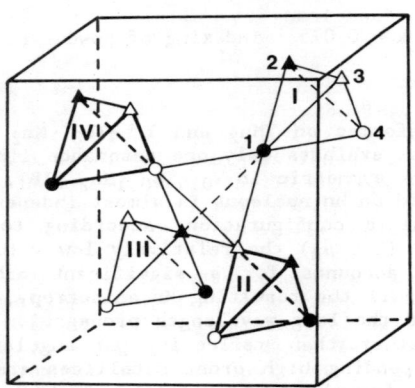

Figure 10 : YMn$_2$: magnetic structure for [001] propagation. The Mn atoms are shared out in two sublattices with different magnetization axes (circles and triangles). Within each sublattice the Mn moments are antiparallel (black and white).

When limited to the nearest neighbors, the surrounding in Mn and R atoms of the 16d Mn atoms in YMn$_2$ is the same as that of the 2a Mn atoms in NdMn$_2$. The reorientation process observed in NdMn$_2$ has put into evidence a large anisotropy of the 2a Mn moment. As opposed to the case of cobalt based similar uniaxial intermetallics, the plane perpendicular to the local threefold axis $\bar{3}m$ is an easy plane of magnetization. Moreover, in this plane the [120] directions are strongly favoured. Translating this in YMn$_2$, each of the 16d Mn atoms has an easy plane of magnetization perpendicular to the local trigonal axis $\bar{3}m$. Within this plane three possible directions of magnetization exist. They are labelled in table 3 for the four atoms of tetrahedron I.

Table 3 : Coordinates of the local easy directions of magnetization of the Mn atoms 1, 2, 3 and 4 at the tops of the tetrahedron I in YMn_2.

	1			2			3			4	
s_1^1	s_2^1	s_3^1	s_1^2	s_2^2	s_3^2	s_1^3	s_2^3	s_3^3	s_1^4	s_2^4	s_3^4
2	-1	-1	2	-1	-1	-2	1	1	-2	1	1
-1	2	-1	1	-2	1	-1	2	-1	1	-2	1
-1	-1	2	1	1	-2	1	1	-2	-1	-1	2

If a multiaxes antiferromagnetic structure associated with the different local easy directions of magnetization is considered, the exchange frustration is increased and leads to a finite intensity for the magnetic peak (001).

In fact, a good fit of the neutron pattern is obtained with a collinear structure in which the axis of the moments is at 45° of the propagation vector. This direction is inconsistent with the local anisotropy favouring the planes perpendicular to the local trigonal axes. Indeed, in collinear structures the direction of the moments that minimizes the total anisotropy is one of the four [111] directions. The best fit should have been then obtained with an angle of 54°.

As mentioned above a two axes magnetic structure in which the four Mn atoms of any tetrahedron are only coupled by pairs does not increase the exchange frustration. Each pair is characterized by one direction of the moment not necessarily parallel to the other. If the planes of easy magnetization were isotropic the two axes structure which minimizes the anisotropy should have its two axes perpendicular to the propagation vector. The corresponding magnetic configuration is reported in figure 11-A. Nevertheless, the calculated magnetic intensities are in poor agreement with the observed ones. In fact, the easy planes are not isotropic. The magnetic anisotropy is minimized when each moment makes the same angle (33°) with one of its three easy directions of magnetization. The resulting magnetic structure, when projected in the plane perpendicular to the propagation vector gives the same arrangement as that reported in figure 11-A. The directions of the moments in tetrahedron I are reported in table 4, together with the closest easy direction of magnetization. This magnetic structure gives the best agreement between the calculated and observed magnetic intensities.

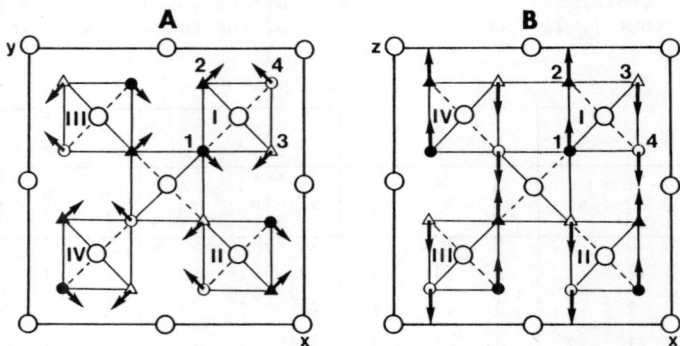

Figure 11 : YMn2 : A - magnetic structure predicted for isotropic easy plane of magnetization perpendicular to the local trigonal 3m axis. It represents also the projection in the plane (001) perpendicular to the propagation vector [001] of the actual structure.
B - projection of the magnetic structure in the plane (010) containing the propagation vector [001].

Table 4 : Directions of the moments and closest easy directions of magnetization of the Mn atoms 1, 2, 3, 4 at the tops of the tetrahedron I in the magnetic structure of YMn$_2$ described with the propagation vector [001].

1		2		3		4	
m_1	s_2^1	m_2	s_1^2	m_3	s_2^3	m_4	s_1^4
1	-1	1	2	-1	1	-1	-2
-1	2	1	1	-1	2	1	1
2	-1	2	1	-2	1	-2	-1

The deduced magnetic structure allows to extract the magnetic scattering amplitude Mf assigned to one Mn atom. Its variation in the reciprocal space is reported as a function of $\sin\theta/\lambda$ in figure 12. The observed magnetic scattering amplitudes are distributed between the two variations calculated using the 3d Mn atomic spherical form factor [20] and assuming a Mn moment of 2.3 μ_B and 2.9 μ_B respectively. This

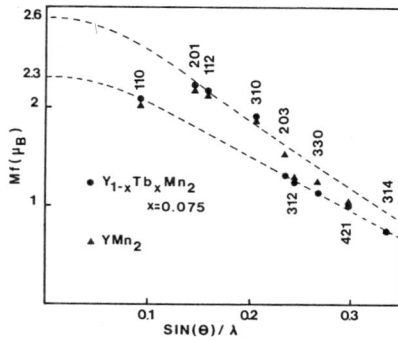

Figure 12 : YMn$_2$ and (Y$_{1-x}$Tb$_x$)Mn$_2$ x = 0.075 : variation in reciprocal space of the observed magnetic scattering amplitude (Mf) of Mn. The dashed lines stand for the magnetic scattering amplitude calculated using the 3d Mn atomic spherical form factor [19] and assuming a Mn moment of 2.3 μ_B and 2.9 μ_B respectively.

dispersion of the observed values cannot be ascribed to experimental inaccuracy. The magnetic intensities have been measured on the D1B diffractometer at the Institut Laue-Langevin. The counting time was sufficiently long to reach an accuracy better than 3 % for each experimental value. The relatively low value measured for the (110) Bragg peak cannot reasonably result from a negatively polarised contribution of the 4d electrons of the Y atoms. These electrons being strongly delocalised the expected contribution would be too large. On the other hand, an orbital contribution cannot be evoked to account for the relatively high values for the (201), (112) and (310) peaks. Indeed the characteristics related to the orbital contribution to the scattering amplitudes appears essentially at much higher values of sinθ/λ. In consequence the dispersion of the magnetic scattering amplitudes accounts for a notable anisotropy of the magnetization spin density not arising from the orbital contribution to the Mn magnetism. In agreement with the phenomenon of Mn moment cancellation in ThMn$_2$, such an anisotropy can be due to a local reduction of the magnetization density along frustrated links. The quantitative study of such inhomogeneities cannot be carried out entirely on powder samples because of the high cubic symmetry which gives rise to mean effects. A single crystal study is hard to perform for the high magnetovolumic effect breaks the crystal at T_N.

The reduction of the magnetization density in frustrated links due to the magnetic ordering affects also the nature of the paramagnetic state. A significant paramagnetic neutron scattering is observed above T_N [12]. It increases with temperature as shown in figure 13. The paramagnetic scattering exhibits a maximum around a wavevector corresponding in real space to the distance between two antiferromagnetic layers of Mn tetrahedron. It is reminiscent of the correlations characteristic of the magnetic ordered state. When the temperature is increased, the thermal disorder weakens the antiferromagnetic correlations and the inherent exchange frustrations associated with these correlations. As a result the deficiency of the

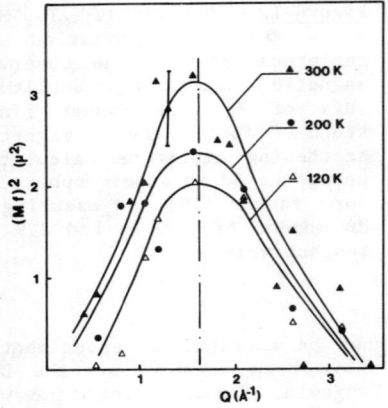

Figure 13 : YMn$_2$: Thermal increase of the paramagnetic scattering $(Mf)^2$.

magnetization in frustrated links in the ordered state is reduced. This effect can lead to the increase of the mean square amplitude of the Mn moments which is at the origin of the thermal increase of the paramagnetic scattering.

6 - CONCLUSION

In short, the study of the RMn$_2$ Laves phases outlines the importance of exchange frustrations inherent to the topology of the atomic packing on the magnetic properties of an itinerant antiferromagnetism close to its instability. The amplitude of the magnetic moment is no more a conserved quantity. Its possible variation induces new and more notable effects of frustration in comparison with localized magnetism. The minimization of the exchange energy in the frustrated links can be more efficient through the reduction of the magnetization density. All these effects are not restricted to the 3d magnetism and may appear in other antiferromagnetic intermetallics with f electrons, such as in cerium or actinides based compounds [21].

REFERENCES

[1] Friedel, J., in Physics of Metals, ed. Ziman J.M., Cambridge University Press, 340 (1969).

[2] Moriya, T., Spin Fluctuations in Itinerant Electron Magnetism, ed. Cardona, M., Fulde, P., Queisser, H.J., Springer series in Solid State Sciences, 56 (1985), and references therein.

[3] Bloch, D., Edwards, D.M., Shimizu, M., Voiron, J., J. Phys. F : Metal Phys., 5, 1217 (1975).

[4] Gignoux, D., Lemaire, R. and Molho, P., J. Appl. Phys., 53, 2087 (1981).

[5] Ballou, R. and Lemaire, R., Proc. ICM88, Paris to appear in J. de Phys.
[6] Smart, J.S., Effective Field Theories of Magnetism, Saunders W.B. Company (1966).
[7] Bertaut, E.F., J. Phys. Chem. Solids, 21, 295 (1961).
[8] Wada, H., Nakamura, H., Yoshimura, K., Shiga, M. and Nakamura, Y., J. Magn. Magn. Mat., 70, 134 (1987).
[9] Janak, J.F. and Williams, A.R., Phys. Rev., B14, 4199 (1976).
[10] Okamoto, T., Nagata, H., Fujii, H., Makihara, Y., J. Magn. magn. Mat., 70, 139 (1987).
[11] Bean, C.P. and Rodbell, D.S., Phys. Rev., 126, 104 (1961).
[12] Deportes, J., Ouladdiaf, B. and Ziebeck, K.R.A., J. Magn. Magn. Mat., 70, 14 (1987).
[13] Yamada, H. and Shimizu, M., Proc. ICM88, Paris, to appear in J. de Phys.
[14] Ballou, R., Déportes, J. Lemaire, R. and Ouladdiaf, B., J. Appl. Phys., 63, 3487 (1988).
[15] Buschow, K.H.J., Sol. Stat. Commun., 21, 1031 (1977).
[16] Deportes, J., Lemaire, R., Ouladdiaf, B., Roudaut, E. and Sayetat, F., J. Magn. magn. Mat., 70, 191 (1987).
[17] Shiga, M., Wada, H. and Nakamura, Y., J. Magn. magn. Mat., 31-34, 119 (1983);
[18] Ballou, R., Déportes, J., Lemaire, R., Nakamura, Y. and Ouladdiaf, B., J. Magn Magn. Mat., 70, 129 (1987).
[19] Berthier, Y., Déportes, J., Horvatic, M. and Rouault, P., Proc. ICM88, Paris, to appear in J. de Phys.
[20] Freeman, A.J. and Watson, R.E., Acta Cryst., 14, 231 (1961).
[21] Rossat-Mignod, J., Burlet, P., Quézel, S. and Effantin, J.M., in Physics of Magnetic Materials, Jadwisin'84 (Poland), World Scientific, Singapore (1985), 1, 411.

MAGNETOELASTIC PROPERTIES OF ZIRCON STRUCTURE COMPOUNDS CONTAINING RARE EARTH IONS WITH ORBITAL DEGENERATE ELECTRONIC STATE

V.I.Sokolov, Z.A.Kazei

Physics Department Moskow State University
Leninskiye Gory, 119899 Moskow
USSR

ABSTRACT

On the example of rare earth (RE) zircon structure compounds the influence of the Jahn-Teller (JT) interactions on the magnetic and magnetoelastic properties is investigated for paramagnetic crystals with different ground electronic states of RE ions. We present the experimental results of heat expansion, magnetostriction and magnetization for monocrystals of RXO_4, where R=Dy, Tb, Tm; X=V,P. In these crystals the JT interaction causes the structural phase transitions: the real ones (RVO_4), or the virtual ones ($TmPO_4$). The mechanisms of magnetostriction for the investigated crystals in the temperature region of the structural phase transitions are discussed. It is shown that the peculiarities of temperature and field dependences are in qualitative accordance with the Vekhter and Kaplan theoretical calculations, basing on the microscopic approach in the scope of the cooperative Jahn-Teller effect theory.

1. INTRODUCTION

The peculiarities of physical properties of crystals with the cooperative Jahn-Teller effect (CJTE) are due to two reasons. Firstly in the JT systems besides the electron-phonon and electron-deformation coupling, which are responsible for the structural phase transitions (SPT), exchange and magnetic interactions exist, which are comparable in magnitude with them. The mutual influence of the distortion and magnetic correlations causes the unusual elastic and magnetic properties of the JT crystals, and strong magnetoelastic effects in particular.

Secondly, if the SPT temperatures T_c are not high (~ 10 K) the elastic molecular fields result in comparatively small splitting of energy levels. Splittings of such magnitudes may be induced by external magnetic fields, which can be realized in experiment. The magnetic field either supresses or enhances the lattice JT deformation depending on the character of the electron spectrum change thus the magnetic field essentially influences the magnetic and anisotropic properties of the JT crystals.

It is clear that the strongest magnetoelastic effects may be expected for the JT compounds with RE ions, characterizing by the strong spin-orbital interaction and weak splitting of the $(2j+1)$-degenerate ground multiplet in the crystal field. Such situation occurs in crystal with zircon structure RXO_4, where R is trivalent RE ion and X is V, P or As. During the last 10-15 years these crystals are subject to intensive experimental and theoretical investigations [1-30].

In this paper the results of the experimental investigation of magnetic and magnetoelastic properties of monocrystals RVO_4 (R=Dy, Tb, Tm) and RPO_4 (R=Tb, Tm) having different schemes of degenerate levels of RE ions are reported. We pay a particular attention to magnetostrictive effects, for which the peculiarities of the coupling of the magnetic and elastic subsystems are revealed for the JT crystals.

2. SAMPLES AND EXPERIMENTAL TECHNIQUES

We have chosen 5 crystals of RE zircons, the characteristics of which are listed in Table 1. The energy spectra of RE ions in these compounds are different and include the most interesting cases of degeneration (or quasi-degeneration) for the electron levels [1]. Due to this fact in the investigated crystals the JT effect shows the specific features: the "real" JT effect ($TmVO_4$), JT pseudoeffect ($DyVO_4$), the JT effect connected with an excited level ($TbVO_4$, $TmPO_4$).

The different case is the crystal of $TbPO_4$. For T_c = 2,13 K (which is 0.15 K below the Neel temperature) the monoclinic lattice deformation appears [2,3]. The nature of this SPT is not established as yet. If we consider according to [4] this SPT in $TbPO_4$ as the JT one, it means that the electron-phonon interaction is close in magnitude to the exchange interaction.

The external magnetic field splits and mixes the electron states of RE ions, and so we can distribute the investigated crystals in two groups, for which the local JT distortions are either enhanced ($DyVO_4$, $TbVO_4$, $TmPO_4$) or supressed ($TmVO_4$) by magnetic field.

The monocrystals of RE vanadates and phosphates were grown at the Magnetic laboratory of the Moscow State University and at the Institute of silicate chemistry (Leningrad). The samples were prepared by the flux method and their typical dimensions are 2 x 1 x 0.5 mm^3. Our X-ray investigations (diffractometer "Geigerflex") are in accordance with the data of the previous works [1], where it was established that the vanadates of Dy, Tb and Tm have the SPT accompaning by the symmetry lowering from the tetragonal one (space group D_{4h}^{19}) to the orthorhombic one (D_{2h}^{24} or D_{2h}^{28}) for values of T_c listed in Table.

Below the T_c a tetragonal unit cell of RE zircons is elongated in either the $\langle 100 \rangle$ direction or $\langle 110 \rangle$ one. It corresponds to the orthorhombic deformation of crystals of B_{1g}-type or B_{2g}-type. In crystals where the JT interactions are small in comparison with the splitting of the lower-lying electronic levels in a crystal field, the SPT either is absent or is in the vicinity of the Neel temperature (see Table). Excepting $TbPO_4$ the magnetic ordering for investigated crystals is observed at the temperatures much below the T_c [1].

The magnetostriction (U) and thermal expansion ($\Delta l/l_o$) were measured using the capacity technique for temperatures between 1.7 and 50 K. The capacitance trasducer was connected in the oscillatory circuit of a high-stability cryogenic generator, operating at a 1.5 MHz frequency. The construction of the apparatus allowed measurement of the longitudinal magnetostriction $U_\parallel$ in a superconducting solenoid to 50 kOe and the transverse $U_\perp$ in a superconducting magnet, performed in a Helmholtz coils geometry (fields up to 30 kOe). The crystals were oriented by the X-ray method to 0.5° accurancy. The corresponding quantities, characterizing the change of crystal size, were determined as $U = [l(H) - l(H=0)]/l_o$ and $\Delta l(T)/l_o = [l(T) - l_o(T_o)]/l_o$, where l_o is the crystal length along the direction of measurement for the magnetic field H=0 and $T_o \gg T_c$.

Our apparatus allows to register the magnetostriction 10^{-6} and controll it for every experiment by the calibrated piezocrystal $Sr_3Ga_2Ge_4O_{14}$, mounted directly in the measuring cell. The absolute error for the deformation measurement is not higher than 5%.

The magnetization (M) was measured by a vibrating sample magnetometer in the fields up to 60 kOe; the sensitivity is not worse than 10^{-4} emu. In our experiments the temperatures were measured by calibrated Allen-Bradly resistors (100 Ω, 1/8 Wt), which also served as sensors for the temperature controller.

3. EXPERIMENTAL RESULTS AND DISCUSSION

3.1 $DyVO_4$

According to [1] for $T > T_c$ the lower-lying electronic states of Dy^{3+} are the Kramers doublets Γ_7 (E_1) and Γ_6 (E_2), separated by the gap $2\Delta = 9$ cm^{-1} and well separeted from the higher-lying excited states.

The mean-field Hamiltonian of the $DyVO_4$ crystal in the magnetic field $h_x = g_\perp M_B H_x$ can be written [5]

$$H = -A \bar{\sigma}_z \sigma_z - \Delta \gamma \sigma_x - h_x S_x \qquad (1)$$

where A is the molecular field parameter, resulting from the correlations of the JT distortions, $\sigma_{x,z}$ and S_x are the operators of the interactions with the molecular, crystalline and magnetic fields, correspondingly, γ is the vibronic reduction factor. In [6] from (1) the order parameter $\bar{\sigma}_z$, characterizing the uniform deformation U is calculated and in [7] the magnetic moment is calculated.

The experiments [6-10] show that below the T_c in $DyVO_4$ (in other JT crystals as well) the crystallographic twins (domains) appear. The reason of this fact is the existence of several equivalent directions of distortions in the elementary cell. The crystallographic (JT) domains influence essentially on the physical properties of the JT crystals.

In the Fig.1 the thermal expansion is shown for $DyVO_4$ along [100] in H=0 and H=20 kOe, parallel and perpendicular to Δl. The external magnetic field H ‖ [100] transforms a crystal to the single-domain state with the easy axis (the axis with the largest g-factor and magnetic susceptibility χ) along the field. If we adopt that in the rhombic phase it holds a < b, then for H ‖ [100] ‖ Δl we measure the thermal expansion along the rhombic a-axis, and for $H \| [010] \perp \Delta l$ - along the b-axis. Consequently, in $DyVO_4$ for $T < T_c$ $\chi_a > \chi_b$, i.e. the easy axis is the short axis of the second order in the basal plane.

The anomaly magnitude $\Delta l/l_o$ in the field is equal $\Delta a/a = -\Delta b/b = 26 \cdot 10^{-4}$; it is the half of the spontaneous deformation of the lattice $(b-a)/a = 52 \cdot 10^{-4}$, determined from the X-ray diffraction in [11]. For $T > T_c$ the field $H \| [100]$ induces the deformation caused by magnetostriction which is of the same symmetry as the spontaneous deformation. Thus the T_c is increased and the SPT is spread. This is in accordance with the numerical calculations by Pytte [12]. Our investigations show that the character of the division into the JT domains depends strongly on the thermal history of a sample and mechanical strains, defects, etc. The crystals being cooled below the T_c in the magnetic field ∼20 kOe do not remain monodomain when the field is switched out at 4.2 K. The cooling in the magnetic field forms only the small prefered orientation of the JT domains.

Fortunately it was found that one of $DyVO_4$ crystals, for which we measured the magnetostriction, was practically in single domained state below the T_c (H=0). In Fig.2 isotherms of the longitudinal magnetostriction for this crystal $U_\|(H)$ in $H \| [100]$ are shown. In these dependences the domain striction shows pronounced jumps in fields ∼5 kOe for T = 1.8 and 4.2 K. It is of principal importance that the $U_\|$ increases essentially when the temperature increases. It reaches the same value as those of the spontaneous lattice deformation near 13 K in the field 40 kOe.

The other peculiarity of the $U_\|(H)$ is the growth of the maximum value of the magnetostriction coefficient $D = 1/U_o \cdot dU/dH$ (U_o is the saturation magnetostriction) near the T_c. This coefficient can be calculated in theory for Hamiltonian (1) [6] and it should increase abnormally for $T \to T_c$ (for $T=T_c$ $D \to \infty$). The data presented in Fig.3 evidence the qualitative agreement of the theory and experiment.

Theoretical analysis shows [13] that the JT interactions give the contribution to the magnetization $\bar{S}_x$ as well. In particular for weak magnetic fields the magnetization of the single domained $DyVO_4$ along [100] has a form [7]:

$$M/M_o \equiv \bar{S}_x = h_x/4kT \,(1 + {}^{kT}/_{\Delta\gamma}\, th^{\Delta\gamma}/kT +$$
$$+ {}^{Ah_x^2}/(4\Delta^2\cdot\gamma^2 kT)\, th^{\Delta\gamma}/_{kT}), \quad T > T_c$$

$$M/M_o \equiv \bar{S}_x = h_x/4kT \,(1 + {}^{kT}/_{\Delta\gamma}\, th^{\Delta\gamma}/kT +$$
$$+ A\bar{\sigma}_z/_{\Delta\gamma}\cdot th^{\Delta\gamma}/kT\,), \quad T \leqslant T_c$$

(2)

From the formulas (2) it is possible to conclude that the JT correlations ($A \neq 0$) give a positive contribution to the M, both above and below the T_c. The M(H) is saturated in high fields, so this contribution causes the bend on the magnetization isotherms. The theory predicts the paramagnetic saturation according to the law $th(h_x/kT)$ when the H ∥ [110] induces the B_{2g}-deformation, correlations of which are small for $DyVO_4$.

These effects we have observed when measuring the magnetization of $DyVO_4$. It is seen from Fig.4 (where the saturation magnetization $M_o=5,35 \cdot 10^4$ emu/mol for 5 K) that the isotherms $M/M_o(H)$ reveal the bend points, which displace into the region of strong fields H when increasing the temperature. The magnetoelastic contribution to the M(H) reaches its maximum value near the T_c. It is evident that it is a consequence of the following fact: for low temperature the striction is supressed by the JT field ($A \gg h_x$), and for $T \gg T_c$ the striction is small because of the smallness of the h_x/kT.

We should remark that in experiment we can observe the

contribution of the JT interactions for $T > T_c$ only. Below the T_c the processes of the domain reconstruction under the influence of magnetic field mask this contribution. To compare with the theory the results of the numerical calculations for $\bar{S}_x$ according to (2) are given in Fig.4. The best agreement with the experiment is reached for the following values of parameters: $A = 11.2$ cm^{-1}, $\Delta = 4.5$ cm^{-1} and $g = 10$.

3.2 TmVO$_4$

The ground electronic state of Tm^{3+} ion is non-Kramers doublet separated from the first excited singlet by 50 cm^{-1}. In this basis the only z-component of the magnetic moment ($g_\perp \approx 0$, $g_{\|} = 10$) is not equal to zero and for the $T_c = 2.15$K the spontaneous deformation appears which is B_{2g}-type [1]. In the magnetic field H$\|$[001] the doublet is splitting, thus resulting in the decreasing of vibronic effects. As a result the external magnetic field reduces the orthorhombic deformation and the temperature T_c, i.e. SPT is supressed by the magnetic field.

The supression effect for the SPT in the field H$\|$[001] is illustrated by the Fig.5. These curves are characterized by the existence of noticeable distortions of the tetragonal crystal lattice up to $T = 3T_c$. As far as the character of the λ-anomaly of the heat capacity of TmVO$_4$ [21] indicates the small contribution of the short-range order for $T > T_c$ it is possible to suppose that the small uniaxial strains serve as a reason for the observed "tail" of $\Delta l(T)/l_0$ in TmVO$_4$, these strains appear due to glueing of the sample in the measuring cell. According to [14] near T_c even the negligible mechanical stresses, applied to the JT crystal, influence considerably on its physical properties due to the softening of the elastic modules.

The isotherms of the transversal magnetostriction

($U_\perp$) for $TmVO_4$ in $H\|[001]$ ($\Delta l \| [110]$) are shown in Fig.6. For $T < T_c$ the increase of the $U_\perp$ is observed. This increase is related to the rehabilitation of the tetragonal phase in the magnetic field $H_c(T)$. The signs of changes of the magnetostriction when increasing H and the thermal expansion when cooling below the T_c are different, i.e. the H gives rise to the rhombic-tetragonal phase transition.

The noticeable value of magnetostriction is the peculiar feature of the $U_\perp(H)$ curves for $T > T_c$ and $H > H_c$, meanwhile the theory gives $U \neq 0$ for rhombic phase only [15]. The most probable reason for the discrepancy of the theory and experiment is the mentioned above elastic strains in $TmVO_4$. The calculation, performed recently in [16], confirms this assumption.

According to the theoretical analysis [17] for $H \perp c$ ($g_\perp = 0$) the magnetostriction of $TmVO_4$ is absent. However our experiment (Fig.7) shows the large striction $U_\|$ for $H \| [110] \| \Delta l$. In the field $H = 40$ kOe for $T = 1.8$ K the value of $U_\| \sim 10^{-3}$. As the temperature increases the striction decreases, but remains noticeable even for $T > T_c$.

It is possible to propose two mechanisms of magnetostriction $U_\|$ in $TmVO_4$. Firstly, this is a reorientation of the JT domains in the magnetic field; the reason for this reorientation is an anisotropy of χ in the basal plane, related to the CJTE. For the Tm^{3+} ion the χ along the $a^{'}$-, $b^{'}$-axes is of the Van Fleck nature, i.e. is caused by the mixing of the excited levels. As far as the value of the χ in the basal plane is small, the domain reorientation field is essentially greater than that in $DyVO_4$. For H=40 kOe the sample is not single domained as yet.

The second mechanism is a "true" striction of a single domained crystal along the easy direction in the basal plane. This striction is of the Van Fleck nature and, according to the calculation [16], reaches the maximum value near the T_c, where the elastic module C_{66} is softening. However the decreasing of $U_{\parallel}$ when T is increasing (Fig.7) shows that the main contribution to the striction of $TmVO_4$ is connected with the JT domain reorientation.

3.3 $TbVO_4$

For the tetragonal crystal $TbVO_4$ the singlet-doublet-singlet scheme of the lower-lying electronic levels is realized[1]. The JT interaction is comparatively great here ($A > 2\Delta$) and the SPT to the orthorhombic phase occurs for the T_c=34 K[18,19]. The B_{2g}-type distortions not only remove the doublet degeneration but enlarge the gap between the singlets below the T_c.

According to [13] the magnetic and magnetoelastic properties of $TbVO_4$ can be described by the Hamiltonian

$$H = -A \bar{\sigma}_z \sigma_z - \Delta \cdot \mathcal{T} - g \cdot \mu_B (H_x S_x + H_y S_y) \quad (3)$$

Here Δ - is the energy gap between the singlet and doublet, σ, $\mathcal{T}$, S - are the electronic operators, given in the singlet-doublet-singlet basis of the Tb^{3+} ion[7]. Using the Hamiltonian (3) in the papers [7,13] the numerical calculation of the quantities $\bar{\sigma}_z$ and $\bar{S}_x$ was performed. The best agreement with the experiment was received for the parameters: $A = 25$ cm^{-1}, $\Delta = 9$ cm^{-1} and $g = 10$.

According to the theoretical calculations and our experimental data the curves M(H) in $H \parallel [110]$ for $TbVO_4$ have the bend points analogously to the case of $DyVO_4$. The temperature interval, where the JT interaction contributes to the M, and the value of this contribution are noticeably smaller then those for $DyVO_4$. The estimate of this contribution (using the linear parts of the M(H)) is equal to ~8% for T = 36 K and H = 50 kOe.

The JT interaction contribution to the magnetostriction has several peculiarities in $TbVO_4$ in comparison with the other RE zircons. In the curves $U_{\parallel}(H)$ (Fig.8) for $T \leqslant 24$ K and $H \sim 5$ kOe the rapid increase of the striction due to the JT domains reorientation is observed. From the comparison with the case of $DyVO_4$ (Fig.2) it is seen that the domain striction for $TbVO_4$ has the opposite sign. This indicates that the easy magnetization axis for $TbVO_4$ is the long axis of the second order (b) in the basal plane. The magnetostriction caused by the JT domains reorientation in this crystal reaches the huge value (10^{-2}).

As it is seen from Fig.8 for $T \sim 25$ K (i.e. for $T \ll T_c$) the characters of the $U_\parallel(H)$ curves are changed. It is essential that according to our X-ray data (for the same crystal of $TbVO_4$) the $T_c = 34$ K. Our opinion is that it may be two reasons for such a behavior of the $U(T,H)$. The change of the character of the JT domains reorientation may serve as a first reason: for $T < 25$ K this process happens by jump and continuously for $T > 25$ K when the H increases. Secondly, according to our measurements of the elastic properties for the polycrystal of $TbVO_4$, for $T \sim 25$ K the minimum of the Young's module takes place, meanwhile the sound attenuation (the frequency - $2 \cdot 10^5$ Hz) has an anomaly in the region of $T_c = 34$ K. The reason of such a behavior of the elastic characteristics is not evident (we should note that the same situation occurs for $DyVO_4$).

The influence of the JT domains on the striction of $TbVO_4$ is illustrated by Fig.9, where the dependences of $U_\perp$ on the H orientation in the (110) plane are presented. It is seen that the $U_\perp$ has the maximum value for $H \parallel [1\bar{1}0]$ and vanishes when $H \parallel [001]$. Such a behavior of the magnetostriction may be caused by the single domained state of the crystal $TbVO_4$ when there is a projection of H on the $[1\bar{1}0]$ axis.

3.4 $TmPO_4$

For the $TmPO_4$ crystal the scheme of the lower-lying electronic levels of the Tm^{3+} ion is analogous to that of the case of Tb^{3+} in $TbVO_4$. But the value of the interaction for $TmPO_4$ is smaller than the Δ: $A=20$ cm^{-1}, $\Delta = 30$ cm^{-1} [20]. For this reason the SPT does not take place. As far as in $TmPO_4$ there exist the strong correlations of the local JT distortions near the Tm^{3+} ions, this crystal is considered as a virtual elastic. This means a compound for which the peculiarities of the physical properties are caused by the vibronic interactions even in the absence of the CJTE. In particular the characteristic minimum of the elastic modulus C_{66} for $T \simeq 20$ K [20] and anomalies of the magnetic properties in the liquid helium temperature region [22,23] are related to these effects for $TmPO_4$.

The investigation of the magnetoelastic properties of $TmPO_4$ is of great interest, because contrary to compounds with the CJTE, here the manifestations of the JT interactions are not masked by the crystallographic domain reorientation processes.

In Fig.10 the dependences of $U_{\parallel}(H^2)$ for $TmPO_4$ in the field $H \parallel [110]$ are presented. It is seen that in the region of 5 K the magnetostriction reaches the value 10^{-3} and depends quadratically on the field for $H \leqslant 15$ kOe. The magnetic field value, for which the deviation from the $U_{\parallel} = \alpha H^2$ dependence begins, is a function of T. For $T \geqslant 28$ K the striction changes according to the H^2 law in the all investigated magnetic field interval. The coefficient α has a non-monotonic dependence on T with the maximum near $T_o = 13$ K. Near T_o the field interval with such a law $U_{\parallel} = \alpha H^2$ is decreased essentially.

From the Hamiltonian (3), supposing $g\mu_B H$, kT and $\Delta\gamma \gg A\bar{\sigma}_z$ and taking into account only the H^2 terms, it is possible to get a simple analytical formula for the α [24]:

$$\alpha = \left(\frac{g\mu_B}{\Delta\cdot\gamma}\right)^2 \cdot \left(ch\frac{\Delta\gamma}{KT} - 1\right) \cdot \left(ch\frac{\Delta\gamma}{KT} + 1 - \frac{A}{KT}\right)^{-1} \quad (4)$$

From (4) it follows that for $A=22$ cm^{-1}, $\Delta\gamma=30$ cm^{-1} and $g=8.0$ the $\alpha(T)$ dependence has a maximum for $T \simeq 14$ K. As it is seen from Fig.11 the JT local distortion correlations not only enhance the magnetostriction in TmPO$_4$ for low temperatures but also result in the appearance of an extremum on the $\alpha(T)$ for $T_o=14$ K which is noticeably below 20 K, where the minimum of the elastic modulus C_{66} is observed. The characteristic features of the $\alpha(T)$ curve depend essentially on the value of the parameter A.

3.5 TbPO$_4$

The peculiarities of the physical properties of TbPO$_4$ are caused by the special scheme of the lowest electronic states of the Tb^{3+} ion in the crystal field: ground state is a non-Kramers doublet and first excited state is a singlet 2.2 cm^{-1} away. According to [2,3] at low temperatures TbPO$_4$ shows two magnetic phase transitions which are very close. Between $T_{N1}=2.28$ K and $T_{N2}=2.13$ K it orders as a simple two-sublattice antiferromagnet (AF) along the tetragonal c-axis. Below T_{N2} the moments are canted towards the [110] direction ($\sim 20°$) in a (110) plane and the symmetry is lowered most probably to monoclinic one [26]. The crystallographic distortion is caused apparently by the CJTE [27].

Because the local symmetry of RE ions in zircon structure D_{2d} has no symmetry centre, monoclinic deformation in TbPO$_4$ results in the SPT being an antiferroelectric ordering as well [27]. In the external magnetic field in TbPO$_4$ the anomalous large magnetoelectric effect is observed [28]. The main features of this phenomenon in the TbPO$_4$ can be described correctly in the scope of the CJTE theory [4].

In Fig.12 the thermal expansion for TbPO$_4$ is shown for the different crystallographic directions. The temperature change $\Delta l/l_o$ reveals that for $T_{N2}=T_c=2.13$ K a

lattice deformation appears, which is $\sim 10^{-3}$ along the
<111> and <110> directions. For T_{N1}=2.28 K the anomaly of
$\Delta l(T)/l_0$ along <111> is much smaller than 10^{-3}, along <110>
it is not observed at all. From this it follows that the
phase transition for T_{N1} is of magnetic nature and that
for T_{N2} is a structural one.

The lowering of symmetry in $TbPO_4$ below the T_c is illustrated in Fig.13, where the angular dependences of the transversal magnetostriction $U_\perp(\varphi)$ are drawn in the (110) plane ($\Delta l \parallel [1\bar{1}0]$). For $T > T_{N1}$ (fig.13, curve 2) the $U_\perp(\varphi)$ dependence has a form characteristic for a tetragonal crystal, it has two second order axes: $[001]$ and $[1\bar{1}0]$. For $T < T_{N2}$ (curve 1) due to symmetry lowering in the (110) plane no symmetry axes exist. The maximum of $U_\perp(\varphi)$ is moved by an angle $\sim 20°$ from the $[001]$ axis. This is in agreement with the results of neutron diffraction measurements of $TbPO_4$ [29].

In Fig.14 the isotherms of the magnetostriction $U_\perp$ for $TbPO_4$ in $H \parallel [1\bar{1}0]$ are plotted. They show the following peculiarities: 1) for $T < T_{N2}$ the curves $U_\perp(H)$ exhibit jumps, caused by the spin-flop transitions; 2) for the first applying of the magnetic field (T=1.85 K) the jump of $U_\perp$ is noticeably larger than that for the following applyings; 3) in the region of the spin-flop transition the hysteresis is observed; 4) in strong fields (H > 10 kOe) the magnetostriction increases not saturating.

The last fact is a little unusual, so as in the paramagnetic phase of "traditional" antiferromagnets the magnetostriction changes slightly when H increases. According to the phase diagram of $TbPO_4$ for $H \parallel [001]$ [27] the spin-flip fields are 8.2 - 9.5 kOe at T = 1.9 - 2.0 K. For our curves $U_\perp(H)$ these fields correspond to the fields, in which the striction changes its sign.

One may suppose that in $TbPO_4$ the external magnetic field causes the noticeable electrostriction due to the anomalously large magnetoelectric effect. This electrostriction gives a contribution to the experimental dependence of $U_\perp(H)$, which is of the opposite sign. As far as electrostriction depends quadratically on the electric field as a rule, the resulting deformation ($U_\perp$) increases when H increases.

4. CONCLUSION

Resuming the work we shall summarize the main results:
1. All the presented experimental data confirm the conclusion of the previous theoretical works of the effective influence of the external magnetic field on the degenerate electronic states of RE ions in the JT compounds of the zircon structure. The observed there magnetoelastic properties show that the local JT distortions are either enhanced ($DyVO_4$, $TbVO_4$) or suppressed ($TmVO_4$) by the magnetic field.
2. Due to the strong electron-phonon coupling and softening of the elastic modules in the investigated compounds the huge magnetostriction effects are observed: for the paramagnetic crystal $TbVO_4$ the magnetostriction is $\sim 10^{-2}$ for T = 5 K and H $\sim$ 10 kOe. Considering a transparancy of RE zircons in the optical region their striction properties may be of definite interest for practical purposes.
3. Below the T_c the magnetostriction of the JT crystals of the zircon structure are caused by two mechanisms: 1) the correlation of the local JT distortions; 2) the structural domains (twins) reorientation. In the virtual elastic $TmPO_4$, for which the SPT does not occur, the first mechanism results not only in the increase of magnetostriction, but in the characteristic anomaly on the temperature dependence of the derivative dU/dH for T_o=14 K as well. The processes of the forming and changing in magnetic field of the JT domains

need a special investigation.
4. The peculiarities of the temperature and field dependences of the magnetoelastic properties for the investigated crystals are in agreement in main features with the Vekhter and Kaplan theory, who use the microscopic approach in the scope of the CJTE theory. To our opinion some quantitative descrepancy between theoretical and experimental results is caused by the peculiarities of the CJTE in real crystals.
5. Among the RE compounds of the zircon structure the $TbPO_4$ crystal is of special interest. The nature of the SPT for this compound is not established unambiguously as yet. The large magnetoelectric effect, observed by Rado and Ferrari [30], the closeness of the structural phase transition (which is antiferroelectric simultaneously) and antiferromagnetic phase transition, the peculiar magnetostriction properties make the $TbPO_4$ to be a perspective object of the future investigations.

REFERENCES

1. Gehring G.A. and Gehring K.A., Rep.Progr.Phys. 38, 1 (1975).
2. Suzuki H. and Nakajima T., J.Phys.Soc.Japan 47, 1441 (1979).
3. Nagele W., Hohewein D. and Domann G., Z.Phys. B, 39, 305 (1980).
4. Vekhter B.G. and Kaplan M.D., Izv.Akad.Nauk SSSR, Ser. Fiz. 51, 1674 (1987).
5. Kaplan M.D., JETP Lett. 35, 105 (1982).
6. Vekhter B.G., Kazei Z.A., Kaplan M.D. and Sokolov V.I., JETP Lett. 43, 369 (1986).
7. Vekhter B.G., Kazei Z.A., Kaplan M.D. and Sokolov V.I., Fiz.Tverd.Tela 30, 1021 (1988).

8. Kasten A. and Becker P.J., Int.J.Magn. <u>5</u>, 157 (1973).
9. Daudin B., Chouteau G. and M'Sirdi N., J.Phys.(France) <u>45</u>, 169 (1984).
10. Kasten A., Z.Phys. B, <u>38</u>, 65 (1980).
11. Göbel H. and Will G., Phys.St.Sol.(b) <u>50</u>, 147 (1972).
12. Pytte E., Phys.Rev. B, <u>9</u>, 932 (1974).
13. Vekhter B.G. and Kaplan M.D., Sov.Phys.JETP <u>60</u>, 1020 (1984).
14. Melcher R.L., Pytte E. and Scott B.A., Phys.Rev.Lett. <u>31</u>, 307 (1973).
15. Vekhter B.G. and Kaplan M.D., Fiz.Tverd.Tela <u>16</u>, 1630 (1974).
16. Vekhter B.G. and Kaplan M.D., Fiz.Nizkich Temper. <u>14</u>, 395 (1988).
17. Elliott R.J., Harley R.T., Heyes W. and Smith S.R.P., Proc.Roy.Soc. A<u>238</u>, 217 (1972).
18. Wells M.R. and Worswick R.D., Phys.Lett. <u>42A</u>, 269 (1972).
19. Harley R.T., Lyons K.B., Fleury P.A. and Smith S.R.P., J.Phys.C <u>16</u>, 1407 (1983).
20. Harley R.T., Manning D.I., J.Phys.C, <u>11</u>, L633 (1978).
21. Cooke A.H., Swithenby S.J. and Wells M.R., Sol.St.Commun. <u>10</u>, 265 (1972).
22. Andronenkó S.I. et al., Fiz.Tverd.Tela <u>25</u>, 423 (1983).
23. Ioffe V.A. et al., Zh.Eksp.Teor.Fiz. <u>84</u>, 707 (1983).
24. Bondar I.A. et al., Zh.Eksp.Teor.Fiz. <u>94</u>, 288 (1988).
25. Sivardiere J., Phys.Rev.B, <u>8</u>, 2004 (1973).
26. Üffinger G., Kasten A., Phys.St.Sol.(b) <u>128</u>, 201 (1985).
27. Domann G., J.Mag. and Magn.Mater. <u>13</u>, 163 (1979).
28. Rado G.T., Ferrari J.M. and Maisch W.G., Phys.Rev. B, <u>29</u>, 4041 (1984).
29. Coing-Boyat J., Sayetat F. and Apostolov A., J.de Phys. <u>36</u>, 1165 (1975).
30. Rado G.T. and Ferrari J.M., Proc. of the 18th Ann.Conf. on Magn. Magn.Mater, ed.C.D.Graham, N.Y.,p.1417 (1973).

TABLE 1 : The phase transition temperatures, the symmetry of the lattice distortion and lower-lying electronic states of some RE zircons (according to [1])

Crystal	T_c, K	T_N, K	Type of the deformation	Lower-lying electr. states of RE ions(cm^{-1})
DyVO$_4$	14.0	3.07	B_{1g}	$E_1(0)$ $E_2(9)$
TbVO$_4$	33.0	0.61	B_{2g}	$A_1(0)$ $E(9)$ $B_1(18)$
TmVO$_4$	2.10	-	B_{2g}	$E(0)$ $A(50)$
TbPO$_4$	2.13	2.28 2.13	monoclinic(?)	$E(0)$ $A(2.2)$
TmPO$_4$	-	-	B_{2g} (virtual)	$A_1(0)$ $E(31)$ $B_1(76)$

FIGURE CAPTIONS

Fig.1 Temperature dependence of the thermal expansion for $DyVO_4$ along $[100]$ in the magnetic field H=20 kOe (1 - $H \parallel [100] \parallel \Delta l$, 2 - $H \parallel [010] \perp \Delta l$) and the a, b parameters of a rhombic unit cell.

Fig.2 Isotherms of the longitudinal magnetostriction for $DyVO_4$ in $H \parallel [010]$.

Fig.3 Experimental (the solid curve) and theoretical (the dashed curve) dependences of the magnetostriction coefficient on the reduced temperature for $DyVO_4$ crystal.

Fig.4 Experimental (-o-) and calculated (—·—) field variations of the reduced magnetic moment for $DyVO_4$ in the internal magnetic field H_i along $[100]$.

Fig.5 The thermal expansion for $TmVO_4$ along the $[110]$ axis in H=0 (a) and the external magnetic field $H \parallel [001]$ (b).

Fig.6 Isotherms of the transverse magnetostriction for $TmVO_4$ ($H \parallel [001]$, $\Delta l \parallel [110]$). The inset shows the geometry of the experiment. The temperature variation of $U_\perp$ in H=0 is due to the thermal expansion of the crystal.

Fig.7 Isotherms of the longitudinal magnetostriction ($H \parallel [110] \parallel \Delta l$) for $TmVO_4$.

Fig.8 Isotherms of the longitudinal magnetostriction for $TbVO_4$.

Fig.9 Angular dependence of the transverse magnetostriction for $TbVO_4$ in the (110) plane at 4.2 K in a field H= 15 kOe ($H \parallel [110] \parallel \Delta l$).

Fig.10 The longitudinal magnetostriction $U_\parallel$ versus H^2 (external field $H \parallel [110]$) for $TmPO_4$ at 4.5 - 34 K.

Fig.11 Temperature dependence of the coefficient α ($U_\parallel = \alpha H^2$) for $TmPO_4$. 1 - experiment; 2,3 - theoretical curves calculated with the molecular field parameter $A = 22$ cm^{-1} and $A = 0$, correspondingly.

Fig.12 Temperature dependence of the thermal expansion for $TbPO_4$ along the $[110]$, $[111]$ directions.

Fig.13 Angular dependence of the transverse magnetostriction for $TbPO_4$ in the (110) plane at 4.2 and 1.85 K in $H = 15$ kOe and 12 kOe, correspondingly.

Fig.14 Isotherms of the transverse magnetostriction for $TbPO_4$ in the field applied along the antiferromagnetic axis and the geometry of the experiment. The curve 1 corresponds to the first applying of the field H at 1.85 K, the curve 2 - the following applyings.

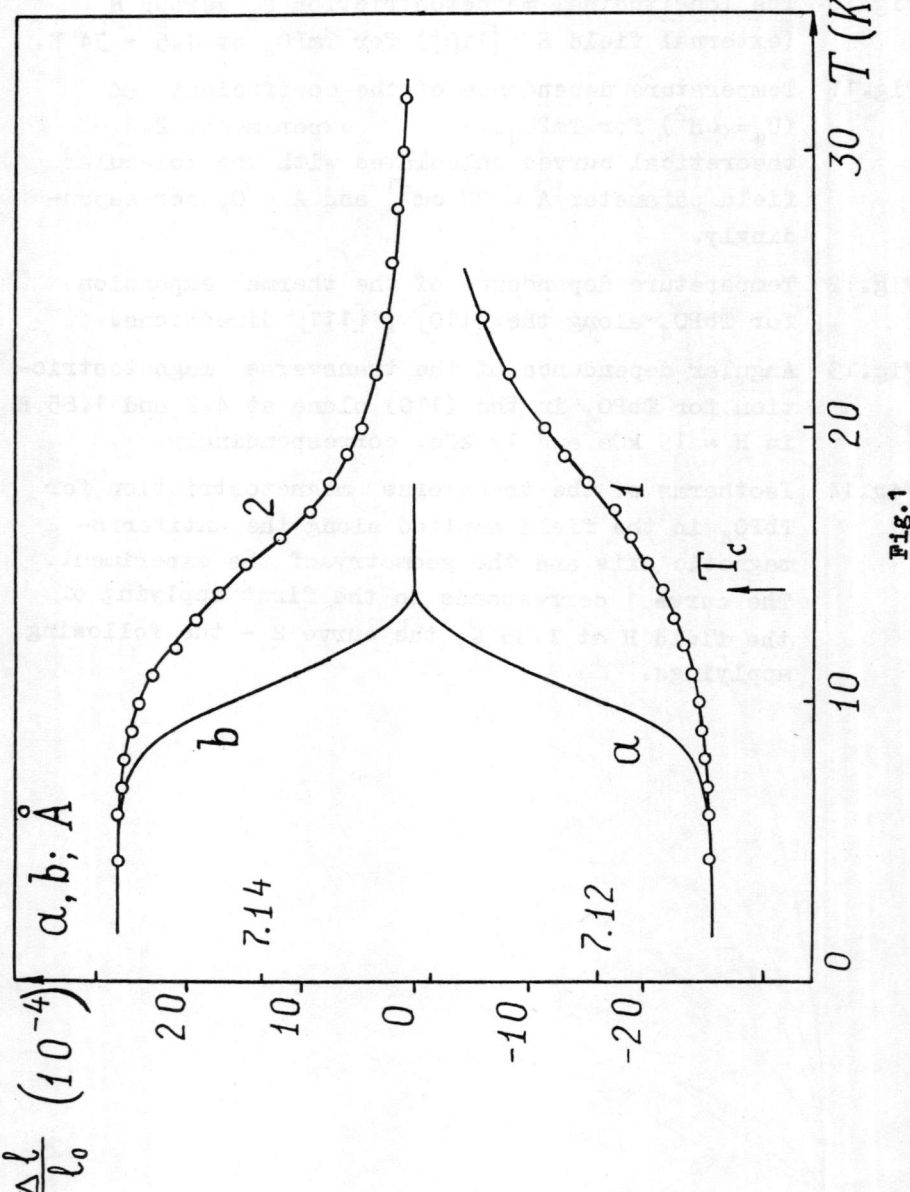

Fig.1

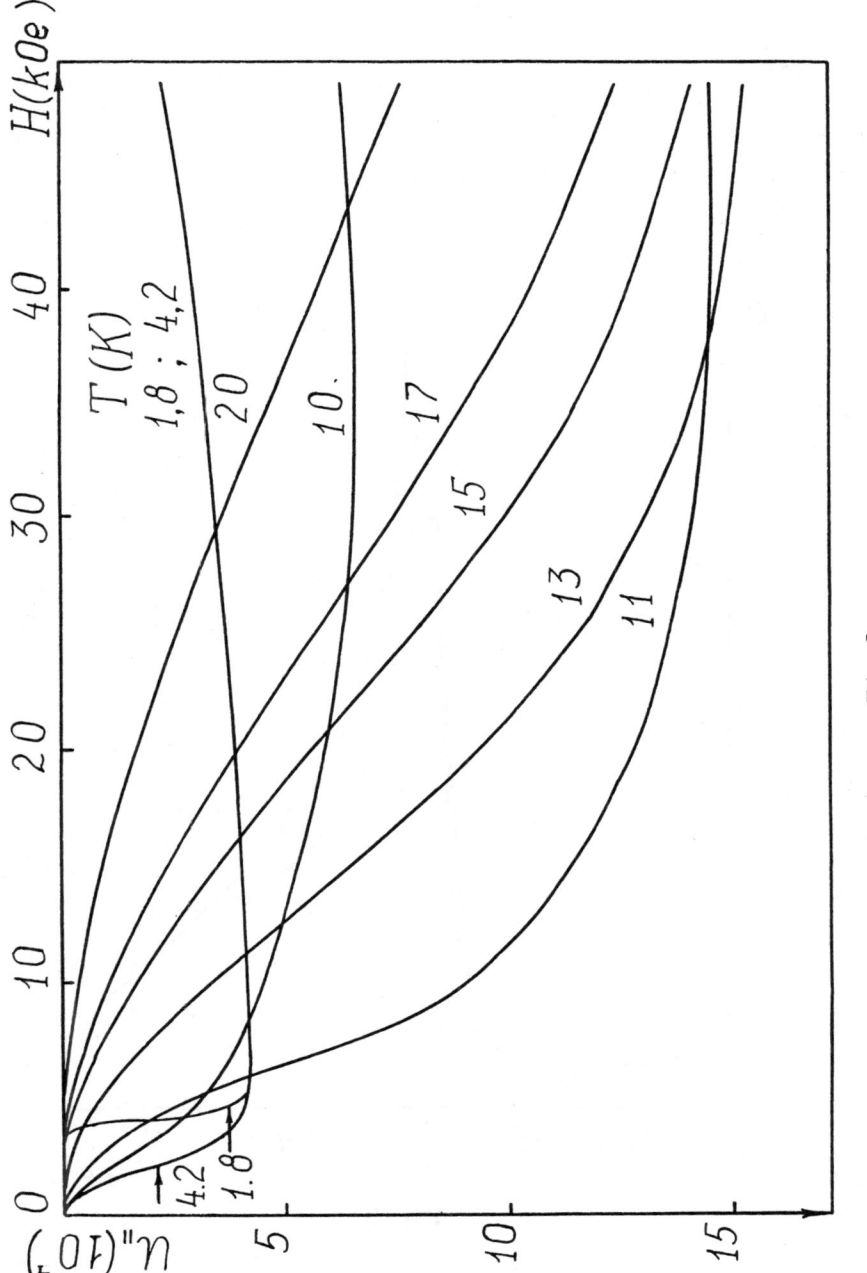

Fig.2

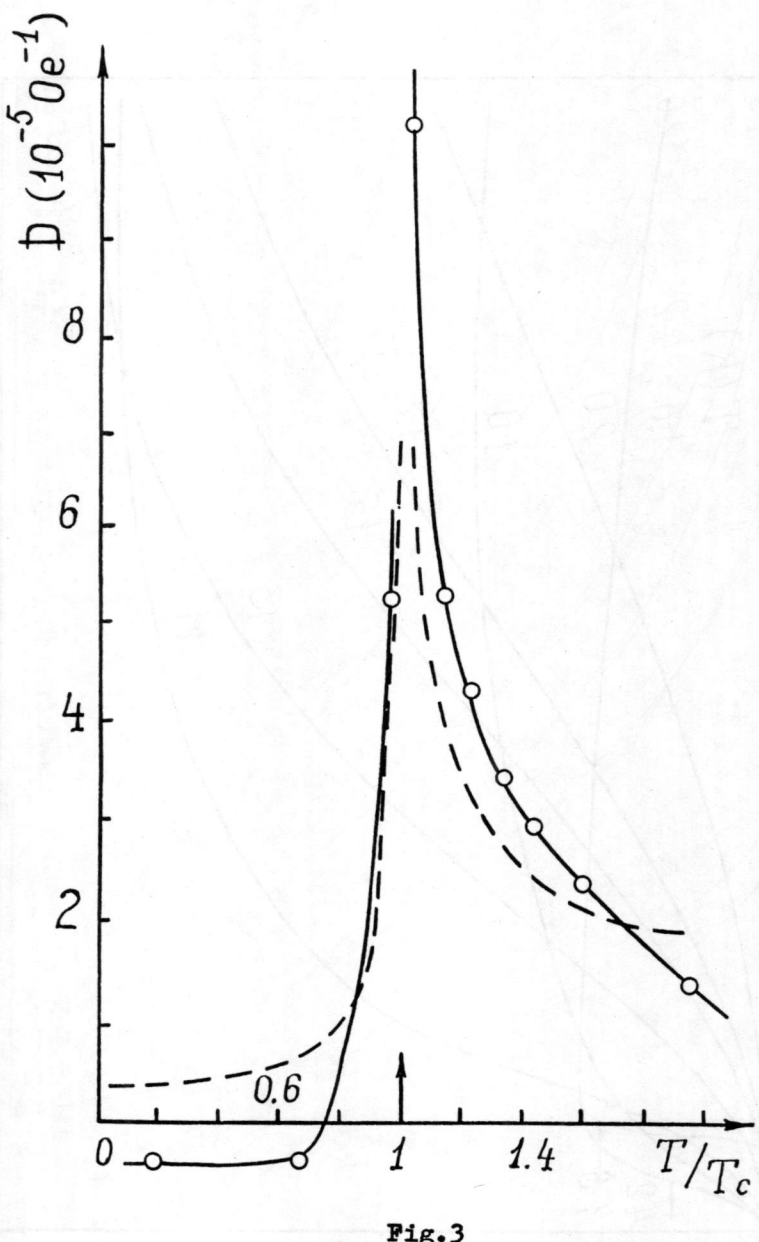

Fig.3

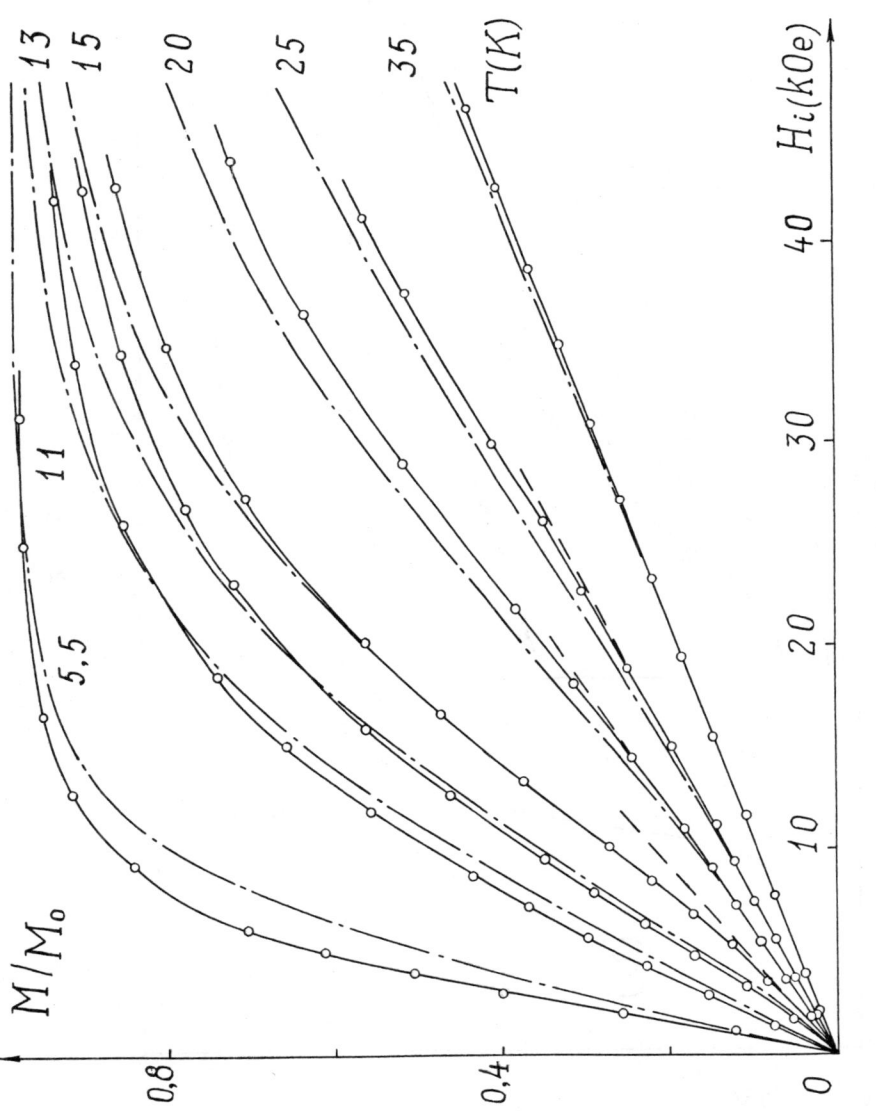

Fig. 4

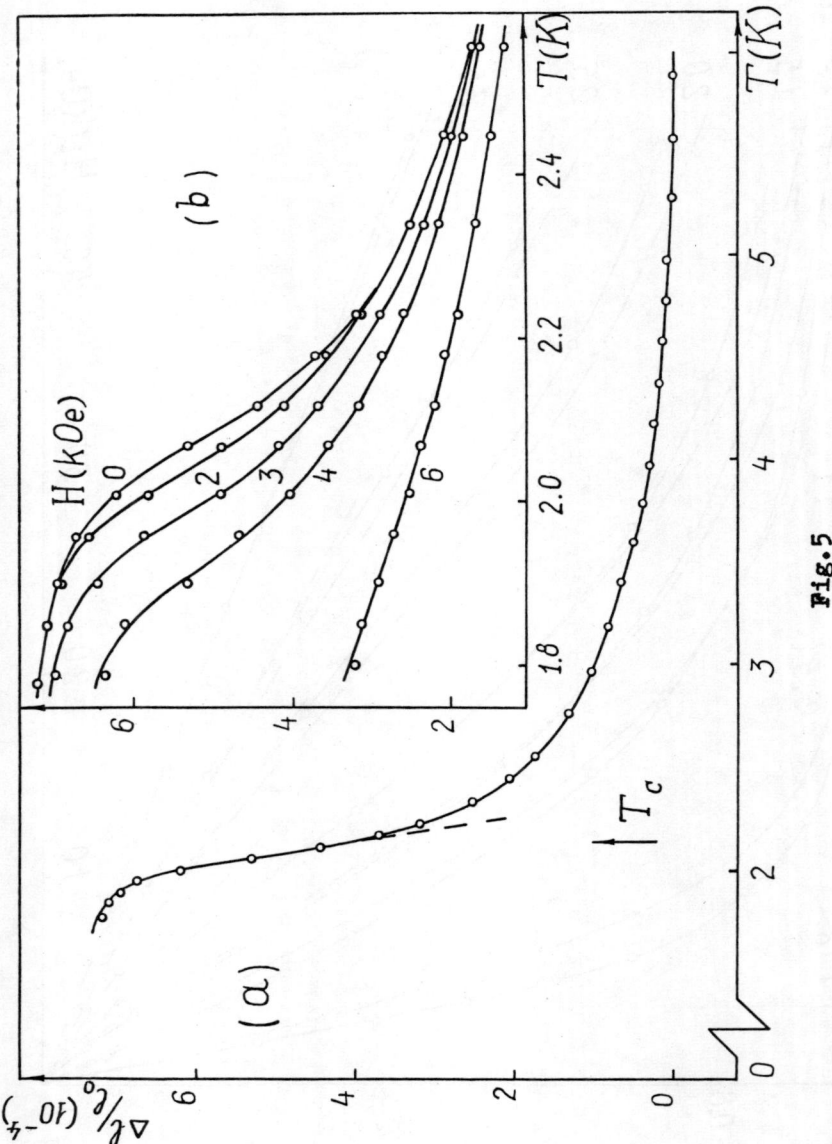

Fig.5

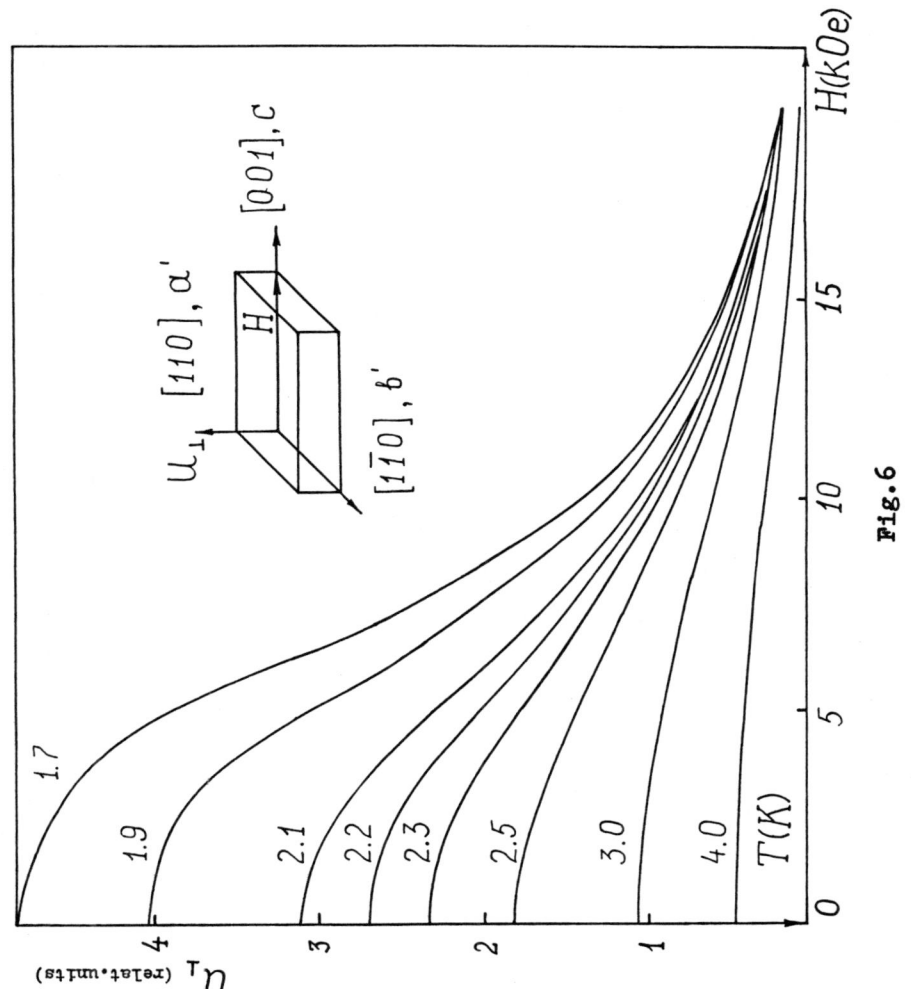

Fig.6

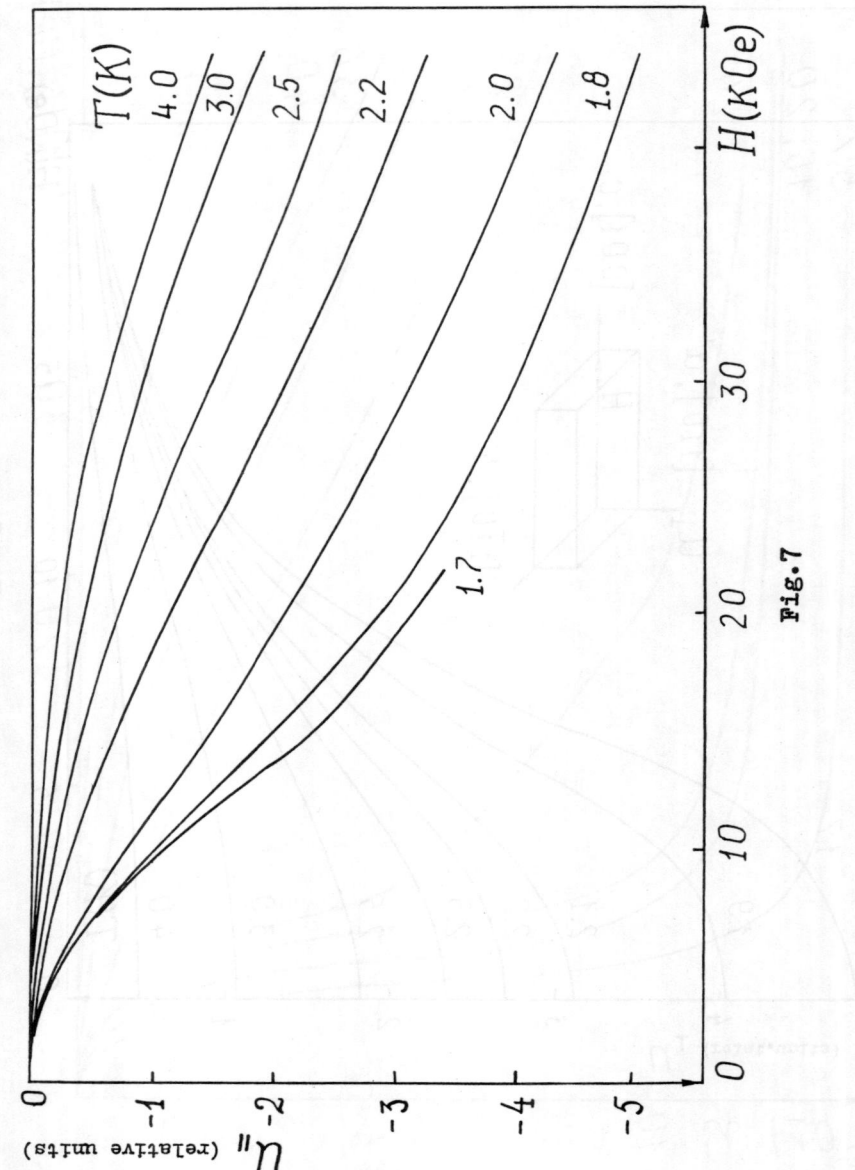

Fig.7

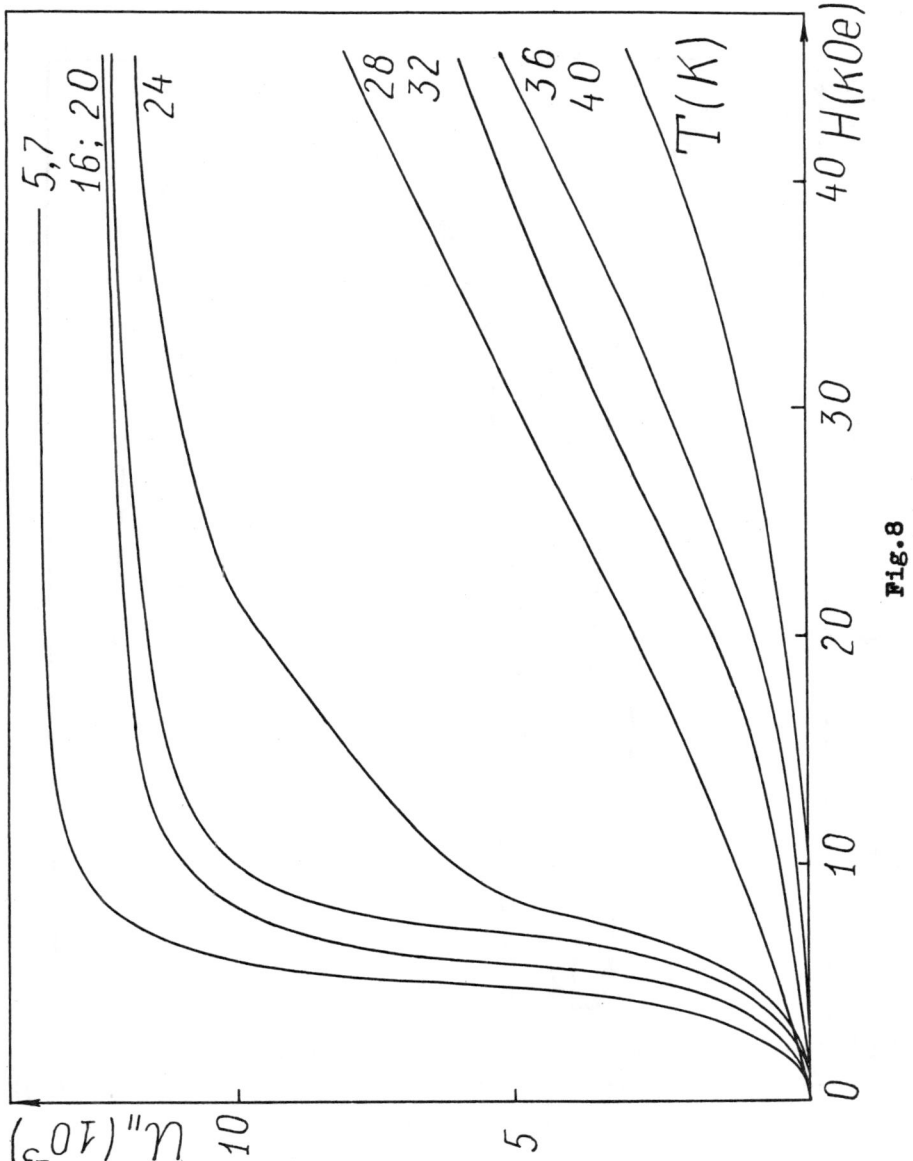

Fig.8

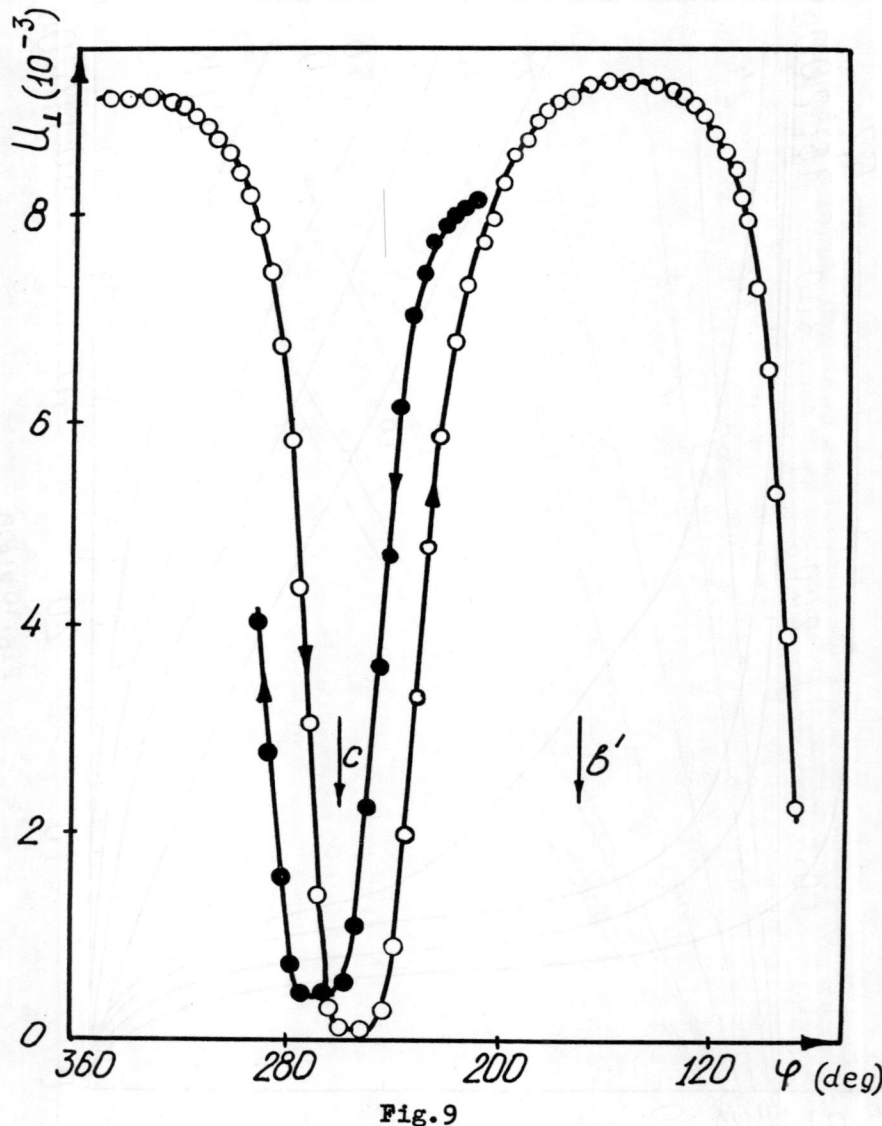

Fig.9

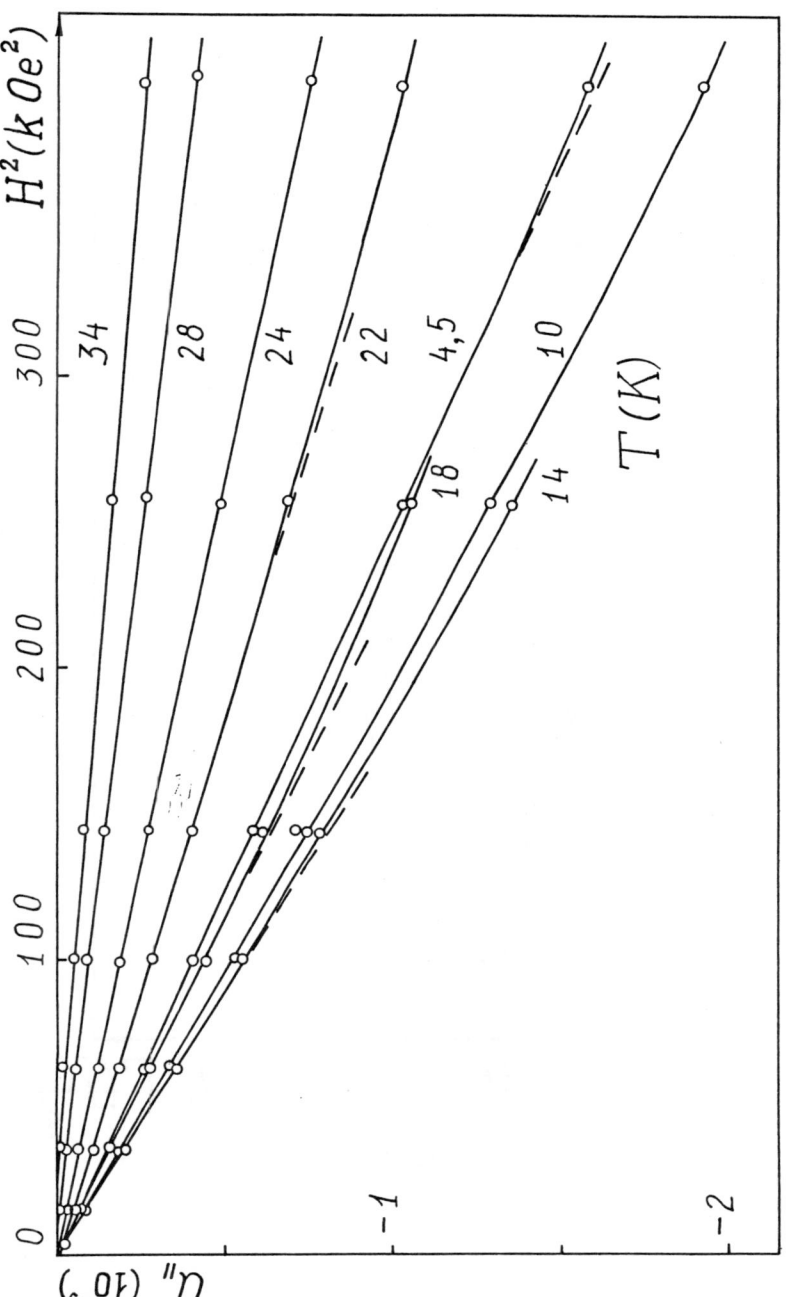

Fig.10

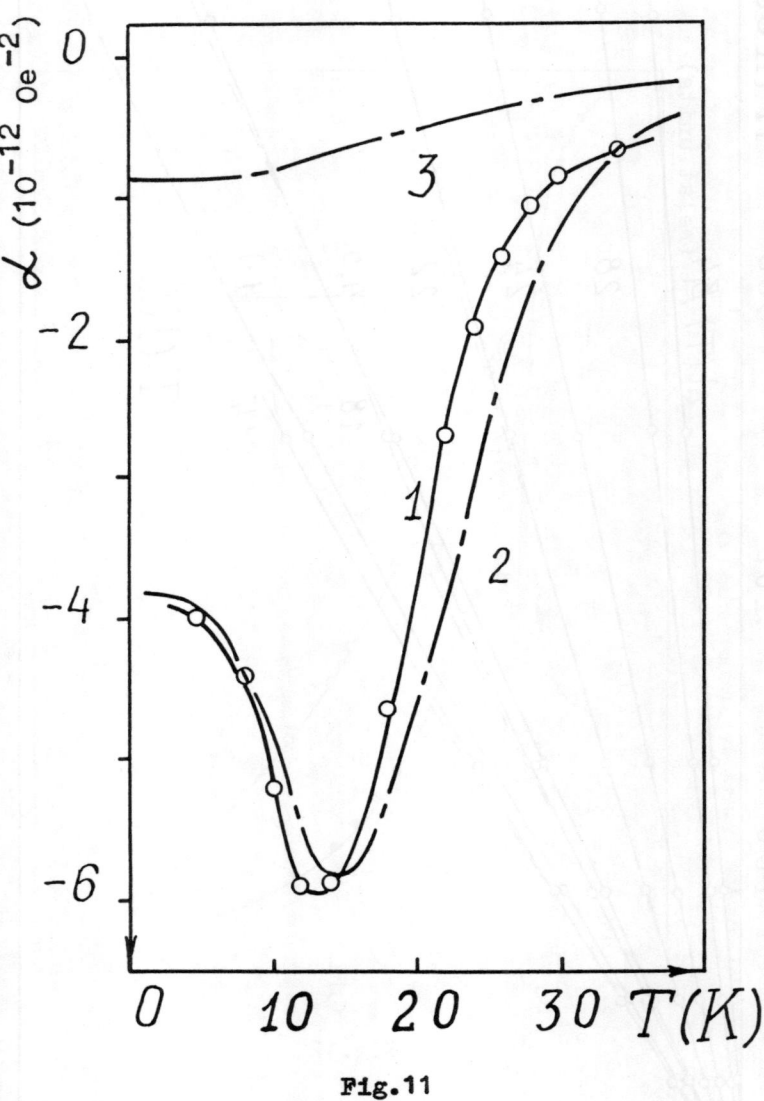

Fig.11

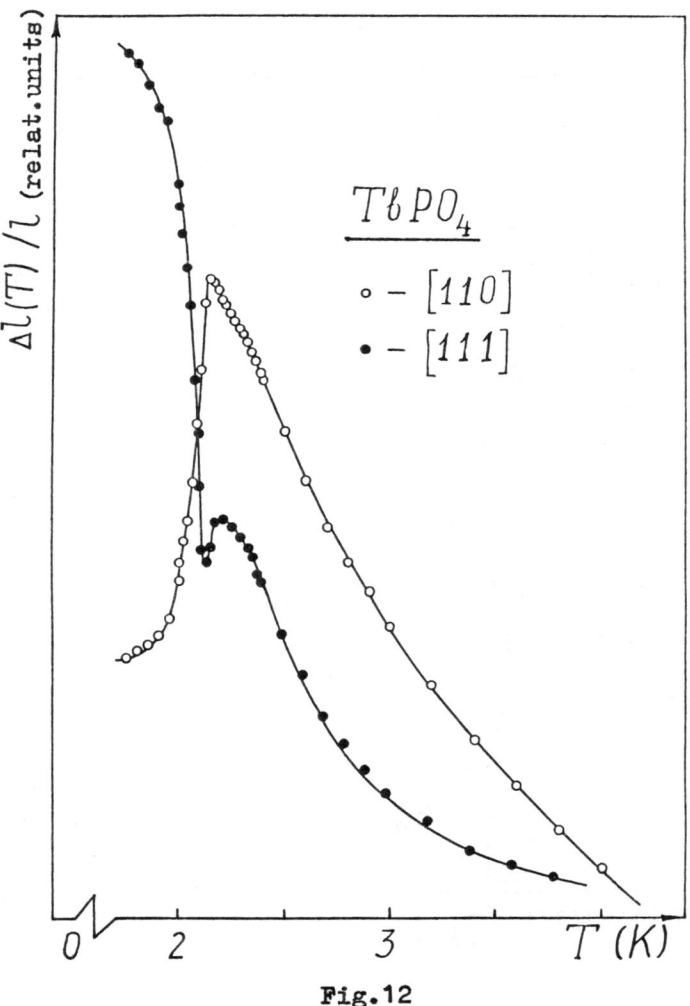

Fig.12

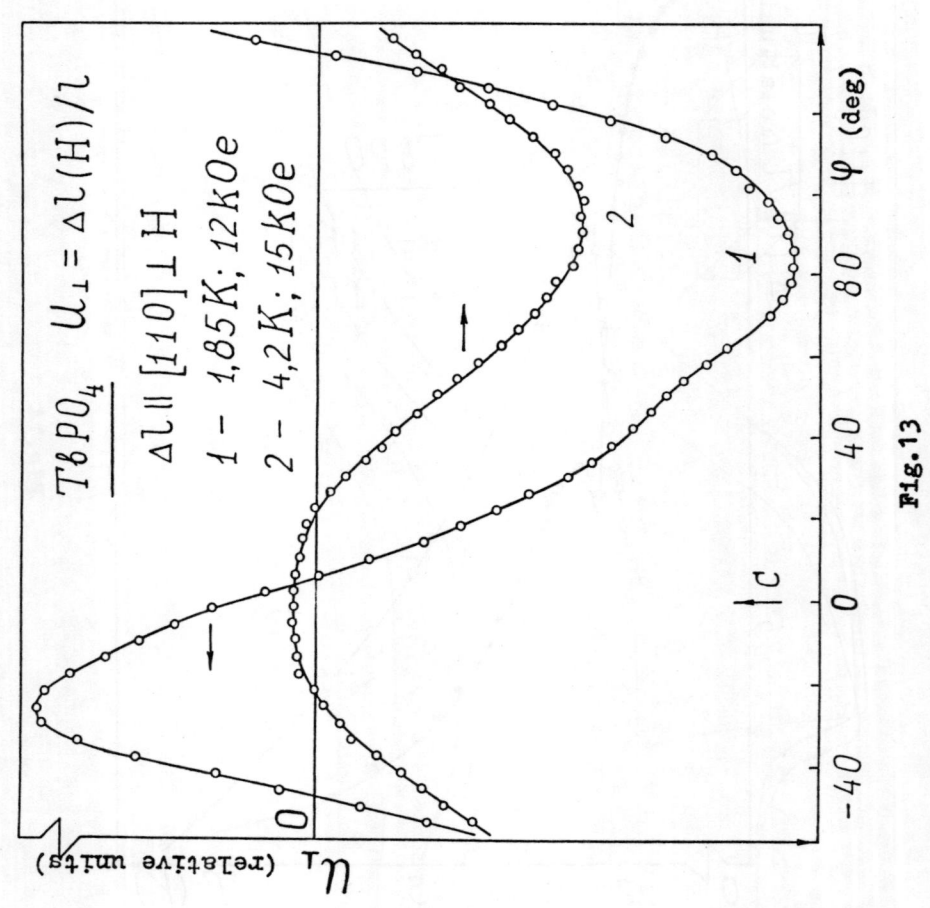

Fig.13

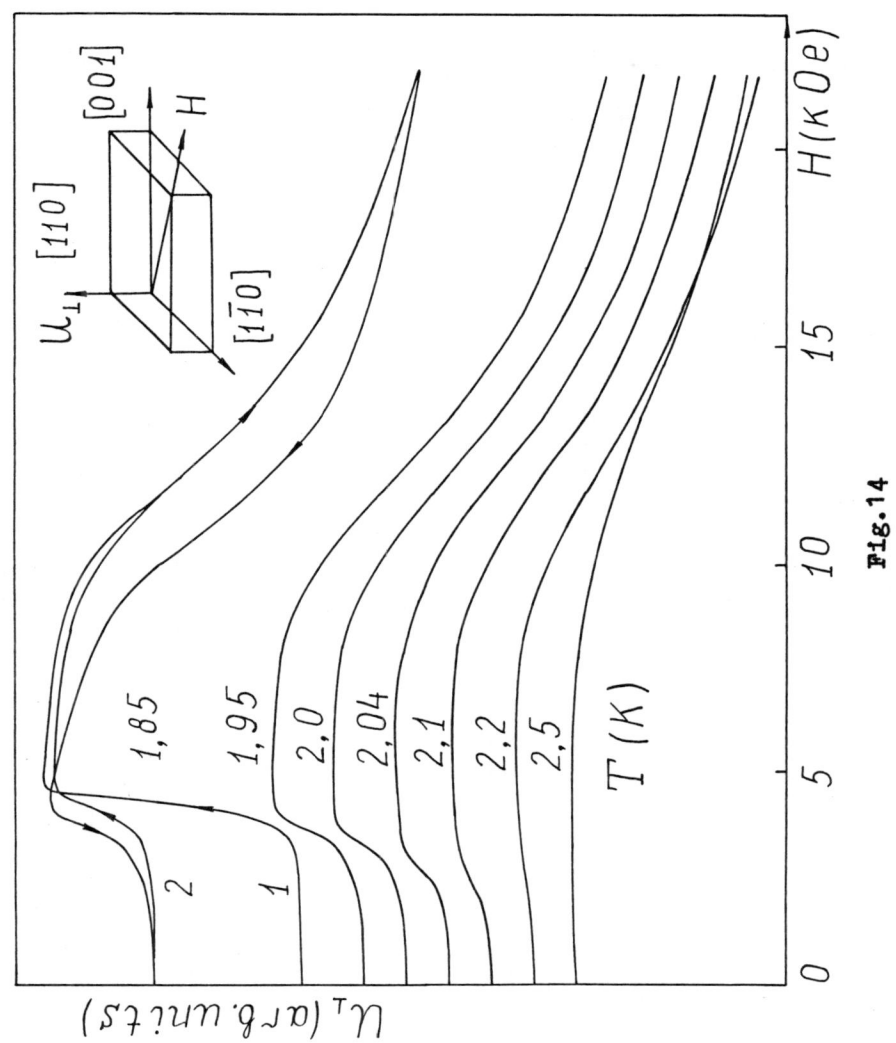

Fig. 14

MAGNETIC ORDER IN ORGANIC COMPOUNDS

Jerzy Pietrzak

Institute of Physics, Adam Mickiewicz University
PL 60-780 Poznań, Poland

ABSTRACT

This paper is review of experimental results on magnetic properties of a new class of low-dimensional molecular crystals. Strongly paramagnetic, antiferromagnetic and ferromagnetic /in high magnetic field/ phases have been observed in them. The analysis of magnetic properties to be described is based on experimental data of the crystal and molecular structure, magnetic susceptibility, magnetization, Mossbauer- and EPR spectroscopy, conductivity and theoretical analysis of the data including newly investigated materials.

1. INTRODUCTION

Various one-dimensional systems /1-D/ constituting a new class of molecular solids have been synthesized from organic[1,2] or organometallic molecules[3] and hydrocarbons[4]. Among others, quasi-one-dimensional organic conductors as well as organic superconductors have been obtained[5]. Optical, magnetic and electric properties of these highly anisotropic materials have been the subject of considerable interest in recent years[2,6-9]. Up to now, however, only a little has been known about the magnetic order in low-dimensional organic materials. At the present stage there are only a few compounds, which belong to the molecular charge-transfer /CT/ complexes and exhibit at low temperatures a cooperative magnetic phenomena, i.e. antiferro-, ferri-, meta- and ferromagnetism. Among them are: $Fe/py/_3Cl_3$-py /py = pyridine/[3,9], $Fe/C_5Me_5/_2^-$ TCNE /Me=CH_3, TCNE=tetracyanoethylene/ and related complexes[10]. We shall present experimental evidence for the existance of magnetic order in these molecular charge-transfer complexes.

2. CRYSTALLINE AND MOLECULAR STRUCTURES

Single crystals of the organometallic complex $Fe/py/_3 Cl_3$-py /referred to as FTPC-P/ obtained in our laboratory make a new class of low-dimensional molecular crystals formed from ferric tripyridine chloride molecules and pyridine molecules. X-ray structure investigations were performed at 190 K and room temperature by Hoser et al.[3], who assigned the complex studied to the orthorhombic system and monoclinic system, respectively. Crystallographic data of this crystal are listed in Table I.

The iron-pyridine ligand molecules /$Fe/py/_3Cl_3$ and pyridine molecules are linked by strong intermolecular coupling between pyridine molecule /III/ and pyridine ligands /I or II/ forming a structural pattern of different linear-chains only observed along the b and c axis of crystal /Fig. 1/.

Table I. Crystallographic data of complexes

Complex	Crystal symmetry	Temperature (K)	Unit cell parameters a (nm)	b	c
Fe/py/$_3$Cl$_3$-py Ref. 3	orthorhombic C222$_1$	190	0.8861	1.6712	1.4280
	monoclinic *)	300	0.4745	1.4395	0.7861
Fe/C$_5$Me$_5$/$_2$TCNE Ref. 10	orthorhombic Cmc2$_1$		1.0598	1.6091	1.5566

*) $\beta = 93.9°$

Preparation of large single crystals facilitating the measurement of their physical properties has to be achieved. Single crystal of FTPC-P used in the studies were stable and of high quality. This was verified by single crystal X-ray diffraction- EPR- and Mossbauer techniques[3,11-13]. This 1-D complex exhibits a very high conductivity along the b- and c-axis only and even becomes superconducting at a low temperature[14].

The 1:1 charge transfer salts of Fe/C$_5$Me$_5$/$_2$-TCNE composition, which have been reported by Miller et al.[10] in the orthorhombic or monoclinic phase can be prepared from tetrahydrofuran or acetonitrile containing solvent molecules, respectively. Both phases show the pattern ...D$^+$A$^-$D$^+$A$^-$... /Fig. 1/, however, the anion is disordered and unrefinable in the orthorhombic phase. All physical measurements were carried out in the orthorhombic phase /Table I/. Preparation of large single crystals has yet to be developed.

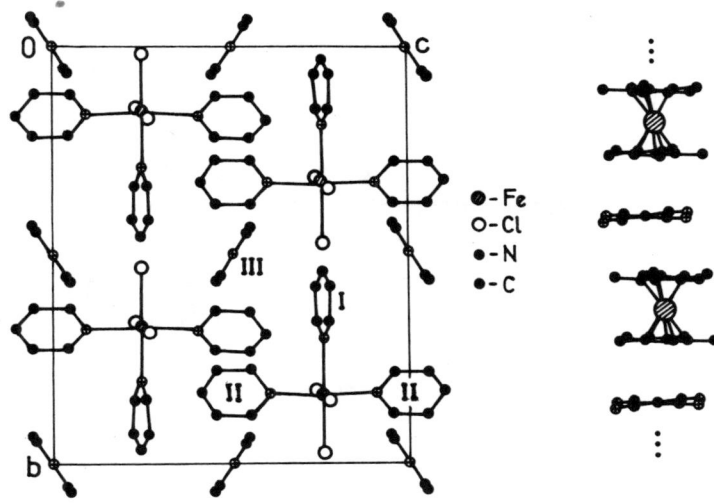

Fig. 1. Crystalline structure of FTPC-P /projection 100/. Stacks of alternated components are parallel to b-/...III II II III.../ or c-axis /...III I III I .../. On the right side - view of Fe/C_5Me_5/$_2$ - TCNE linear--chain. Ref. 3 and 10.

The Fe/C_5Me_5/$_2$-TCNE and related complexes are in general low conducting linear systems. However, this class of compounds exhibit a variety of interesting optical[15,16], and magnetic[10,17,18], properties as well as a spin-Peierls transition[19].

3. MAGNETIC SUSCEPTIBILITY AND MAGNETIZATION

3.1. Fe/py/$_3$Cl$_3$-py Complex

Magnetic susceptibility of oriented single crystals FTPC-P was measured between 1.9 and 280 K with a string magnetometer in a field of 13.3 kOe[9].

The magnetization was also measured in fields up to 100 kOe with a Bitter-type coil at 1.9 to 8 K.

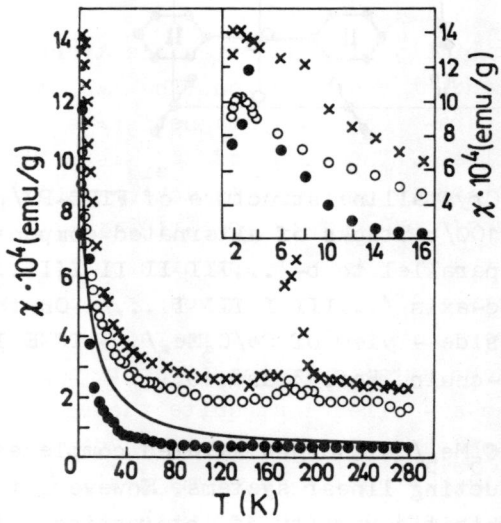

Fig. 2. Temperature dependences of total magnetic susceptibilities, χ_a /•/, χ_b /×/ and χ_c /o/, with the magnetic field parallel to the principal axes of a single crystal of FTPC-P. Inset: The plot at temperatures lower than 20 K is shown on an expanded scale. Solid line - magnetic susceptibility /isotropic/ of disordered single crystal. Ref. 9 and 20.

Fig. 2 shows that the principal magnetic susceptibilities χ_a, χ_b and χ_c obtained for high quality single crystals[9] may be described by Curie-Weiss laws of the type

$$\chi_i = \chi_{ei} + C/T-\theta/^{-1} \qquad (1)$$

i = a,b or c-axis

where

$$\chi_{ei} = \chi_{ep} + \chi_{ed} \qquad (2)$$

is the susceptibility of conduction electrons i.e. the temperature independent contribution. The two terms in Eq. 2 represent the para- /Pauli paramagnetism/ and diacontributions, respectively. The value of χ_{ei} is so small in disordered FTPC-P single crystals /stacking faults/[20] that $\chi_a = \chi_b = \chi_c = C/T-\theta/^{-1}$ /Fig. 2 solid line/. The difference between the observed χ_i for high quality and disordered single crystal therefore gives χ_{ei} immediately. The results are shown in Fig. 3. In the b- and c- directions /along the stacks axis/ only the first term in Eq. 2 is important. In the a - direction, quite the contrary, the second term is dominant. The temperatures dependence of χ_i in 13.3 kOe /Fig. 2/ and in lower magnetic fields shows that FTPC-P is an antiferromagnet with a Néel temperature of 3.8 K. Since Néel temperature decreases with increasing applied magnetic field different for each direction the T_N value at zero field was obtained by extrapolating the T_N readings at several magnetic fields to that at zero field. For instance the peaks of susceptibilities χ_a, χ_b and χ_c appear at transition temperatures of 3.6, 2.1 and 2.6, respectively /Fig. 2/. From the linear part of $\chi_i^{-1}/T/$ above 10 K, the Curie-Weiss temperatures were determined to be $\theta_b = \theta_c = -8$ K and $\theta_a = 1.3$ K and the effective magnetic moments $\mu_{effa} = 3.6 \mu_B$, $\mu_{effb} = 8.2 \mu_B$ and $\mu_{effc} = 7.1 \mu_B$.

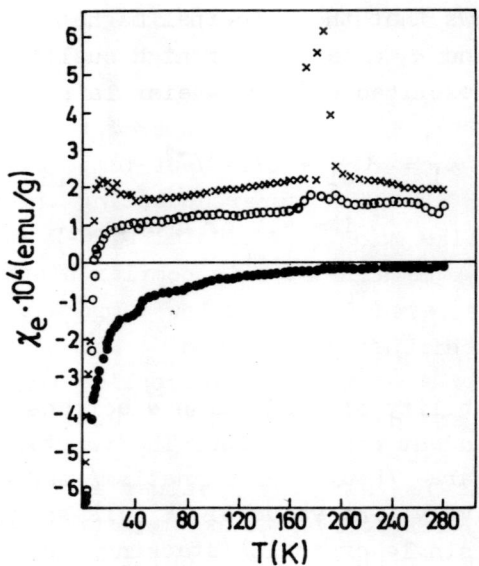

Fig. 3. Temperature dependences of itinerant electron susceptibilities, χ_{ea} /•/, χ_{eb} /×/ and χ_{ec} /o/, for FTPC-P single crystal. Ref. 23

One should note that for the a - axis, the susceptibility deviates from a Curie-Weiss law at low temperatures below 10 K. This indicates the presence of antiferromagnetic interactions coexisting with the ferromagnetic ones as it is shown from the positive Curie-Weiss temperature.

In high temperatures paramagnetic susceptibilities χ_b and χ_c or χ_{eb} and χ_{ec} /Figs. 2 and 3/ only show a bump near 200 K corresponding to the anomaly observed also near 200 K in the conductivity[8] and in the EPR spectra[12]. Furthermore, X-ray investigations[3] have shown that below 200 K an orthorhombic system is observed with no change of the periodicity along b, but with a doubling along a and c /Table I/. This anomaly is related to a static structural distortion /$T < T_p$/ i.e. a Peierls transition accompa-

nied by a static spin density wave of a 1-D electron gas.
In lower temperatures, the onset of the sharp drop in susceptibilities χ_{ea}, χ_{eb} and χ_{ec} occurs at ~27, 12 and 17 K, respectively /Fig. 3/. The midpoints of the all drops are at 8 K and are much higher than Néel temperature. We interpret this sharp drop as a superconducting transition. So, the organic - like molecular crystal FTPC-P is not only a paramagnetic 1-D metal, but this complex also is 3-D antiferromagnetic superconductor at low temperature.

The magnetizations M_a, M_b and M_c are plotted in Fig.4 versus the field H which has been applied to induce than along the a, b and c-axis at 4.2 K[14]. Both the M_b and M_c saturate in the field of ~70 kOe which is considerably higher than M_a. They saturation values are 115.5, 101.5 and 36.0 emu/g, respectively. The magnetization curves cannot be fit to Brillouin function, the deviation at low fields indicated the appearance of pairbreaking in the superconducting state which disappears in high fields.
It should be pointed out that the field dependences of magnetizations above the critical temperature can already be fit to Brillouin functions. At 1.9 K the magnetizations are anisotropic for low fields as shown in Fig. 4 for the a and c-axis the saturated values of 3.6 emu/g do not depend on the orientation of high field H. This field - dependent switching from antiferromagnetic to a high - moment behaviour is consistent with **metamagnetism**. The measured values of the effective magnetic moments in low fields along the a, b and c-axis are 4.1, 9.4 and 8.2 in Bohr magneton units, while the effective moments deduced from Curie-Weiss law are smaller by ~12%.

The temperature - dependent Fe^{3+} ion paramagnetism has been subtracted from the magnetization value:
$M_{ei} = M_i - M_a$ /i=b or c/. The corresponding paramagnetic effective moment of conducting electrons /itinerant/ calcu-

lated from $M_e = NngS\mu_B$, where N is the molecule number per cm^3, and n - the spin number per molecule, assuming a g value of 2.0 and S of 1/2 is $6.2\mu_B$ for both the bonde - axis and the spin number n of 6.

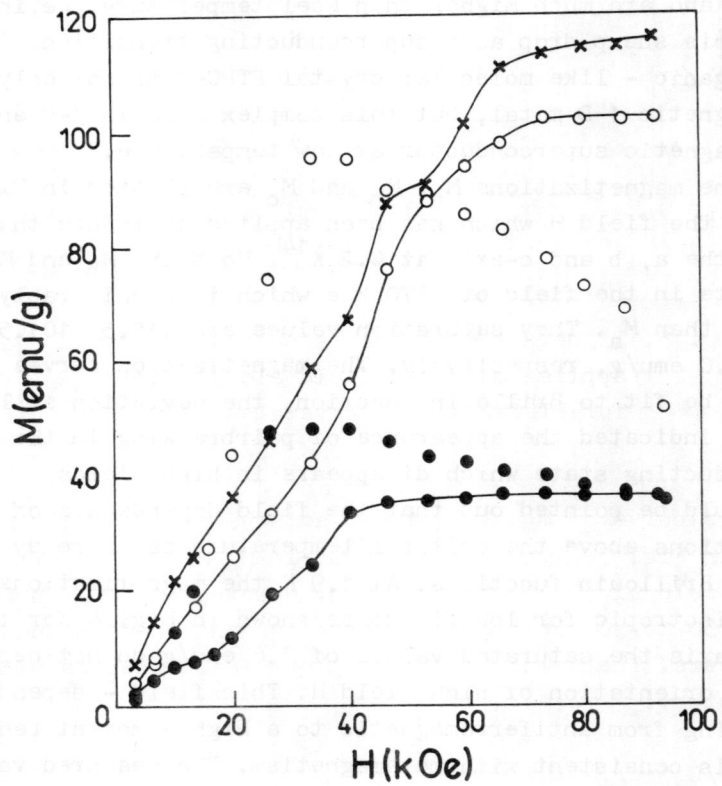

Fig. 4. Magnetization M_a/•/, M_b/×/ and M_c/○/ vs applied magnetic field along a-, b- and c-axis of FTPC-P single crystal at 4.2 K /solid lines/ and 1.9 K /for a- and c-axis only/. The solid lines are guides for the eye. Ref. 14

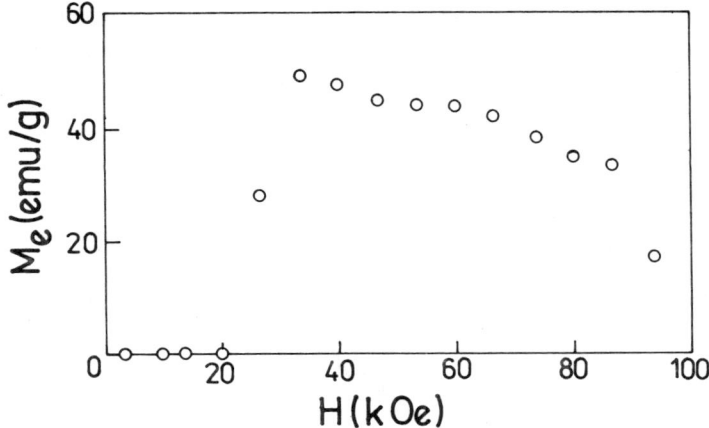

Fig. 5. Itinerant electron magnetization vs applied magnetic field along c-axis for FTPC-P single crystal at 1.9 K. Ref. 14

The curve of M_e along the c-axis vs field at 1.9 K is shown in Fig. 5. A sudden increase in M_e above the critical a field of 20 kOe is observed. However, the M_e in field range ~30 to 90 kOe smoothly decreases and then it suddenly disappears. The value of the critical field changes with temperature.

3.2. $Fe/C_5Me_5/_2$-TCNE Complex

Investigations of magnetic properties of $Fe/C_5Me_5/_2$-TCNE were performed for polycrystalline samples and single crystals[10]. However, the studies of the aligned single crystals give a more detailed understanding[21]. The magnetic susceptibility at 200 K is 0.147×10^{-4} emu/g. The single crystal reciprocal susceptibility can be fit by the Curie-

-Weiss law above 60 K with Θ = 30 K /Fig. 6/, suggesting the dominant ferromagnetic interactions[10,18]. Below 60 K a substantial departure from Curie-Weiss behaviour is evident /Fig. 6/.

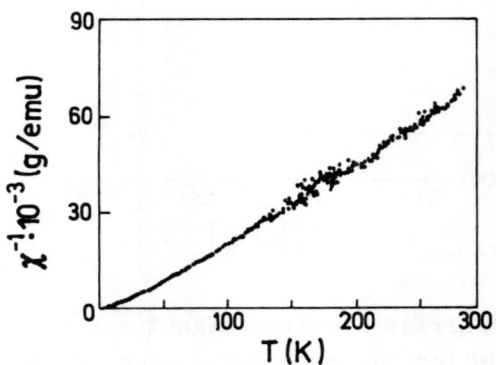

Fig. 6. Reciprocal magnetic susceptibility vs temperature, with the magnetic field parallel to the c-axis of single crystal of $Fe/C_5Me_5/_2$-TCNE. Ref. 10

From the low-field $\chi/T/$ data and from the temperature dependence of spontaneous magnetization, the Curie temperature $/T_c/$ was determined to be 4.8 K /Fig. 7/. Examination of variations of $\chi \propto /T-T_c/^{-\gamma}$ above T_c, $M_s \propto /T_c-T/^{-\beta}$ below T_c, and $M \propto H^{1/\delta}$ at T_c enables a valuation of the γ, β and δ critical exponents[22]. The results obtained by Miller et al.[10] for the magnetic field parallel to the stack axis are 1.2, 0.5 and 4.5, respectively. When these values are compared with model-dependent predictions, the exponents are consistent with a 3-D behaviour and there is no evidence for an intermediate 2-D regime[20].

The magnetization vs applied field measured for $Fe/C_5Me_5/_2$-TCNE exhibits hysteresis loops[21]. At 4.7 K

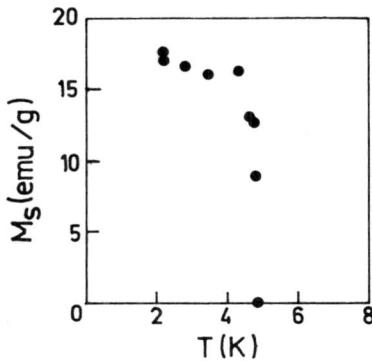

Fig. 7. Spontaneous magnetization of $Fe/C_5Me_5/_2$-TCNE vs temperature in Earth's magnetic field. The field gradient was aligned in parallel to the stacking axis. Ref. 10

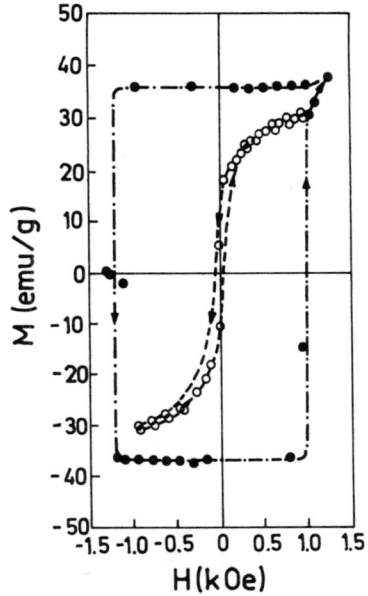

Fig. 8. Hysteresis loops for $Fe/C_5Me_5/_2$-TCNE single crystal oriented with the stacking axis parallel to the applied magnetic field at 2.0 K /●/ and 4.7 K /○/. Ref. 10

a hysteresis with coercive field of 30 Oe is observed. A large coercive field of ~1 kOe is observed at 2 K[20] as shown in Fig. 8. A well-defined remanent magnetization nearly equal to the saturation magnetization of 35.9 emu/g /1.63x10^4/ emu/mol/ is seen. It is noteworthy that the saturation value which was obtained here for $Fe/C_5Me_5/_2$-TCNE is similar to the value obtained for FTPC-P single crystals along the a-axis.

4. MÖSSBAUER SPECTRA

Mössbauer effect measurements were also made to determine the local environment effects.

The high quality single crystals of FTPC-P complex exhibit a single asymmetrical quadrupole doublet at 77 to 290 K[13]. At 4.2 K, i.e., near the Néel temperature, Mossbauer spectra for the a, b and c directions parallel to the gamma /γ/ beam show extreme broadening /Fig. 9/[23].

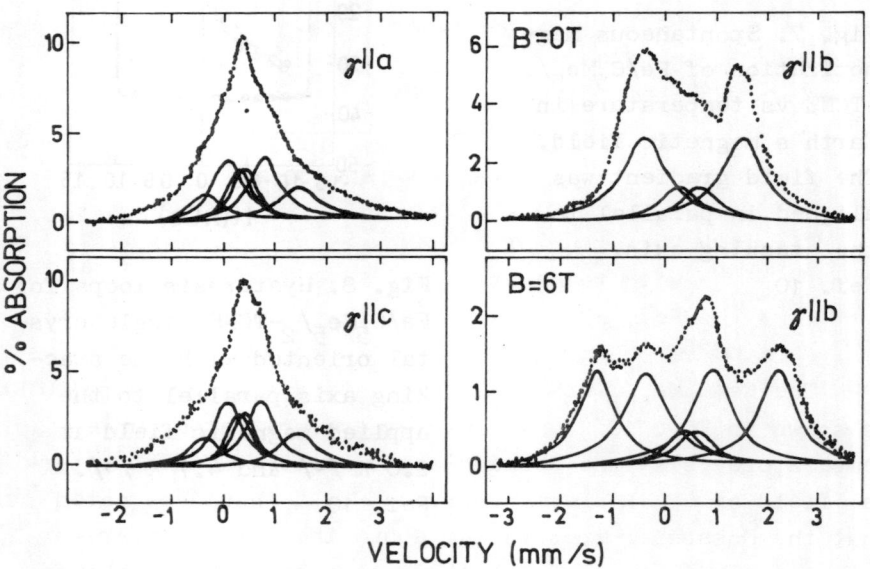

Fig. 9. ^{57}Fe Mössbauer spectra for single crystals of FTPC-P obtained in a, b or c direction parallel to the gamma beam. Ref. 23

These spectra were analysed by a specially RKU-01 computer programme which enhanced the spectral resolution[24]. The six peak magnetic patterns appear at the spectrum zero applied field.

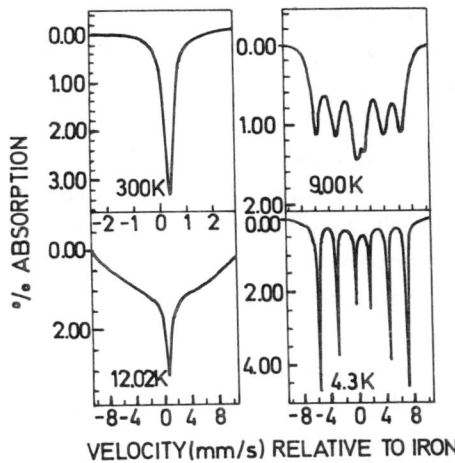

Fig. 10. Variation of ^{57}Fe Mössbauer spectra with temperature for Fe/C$_5$Me$_5$/$_2$-TCNE. Ref. 10

The Fe/C$_5$Me$_5$/$_2$-TCNE complex exhibit a singlet at 300 K is shown in Fig. 10[10]. At low temperature below Curie temperature, six-line Zeemann split spectra with an internal field of 424 kOe are observed[10]. It should be noted that the Mössbauer spectra far above the Curie temperature i.e. $T \cong 2T_c$, show clearly distinguished six peak magnetic patterns with an internal field of ~396 kOe /Fig. 10/.

5. SUMMARY

In this review we have summarized the magnetic studies of some of the key organometallic charge-transfer complexes that have become important in recent years. The magnetic properties of these complexes are important to the development of an understanding of the electronic structure and cooperative magnetic phenomena of the molecular solids.

The results obtained up to now allow to drow some interesting conclusions. A quantative bulk ferromagnetic and antiferromagnetic ordering has been established for the $Fe/C_5Me_5/_2$-TCNE and FTPC-P complexes, respectively. Magnetic interactions along a stack /1-D/ were found to dominante above 30 K for FTPC-P and above 16 K for $Fe/C_5Me_5/_2$-TCNE, whereas near critical temperature and Néel temperature $/T_{cr} > T_N/$ or Curie temperature the presence of 3-D bulk behaviours were evidenced on the ground of the magnetic susceptibility, magnetization and Mössbauer spectra data. Moreover, the strong influence of itinerant electrons along the stacking direction /due to $\pi - \pi$ orbital overlap between succesive pyridine molecules in the one-dimensional stack/ on anisotropic magnetic properties was observed for FTPC-P single crystals. Interaction of both spin subsystem i.e. localized 3d - electrons and itinerant electron leads to a number of new phenomena in organic-like molecular crystals.

REFERENCES

1. Shchegolev, I.F., Phys.Stat.Sol. /a/ 12, 9 /1972/.
2. Andre, J.J., Bieber, A., and Gautier, F., Ann.Phys. 1, 145 /1976/.
3. Hoser, A., Kałuski, Z., Januszczyk, M., Pietrzak, J., and Głowiak, T., Acta Cryst. C39, 1039 /1983/.
4. Chiang, C.K., Fincher, C.R., Park, Y.W., Heeger, A.J., Shirakawa, H., Louis, E.J., Gan, S.C., and Mc Diarmid, A.G., Phys.Rev.Lett. 39, 1098 /1977/.
5. Jerome, D., and Schulz, H.J., Adv.Phys. 31, 299 /1982/.
6. Miller, J.S. Ed: Extended Linear Chain Compounds,

Plenum, New York, 1983, Vol. 1-3.
7. Patil, A.O., Heeger, A.J., and Wudl, F., Chem.Rev. 88, 183 /1988/.
8. Januszczyk, M., Pietrzak, J. and Stecki, A., Phys. Stat.Sol. /a/ 73, K195 /1982/.
9. Janicki, J., Januszczyk, M., Pietrzak, J., and Mydlarz, T., Acta Phys. Pol. A72, 311 /1987/.
10. Miller, J.S., Epstein, A.J., and Reiff, W.M., Isr. J. Chem. 27, 363 /1987/, Chem.Rev. 88, 201 /1988/.
11. Januszczyk, M., PhD Thesis, A.Mickiewicz Univ., Poznań 1982, Poland
12. Pietrzak, J., and Januszczyk, M., Proc.Conf. Radio- and Microwave Spectroscopy RAMIS-83, A.Mickiewicz Univ. Poznań 1985, p. 193, Poland.
13. Janicki, J., Januszczyk, M., Krzyminiewski, R., Pietrzak, J., Suwalski, J., and Śledzińska, I., Acta Phys. Polon. A68, 69 /1985/.
14. Janicki, J., Januszczyk, M., Pietrzak, J., and Mydlarz, T., /in preparation/.
15. Mullikan, R.S., Person, W.B., Molecular Complexes: A Lecture and Reprint Volume: Viley, New York, 1969
16. Soos, Z.G., Annu. Rev. Phys. Chem. 25, 121 /1974/.
17. Candela, G.A., Swartzendruber, L., Miller, J.S., Rice, M.J., J.Am.Chem.Soc. 101, 2755 /1979/.
18. Miller, J.S., Calabrese, J.C., Bigelow, R.W., Epstein, A.J., Zhang, R.W., Reiff, W.M., J.Chem.Soc., Chem. Commun. 1026 /1986/.
19. Bray, J.W., Hart, H.R., Interroute, L.V., Jacobs, I.S., Wothins, G.D., Prober, D.E., Phys.Rev. Lett. 35, 744 /1976/.
20. Pietrzak, J., Januszczyk, M., and Obuszko, Z., Proc. Nat. Conf. Molecular Crystals 79, Błażejewko, Poznań 1979, Poland.
21. Chittapeddi, S.R., Cromack, K.R., Miller, J.S., and Epstein, A.J., Phys.Rev.Lett. 58, 2695 /1987/.

22. Baker, G.A., Rushbrok, G.S., Gilbert, H.E., Phys.Rev. A135, 1272 /1964/.
23. Janicki, J., PhD Thesis, A.Mickiewicz Univ., Poznań 1987, Poland.
24. Koper, A., and Krzyminiewski, R., Acta Magnetica 2, 3 /1985/.

MAGNETIC PROPERTIES OF CERIUM MONOPNICTIDES

Leon Kowalewski

Institute of Physics, A. Mickiewicz University

60-769 Poznań, Poland

ABSTRACT

The origins and some effects of hybridization mediated Coqblin-Schrieffer and Kasuya-Takahashi exchange interactions in cerium monopnictides are discussed. Unusually strong anisotropy and multichannel character of these interactions were of particular interest. Moreover, the mechanism of continuous magnetic phase transition is analysed.

1. INTRODUCTION

Magnetic properties of the face-centered-cubic cerium monopnictides CeX (X=P, As, Sb, Bi) appear to be extremely anomalous when compared with those of other rare-earth monopnictides, or other cerium systems [1-9]. The main origin of this behaviour is the hybridization of quasilocalized 4f electrons with the 5d-conduction or p-valence band charge carriers [3,4,9]. The quasilocalized and wide band electron subsystems can be described by the Anderson model.

The band structure of these monopnictides changes evidently if one goes from the light to heavy ones [3,4,9]. The light cerium monopnictides are narrow gap semiconductors [3] which have neither valence holes nor conduction electrons. Some authors however, consider them to be weakly overlapping semimetals[1,3,11] with a very small number of valence holes and conduction electrons. Heavy monopnictides are always semimetals with several per cent of valence holes and the same per cent of conduction electrons per each cerium atom[3].

The hybridization mediated exchange interactions between cerium ions occur via different charge carrier states[4,9,11] and depend on the symmetry of the cerium crystal field states. The relative strength of hybrydization mediated channels of the exchange interactions varies depending on an actual cerium monopnictide.

The aim of our paper is to review the anomalous properties of cerium monopnictides and discuss their main origins. Especially, we will discuss the fundamental problem of the cerium monopnictides, which is the origin and effect of the unusually strong anisotropy of exchange interactions. We will also discuss the multichannel character of these interactions.

The paper is organised as follows. In Section 2 we review the properties of cerium monopnictides. In Section 3 we discuss two models of anisotropic indirect f-f exchange interactions with intermediating either conduction or valence band states, respectively. In Section 4 we analyse a mechanism of magnetic phase transitions in cerium monopnictides on the basis of both the Landau theory and an account of the critical behaviour of longitudinal magnetic fluctuations. An experimental evidence of the anisotropy of the effective exchange interactions as well as that of their multichannel character are given in Section 5. Final remarks and conclusions are presented in Section 6.

2. ANOMALOUS PROPERTIES OF CERIUM MONOPNICTIDES

In order to discuss the anomalous behaviour of cerium monopnictides we start with a brief review of the necessary information on their properties [1-11]:

1. The calculated effective 4f level in CeSb and CeBi is about 1eV below the Fermi level while the one observed in the X-ray photoemission spectroscopy is found to be 3eV below the Fermi level[3].

2. The crystal field splitting between the ground doublet Γ_7 and excited quartet Γ_8 is much smaller in the light cerium monopnictides than in other rare-earth monopnictides, and is found to be extremely small in CeSb and CeBi[3,5,12,13]. The experimental values of the crystal field splitting and those extrapolated from other rare-earth monopnictides are given in Table I.

TABLE I

crystal field splitting Δ	CeP[1,5]	CeAs[1,5]	CeSb[1,5,12]	CeBi[1,5,12]
extrapolated Δ	390	345	264	247
measured Δ	170	160	37	8

3. In contrast to the small splitting Δ, and consequently the small single ion anisotropy, the total anisotropy field in CeSb and CeBi is very large[3,8,9,14-17]. Moreover this anisotropy field is directed along the [001] axis, while the easy direction of the ground doublet Γ_7 is along the [111] axis.

4. The appearence of the strong total magnetic anisotropy together with the small single ion one, suggests that the exchange interactions must be highly anisotropic. However, from the Néel temperatures[1] (16.2K for CeSb[12], 25.5K for

CeBi[12], 8.5K for CeP[12] and 7.5K for CeAs[12])they seem to be rather small in magnitude.

5. Anisotropy of the dispersion of transverse magnetic excitations and of longitudinal magnetic fluctuations above T_N is unusual for compounds[7,8,18] with cubic symmetry.

6. Magnetic excitations are not described by the same exchange parameters as the static properties[1,18].

7. The magnetic moment induced in an external magnetic field in the temperatures far above T_N exhibits strong anisotropy[19].

8. The magnetic structures of cerium monopnictides consist of various antiparallel sequences of the ferromagnetic (001) planes with the the magnetic moments aligned along [001] direction[2,3,20-22].

9. In CeSb the higher temperature ordered structures involve paramagnetic (001) planes which coexist with the magnetically ordered planes with their moments aligned along the [001] axis[23].

10. In CeSb[24,25] and CeBi[25] a large tetragonal distortion occurs at the Néel point T_N.

11. In CeSb and CeBi strong short-range ordering effects exist well above T_N[26,27].

3. ANISOTROPIC BAND MEDIATED EXCHANGE MODELS

In cerium monopnictides there exist two band channels of the hybridization between the localised 4f electrons and charge carrier states[3,28,30]:

1) the first channel via the conduction $5d-t_{2g}$ electrons leads to the long-range, anisotropic RKKY-type f-f exchange interaction. It is the so called Coqblin-Schrieffer model[28].

2) the second channel via the p-valence states, especially the p-valence holes leads to the short range, strongly

anisotropic f-f exchange interactions. It is the so called Kasuya-Takahashi model[3].

The main difference between these two models consists in the choice of the charge carriers in spin-orbit states[30].

Coqblin and Schrieffer consider the free electron limit - the Bloch states with freely moving electrons.

Kasuya and Takahashi consider the localized charge carriers in atomic orbitals around each site.

In both models the symmetry of 4f crystal field states plays the fundamental role in description of the effective interionic f-f interactions. Each of the hybridization channels can be split into subchannels according to the symmetry of the crystal field states. In order to discuss the different channels of the hybridization mediated exchange f-f interactions, we give a very short review of intermediating processes of the f-f indirect exchange.

3.1. Conduction 5d-Band Channel

Coqblin and Schrieffer[28] assume the band conduction states in the free electron limit. They decompose the free electron states into the partial spherical wave states.

In order to discuss the hybridization of localized f electrons and band d-electrons it is convenient to take into account the spin-orbit states from these partial wave states and consider only the states of a definite angular momentum J. Let us also restrict the states of the localized 4f electrons to the ground manifold $4f_{5/2}$ and neglect the manifold $4f_{7/2}$ because the spin-orbit coupling is relatively large in comparison with the 4f-5d interaction.

In Section 3.1 we discuss the simplified two-cerium ion interaction neglecting the crystal field effects. By including the crystal field effects however, we obtain no essentially new qualitative results on the exchange interactions.

The electronic system is described by the basic Anderson Hamiltonian[4,28]

$$H = H_0 + H_1$$

where

$$H_0 = \sum_{|k|,M} \varepsilon_k \hat{n}_{|k|M} + \sum_{i,M} E_M \hat{n}_M(i) + 1/2\, U \sum_{i,M,M'} \hat{n}_M(i)\, \hat{n}_{M'}(i)$$

$$H_1 = \sum_{i,|k|,M} [V_{|k|Mi}\, c^+_{|k|M}\, c_M(i) + V^*_{|k|Mi}\, c^+_M(i)\, c_{|k|M}] .$$

The creation and annihilation operators of electrons in the hybridization term H_1 create (or annihilate) the spin-orbital states, especially: $c^+_M(i)$ – creates an electron with a total angular momentum $J = 5/2$ and its z-component M; whereas $\hat{c}^+_{|k|M}$ – creates a band electron with a wave vector $|k|$ and the same component of the total angular momentum J.

If the mixing potential is assumed to be spherical the cerium 4f electron states interact with the appropriate partial spherical wave states with the same quantum numbers $l=3$, $J = 5/2$.

E_M denotes the unperturbed energy of the localized 4f electron with respect to the Fermi level E_F. The X-ray photoemission spectroscopy[3] shows that the bare energy of the 4f electrons is relatively far from E_F. However, effective level of a 4f electron, shifted due to the hybridization is rather close to the Fermi energy.

In the limit of small 4f-5d mixing the canonical Schrieffer-Wolff transformation can be used to replace the hybridization term H_1 by the effective resonant exchange scattering term

$$H_{4f-band} = -\sum_{\substack{|k||k'| \\ M,M'}} J^{MM'}_{|k||k'|} c^+_{|k'|M'} c_{|k|M} \left(c^+_M c_{M'} - \frac{\delta_{MM'}}{2J+1} \sum_{M''} \hat{n}_{M''} \right)$$

The coefficient $J^{MM'}_{|k||k'|}$ describes an antiferromagnetic exchange and depends on the energy of the 4f electrons measured from the Fermi level as seen from the formula:

$$J^{MM'}_{|k||k'|} = \frac{1}{2} V_{|k|M} V^*_{|k'|M'} U \left[\frac{1}{(\varepsilon_k - E_M)(\varepsilon_k - E_M - U')} + \frac{1}{(\varepsilon_{k'} - E_{M'})(\varepsilon_{k'} - E_{M'} - U)} \right]$$

For $|k|$ and $|k'|$ equal to the Fermi wave vector k_F we can assume $\varepsilon_k \longrightarrow \varepsilon_F = 0$. If additionally $E_M = E_0$ all the scattering coefficients are approximately the same and equal to[4]

$$J_{k_F k_F} \cong \frac{|V_{k_F,\sigma}|^2 U}{E_0 \cdot (E_0 + U)} < 0 .$$

The scattering coefficient is determined by the single-site hybridization of the 4f and 5d electron states.

It is noteworthy that the existence of a localized moment of cerium ion depends on the strength of the effective exchange interaction $J_{kk'}$.

As known[31] the same Anderson model enables us to discuss the two competing processes, i.e. the hybridization mediated f-f exchange interaction of magnetic cerium ions and the effective suppression of cerium magnetic moments due to the Kondo spin fluctuations. The first of the processes dominates in the cerium monopnictides.

In the case of n cerium ions in a Fermi bath all scattering terms can be treated as a perturbation and in the second order of perturbation calculus lead to the effective 4f-4f interactions. The Coqblin-Schrieffer exchange interaction takes the form[4]:

$$H_{ij}(R) = \sum_{MM'} E_{ij}^{MM'}(R) \left[c_{M'}^+(i) c_M(i) - \frac{\delta_{MM'}}{2J+1} \sum_{M''} n_{M''}(i) \right] \times$$

$$\left[c_M^+(j) c_M(j) - \frac{\delta_{MM'}}{2J+1} \sum_{M''} n_{M''}(j) \right]$$

where the bonding axis is assumed to be along the quantization axis. The Coqblin-Schrieffer two-ion interaction is of the long range RKKY type. It displays also a high anisotropy with respect to the bonding axis.

The strongest two-ion coupling with energy $E^{\pm 1/2, \pm 1/2}(R)$ corresponds to the orbital moments perpendicular to the bonding axis and the charge distributed along this axis. The leading term of this interpretation is proportional to $(k_F R)^{-2}$.

The coupling is weaker if the charge is further away from the bonding axis.

However, all the two-ion bonding axes are not parallel and therefore the arrangement of orbital moments depends on the distance between cerium ions and on the number and geometry of their nearest neighbours.

The interaction in the common crystal-lattice coordinate system takes the form[4]

$$\mathcal{H}_{ij} = -E_{ij} \sum_{\mu\nu\varepsilon\sigma} B_{\mu\nu}^{\varepsilon\sigma}(\theta_{ij}) e^{-i(\mu-\nu+\varepsilon-\sigma)\phi_{ij}} L_{\mu\nu}^{(i)} L_{\varepsilon\sigma}^{(j)}$$

where θ and ϕ are angles between the interionic bonding and crystalline quantization axes. The indices $\mu,\nu,\varepsilon,\sigma$ are the magnetic quantum numbers of cerium eigenstates. The factors $B_{\mu\nu}^{\varepsilon\sigma}(\theta)$ determine anisotropy of the two-ion interaction. In particular the maximum of $B_{5/2\;5/2}^{5/2\;5/2}(\theta=\pi/2)^{32}$ (see Fig.1) explains the preference of the fully saturated magnetic moments for the planes transverse to the moment direction.

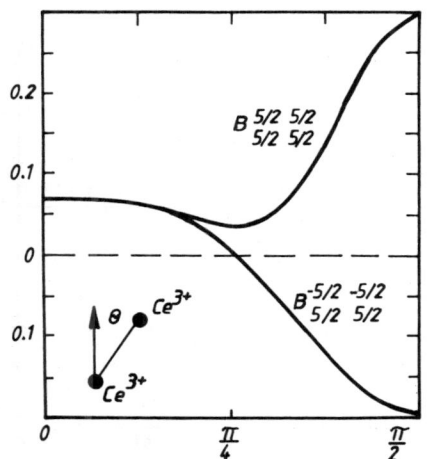

Fig. 1. Variation of diagonal $B^{\varepsilon\sigma}_{\mu\nu}$ element versus the angle θ between the interionic bonding and magnetic ordering axes [after R. Siemann and B.R. Cooper[3]].

The anisotropy of the Coqblin-Schrieffer interactions results also from the behaviour of their Fourier transforms. For example $J^{5/2,5/2}_{5/2,5/2}(\mathbf{q})$ (see Fig.2) shows a flat dispersion in the [001] direction unlike in the [110] and [111] directions. It corresponds to a weak coupling between and a rather strong one within the [001] planes.

Finally, we can say that the hybridization of the 4f electrons with band 5d electrons results in the strongly anisotropic f-f interactions of cerium ions.

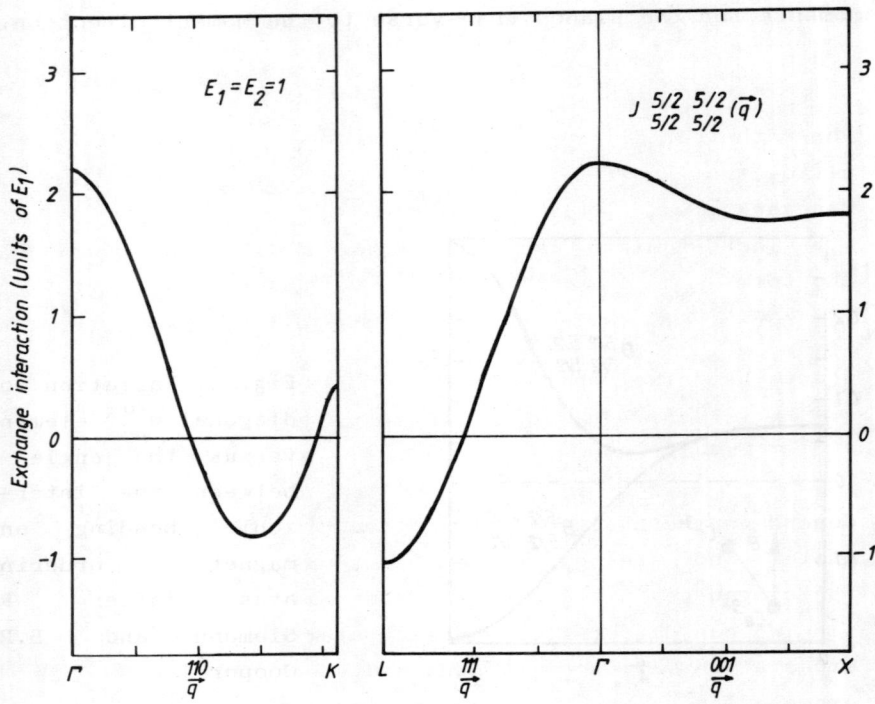

Fig. 2. Dispersion of the $J^{5/2\,5/2}_{5/2\,5/2}(\mathbf{q})$ - component of the Fourier transformed C-S exchange energy for the same nn and nnn interactions [after B.R. Cooper et al.[4]].

3.2. Valence Band Channel

The valence band channel in the p-f mixing model of Kasuya and Takahashi[3] also explains the strongly anisotropic indirect exchange interaction of cerium ions.

The p-valence band term
$$H_p = \sum_{k\mu\sigma} \varepsilon_{k\mu\sigma} |k\mu\rangle \langle k\mu|$$
is mainly described by the p states $|s=1/2,\ m_s,\ l=1, m_l; i\rangle$

p-band take the form:

$$|k\mu\rangle = 1/N \sum_p \sum_{jm_j} \sum_{lm} B(k\mu,jm_j) e^{ik\cdot R_p} \langle sm_s 1m_l | jm_j\rangle |sm_s,1m_l;p\rangle$$

where the parameters $B(k\mu,jm_j)$ are fitted by Kasuya et al. to the known valence band of LaSb calculated by Hasegawa[33].

The p-f mixing contribution to the Hamiltonian takes the form

$$H_{pf} = \sum_i \sum_{\Gamma\nu} \sum_{k\mu} |k\mu\rangle\langle k\mu|V|J\Gamma\nu;i\rangle\langle J\Gamma\nu;i|$$

where the anisotropic matrix elements

$$\langle k\mu|V|J\Gamma\nu;i\rangle = e^{ik\cdot R_p} \sum_{jm_j} B^*(k\mu,jm_j)\langle jm_j;k|V|J\Gamma\nu;i\rangle$$

depend on the charge distribution in the crystal field states and are expressed in terms of the two-center (pnictogen-cerium) integrals $(pf\sigma)$ and $(pf\pi)^3$.

We can differentiate the crystal field states by the expectation values of the associated angular momenta. The expectation value of the z-component of the total angular momentum in the two states $(\Gamma_8 1)$ of the quartet Γ_8 is nearly fully saturated and equal to $\pm 11/6$. In the remaining states $(\Gamma_8 2)$ of the quartet Γ_8, the expectation value of J^z is small and equal to $\pm 1/2$. In the states of the ground doublet Γ_7, the $\langle J^z\rangle$ value is equal to $\pm 5/6$. For the nearly fully saturated moment along the [001] direction, the 4f wave function is extended on the (001) plane. We will see later that in the p-f mixing model the states $\Gamma_8 1$ play the dominant role in explaining the anomalous magnetic properties of cerium monopnictides.

Hasegawa has shown for the La monopnictides[33] that the top of the p-valence band is at the Γ point, while the bottoms of the conduction $5d-t_{2g}$ bands are at the three X points of the Brillouin zone.

When the energy gap between the top of the valence band and the bottom of the conduction band is narrow, or the bands are slightly overlapping, the anisotropic p-f mixing pushing up some valence states, either creates the p-valence holes or increases their number. If the number of the p-holes is small, they are restricted to the Γ point only. Consequently the p-f mixing matrix elements are nonzero only between the crystal field states of the quartet Γ_8 and the analogous quartet spin orbital states $|P\ 3/2\ \Gamma\rangle$.

The energy shift of the $|P\ 3/2\ \Gamma\rangle$ states, which are pushed up due to mixing with the states of the quartet Γ_8, is given by[3]

$$\Delta E(P\ 3/2;i) = \frac{|\langle \Gamma_8;i|V|P\ 3/2;\Gamma\rangle|^2}{\Delta E_{4f}} X_0 X(\Gamma_8;i)$$

where X_0 denotes the concentration of Ce atoms and $X(\Gamma_8;i)$ - the thermal population of the Γ_8 states, ΔE_{4f} - the energy interval between the 4f level and the top of the valence band at the Γ point.

Kasuya and Takahashi[3] treating the p-f mixing as perturbation, have shown that the dominating Γ_8-Γ_8 interaction between nearest neighbours depends on two different charge distributions, i.e. $\Gamma_8 1$ or $\Gamma_8 2$, and moreover on the geometrical orientation of the pair — namely the ions of the pair can lie either in the same (001)-layer or in different (001)-layers.

The effective symmetry exchange interaction of cerium ions in the p-f mixing model approximately can be written in the form[3]

$$H_{f-f} = -\sum_{i,j}\sum_{\mu\nu} |\Gamma_8\mu;i\rangle|\Gamma_8\nu;j\rangle J^{\mu\nu}(R_{ij})\langle\Gamma_8\mu;j|\langle\Gamma_8\nu;i|$$

The strongest f-f interaction occurs between the quartet states $\Gamma_8 1$ which exhibit nearly fully saturated moment along the [001] direction. These crystal field wave functions are extended in the (001) plane and strongly mixed with the p states of pnictogens in the same (001) plane. The occupation of the particular cubic crystal field state $\Gamma_8 1$ which ensures the strongest p-f mixing, is one of the origins of the strong magnetic anisotropy with the [001] easy axis.

For a large number of the p-holes, there are some p-hole states in the vicinity of the Γ point. They can be responsible for the relatively small $\Gamma_7-\Gamma_7$ and $\Gamma_7-\Gamma_8$ exchange interactions.

The unusual anisotropy of f-f interactions results also from the anisotropic dispersion in the valence bands. The anisotropic p-f mixing pushes up a special type of valence bands and leads to their anisotropic dispersion.

The perturbed valence bands along the two Δ axes are shown in Figure 3. The shift of the valence bands is associated with an increase in the number of the appropriate holes and with a shift of the Fermi level from E_{F1} to E_{F2}. The dispersion of these valence bands is strongly anisotropic. For example, the Fermi wavevector K_{F_z} along the Δ_z axis, shown in Fig.3 is larger than the K_{F_x} along the Δ_x axis.

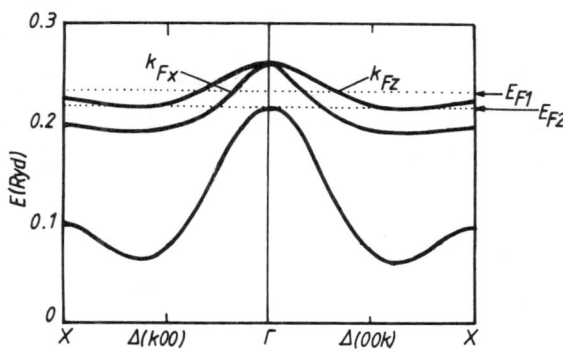

Fig.3. Valence bands for LaSb[3,33]. The Fermi levels E_{F1} and E_{F2} correspond to the 0.03 or 0.1 holes per cerium atom respectively [after H.Takahashi and T.Kasuya[3]].

Especially, in the limit of strong anisotropy, the Fermi

surface becomes cylindrical along [001] axis and we expect no interaction along the z axis.

Takahashi and Kasuya[3] examined also the dependence of the anisotropic $\Gamma_8 1 - \Gamma_8 1$ interaction on the number of the p-holes (n_h). They show that the ferromagnetic, intralayer interaction in the (001) plane becomes larger, while the interlayer nearest neighbour coupling J_1 becomes smaller with an increase in the number of the p-holes. Moreover, the interlayer next nearest neighbour interaction J_2 can even change its sign and become antiferromagnetic for a certain number of the p-holes. The tendency of changes of the coupling parameters illustrate the Table II.

TABLE II. The number of holes (n_h), f-f coupling parameters (J_0, J_1, J_2), for CeBi with crystal field splitting $\Delta=8K$ and $\langle J_z \rangle = 11/6$ ordering [after Takahashi et al.[3]]

n_h/Ce	J_0(k)	J_1(k)	J_2(k)
0.03	246.8	73.6	15.1
0.10	599.0	19.4	−0.4

The valence band distortion due to the p-f mixing, especially the anisotropy of the dispersion in the valence band, is much more profound in the ordered states than in the paramagnetic one. Finally we come to the conclusion that the shape of the crystal field states of 4f electrons and the anisotropic dispersion in the p-valence band and consequently the strong anisotropy of the Fermi surface are the essential origins of the strong magnetic anisotropy in the cerium monopnictides.

4. MICROSCOPIC MECHANISM OF A SECOND ORDER PHASE TRANSITION IN CERIUM MONOPNICTIDES.

The multichannel hybridization mediated exchange interactions determine also a mechanism of the magnetic spontaneous polarization in cerium monopnictides. At first, we discuss the microscopic aspects of the thermodynamical approach of Kasuya et al., who used the Landau theory. Next, we analyse the critical behaviour of the longitudinal bulk and surface magnetic excitations in vicinity of the Néel point.

4.1. Kasuya's Approach To The Magnetic Phase Transition In CeBi.

Kasuya et al. pointed out to the distinguished role of the $\Gamma_8-\Gamma_8$ channel of the p-valence holes in an analysis of the magnetic phase transition in CeBi[3].

In the paramagnetic phase there occurs degeneracy of the thermal populations of the ground doublet states Γ_7 and the excited quartet states Γ_8, respectively. The Néel point, is defined as a temperature at which the degeneracy in the Γ_8 quartet is removed and simultaneously a longitudinal magnetic moment appears in the system. The order parameter is defined as a deviation of the thermal populations $\chi(Q_i)$ of the Γ_8 states from their equilibrium value χ_Q^0. The quartet Γ_8 is distinguished by the particular $\Gamma_8-\Gamma_8$ channel of the indirect exchange coupling.

In order to find a temperature of the second order phase transition T_{cr}, Kasuya et al.[3] expanded the free energy around its equilibrium value χ_Q^0 up to the second order of the ordering parameter $\chi(Q_i)-\chi_Q^0$.

The critical point is defined as that at which the coefficient in the second order term of the free energy

vanishes.
Then[3]

$$k T_c = \frac{M^2 \chi_Q^0}{\Delta E_{4f}^2 \cdot a} \left[1/4\, D_{P\,3/2}(E_F) - \frac{2n(Pi)}{\Delta E_{4f}\sqrt{a}} \right].$$

where $M = \langle J=5/2, \Gamma_8 i|V|P3/2,i\rangle$; $D_{P3/2}(E_F)$ is the total density of states of the $|P\,3/2\rangle$ bands at the Fermi level; $n(P_i)$ is the number of holes for the $|P\,3/2\,i\rangle$ state, whereas the parameter

$$a = 1 + 4M\chi_Q^0\,(\Delta E_{4f})^{-2}.$$

The critical point increases with increasing value of p-f mixing matrix element M and thermal population χ_Q^0 of the quartet Γ_8. Moreover, to get a positive T_c, the total density of the quartet states $|p\,3/2\rangle$ at the Fermi level, has to be sufficiently great in comparison with the total number of holes in the $|p\,3/2\rangle$ states.

From the thermodynamical approach of Kasuya et al. it follows that the symmetry breaking in the Γ_8 states at temperature T_c leads to a simultaneously appearence of longitudinal magnetic moment. Therefore, the Landau type theory of a second order phase transition turns out to be associated with instability of the longitudinal magnetic excitations.

4.2. Low-Frequency Magnetic Dynamics

The Kasuya's approach presented in Section 4.1. indicate that the $\Gamma_8-\Gamma_8$ channel of the effective exchange interactions via the quartet p-hole spin-orbit states, plays a fundamental role in model interpretation of the magnetic phase transition in CeBi.

However in order to explain different magnetic properties of other cerium monopnictides it may be necessary to include in our model also other channels of the f-f exchange interactions. Especially the dynamical mechanism of magnetic phase transitions in other cerium monopnictides may be governed by other channels of exchange interactions.

Therefore, it seems quite natural to propose a phenomenological approach, which would enable us to take into account all the channels of the exchange interactions. To this purpose we should decompose the total angular momentum in the cubic crystal-field basis into the Curie-Langevin and Van Vleck components. For example the matrices of the total angular momenta; namely:

$$J^z = \begin{pmatrix} a & 0 & d & 0 & 0 & 0 \\ 0 & -a & 0 & -d & 0 & 0 \\ d & 0 & b & 0 & 0 & 0 \\ 0 & -d & 0 & -b & 0 & 0 \\ 0 & 0 & 0 & 0 & c & 0 \\ 0 & 0 & 0 & 0 & 0 & -c \end{pmatrix}, \quad J^+ = \begin{pmatrix} 0 & -a' & 0 & b' & 0 & 0 \\ 0 & 0 & 0 & 0 & c' & 0 \\ 0 & b' & 0 & a' & 0 & 0 \\ 0 & 0 & 0 & 0 & d' & 0 \\ 0 & 0 & 0 & 0 & 0 & e' \\ c' & 0 & d' & 0 & 0 & 0 \end{pmatrix}$$

(where $a = -5/6$, $b = 11/6$, $c = 1/2$, $d = \sqrt{20}/3$, $a' = 5/3$, $b' = -2/3\sqrt{5}$, $c' = -2\sqrt{5/3}$, $d' = 2/\sqrt{3}$, $e' = 3$ and $J^- = (J^+)^+$) can be decomposed into the following components

$$J^\alpha = j_\alpha^{\Gamma_7} + j_\alpha^{\Gamma_8} + j_\alpha^{vv}, \quad \alpha = z, +, -$$

Resorting to this decomposition we propose a Hamiltonian, which distinguishes various channels of exchange interactions by appropriate values of the parameter $F_{i\alpha,j\alpha'}^{mn}$, ($m,n = \Gamma_7, \Gamma_8, vv$; $\alpha = z, +, -$)

$$\mathcal{H} = \sum_i H_{CF}(i) - 1/2 \sum_{ij} \sum_{\alpha,\alpha'} \sum_{m,n} F_{i\alpha,j\alpha'}^{mn} (j_{i\alpha}^m)^+ j_{j\alpha'}^n - \sum_i h_i J_i .$$

This Hamiltonian enables us to discuss magnetic properties of system depending on the strength of different channels of the exchange interactions.

Especially, in order to discuss the mechanism of

magnetic phase transition we assume a set of dynamical variables, which generate apropriate longitudinal magnetic excitations and treated as slow variables in the vicinity of a critical point.

These excitations are the poles of the extended longitudinal dynamical susceptibility defined by the relation:

$$\langle J^z_{q,\omega} \rangle = \chi^{J^z J^z}_{q,\omega} h_{q,\omega}.$$

This total susceptibility can be decomposed into appropriate partial susceptibilities:

$$\chi^{J^z J^z}_{q,\omega} = \sum_{\Gamma_7, \Gamma_8, vv} \chi^{mn}_{q,\omega}$$

which are determined in MFA, by the following set of equations

$$\chi^{mn}_{q,\omega} = \chi^{mn}(\omega) \delta_{mn} + \chi^{mn}(\omega) \sum_k F^{mn}_q \chi^{kn}_{q\omega}.$$

Next, using the Mori's technique[35-38], we can discuss the low-frequency dynamics of the longitudinal excitations in the paramagnetic vicinity of the Néel point. As known, the critical dynamical behaviours of a system can be described by the spectral intensity $S_{q,\omega}$ expressed by

$$S_{q,\omega} = \text{cth} \frac{\omega}{2KT} \text{Im} \chi^{J^z J^z}_{q,\omega}$$

If a continuous phase transition occurs and if the order parameter is ergodic, the static limit of the dynamical susceptibility $\chi^{J^z J^z}_{q\omega}$ tends to infinity. In other words, at least one of the spectral intensity peaks moves towards $\omega=0$, and its half-width reaches zero, as well. The problem of dynamical mechanism of magnetic phase transition resolves into the problem of low-frequency dynamics of longitudinal magnetic excitations and fluctuations.

The final results[36-37] show evidently that the mechanism of magnetic phase transition essentially depends not only on such parameters as the crystal field splitting Δ and the Néel point but also on the relative strength of different channels of exchange coupling.

If the Γ_8- Γ_8 channel is distinguished, as it is in the case of heavy cerium monopnictides, we get for the same order of the crystal field splitting Δ = 10 K and the Néel point T_N= 10 K, the central mode mechanism of the phase transition. The magnetic order appears due to the slowing down of the intraquartet Γ_8 fluctuations - the width of the Γ_8 - central mode tends to zero if T approaches the Néel point[36-37].

In the case of CeP and CeAs the crystal-field splitting is of the order of 200 K, while the Néel point is of the order of 10 K (strictly 8,5 K for Ce and 7,5 K Ce). Moreover, it seems that any channel of interactions is distinguished and the appearance of magnetic order is caused by the slowing down of the intradoublet Γ_7 collective fluctuations[35-37].

It is also noteworthy that in the case of sufficiently large negative defects of the crystal field splitting Δ at the boundary, a surface longitudinal magnetic central mode appears for temperature T_s slightly higher than T_N. The half width of that surface central mode tends to zero if $T \longrightarrow T_s$ from above. It means that a spontaneous magnetic polarization at the surface occurs for the transition temperature which is slightly higher than the Néel point of the bulk[39-40].

5. EXPERIMENTAL VERIFICATIONS OF UNUSUAL f-f INTERACTION

The unusual anisotropy of indirect exchange interactions as well as their peculiar multichannel character can be confirmed and observed in inelastic and diffuse critical neutron scattering[1,7,8,41].

Let us discuss for example an anisotropic spectrum of magnetic excitations in CeBi which is unusual for a cubic systems. J. Rossat-Mignod et al.[1,41] show that the transverse excitations with the wave vector **q** perpendicular to the ferromagnetic (001) plane (**q** ∥ **k**), are dispersionless with the energy of about 1 THz. These excitations indicate that the coupling between the (001) planes is very weak (about 1K).

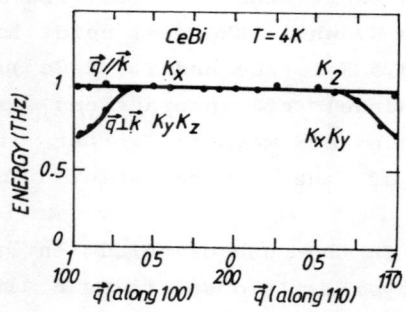

Fig.4. Magnetic exci - tation spectrum of CeBi at T = 4K [after J.Rossat-Mignod et al.[1]]

However, the same excitations with a wave vector **q** lying within the (001) plane exhibit a dispersion and moreover there is a minimum in their energy at the X-point of the Brillouin zone. Since the coupling between the ferromagnetic (001) planes in CeBi is relatively weak, we can consider these planes as quasi 2D systems. Then, the dispersion relation for magnetic excitations with a wave vector **q** lying within the (001) planes takes the form

$$\omega(q) = \Delta - |<1|J^-|0>|^2 J(q) .$$

The minimum of $\omega(q)$ corresponds to the maximum of the Fourier transform of the exchange integrals between ions in the (001) plane.

On the other hand the ferromagnetic ordering within the (001) planes correspond to the maximum of $J(q)$ for $q = [0,0,0]$.

This contradiction can be solved by assuming that the exchange parameter in the high frequency (Van Vleck) magnetic excitations are not described by the same exchange parameter as that which describe the static configurations. The spontaneous magnetic polarization is associated with the $\Gamma_8 - \Gamma_8$ channel of the exchange interactions, whereas the Van Vleck excitations should be associated with the $\Gamma_7 - \Gamma_8$ channel of the exchange interactions.

The anisotropy of the exchange interactions is observed also in the diffuse critical neutron scattering (DCNS) - elastic scattering of neutrons on short-range ordered spin fluctuations.

As known, above the antiferromagnetic ordering temperature, nearly total intensity of magnetic fluctuations is due to the longitudinal fluctuations. Hälg and Furrer[7,8] have shown that these fluctuations above the antiferromagnetic ordering temperature are highly anisotropic in cerium monopnictides. They performed the DCNS experiments on CeSb and CeAs. The appropriate cross section is proportional to the static limit of the longitudinal extended susceptibility:

$$\frac{d\sigma}{d\Omega} \sim kT\, F^2(q)\, [1-(q_z/q)^2]\, \chi^{J^z J^z}(q,\omega = 0)$$

The latter is related to the Fourier transform of the exchange interactions and displays the same symmetry, so, we can write

$$\chi^{J^z J^z}_{q,\omega=0} = \frac{\chi^{zz}}{1 - F^{zz}_q \chi^{zz}}$$

where χ^{zz} is the single-ion static susceptibility. Next, if we expand $F^{zz}(q)$ around the ordering wave vector Q and put

$$F^{zz}_{Q+q} \sim F^{zz}_Q - \left(\frac{d^2 F^{zz}_q}{dq^2}\right)_{q=Q} q^2$$

we arrive at the Ornstein Zernike formula

$$\chi^{J^z J^z}_{Q+q} \cong \frac{c_z}{x_z^2 + q^2} \quad ,$$

where $x_z^2 \sim \left(\frac{d^2 F^{zz}_q}{dq^2}\right)^{-1}_{q=Q}$.

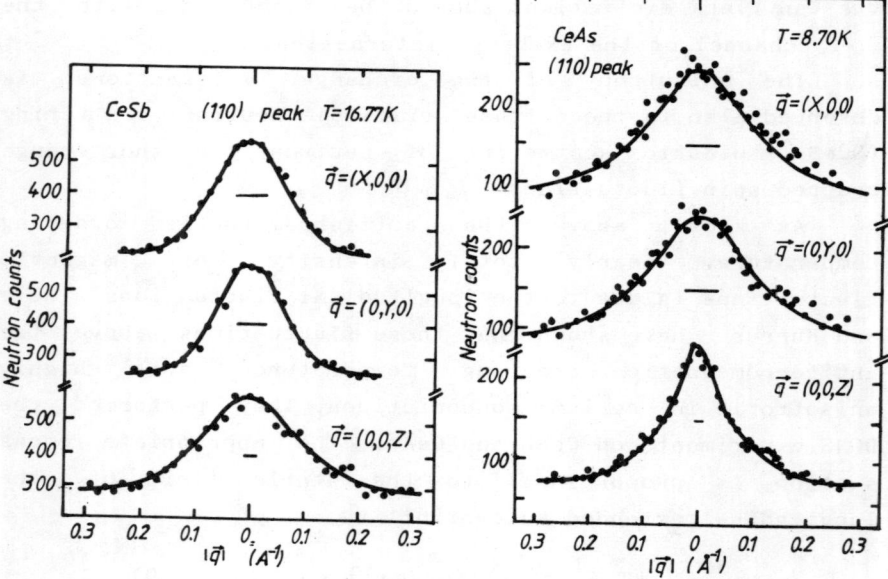

Fig.5. DCNS intensity distributions in CeSb and CeAs [after B.Hälg and A.Furrer[8]].

Finally, the cross secion takes the Lorentzian form in the q-space with its peak at the antiferromagnetic ordering wave-vector Q :

$$\frac{d\sigma}{d\Omega} \sim \frac{\varkappa_z}{\varkappa_z^2 + q^2} .$$

The halfwidth of this Lorentzian curve is equal to the inverse of the correlation length ($\varkappa_z = 1/\xi_z$). DCNS intensity distributions in CeSb and CeAs are shown in Fig.5 It is easy to see that the half-width of these Lorentzians depends on the direction of wave-vector q. We see also that the correlation length in CeSb in the plane (001) is greater than that in the direction [001]. It means that the exchange interactions within the (001) plane are greater than the interactions between them. For CeAs the ratio of the couplings is reversed.

Finally, we can say that the experimentally observed unusual anisotropy of the dispersion of transverse magnetic excitations and the anisotropy of the longitudinal fluctuations above T_N, confirm the hybridization mechanisms of the indirect exchange interactions.

6. CONCLUSIONS

Our review is confined to the discussion of the fundamental origins of anomalous properties of cerium monopnictides, namely the mechanisms of strongly anisotropic indirect exchange interactions. However, we also discuss some of the anomalous phenomena, such as the unusual process of magnetic spontaneous polarization, and the extraordinary anisotropy of the dispersion of magnetic excitations as well as that of magnetic longitudinal fluctuations above the Néel point.

Both models of hybridization mediated exchange interactions lead to similar results. Both of them show that the exchange interactions strongly depend on the shape of the 4f charge distribution and moreover, on the anisotropy of the dispersion of the band charge carriers and consequently the anisotropy of the Fermi surface.

It seems however, that in spite of a great progress which has been made recently in understanding of the mechanisms of the hybridization mediated exchange interactions a general theory explaining the all aspects of opening or blocking of distinct channels of the indirect exchange interactions is still needed.

Many problems still remain to be solved before we fully understand the unusual electronic and magnetic properties of cerium monopnictides. For example, as already known an important role is played by a shift of the special type of valence states due to the p-f mixing. It seems that assuming a non-spherical form of the mixing potential, we may obtain some further effects in relation with that aspect.

The influence of the single-site correlations U on the f-f exchange interactions should also be discussed. A full symmetry analysis of exchange interactions for the case of large number of the p holes, anisotropically distributed in the valence band, and for the case of different occupation of the $5d-t_{2g}$ conduction band states, should also be desired.

In our brief review we have not discussed some essential effetcs resulting from the f-electron delocalization without hybridization and also some effects of the hybridization induced distortion.

REFERENCES
1. J. Rossat-Mignod, P. Burlet, S. Quezel, J.M. Effantin, . D. Delacôte, H. Bartholin, O. Vogt, D. Ravot, J. Magn. Magn. Mat. 31-34, 398 (1983)
2. J. Rossat-Mignod, P. Burlet, S. Quezel, O. Vogt, Physica 102B, 237 (1980)
3. H. Takahashi, T. Kasuya, J. Phys. C18, 2697, 2709, 2721, 2731, 2745, 2755 (1985)
4. B.R. Cooper, R. Siemann, D. Yang, P. Thayamballi, A. Banerjea: in Handbook on the Physics and Chemistry of the Actinides, eds. A.J. Freeman and G.H. Lander, Elsevier Science Publishers B.V. 1985, p. 435
5. R.J. Birgenau, E. Bucher, J.P. Maita, L. Passel, K.C. Turberfield, Phys. Rev. B8, 5345 (1973)
6. B.R. Cooper, J. Less-Common Mat. 133, 31 (1987)
7. B. Hälg, A. Furrer, J. Appl. Phys. 55, 1860 (1984)
8. B. Hälg, A. Furrer, Phys. Rev. B34, 6258 (1986)
9. J.M. Wills, B.R. Cooper, Phys. Rev. B36, 3809 (1987); J. Magn. Magn. Mat. 54-57, 1049 (1986)
10. F. Hulliger, H.R. Ott, Z. Phys. B29, 47 (1978)
11. B.R. Cooper, J.M. Wills, N. Kioussis, Q.G. Sheng, ICM, Paris 1988.
12. P. Thayamballi, B.R. Cooper, Phys. Rev. B30, 2931 (1984)
13. H. Heer, A. Furrer, W. Hälg, O. Vogt, J. Phys. C12, 5207, (1979)
14. G. Busch, O. Vogt, Phys. Letters 25A, 449 (1967)
15. B.R. Cooper, O. Vogt, J. de Physique 32, C1-1026 (1970)
16. H. Bartholin, D. Florence, Tcheng-Si Wang, O. Vogt, Phys. Stat. Solidi a 24, 631 (1974)
17. P. Burlet, J. Rossat-Mignod, H. Bartholin, O. Vogt, J. Physique 40, 47 (1979)
18. J. Rossat-Mignod, D. Delacôte, J.M. Effantin, C. Vettier, O. Vogt, Physica 120B, 163 (1983)

19. T. Suzuki, M. Sera, H. Shida, K. Takegahara, H. Takahashi, A. Yanase, T. Kasuya, Valence Fluctuations in Solids, 1981, Amsterdam, North Holland, p. 255
20. J. Rossat-Mignod, P. Burlet, J. Villain, H. Bartholin, Tcheng -Si-Wang, D. Florence, O. Vogt, Phys. Rev. B16, 440 (1977)
21. J.W. Cable, W.C. Koehler, AIP Conf.Proc.5, (1972) p. 1381
22. B. Rainford, K.C. Tuberfield, G. Busch, O. Vogt, J. Phys. C1, 679 (1968)
23. N. Kioussis, B.R. Cooper, A. Banerjea, Phys. Rev. B (1987)
24. F. Lévy, Phys. Condens, Matter 10, 85 (1969)
25. F. Hulliger, M. Landolt, H.R. Ott, R. Schmelzer, J. Low Temp. Phys. 20, 269 (1975)
26. B. Hälg, A. Furrer, W. Halg, O. Vogt, J. Phys. C14, L961 (1981)
27. M. Sera, T. Fujita, T. Suzuki, T. Kasuya, Valence Instabilities, Amsterdam, North Holland, p. 435
28. B. Coqblin, J.R. Schrieffer, Phys. Rev. 185, 847 (1969)
29. B. Cornut, B. Coqblin, Phys. Rev. B5, 4541 (1972)
30. D. Ravot, A. Mauger, J.C. Achard, M. Bartholin, .J. Rossat -Mignod, Phys. Rev. B28, 4558 (1983)
31. J.R. Schrieffer, P.A. Wolff, Phys. Rev. 149, 491(1966)
32. R. Siemann, B.R. Cooper, Phys. Rev. Lett. 44, 1015(1980)
33. A. Hasegawa, J. Phys. C13, 6147 (1980)
34. K.W. Becker, P. Fulde, J. Keller, Z. Physik B28, 9 (1977)
35. M. Thomas, L. Kowalewski, J. Magn. Magn. Mat. 76-77, (1988)
36. L. Kowalewski, M. Thomas, J. Magn. Magn. Mat. 63-64, 15 (1987)
37. M. Thomas, L. Kowalewski, to be published
38. H. Mori, Prog. Theor. Phys. 33, 423 (1965)
39. R. Wojciechowski, L. Kowalewski, Physica B (1988)
40. R. Wojciechowski, L. Kowalewski, J. Magn. Magn. Mat. 76-77, (1988)
41. J. Rossat-Mignod, D. Delacôte, J.F. Effentin, C.Vettier, O. Vogt,Physica 120B, 163 (1983)

PROSPECTS FOR APPLICATIONS OF OXIDE SUPERCONDUCTORS
(EXTENDED ABSTRACT)

A. I. Braginski[*]
Westinghouse R&D Center
1310 Beulah Road
Pittsburgh, Pennsylvania 15235
USA

ABSTRACT

Critical parameters and superconducting characteristic lengths are the criteria for superconducting material usefulness. It is shown that oxide superconductors do not meet, at present, some of the criteria.
Projections for attainable properties indicate that large scale, HTS power apparatus is less feasible than small scale electronic applications.

1. CRITICAL PARAMETERS AND CHARACTERISTIC LENGTHS

The recent discoveries of superconductivity at relatively high cryogenic temperatures, attaining T_c = 125 kelvin in early 1988, resulted in predictions of an enormous and almost immediate impact on the electric power and electronic technologies due to the reduction in cryocooling requirements when operating at the temperature of liquid nitrogen, T = 77K, instead of liquid helium, 4.2K. The purpose of this paper is to assess the feasibility of applications in these two areas by comparing the requirements for a useful superconductor's critical parameters, characteristic lengths scales and physico-chemical, mechanical etc. characteristics, with the presently known properties of the new high-T_c oxide compounds.

In the large-scale, electric power area, type II superconductors are and will be used predominantly in the form of composite, multifilamentary wires/cables of which electromagnet windings can be fabricated. In the case of transmission cables and some dc magnets, composite tape conductors are also acceptable. At the temperature of operation, which from thermal stability and current density considerations should not exceed T_{op} = 0.5 to 0.7 T_c, the essential electrical requirement is that of a highest possible critical current density, J_c between 10^5 and 10^6 A/cm^2 in the superconductor, at the intended magnetic field of operation in the mixed state, $H < H_{c2}$, where H_{c2} is the upper critical field. A highest possible $H_{c2}(T_{op})$ is desired

---

[*]Supported in part by AFOSR Contract No. F49620-88-C-0039.

since one of the principal benefits is the ability to generate strong magnetic fields. To attain very high J_c in the superconductor, fluxoids must be pinned in the condensation energy minima formed by a dense network of random defects of a size comparable to the superconducting coherence length, ξ. This characteristic length must be large enough for the atomic-scale defects not to disrupt superconductivity and thus impede the flow of supercurrent. The most essential mechanical requirement is to have a strong and ductile wire capable of withstanding enormous Lorentz forces without damage and deterioration of the critical surface (T_c-H_{c2}-J_c) under mechanical load and deformation. Stable, low ac loss, composite conductors can be obtained by proper geometrical design using low-temperature superconductors (LTS) with acceptable critical and mechanical properties and a high-conductivity, stabilizing metal matrix (Cu, Al).[1] In radio-frequency power applications, e.g. accelerator cavities, a highest possible lower critical field, H_{c1}, is required to minimize rf losses.

In small scale, low power, electronic applications, Josephson tunnel junction devices and circuits are patterned from films and multilayered film structures. High fields are not required, and $T_{op} < 0.5$ to $0.7 \, T_c$ is again the rule, so that the critical parameter of importance is the highest possible self-field $J_c(T_{op}) = 10^6$ to 10^7 A/cm^{2}s. Longest possible ξ minimizes sensitivity of T_c and energy gap, Δ, to surface/interface disorder and degradation while shortest possible penetration depth, λ, minimizes flux trapping in Josephson junctions and maximizes the velocity of signal propagation in transmission lines.[2] Type I superconductors with $\xi \gg \lambda$ would be ideal, if not for their low T_c and mechanical softness. A hard, high-melting point refractory material having a crystalline morphology and low-temperature properties resistant to thermal cycling is required for long-term device stability.

Low temperature superconductors (LTS) used in both classes of applications have cubic crystal structures, are electrically nearly isotropic and polycrystalline. Niobium and some of its alloys (Nb-Ti) are ductile while niobium and vanadium compounds are brittle.

2. PROPERTIES OF HIGH-TEMPERATURE SUPERCONDUCTORS (HTS)

All materials in this category which have $T_c > 30K$ are cuprates i.e. oxide compounds of copper, an alkaline earth metal and other elements.[3] The number of materials and their confirmed T_c's are growing due to a very high level of the worldwide research effort. The confirmed record T_c is presently 125K, in $Tl_2Ca_2Ba_2Cu_3O_y$,[4] but the most researched material to date is $Y_1Ba_2Cu_3O_7$ (YBCO) and its derivatives where Y is replaced by other rare-earth elements. Their common T_c is 90-95K. The Bi- and Tl-based cuprates (BSCCO and TBCCO) crystallize in several structures where T_c increases with the number of stacked Cu-O planes. All HTS cuprates known are ceramic materials, very brittle and environmentally rather unstable. All have orthorhombic or tetragonal crystal structures which are perovskite-related. They exhibit strongly anisotropic electrical conductivity, J_c, and H_{c2} estimated at 1000 to

3000 kOe in the favorable a-b crystal plane at T = 0K and, in YBCO, up
to 300 kOe at 77K. The coherence length is also anisotropic and
extremely short, comparable to the interatomic spacing in the direction
normal to the a-b plane. Due to this short range of superconducting
interactions, atomic-scale defects disrupt superconductivity. For
example, the transfer of current between crystallites in
polycrystalline materials is severely limited by imperfect grain
boundaries. This very short ξ is perhaps the single most fundamental
limiting parameter for most applications. However, the intrinsic
(depairing) J_c limit is high, in the 10^8 A/cm^2 range, and self-field J_{cs}
$\leq 5 \times 10^6$ A/cm^2 has been obtained at 77K in the a-b plane of nearly
single-crystalline films.

3. LARGE-SCALE APPLICATIONS

Table 1 is a listing of large-scale applications of
superconductivity. These application concepts were developed for LTS
and a standard temperature of operation at or in the vicinity of 4.2K.
Most cases were developed to a successful prototype demonstration and
some to manufacturing for the scientific, medical and industrial
markets. The range of current densities in the superconductor was
1×10^5 to 1×10^6 A/cm^2 (a factor of 2 to 5 less in stabilized wires)
in fields between self and H_{op} = 120 kOe, most often about 50 kOe. In
most cases, field applications did not follow the prototype development
due to economic rather than technical constraints, due in part to the
necessity of operation in liquid helium.

Due to the extremely short ξ, the current transfer in bulk
polycrystalline aggregates of HTS materials is hampered by Josephson
weak-links occurring at imperfect grain boundaries and other defects.
In wire-shaped ceramic cores, best J_{cs} at 77K are typically in the
range of 1 to 5×10^3 A/cm^2 and dwindle to negligible values in even
weak magnetic fields. Melt-textured growth of YBCO used to minimize
the effect of grain boundaries by reducing their number, the impurity
content and by partial alignment[5] was claimed to produced some
improvement: at 77K J_{cs} = 1.7 $\times 10^4$ A/cm^2 and J_c = 10^3 A/cm^2 at 10 kOe.
A related, perhaps more promising approach to a practical conductor
fabrication is the pulling of textured or single-crystalline fibers by
the laser-heated pedestal growth method which[6] achieved
J_{cs} = 4×10^4 A/cm^2 near 70K in BSCCO. These values fall still far
short of what is needed and the approaches do not appear readily
extendable to economical wire fabrication. Conceptual designs of some
apparatus of Table 1 were developed[7] to demonstrate practicality of HTS
use with $J_c(H_{op})$ of only 10^4 A/cm^2 at $H_{op} \geq 20$ kOe but even such HTS
wire performance does not yet appear realistic. The HTS brittleness
will also impose strenuous requirements on the conductor design,
manufacturing and handling.

In conclusion, until the feasibility of practical HTS composite
conductors is demonstrated, the large scale applications of
superconductivity will still be based on LTS and liquid helium
cryogenics. Consequently, their impact on economy will remain limited.

Table 1 — Most Viable Large Scale Applications of Superconductivity (LTS)

Energy Systems:	Status
• Electric power generators	p
• Electric energy storage	p
• Electric power transmission lines and other TDS gear	p
• Nuclear fusion magnets	p
• High energy physics magnets and cavities for particle accelerators	m/u
• Magnetohydrodynamic power plant magnets	p/u
Transportation:	
• Motors for ship propulsion	p
• Levitated suspension and linear motors for trains	p
Medicine:	
• High field magnets for magnetic resonance imaging	m/u
Other Fields:	
• Magnetic ore separation	m/u
• Compact power supplies/generators for airborne and space platforms	p
• Electromagnetic projectile launchers	c

Legend: c - concept, p - prototype, m/u - manufactured and used.

4. SMALL SCALE APPLICATIONS - ELECTRONICS

Table 2 contains a listing of viable LTS electronic components and functions with examples of feasible applications. Of these, only instrumentation apparatus such as SQUID (superconducting quantum interference device) magnetometers, volt standards and, recently, sampling oscilloscopes/reflectometers are manufactured and in use. The perceived obstacles to broader field use have been many, with the helium cryogenics dominant, especially in portable equipment.

Prospects for HTS electronic components are rather bright since the devices operate in self- or low magnetic fields and the use of epitaxial HTS thin films having (in one plane) the required high J_{cs} of $\geq 10^6$ A/cm^2 up to 77K is possible. Cryocooling to temperatures below 77K but within the range of efficient, single-stage cryocoolers ($T \geq 45K$) makes it possible to meet the $T_{op} \leq 0.5\ T_c$ requirement. The material problems which must be solved, however, are: the observed high radio frequency surface losses (in the gigahertz range), the high level of low-frequency 1/f electronic noise seen in both bulk and thin film materials (attributed mostly to flux motion or flux trapping/untrapping) and degradation of superconductivity within several coherence lengths from film surfaces and interfaces with insulators and semiconductors. Mechanisms of excess rf loss and noise in HTS are not yet understood sufficiently but the magnitude of these undesirable effects appears to be related to the presence of grain boundaries and to decrease with improved crystalline quality of epitaxial films. It can be expected that truly single-crystalline, defect-free films will perform adequately and will also have undegraded surfaces/interfaces.

Table 2 — Viable Electronic Applications
of Conventional Superconductors

Generic Circuit Components/Functions:
- Magnetic field (SQUID) and radiation detector
- Logic gate/switch
- Shift register
- A/D converter
- Phase shifter
- Mixer
- Antenna
- Oscillator
- Parametric amplifier
- Transmission line
- Delay line
- Chirp transform filter
- Convolver
- Correlator

Applications:
- Ultrafast- and super-computers
- Medical — contactless sensing, encephalography
- Instrumentation — sampling oscilloscope/reflectometer, spectrum analyzer, volt standard, magnetometer
- Conventional radar
- Gigabit rate communication
- Infrared imaging systems
- Magnetotellurics, magnetic anomaly detection.

Low losses at microwave frequencies, two orders of magnitude below copper at the same f and T, will make it attractive to develop and use HTS passive devices: antennae, transmission and delay lines, filters, resonators etc. (Table 2). In these applications λ = 100 to 200 nm sets the depth scale for film perfection, a manageable requirement. Sufficiently low 1/f noise energy at T_{op}, below 10^{-29} to 10^{-30} Joule/Hz at 1 Hz will be required to make rf and dc SQUID detectors viable. Improved flux pinning and elimination of electron traps appear possible. Undegraded superconductivity at the film interface with insulators will make Josephson tunnel junctions and digital circuits of Table 2 practical. However, this goal will be much more difficult to attain than the loss and noise reduction since short ξ sets the depth scale. Equally or even more difficult will be to attain good interfaces with semiconductors. Progress in this area would lead to hybrid circuits of passive superconductor elements and interconnect lines with active devices such as MOSFET, MODFET and CMOS which operate advantageously at low temperatures of 45 to 77K due to enhanced carrier mobility and have a potential for high gain, wide bandwidth and reduced noise.[8] It is also possible that novel transistor-like 3-terminal devices incorporating superconductors and semiconductors would emerge.

It should be noted that HTS tunnel junction circuits, even if feasible, will not be particularly suitable for high-density digital computer circuits. While LTS circuits are fast (a few picosecond switching is typical of Josephson device gates) their biggest advantage over semiconductor technology is in the power dissipation being reduced by two to three orders of magnitude. This is so since the LTS energy gap Δ(LTS) is only a few millivolts, compared with the order of one volt in the semiconductor. Consequently, for the presently used logic, the low power density advantage will largely disappear when substituting HTS since the power dissipation is expected to increase with K^2 where $K = Δ(HTS)/Δ(LTS) > 10$.[9] The prejudice against using liquid helium cryogenics relates more to psychological than to technical or economic obstacles, especially in the case of large, stationary systems. The potential of LTS digital electronics is considerable and yet to be exploited.[10]

In conclusion, early electronic applications of HTS can be expected, especially as passive elements and SQUID's. Circuits using Josephson tunnel devices and hybrids appear, at present, less feasible. The economic impact of HTS electronics is likely to remain limited until integration with semiconductors becomes practical. Current interest in superconductivity may also lead to a broader use of LTS electronics.

5. REFERENCES

1. Wilson, M. N. in "Superconductor Materials Science", S. Foner and B. B. Schwartz, editors, Plenum, NY 1981, 63-131.

2. Beasley, M. R. and Kircher, C. J., ibidem, 605-684.

3. Bednorz, J. G. and Muller, K. A., Z. Phys. B. - Condensed Matter **64**, 189 (1986); Wu, M. et al., Phys. Rev. Lett. **58**, 908 (1987).

4. Parkin, S. S. et al., Phys. Rev. Lett. **60**, 2539 (1988).

5. Jin, S. et al., Proc. Symp. HTS (Cincinnati), Am. Cer. Soc., May 1988 (in print).

6. Feigelson, R. S. et al., Science **240**, 1642 (1988).

7. Wolsky, A. M. et al., "Advances in Applied Superconductivity: A Preliminary Evaluation of Goals and Impacts," Argonne National Laboratory Report ANL/CNSV-64, January 1988.

8. Van Duzer, T., Cryogenics **28**, 527 (1988).

9. Hasuo, S., Presentations at the 1988 FED Workshop, Miyagi-Zao, June 1988 and ASC, San Francisco, August 1988 (unpublished).

10. Hasuo, S., "High-Speed Josephson Integrated Circuit Technology," IEEE Trans. **MAG-25**, (1989) - to be published.

MAGNETIC ORDER AND SUPERCONDUCTIVITY IN $RBa_2Cu_3O_z$

I. Felner, Y. Wolfus,* E.R. Bauminger and I. Nowik
Racah Institute of Physics, The Hebrew University, Jerusalem, Israel
*Department of Physics, Bar-Ilan University, Ramat Gan, Israel.

Magnetic susceptibility, powder x-ray diffraction and specific heat measurements were used to study the high-temperature superconductor $YBa_2(Cu_{0.91}Fe_{0.09})_3O_{7.1}$ and the nonsuperconductor quenched $YBa_2(Cu_{0.91}Fe_{0.09})_3O_{6.1}$ samples. The crystal structure for both samples is tetragonal with almost the same lattice parameters. No magnetic spin glass state of iron is observed in the two samples at low temperatures and the quenched sample exhibits antiferromagnetic behaviour with $T_N=415$ K. For the superconducting sample glassy features appear below a temperature, Tg, whose field dependence differ substantially from that observed in spin glasses. Tg scales with applied field as $H^{\frac{1}{4}}$. The linear term γ obtained from the specific heat studies is extremely high: $\gamma = 73$ mJ/mole K^2.

Mossbauer studies of Fe^{57} in $RBa_{2-y}K_y(Cu_{1-x}Fe_x)_3O_z$, with R=Y and Pr; y=0 and 0.5; x=0.01, 0.05 and 0.09 and z between 5.9 and 7.1, have been performed. The majority of iron ions enters the Cu(1) site and yield six quadrupole doublets, following the six oxygen coordination around the iron ions. A minority of the iron ions enter the Cu(2) site and reveal its magnetic order. For R=Y, y=0, x=0.1, T_N equals 280 K and 415 K for z=6.5 and 6.1 respectively. The magnetic moments lie in the basal plane. In tetragonal, oxygen rich $PrBa_2(Cu_{0.9}Fe_{0.1})_3O_{6.9}$, $T_N=325$ K and in superconducting

$YBa_2(Cu_{0.91}Fe_{0.09})_3O_{7.1}$ there is no magnetic order. In nonsuperconducting $YBa_{1.5}K_{0.5}(Cu_{0.95}Fe_{0.05})_3O_{6.1}$, two distinctly inequivalent magnetic iron sites are observed, probably corresponding to iron in the Cu(2) site with different Ba-K neighbours.

Introduction

Recent susceptibility and magnetization measurements on the high T_c materials, have shown certain similarities to well known spin glass materials.[1-2] For example the shielding zero field cooled (ZFC) branch of the magnetization is irreversible, metastable and lower than the field cooled (FC) Meissner branch, i.e. it is more strongly diamagnetic. The temperature T_g which is defined as the inflection point of the two branches, is magnetic field dependent and scales with the applied field as $H^{2/3}$.[3] It soon became clear that the granular nature of these materials and the short coherence length are responsible for the existence of internal Josephson junctions at the twin boundaries, and are the origin of the so called, superconducting glassy state.[4]

It has been reported by several groups[5-7] that partial substitution of Cu in the orthorhombic superconductor $YBa_2Cu_3O_7$ by Fe, progressively decreases the T_c of the system, and for Fe concentration higher than 2% the crystal structure transforms to tetragonal. Mossbauer[8] and neutron diffraction[9] measurements show definitely that the Fe ions substitute preferentially for the Cu(1) sites located on the basal plane of the structure and only about 10-20% of Fe substitute for Cu in the Cu(2) sites on the CuO_2 planes. Moreover, low temperature Mossbauer studies of samples doped with more than 5% Fe, show[8, 10-12] magnetic broadening (around 10-20K) at temperatures which depend on Fe concentration, an indication of magnetic ordering. At 4.1K the spectra become more complicated showing a distribution of magnetic hyperfine fields which are reminiscent of spectra for spin glass systems. It was assumed that spin glass short range magnetic order of Fe ions coexists with superconductivity in the materials at 4.1K[12-13]. These two independent spin glass features in the Fe doped samples motivated the present work.

In order to clarify, whether the two spin glass phenomena are connected or not, extensive magnetic susceptibility measurements at low temperatures and low magnetic fields have been carried out on both the fully oxygenated superconducting (S.C.) $YBa_2Cu_{2.73}Fe_{0.27}O_{7.13}$ and on the quenched nonsuperconducting $YBa_2Cu_{2.73}Fe_{0.27}O_{6.1}$; both samples were taken from the same batch. The Mossbauer spectra of the quenched sample at 4.1K also show the same complicated shape mentioned above[8]. The bulk susceptibility measurements on both samples show no evidence for spin glass or long range magnetic ordering of the Fe ions at low temperatures. In addition, the glassy behaviour of the superconducting state was studied. It was found that T_g scales with the applied magnetic field as $H^{1/4}$ in contrast with the 2/3 power law found in other systems[3]. The specific heat measurements showed no evidence of any magnetic transitions and the γ value obtained from the linear region of $C_P(T)/T$ vs T^2 was 73 mJ/mole K^2, remarkably larger than in the pure $YBa_2Cu_3O_7$ sample.

Mossbauer studies of Fe^{57} in $RBa_{2-y}K_y(Cu_{1-x}Fe_x)_3O_z$, with R=Y or Pr, y=0.0 and 0.5, x=0.01, 0.05 and 0.09, and z=5.9 to 7.1 at temperatures between 4.1 K and 480 K are also reported. These studies show that the fraction of iron ions which enter the Cu(2) site order magnetically in the non superconducting compounds, with different ordering temperatures. In $YBa_2Cu_3O_z$ the iron spectra reveal magnetic order for $z \leq 6.5$, with $T_N \leq 415$ K. There is magnetic order in the Cu(2) site in tetragonal nonsuperconducting $PrBa_2Cu_3O_{6.9}$, with $T_N = 325$ K. There is no magnetic order in superconducting $YBa_2Cu_3O_{6.9}$. In $YBa_{1.5}K_{0.5}Cu_3O_z$ magnetic order in the Cu(2) site exists up to 450 K. Two inequivalent magnetic Cu(2) sites are observed in this case. They are explained in terms of the different local environment of the Cu(2) ions, according to the number of K neighbours. The results indicate that the Ba-K ions play either an active or passive (rejecting oxygen) role in the magnetic exchange within the Cu(2) planes. Iron ions in the Cu(1) site are not ordered magnetically in either the superconducting or the quenched nonsuperconducting sample. The magnetic hyperfine structure observed in the Fe ions replacing Cu(1) in the nonsuperconducting samples with both x=0.01 and x=0.09 as well as in the superconducting samples with x=0.09, may be due to long paramagnetic spin relaxation times, consistent with the magnetic and specific heat measurements mentioned above. It seems that the lower the superconducting transition temperature, the longer the spin relaxation times.

Experimental Details

The samples were prepared by conventional methods[5]. Four variations for the preparations of $YBa_2(Cu_{1-x}Fe_x)_3O_z$ were employed: (a) When slow cooling from 950°C in an oxygen atmosphere was applied, fully oxygenated superconducting samples were obtained. Based on the oxygen content measured in iron free samples, z=6.9 was assumed for samples with x=0.01. In samples with x=0.1, the exact iron and copper concentrations were determined by chemical analysis using atomic absorption spectrometry. The Fe concentration was slightly less than the nominal composition. According to neutron diffraction studies[9] the oxygen concentration for the fully oxygentated $YBa_2Cu_3O_x$ doped with 10% Fe was 7.13. The chemical formula deduced for this tetragonal S.C. sample is therefore $YBa_2(Cu_{0.91}Fe_{0.09})_3O_{7.1}$. (b) When the samples were quenched from 800°C, the tetragonal phase was obtained and according to independent oxygen content estimates (based on weight loss) z=6.1 in these samples. Surprisingly, no measurable differences were observed between the lattice parameters of the two tetragonal samples (a) and (b) which are a=3.874(2)Å and c=11.70(3)Å. (c) For the samples quenched from 600°C, which are tetragonal and not superconducting for x=0.09, z=6.5[14] was estimated, and (d) for samples which were cooled in a vacuum furnace z=5.9[15] was estimated. The $PrBa_2(Cu_{1-x}Fe_x)_3O_z$ samples were cooled slowly and fully oxygenated and z=6.9 was estimated. For the $YBa_{1.5}K_{0.5}(Cu_{1-x}Fe_x)_3O_z$ samples, which were cooled slowly, iodometric analysis was used to determine the total amount of Cu and the concentration of Cu^{+3}, and allowed the calculation of the total amount of oxygen.[16] This chemical analysis yielded z=6.5 in the superconducting samples and z=6.1 in the samples quenched from 800°C. In all samples, the absolute z values are estimated within an accuracy of 0.1. All the samples were analyzed by X-ray diffraction measurements and their lattice parameters were obtained. These studies showed that all samples were single phase.

The dc susceptibility measurements were carried out in a commercial S.H.E. SQUID magnetometer and in a 155 PAR vibrating sample magnetometer in various fields between 10 Oe and 40 kOe as a function of temperature in the range 4.2-300K. The magnetization was measured by two different procedures: (a) The sample was zero field cooled (ZFC) to 5K, a field was applied, and the magnetization of the shielding branch was measured as a function of temperature. (b) The sample was field cooled (FC) from above T_c in a field and the Meissner branch was measured. The heat capacity was measured over the temperature range 1.5-70K in an automated adiabatic calorimeter employing the Nernst step heating method.

A conventional, constant acceleration, Mossbauer drive with a 100 mCi ^{57}Co(Rh) source at room temperature and a Harwell proportional counter were used. Velocity calibration was performed with a metallic iron foil at room temperature. Isomer shifts are quoted with respect to this absorber. Mossbauer spectra were recorded between 4.1 K and 450 K.

Experimental results and discussion

1) $YBa_2(Cu_{1-x}Fe_x)_3O_z$; x=0.01 and 0.09; z=5.9-7.1

On these compounds, Mossbauer and detailed magnetic studies were performed:

I. Mossbauer Studies:

a) $YBa_2(Cu_{.99}Fe_{0.01})_3O_z$ with $z=5.9$ to 6.9.

In all these compounds the iron replaces Cu(1) ions[8] and no well defined magnetic subspectrum is found in most of these compounds at any temperature above 4.1 K. The spectra obtained at 90 K and above are all composed of 6 quadrupole doublets, corresponding to 6 inequivalent iron sites, due to different oxygen neighbour configurations in the Cu(1) site[8]. The relative sizes of the quadrupole splittings in the six sites agree with point charge calculations for Fe^{3+}[8]. At 4.1 K, the spectra for $z \leq 6.5$ show complicated structure. These spectra may be due to spin-glass-like order of the iron in the Cu(1) site, but can also be interpreted as due to paramagnetic slow spin relaxation. Some of the spectra obtained at 4.1 K with different z values were shown in Fig. 2 of Ref. 8. In the spectra obtained with z=5.9 samples, a weak but well defined magnetic sextet was observed up to 430 K (Fig. 1). In these samples a larger percentage of the iron enters the Cu(2) site, and thus yields the sextet as explained below.

b) $YBa_2(Cu_{0.91}Fe_{0.09})_3O_z$ $z=6.1; 6.5$ and 7.1

Recent neutron-diffraction measurements[17] on $YBa_2Cu_3O_{6.1}$ show the existence of long-range three dimensional antiferromagnetic order of the Cu spins at the Cu(2) sites. The magnetic moments are within the CuO_2 plane and the planes are coupled antiferromagnetically along the c axis. T_N is about 410K.[17]

In the samples with $z \leq 6.5$, which are not superconducting, Mossbauer measurements show a well defined magnetic sextet, which accounts for about 25% of the spectral area. This spectrum is attributed to iron,

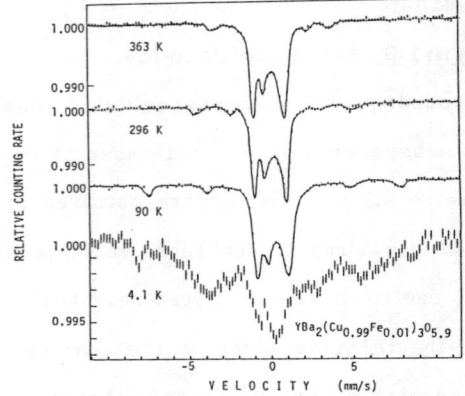

Fig. 1 Several Mossbauer spectra of Fe^{57} in $YBa_2(Cu_{0.99}Fe_{0.01})_3O_{5.9}$.

which now also replaces copper in the Cu(2) site and orders antiferromagnetically.[17] All iron ions in the Cu(2) site are equivalent in terms of oxygen environment and yield a well defined magnetic spectrum. At 4.1 K the magnetic hyperfine field is 514 kOe and the quadrupole interaction is $eqQ/4=-0.17$ mm/s. The same values are obtained for all other samples discussed in this paper. As the temperature is raised, the magnetic splitting decreases and the Neel temperature in each of the compounds can be obtained. The Neel temperature depends on the oxygen content, increasing as z decreases. No magnetic sextet is observed in superconducting $YBa_2(Cu_{.91}Fe_{0.09})_3O_{7.1}$. Fig. 2, Fig. 3 and Fig. 4 show the spectra obtained in $YBa_2(Cu_{1-x}Fe_x)_3O_z$ with z=6.1, z=6.5 and z=7.1 at various temperatures. The well defined magnetic subspectrum is clearly seen and the change of the size of the hyperfine field can easily be followed. A Neel temperature of 415 K is obtained for $YBa_2(Cu_{.91}Fe_{0.09})_3O_{6.1}$. In $YBa_2(Cu_{.91}Fe_{0.09})_3O_{6.5}$ similar spectra are observed, yet the magnetic ordering temperature in this compound is $T_N=280$ K. The electric field gradient at the Cu(2) site according to point charge calculations is positive and points along the c axis. The measured negative effective quadrupole

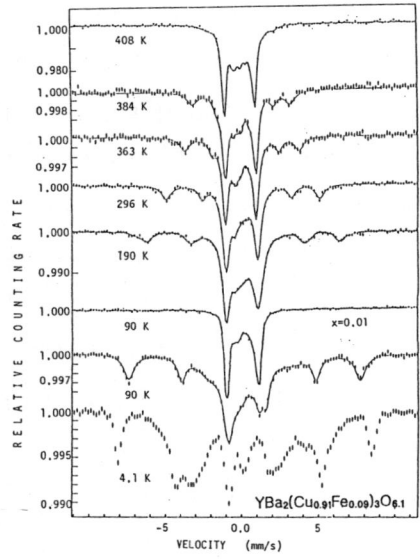

Fig.2 Mossbauer spectra of Fe^{57} in $YBa_2(Cu_{0.91}Fe_{0.09})_3O_{6.1}$ at several temperatures. Note that x=0.09 in all spectra, except at 90 K, where the spectrum with x=0.01 is also shown.

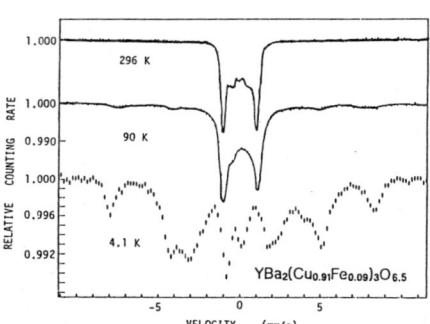

Fig.3 Mossbauer spectra of Fe^{57} in $YBa_2(Cu_{0.91}Fe_{0.09})_3O_{6.5}$ at several temperatures.

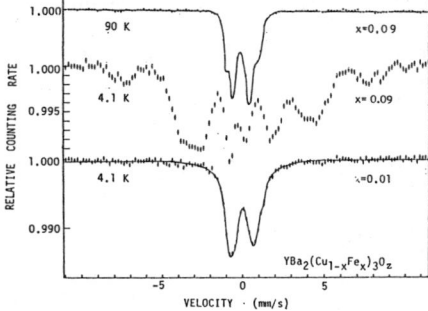

Fig.4 Mossbauer spectra of Fe^{57} in $YBa_2(Cu_xFe_{1-x})_3O_z$ (with z=6.9 for x=0.01 and z=7.1 for x=0.09).

interaction proves that the magnetic moments lie in the basal plane.[17] No magnetic subspectrum could be seen in the $YBa_2(Cu_xFe_{1-x})_3O_{7.1}$ samples. Fig. 5 shows the hyperfine fields as function of temperature in the different samples. The lines were extrapolated to give an estimate of the ordering temperatures. The fact that the ordering temperature of Fe in $YBa_2(Cu_{0.91}Fe_{0.09})_3O_{6.1}$ agrees so well with the neutron diffraction results[17] proves that through the iron, the magnetic ordering of copper in the Cu(2) site is indeed measured. The ordering temperature is not affected by the presence or concentration of the iron impurity. All spectra at 4.1 K show, besides the well defined sextet, a complicated central subspectrum, similar to those observed in the $YBa_2(Cu_{0.99}Fe_{0.01})_3O_{6.1}$ samples, which may be attributed to magnetic spin-glass-like order of the iron in the Cu(1) site or to long spin relaxation time phenomena. The best way to distinguish between the two assumptions is to perform detailed magnetic measurements on both superconducting and quenched samples at low temperatures and the detailed magnetic studies are described below.

II. Magnetic studies:

(a) The quenched, non-superconducting $YBa_2(Cu_{0.91}Fe_{0.09})_3O_{6.1}$

In Figure 6 the FC and ZFC magnetic curves of the quenched sample are presented, and it seems clear that there is no sign of spin glass magnetic behaviour of Fe ions at low temperatures, although the Mossbauer spectra show broadening and magnetic hyperfine structure at low temperatures. The coincidence of the ZFC and FC curves also

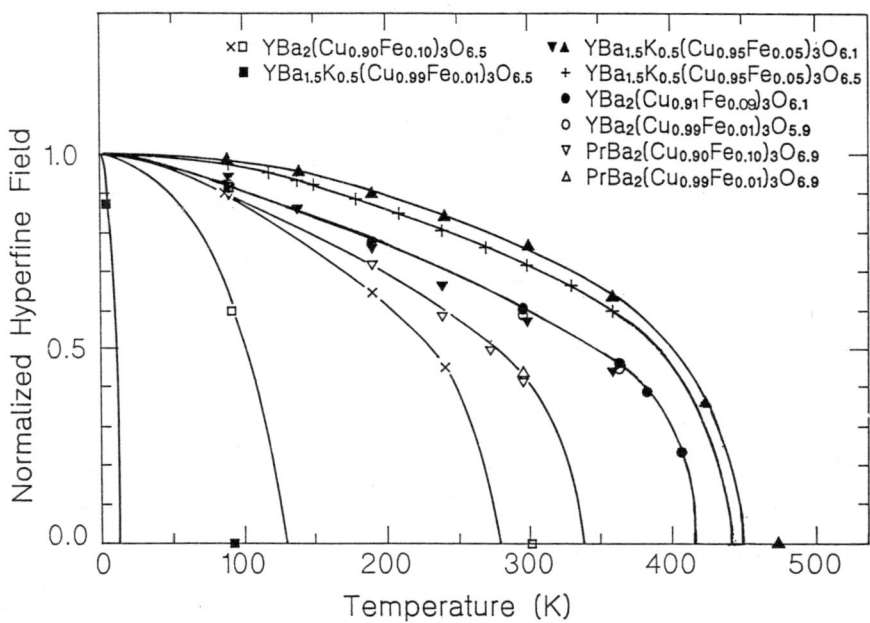

Fig.5. Temperature dependence of the magnetic hyperfine field acting on Fe^{57} nuclei in the Cu(2) site in $RBa_{2y}K_y(Cu_{1x}Fe_x)_3O_z$.

occurs at low magnetic fields. Suprisingly, the magnetization curve is non-linear as a function of the applied field, and a sharp upturn

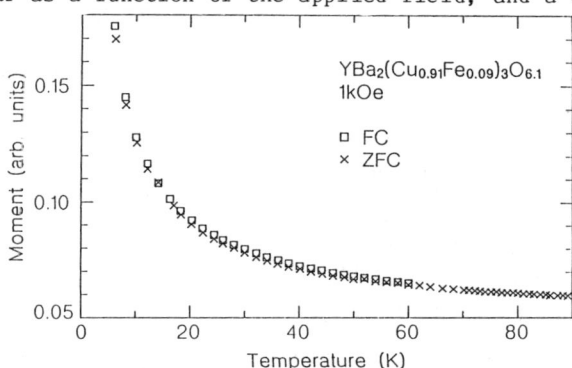

Fig.6 ZFC and FC magnetic susceptibility curves of the quenched sample $YBa_2(Cu_{0.91}Fe_{0.09})_3O_{6.1}$.

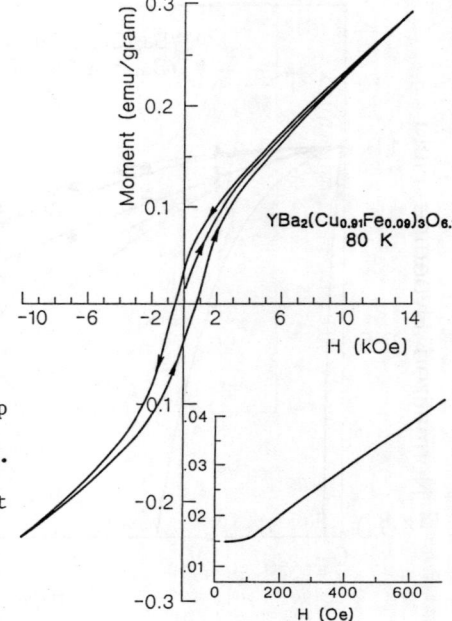

Fig.7 Magnetic hysteresis loop for the quenched sample at 80 K. In the insert the magnetization at low magnetic field is shown.

is observed. This upturn can be referred to a spin flip transition, and is displayed together with the hysteresis loop, both measured at 80K, in Figure 7. Both the spin flip transition (in the insert) and the full hysteresis loop with a coercive field of 600 Oe are attributed to field dependence of the antiferromagnetic nature of Cu and Fe in the Cu(2) planes.

In Figure 8 the different susceptibility curves obtained below and above the spin flip transition are shown. The difference between the susceptibility curves obtained at low (50-100 Oe) and high (500-15000 Oe) magnetic fields is obvious. In this figure the susceptibility curve for the S.C. sample at 15 kOe is also shown, and the higher values of this sample relative to the quenched sample will be discussed later. The temperature dependence of the susceptibility for the

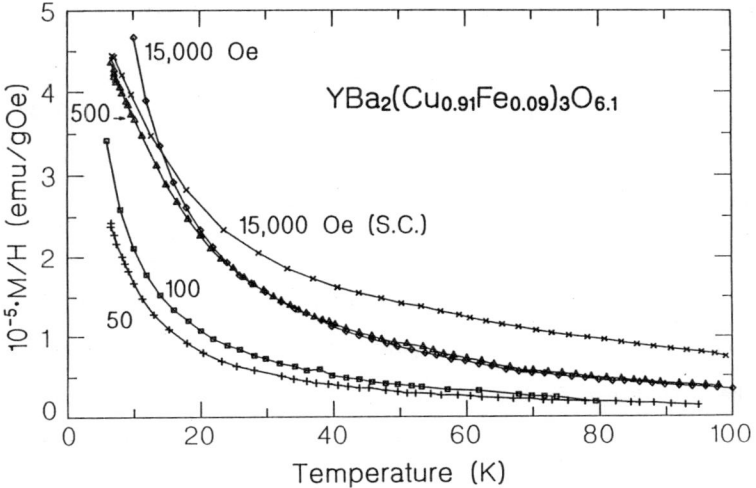

Fig.8 Susceptibility curves vs. temperature for the quenched sample at 50, 100, 500, 15000 Oe. Note the difference between the low and high field curves. The susceptibility of the S.C. sample at 15000 Oe is also shown.

quenched sample in the two regions is well characterized by the Curie-Weiss law

(1) $$\chi = \chi_0 + C/(T-\theta)$$

where χ_0 is the temperature independenct susceptibility, C and θ are the Curie-Weiss constant and temperature respectively. The molar values obtained by least-squares fits for 50 Oe and 100 Oe are $\chi_0 = -6.2 \times 10^{-3}$ and -1.4×10^{-3} emu/(mole.Oe), C=0.16 and 0.35 emu.K/(mole.Oe) and $\theta = -1.0$ and 1.3K respectively. In this compound the Cu ions in the Cu(1) sites are predominantly monovalent and nonmagnetic. The contribution to the magnetic moment arises from two sources; 1) Fe^{+3} and

Cu^{+2} ions which are antiferromagnetic coupled within the CuO_2 planes and 2) Fe^{+3} ions (about 80%) which occupy the Cu(1) sites. There are too many unknown parameters to allow us to offer any physically meaningful interpretation to the difference in these molar Curie-Weiss constants.

All the magnetic features mentioned above are attributed to the antiferromagnetic nature of Cu and Fe in the Cu(2) sites of the quenched sample and not to other magnetic phases such as unincorporated Fe. Otherwise these features would be also detectable in the superconducting sample above T_c, but there is no sign for magnetic phases in this sample (see below), and no unincorporated iron is seen in the Mossbauer spectra.

(b)   Superconducting $YBa_2(Cu_{0.91}Fe_{0.09})_3O_{7.1}$

T_c for this compound (obtained at 5 Oe) is 70(1)K and the reduction of T_c from that in $YBa_2Cu_3O_7$ (T_c=92K) appears to be not a magnetic effect (since Cu/Zn substitution has a much larger effect upon the suppression of T_c[18]) but due to the introduction of structural disorder in the crystal structure. Half the hysteresis loop at 4.1K for applied fields up to 40 kOe is shown in Figure 9. The sample was cooled in zero field to obtain the virgin curve. From the insert in Figure 9, which shows the virgin curve on an extended scale, it can be seen that M vs H begins to curve (flux begins to penetrate the sample) at H_{c1}=200 Oe. The magnetization curve is reversible as long as one never exceeds H_{c1} and the value of the volume susceptibility is 26% of $1/4\pi$ at 4.1K. From the width of the hysteresis loop, it is possible to obtain a crude estimate of the critical current density J_c using

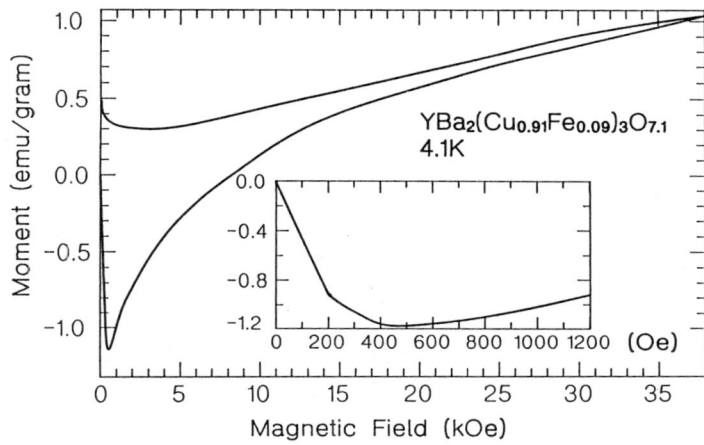

Fig.9 The magnetization M vs. the applied field H for the S.C. YBa$_2$(Cu$_{0.91}$Fe$_{0.09}$)$_3$O$_{7.1}$ at 4.1 K. The insert shows the magnetization values at low H in an extended scale.

$J_c = 10\,\Delta m/d$ where Δm denotes the difference between the negative and positive parts of the magnetization and d is the thickness of the sample. J_c obtained is of the order 300 A/cm^2 at 4.1K and 10 kOe with an error of 20% due to the uncertainty of d. This value is three orders of magnitude smaller than values obtained for undoped YBa$_2$Cu$_3$O$_7$ samples.

The temperature dependence of the susceptibility M/H for the S.C. material at 15 kOe is shown in Figure 8. In the normal state at T>20K the susceptibility is higher than that of the quenched sample. This reduction for the latter compound is attributed either to the general decrease of the Curie-Weiss component with the decreasing oxygen content[19] leading to the decrease of M/H in the quenched sample, or to antiferromagnetism discussed above, which would mean that even at that magnetic field some spins remain coupled.

Attention will now be focussed on the glassy behaviour of the superconducting features. Figure 10 exhibits several typical ZFC-FC curves with different applied magnetic fields. The FC magnetization vs temperature at 50 Oe reaches 54% of the ZFC magnetization. This indicates that the Meissner effect (the flux expulsion) is incomplete and flux is pinned during the cooling process. When the magnetic field is increased to above H_{c1} an upturn in the shielding curve is observed at low temperatures as shown for the 500 Oe curve. For high magnetic fields, both FC and ZFC are positive although the sample is still in the S.C. state. Above T_c all the susceptibility curves, measured either <u>at low or high</u> magnetic fields coincide into one curve which is well characterized by the Curie-Weiss law, Equation 1. This is in contrast to different susceptibility curves obtained for the quenched sample (Figure 8). At 300 Oe $\chi_o = 1.4 \times 10^{-5}$ emu/(mole.Oe) and $C = 0.40$ emu·K/(mole.Oe) and $\theta = 2.8K$ are derived. These values agree perfectly with Sakagami et al[21], and if the Curie constant of pure $YBa_2Cu_3O_7$ ($C = 0.14$ emu·K/(mole.Oe))[20] is subtracted from the above value, one obtains an effective moment $P_{eff} = 3.2 \mu_B$ /Fe ion.

To explain the upturn at magnetic fields higher than H_{c1} and the positive susceptibility obtained at magnetic fields higher than 2000 Oe, it was assumed that the susceptibility M/H is composed of diamagnetic contribution due to part of the sample which is S.C., and a paramagnetic contribution from the rest of the sample due mainly to

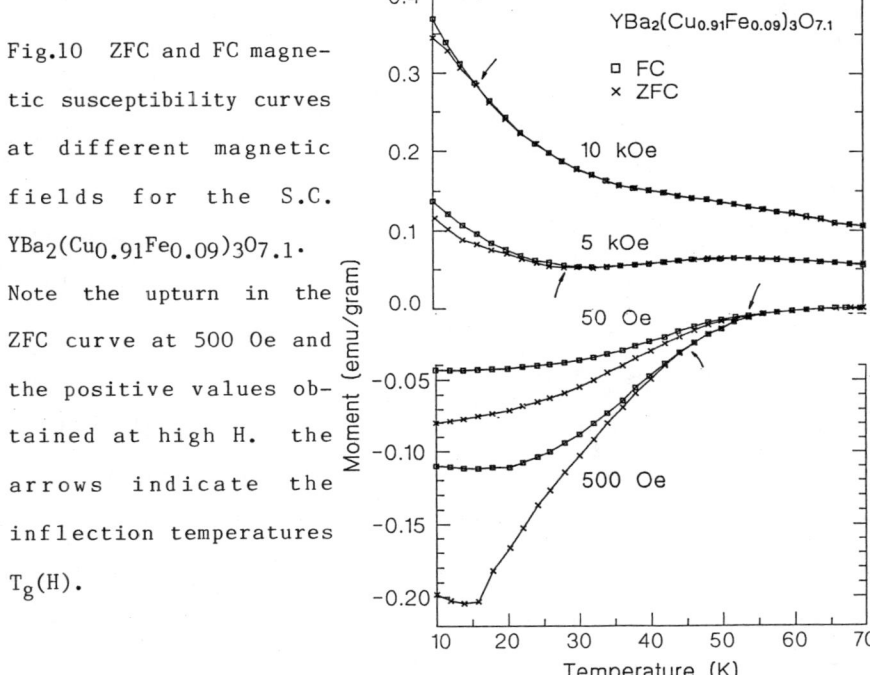

Fig.10 ZFC and FC magnetic susceptibility curves at different magnetic fields for the S.C. $YBa_2(Cu_{0.91}Fe_{0.09})_3O_{7.1}$. Note the upturn in the ZFC curve at 500 Oe and the positive values obtained at high H. the arrows indicate the inflection temperatures $T_g(H)$.

Fe^{+3} ions. The effective magnetic field acting above H_{c1} on the iron ions is just the external applied field, and the magnetic data for the paramagnetic component are those obtained from the temperature dependence of the susceptibility at $T>T_c$ given above.

The temperature dependence of the diamagnetic susceptibility of the S.C. part in the ZFC branch is proportional to an empirical volume equation:

(2) $$V_{S.C.} = V_o(1-T/T_c(H))^\alpha$$

and the total susceptibility is given by:

(3) $$M/H = -V_{S.C.}/4\pi d + (1-V_{S.C.}) \cdot (\chi_o + C/T - \theta)$$

where V_0 is the volume fraction of the sample which is superconducting at OK, d is the measured density and V_0, T_c and α are free parametes. The ZFC curve measured at 1000 Oe is shown in Figure 11 together with the calculated susceptibility (solid line) and the diamagnetic contribution (dashed line) calculated by least square fits of Eq. 3 and Eq. 2 respectively. The free parameters obtained, are $V_0=0.017(1)$, $T_c=64(1)$ and $\alpha=1.9(1)$. T_c obtained here as a free parameter of Eq. 3 is lower than 70(1)K measured at 5 Oe; it is therefore impossible to calculate from these values the field dependence of the higher critical field Hc_2. The positive susceptibility curves obtained at higher magnetic fields (Figure 10) are due to the reduction in the diamagnetic contribution. Equation 3 holds only for the ZFC branch, since, in the FC process additional unknown factors (as the amount of trapped flux) have to be considered. It should be mentioned that the present treatment differs slightly from that given in ref. 2.

The difference between the FC and the ZFC branches vanishes at T_g (indicated by arrows in Figure 10). The field dependence of T_g is shown in Figure 12, yielding a rather weak dependence of T_g on H at low fields. From the linearity of the $\log(T_g(0)-T_g(H))$ versus log H plot (insert of Fig. 12) the slope obtained is 0.25±0.02. The fact that T_g scales with the applied field as $H^{1/4}$ stands in contrast to other high T_c superconductors which exhibit superconducting glassy phenomena[3] and clearly differs also from the regular magnetic spin glass state where the temperature of the irreversibility shows a strong field dependence[22] of the form $T_g(0)-T_g(H) \propto H^{2/3}$.

The specific heat of the S.C. sample over the temperature range 4-70K

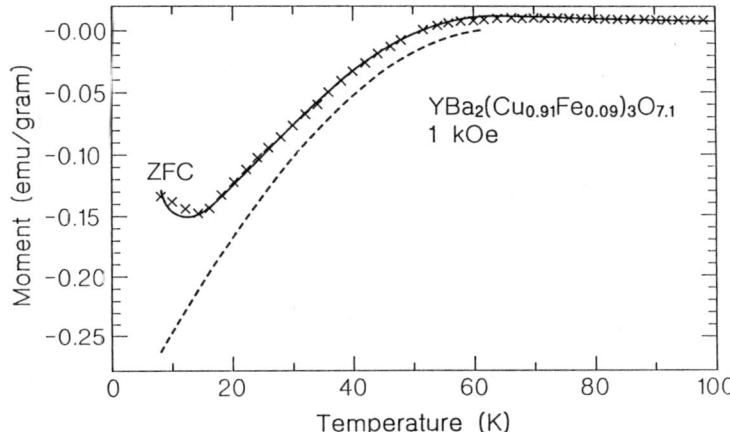

Fig.11 The temperature dependence of the ZFC susceptibility of the S.C. sample at 1000 Oe. The solid line is the fitted susceptibility (see text) and the dashed line is the pure diamagnetic contribution to the susceptibility.

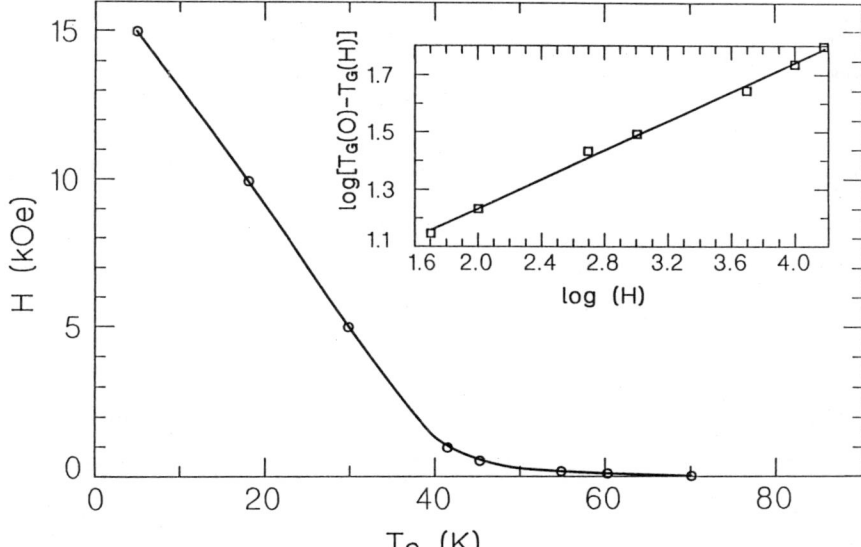

Fig.12 The glassy temperatures (T_g) as a function of the applied field H. The insert shows the linear fitting to the log log plot.

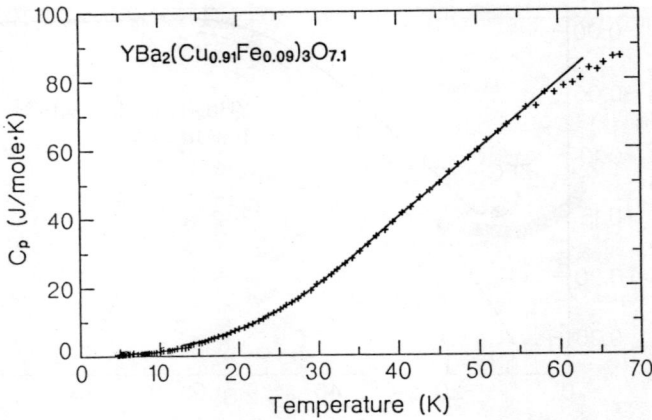

Fig.13 C_P vs. T curve for the S.C. Fe doped sample.

is shown in Figure 13. Figure 14 shows the specific heat of the 10% Fe doped sample together with the pure $YBa_2Cu_3O_7$ phase in the usual presentation of C_P/T vs T^2. At very low temperatures, the strong upturn in C/T, common for these materials, is observed in both samples.

From the nearly linear variation of the data in Figure 14 a finite C/T value at T=0 was extrapolated and denoted with γ^* in order to distinguish between this term and the normal electronic specific heat coefficient γ. For the undoped samples γ^* varies between 2.5 and 17 mJ/mol.K^2, while T_c remains almost constant. In this connection it should be noted that γ^* values of 5 and 9 mJ/mol.K^2 for tetragonal and orthorhombic $YBa_2Cu_3O_z$ single crystals have been observed.[23]. Substitution of Cu by 10% Fe significantly increases γ^* up to 73 mJ/mol.K^2. This appears to be a systematic trend in $YBa_2Cu_3O_7$ when Cu is replaced by either Zn[18] or Ni[24-25], and cannot simply be attributed

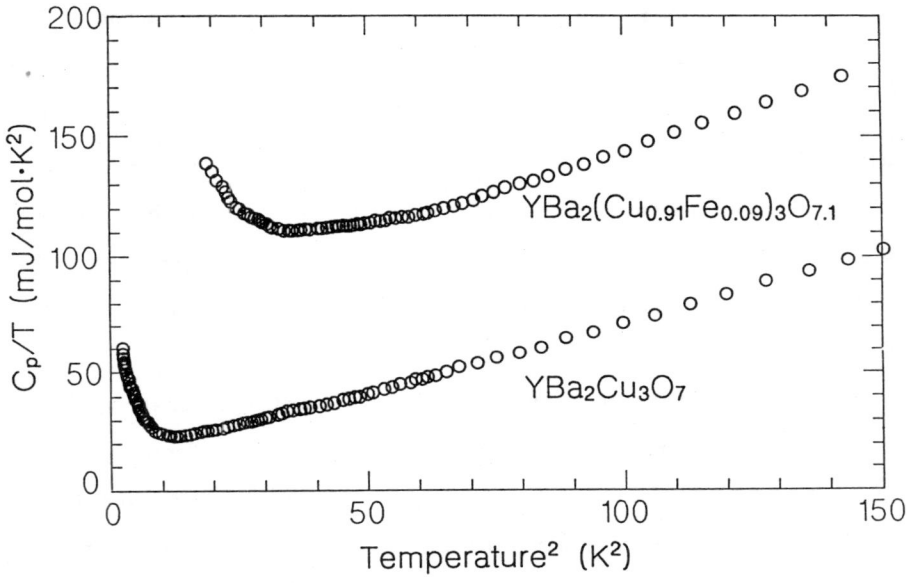

Fig.14 C_p/T vs. T^2 for $YBa_2Cu_3O_7$ and S.C. $YBa_2(Cu_{0.91}Fe_{0.09})_3O_{7.1}$ at low temperatures.

to a normal state electronic heat capacity of an impurity phase. Otherwise the second phase containing Fe would be detectable in the Mossbauer or x-ray measurements and should exhibit an unreasonably high γ value, comparable to that of a heavy fermion system. This systematic trend is also observed in $La_{1.85}Sr_{0.15}CuO_4$ on substituting Cu by Zn: γ^* rises by a factor of 4 upon 2.5% Zn doping[24]. Under the assumption of almost linear increase of γ^* upon replacement of Cu by Zn or another transition metal, $d\gamma^*/dx = 3$ and 2.1 mJ/Cu-mol.K^2 per mol%Zn were derived for the LaSrCuO and the YBaCuO systems respectively. Here mJ/Cu-mol.K^2 was used in order to have comparable values for both systems. With the latter value (multiplied by a factor of 3 to obtain mJ/mol.K^2), γ^* was estimated for the 9% Fe doped sample to be 73.7 mJ/mol.K^2 which is in good agreement with the experimental value obtained. These findings together with arguments concerning the Zn substitution presented recently[25] indicate that the finite γ^* term is an intrinsic property of these materials and may arise from unpaired non-superconducting carriers and/or from a vanishing gap at certain regions of the Fermi surface.

In conclusion of this section we would like to stress, that we provided the first static magnetic measurements of the antiferromagnetic nature found in the nonsuperconducting system using Fe dopant as a microscopic probe. We did not find magnetic spin glass behaviour of Fe ions at low tmeperatures in both S.C. and quenched samples, which was assumed from Mossbauer studies[10-12]. The broadening in the Mossbauer spectra at low temperatures may be due to slow spin relaxation rate phenomena in the superconducting samples, and due to magnetic order in the Cu(1) site induced by the magnetically ordered Cu(2) ions in the non superconducting samples. Regarding the glassy phenomena of the superconducting transition we show that T_g scales with the applied field as $H^{1/4}$. The linear term in the specific heat for the S.C. sample is extremely high.

2) $PrBa_2(Cu_{1-x}Fe_x)_3O_{6.9}$ and $YBa_{1.5}K_{0.5}(Cu_{1-x}Fe_x)_3O_z$ - Mossbauer Studies:

The nonsuperconducting samples of $YBa_2(Cu_{1-x}Fe_x)_3O_z$ with z=6.1 and z=6.5 in which the magnetic site is seen, differ from the metallic superconducting sample with z=7.1 in several respects: 1) They are semiconductors with different electronic properties and band structure ; 2) They are deficient in their oxygen concentration; and 3) they have a tetragonal structure. In order to decide whether all these conditions are essential for magnetic order, or whether any one of them is sufficient, we performed Mossbauer studies on the following two systems: $PrBa_2Cu_3O_{6.9}$ which is well known not to be a superconductor, although it is fully oxygenated; on the other hand, the low oxygen-content $YBa_{1.5}K_{0.5}Cu_3O_{6.5}$ is superconducting with $T_c=88$ K[26].

a) $PrBa_2(Cu_{1-x}Fe_x)_3O_{6.9}$ with x=0.01 and x=0.1

These tetragonal compounds, with a=b=3.900A and c=11.66Å for x=0.01 and a=b=3.927A, c=11.69Å for x=0.10, are rich in oxygen yet not superconducting. In these compounds a well defined magnetic subspectrum is seen in the Mossbauer spectra even for x=0.01 (Fig. 15). It thus seems that in these compounds a larger fraction of the Fe ions enter the Cu(2) site. The Neel temperature of the iron ions in the Cu(2) site is 325 K for both x=0.1 and x=0.01, (Figs. 5 and 15) proving again that T_N does not depend on iron concentration. Thus, though this compound is rich in oxygen, like the orthorhombic superconducting compounds, it is nevertheless nonsuperconducting, and tetragonal, and its magnetic behaviour resembles that of oxygen-poor, tetragonal, nonsuperconducting compounds. This shows that the number of oxygen ions alone cannot account for the disappearance of the magnetic interactions in the Cu(2) site.

b) $YBa_{1.5}K_{0.5}(Cu_{1-x}Fe_x)_3O_z$, x=0.01, 0.05 and z=6.1 and 6.5

These compounds exhibit superconductivity for z=6.5, with $T_c=88$ K for x=0.01 and $T_c=60$ K for x=0.05. The magnetization curves are shown in

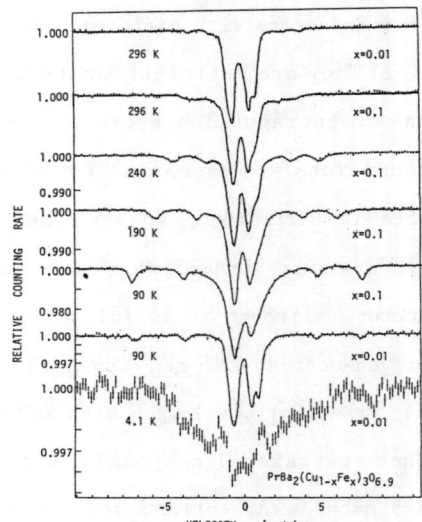

Fig.15 Mossbauer spectra of Fe^{57} in $PrBa_2(Cu_{1-x}Fe_x)_3O_{6.9}$ with x=0.01 and 0.1 at several temperatures.

Fig. 16. The structure is orthorhombic for x=0.01 and z=6.5, and tetragonal for all other samples. The Mossbauer spectra at different temperatures obtained in samples with x=0.05 are shown in Figs. 17 and 18. One observes magnetic subspectra both in the samples with z=6.5 (which are superconducting) and in those with z=6.1 (which are not).

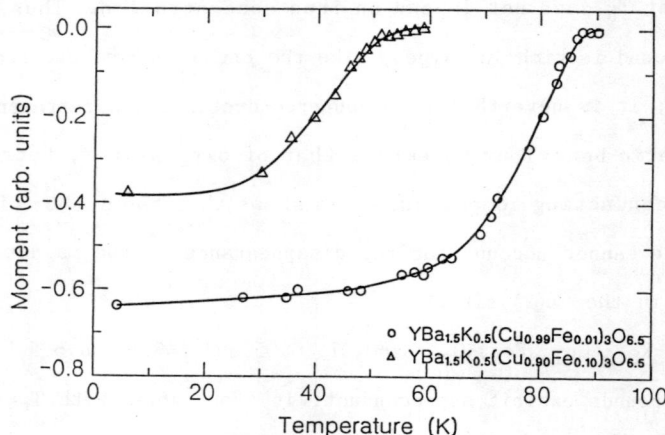

Fig.16 The magnetic suceptibility of $YBa_{1.5}K_{0.5}(Cu_{1-x}Fe_x)_3O_z$.

In the latter, two magnetic sites are clearly observed. One of these sites follows the magnetic behaviour of $YBa_2(Cu_{0.91}Fe_{0.09})_3O_{6.1}$ with $T_N=415$ K, the other shows a higher T_N of about 450 K (Fig. 5). The relative intensity of the magnetic subspectra is higher in samples containing K than in samples which do not contain potassium. Since we attribute the magnetic sub-spectra to iron in the Cu(2) site, this would imply that K attracts iron into this site. It is reasonable that K^+, which replaces Ba^{2+}, prefers to have as its neighbour Fe^{3+} instead of Cu^{2+}. In the sample with z=6.5, which is superconducting with $T_c=60$ K, only one magnetic site with $T_N=450$ K is observed. It thus seems that in this compound superconductivity and magnetic order in the Cu(2) site coexist. A reasonable explanation for the appearance of the two different magnetic sites in the z=6.1 samples, with different T_N and different quadrupole interactions, is to assume that the Fe ions in the different magnetic sites differ in their local

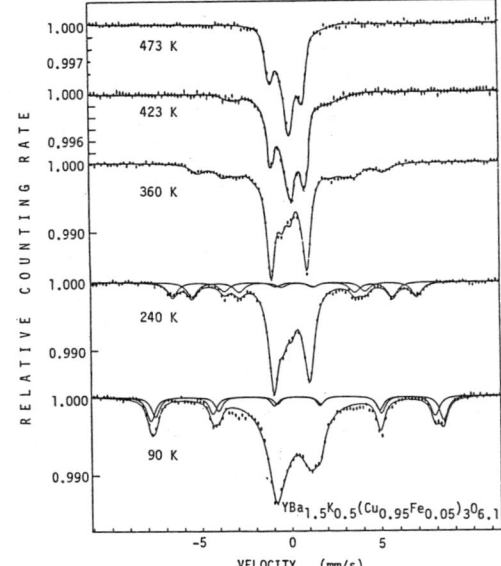

Fig.17 Mossbauer spectra of Fe^{57} in $YBa_{1.5}K_{0.5}(Cu_{0.95}Fe_{0.05})_3O_{6.1}$ with x=0.01 and 0.1 at several temperatures.

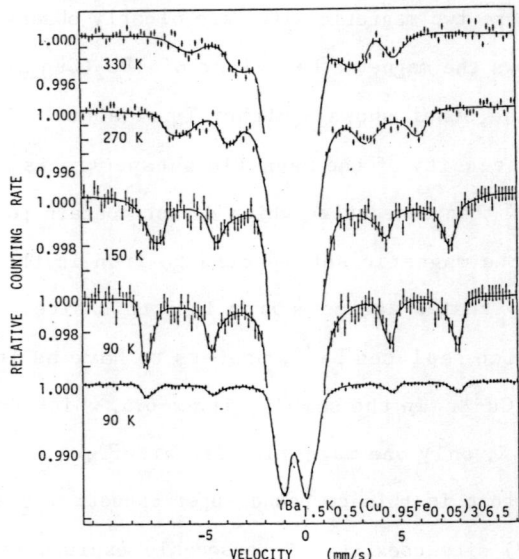

Fig.18 Mossbauer spectra on an expanded scale of Fe^{57} in $YBa_{1.5}K_{0.5}(Cu_{0.95}Fe_{0.05})_3O_{6.5}$ at several temperatures. The spectrum at 90 K is shown in normal and expanded scale.

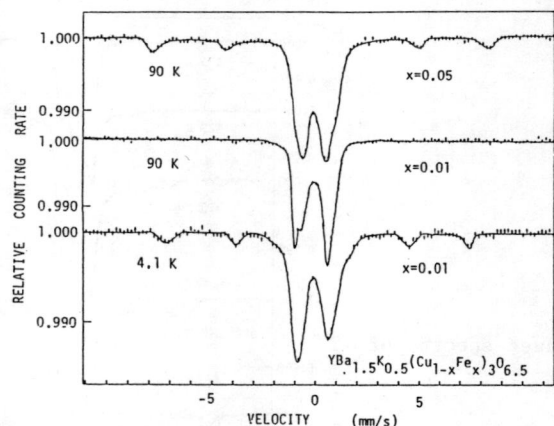

Fig.19 Mossbauer spectra of Fe^{57} in $YBa_{1.5}K_{0.5}(Cu_{1-x}Fe_x)_3O_{6.5}$. Note that the spectra with x=0.01 show a magnetic subspectrum at 4.1 K, but not at 90 K.

environments. The samples with K contain less oxygen[26]. In order to compensate for the lower valence of K, oxygen may be expelled from the Cu(2) plane and/or change the valence of Cu(2) near it. Those iron ions which have no oxygen vacancy and have regular Cu(2) neighbours show the same magnetic behviour as those in $YBa_2(Cu_{0.9}Fe_{0.1})_3O_{6.1}$ with $T_N=415$ K, losing their magnetic interactions (in the compounds with and without K) as more oxygen is added in the Cu(1) chains. Yet oxygen is not added to the Cu(2) plane, so that those iron ions with less oxygen near them show different magnetic behaviour. As the iron probes the magnetic behaviour only of the Cu(2) ions, only those Cu(2) ions which lack oxygen near them remain magnetic even in the superconducting samples, with $T_N=450$ K. Thus in the samples with K and z=6.1 there are two magnetic sublattices. It may be that in these compounds there are regions which are potassium rich and others which are potassium poor. In the samples with x=0.01, Fe enters the Cu(2) site only in the potassium-rich region and thus stays magnetic in the superconducting state (Fig. 19). Here we find $T_N=15$ K, whereas $T_c=88$ K. Recent data show that the magnetic transition temperature is very sensitive to the oxygen content (z) of the sample[27]. There is a sharp transition region where the Neel temperature changes drastically with z: $dT_N/dz=-4000$ K. The low T_N is therefore probably due to a slightly higher oxygen content in this sample.

Conclusions

The magnetic ordering temperatures found here are independent of the iron content and for $YBa_2(Cu_{0.91}Fe_{0.09})_3O_{6.1}$, T_N is identical to that found in $YBa_2Cu_3O_{6.1}$[17], which does not contain any iron. This proves that iron serves here only as a probe for the magnetic properties of Cu in the Cu(2) site.

The results show that K attracts iron into the Cu(2) site and also that at very low oxygen content (z=5.9), prepared by slow cooling in a vacuum furnace, a relatively larger percentage of the iron enters the Cu(2) site.

The relative values of the quadrupole splittings for iron with different numbers of oxygen neighbours in the Cu(2) site are consistent with point charge calculations [8], according to which the electric field gradient (EFG) in the Cu(2) site should be positive and point along the c-axis. The measured negative effective quadrupole interaction indicates therefore that the magnetic moments lie in the basal plane[17]. All Fe ions in the Cu(1) site have similar isomer shifts. The isomer shift of Fe in the Cu(2) site is greater, pointing to a smaller electron density at the Fe nuclei in this site. The broadening in the Mossbauer spectra at low temperatures is probably due to slow spin relaxation rate phenomena. Regarding the glassy phenomena of the superconducting transition we show that T_g scales with the applied field as $H^{1/4}$.

The collected evidence indicates that superconductivity and Cu(2) antiferromagnetism in the $RBa_2Cu_3O_z$ compounds do not generally coexist. Magnetic order is enhanced with decreasing oxygen content in the sample. Nevertheless magnetic order appears in oxygen-rich $PrBa_2Cu_3O_{6.9}$. Superconductivity is enhanced with increasing oxygen concentration, but $PrBa_2Cu_3O_{6.9}$ is again a counter example. Only in

$YBa_{1.5}K_{0.5}(Cu_{1-x}Fe_x)_3O_z$ do we find coexistence of a magnetic sublattice with superconductivity. Even in this case one may claim that the compounds contain phases with different amounts of oxygen and/or potassium. All the results can be explained by the valence of Cu mainly in the Cu(2) site. In order to preserve charge neutrality in a sample of $YBa_2Cu_3O_6$, Cu(1) has to be predominantly monovalent, and Cu(2) predominantly divalent. In this case, Cu(2) is magnetic as observed. As more oxygen is added, both Cu(1) and Cu(2) become of mixed valencies. Adding oxygen mainly changes the valence of Cu(1) due to the proximity of Cu(1) to these oxygens - but it also changes somewhat the valence of Cu(2), which becomes of mixed divalent-trivalent valence and thus, in the samples with $z \geq 6.5$, which are superconducting, it loses its magnetic moment. If Y is replaced by Pr, since Pr is not purely trivalent (but rather mixed trivalent-tetravalent)[28] it keeps the Cu in Cu(2) more divalent, and, therefore, it is magnetic even for high oxygen content. K^+ on the other hand, when it replaces Ba^{2+}, makes Cu(2) more trivalent, even for lower oxygen content. It thus seems that for superconductivity to occur, Cu ions both in the Cu(1) and Cu(2) site have to have mixed valencies, while for magnetic order to appear, Cu(2) has to be predominantly divalent. This picture is consistent with extended x-ray-absorption fine-structure spectroscopy results,[29] which leads to the conclusion that both Cu(1) and Cu(2) atoms have a local oxygen coordination which favors a mixed divalent-trivalent character.

Acknowledgement:

This research was supported in part by a grant from the US-Israel Binational Science Foundation (BSF), Jerusalem, Israel.

References

1. K.A. Muller, M. Takashige and J.G. Bednorz, Phys. Rev. Lett. $\underline{58}$, 1143 (1987).
2. Y. Yeshurun, I. Felner and H. Sompolinsky, Phys. Rev. B$\underline{36}$, 840 (1987).
3. K.A. Muller, K.W. Blazey, J.G. Bednorz and M. Takashige, Physica $\underline{148B}$, 149 (1987).
4. G. Deutscher and K.A. Muller, Phys. Rev. Lett. $\underline{59}$, 1745 (1987).
5. I. Felner, I. Nowik and Y. Yeshurun, Phys. Rev. B$\underline{36}$, 3923 (1987).
6. M. Mehbod, P. Wyder, R. Deltour, Ph. Duvigneaud and G. Naessens, Phys. Rev. B$\underline{36}$, 8819 (1987).
7. Y. Maeno, T. Tomita, M. Kyogoku, S. Awaji, Y.A. Aaki, K. Hoshino, A. Minami and T. Fujita, Nature $\underline{328}$, 512 (1987).
8. E.R. Bauminger, M. Kowitt, I. Felner and I. Nowik, Solid State Comm. $\underline{65}$, 123 (1988).
9. P. Bordet, J.L. Hodeau, P. Strobel, M. Marezio and A. Santoro, Solid State Comm. 1988 (in press).
10. X.Z. Zhou, M. Raudsepp, Q.A. Pankhurst, A.H. Morrish, Y.L. Luo and I. Mattrtense, Phys. Rev. $\underline{36}$, 7230, (1987).
11. S. Nasu, H. Kitagawa, Y. Oda, T. Kohara, T. Shinjo, K. Asayama and F.E. Fujita, Physica $\underline{148B}$, 484 (1987).

12. Z.Q. Qiu, Y.W. Du,. Tang, J.C. Walker, W.A. Bryden and K. Moorjani, J. Magn. Magn. Matter, 1988 (in press).
13. Q.A. Pankhurst, A.H. Morrish, M. Raudsepp and X.Z. Zhou, J. Phys. C21, L7 (1988).
14. J.D. Jorgensen, B.W. Veal, W.K. Kwok, G.W. Crabtree, A. Umezawa, L.J. Nowicki and A.P. Paulikas, Phys. Rev. B36, 5731 (1987).
15. I. Felner and B. Barbara, Solid State Comm. 66, 205 (1988).
16. D.C. Harris, M.H. Hills and T.A. Hewston, J. of Chem. Educ. (1987).
17. J.M. Tranquada, D.F. Cox, W. Kunnmann, H. Moudden, G. Shirane, M. Suenaga, P. Zolliker, D. Vaknin, S.K. Sinha, M.S. Alvarez, A.J. Jacobson and D.C. Jonston, Phys. Rev. Lett. 60, 156 (1988).
18. K. Remschnig, P. Rogl, E. Bauer, R. Eibler, G. Hilscher, H. Kirchmayr and N. Pillmayr, Proc. Int. Discussion Meeting, Manterndorf, Plenum Press.
19. M. Tokumoto, H. Ihara, M. Hirabayashi, K. Murata, N. Terada, T. Matsubara and Y. Kimura, Physica, 148B, 436 (1987).
20. J.R. Thompson, D.K. Christen, S.T. Sekula, B.C. Sales and L.A. Boatner, Phys. Rev. 36, 836 (1987).
21. E. Sakagami, Y. Oda, Y. Yamada, T. Kohara and K. Asayama (preprint).
22. P. Monod and H. Bouchiat, J. Phys. (Paris) Lett. 43, L-45 (1982).
23. S. von Molnar, A. Torreseen, D. Kaiser, F. Holtzberg and T. Penny, Phys. Rev. B37, 3762 (1988).

24. Chan-soo Jee, S. Rahman, A. Kebede, D. Nichols, J.F. Crow, T. Mihalisin and P. Schlottmann, Bull. A. Phys. Soc. 33, 465 (1988).
25. G. Hilscher, N. Pillmayr, R. Eibler, E. Bauer, K. Remschnig and P. Rogl, Z. f. Physik (in press).
26. I. Felner, M. Kowitt, Y. Lehavi, L. BenDor, Y. Wolfus and I. Nowik, Physica C 153-155, 898 (1988).   27. J.H. Brewer et al., Phys. Rev. Letters, 60, 1073 (1988).
28. I. Felner, M. Kowitt, Y. Lehavi, D. Edery, L. BenDor, Y. Wolfus and I. Nowik, Modern Phys. Lett. B 2, 713 (1988).
29. G.M. Antonini, C. Calandara, F. Corni, F.C. Matacotta and M. Sacchi, Europhys. Lett., 4, 851 (1988).

ENERGY GAP IN HIGH-T_c SUPERCONDUCTORS STUDIED BY MEANS OF ELECTRON TUNNELING SPECTROSCOPY

Jerzy Raułuszkiewicz

Institute of Physics, Polish Academy of Sciences
al. Lotnikow 32, 02-668 Warszawa
POLAND

ABSTRACT

Electron tunneling spectroscopy, so effective in studying of conventional superconductors, has encountered many difficulties when applied to the high-T_c superconducting ceramics. The reasons leading to uncertainty of tunneling data are discussed with special regard on surface degradation and small particle charging effects.

1. INTRODUCTION

Electron Tunneling Spectroscopy (ETS) is a widely recognized and applied diagnostic tool in the field of superconductivity. Since its discovery by Giaever in 1960 [1] ETS has provided the most complete characterization of the superconducting state that is presently available and has contributed in the most important way to our present understanding of conventional superconductors.

ETS is realized in tunnel junctions. A tunnel junction consists of two conducting electrodes separated by thin (20 - 50 Å) insulator (or vacuum). The tunneling current depends on tunneling probability which characterizes the barrier, and on electronic densities of states in both electrodes. Thus various density of states (DOS) effects in both or in one electrode only, can be investigated in tunneling experiments, like the just mentioned quasiparticle density of states in superconductors, first and foremost the superconducting gaps Δ with all their features, electron-phonon coupling strength, phonon spectra etc. All ETS information, some of fundamental importance, can be deduced from the analysis of I-V, (or dI/dV, d^2I/dV^2) characteristics.

Thus it was obvious that accordingly, since the discovery of high-T_c superconductors (HTS) [2-5] the ETS techniques have been widely applied to cast some light on various aspects of these new materials. Some of the enthusiastic tunnelists claimed: give us your mysterious superconductor, our excellent ETS technique will resolve all your problems related to the superconducting gap...However, it turned out very soon that in spite of the great amount of ETS experiments the results obtained in many laboratories all over the world, and in some cases even in the same laboratory, were not convincing, not consistent with each other, not reproducible and difficult to interpret in terms of old, known pairing mechanisms.

In general, the tunneling dI/dV (or dV/dI) spectra measured on the HTS exhibit some typical peculiarities: i) great zero-bias anomalies, ii) many peaks in dV/dI symmetric in bias, iii) temperature dependances of positions of these peaks do not follow the Δ(T) dependence of BCS superconductors.

2. EXPERIMENTAL REMARKS ON TUNNELING JUNCTIONS

From the experimental point of view the ETS measurements can be performed using planar or point contact junctions. *Planar junctions* consist of bulk ceramic, (as well as single crystal or evaporated thin superconducting film) as the base electrode and of evaporated conventional superconducting thin film (Pb, Nb) or simply Au or Ag paints used as the counter electrodes. In conventional tunneling junctions either the natural oxides of one of electrodes, or artificially fabricated thin insulating films are usually used as the barriers for tunneling. In the case of HTS the degradated, non superconducting surface layers act as the barriers for tunneling. In case of HTS the conventional thin film sandwich type (or Giaever type) junctions are by far the most difficult to be realized.

Some experiments were also done using the *break junctions*, where tunneling occurs across a fracture produced in UHV or under liquid helium. All these structures, differently from point contacts, cannot be adjusted during experiments.

In the *point contact* junctions a very sharp metallic tip either touches the surface of investigated sample, or is kept above the surface within a tunneling distance. This technique when used together with the scanning tunneling microscope (STM) is the most promising in ETS. It makes it possible to scan the surface, to check its configuration and electrical properties, and next, to select the place on surface where tunneling characteristics will be measured. Therefore, practically all tunneling experiments have recently adopted the point contact which is pressed through the surface layer of the material, so that electron tunneling occurs between the tip and a few grains in the bulk ceramic.

3. ONE- AND TWO-ELECTRON TUNNELING

Interesting information on the superconducting properties of the investigated material can be drawn from tunneling experiments by referring to one-electron tunneling or to two-electron (Cooper pairs, Josephson) tunneling. For these purposes the ETS should include barrier junction structures, or the weakly coupled superconductor configurations. A bridge type film junctions imply the availability of very reliable HTS films, because the link dimensions L should be smaller than the coherence length ξ_o, which is of the order of 10 Å in HTS materials. The condition of $L < \xi_o$ is not possible to fulfill, but in spite of this condition the Josephson weak links, in the best case 17 μm wide, seem to be achieved up to 68 K [6]. The observation of Cooper-pair tunneling in investigated ceramics is the best demonstration of superconductivity as well as of the possibilities of practical applications of HTS in the SQUID systems.

Since the beginning of investigations of HTS the one-electron tunneling appeared to be the most straightforward way to detect and evaluate from the current-voltage characteristics and their derivatives the energy gap Δ (if any). The point contact technique, either combined with scanning tunneling microscope or not, was widely applied. A large number of experiments both on La-Sr-Cu-O and Y-Ba-Cu-O [6-8] superconductors was performed. The metal tip made by conventional superconductors, like Pb or Nb, was employed in contact with ceramic pellets. Also the point contact Y-Ba-Cu-O to Y-Ba-Cu-O and break junction [8] configurations were investigated. A large variety of gap values was reported leading consequently to different values of the ratio $\gamma = 2\Delta(0)/k_B T_c$, a measure of the strength of the coupling between the electrons, whatever excitation mediates the pairing. This well known fundamental parameter underlays the theoretical interpretation of the mechanisms of superconductivity. The results of many measurements and estimations of γ are presented after Prof. Barone [9] in Fig. 1 as the statistical sketch. For conventional superconductors with phonon-induced pairing γ varies from the weak-coupling limit of 3.5 to the strong-coupling being almost 5.

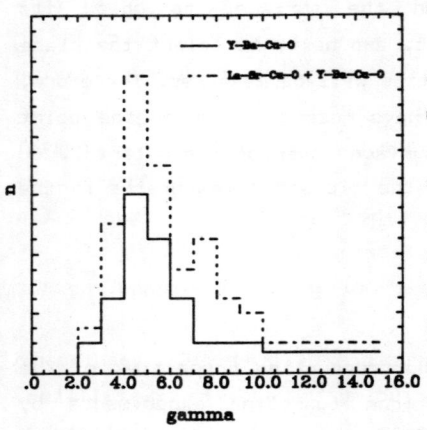

Fig.1 Statistical sketch of γ measurements [9]

4. RELIABILITY OF TUNNELING DATA

The comparison of precisely measurable values of γ for conventional superconductors with those for HTS, sketched in Fig. 1, leads to a question, to what extent the later can be reliable and consequently - to what extent ETS can be reasonably applied to HTS. In HTS the ETS measurements usually do not give very clear I-V curves with well defined gap structure. The most explored dI/dV characteristics, however, have rather *to many structures* and are not similar to those obtained on conventional superconductors, therefore they are difficult to interpret unambiguously.

Several reasons leading to uncertainty in determination of energy gaps in HTS from ETS measurements can be listed as follows:

1. Porosity of ceramic pellets. Because of its ceramic structure the surface is not well defined, not flat and regular. Possible leakage currents and edge effects.

2. Degradation of the superconducting properties of HTS when in contact with another material or with air by which the oxygen stoichiometry is easily affected. The oxygen content in the surface layer described by the index δ in the formula $Y_1Ba_2Cu_3O_{7-\delta}$ differs greatly as compared to the bulk of material. Thus a nonconducting or semiconducting surface layer is formed on the surface of samples making it difficult even to prepare good ohmic contacts [10].

3. Inhomogeneity of material with respect to stoichiometry, crystal structure, oxygen content etc., which would lead to local changes of all superconducting parameters. Thus on the same sample the tunnel tip probing different places gives quite different results.

4. Grain structure resulting in Giaever-Zeller type of tunneling not reviling the gap.

5. Anisotropy in crystallographic structure of unit cell leading to anisotropy of physical properties , including the gap, in crystallites.

6. Edge effects which can occur in the case of too artless electrode pattering technology adopted.

7. Magnetic flux trapping which can produce occasionally normal regions in electrodes.

8. Heating effects acting locally against superconductivity in point contacts or narrow bridges, where the local densities of currents may be very high.

9. Pressure effect occurring locally in point contact junctions as well as non uniform strains even in planar junctions.

10. Ambiguous mechanism of superconductivity and unidentified effects which could be observed.

11. Lack of calculations of tunneling currents and expected structures for comparison with an experiment.

12. Finite voltage structures due to self coupling in Josephson junctions.

13. Capacity effects connected with the small particle granular structure leading to additional smearing of gap and introducing in some cases additional structures.

This list could be continued, but even in the length given above, it can discourage tunnelists. However, this list should be always kept in mind when making the diagnostic of ETS. Some of items listed above are related to the specific experimental method and procedure or to particular material, some are of more intrinsic nature, but all can have their specific implications on study of HTS. For instance, the effect of pressure on the La-Sr-Cu-O material is very dramatic, whereas the pressure independent gap was observed in the Y-Ba-Cu-O. I would like to discuss a little more the problems of surface degradation and of small particle capacity effects.

5. DEGRADATION OF SURFACE

The most serious difficulties in the application of ETS diagnostic to HTS are connected with the degradation of surface. Thus practically all reliable tunneling experiments were performed using point contacts which were pressed through the surface layer of the material [11] so that electron tunneling occurs between a tip and a few grains in the bulk of superconductor. When these grains are in the superconducting state, one obtains a tunnel gap $\Delta(0)$ which is about 15 - 20 meV [12,13] or even 50 meV [14], which corresponds to γ = 3.8 - 5.0 or 13 for T_c = 92 K. An analogous result is obtained for the break junctions where tunneling occurs across a fracture of the material produced under liquid He [15]. The common future of both point contact and break junctions tunneling is that one observes in most experimental trials a tunneling characteristics of a normal metal, and further, that in relatively few case when superconducting tunneling does occur, a variable gap is found. This is an indication that the sample is not a uniform, well defined superconductor, but it consists of a mixture of both normal and superconducting regions with variable superconducting properties.

The infrared reflectivity measurements [12,16] reveal much lower energy gaps when compared with tunneling technique: 6 meV < Δ < 13 meV. This indicates, that the superconducting properties of the surface, at which the reflection occurs can differ considerably from that of the bulk of sample. On the other hand, infrared measurements performed on a single crystal result in a gap Δ = 31 meV (for electric field vector in the a-b plane of crystal).

Let me describe two examples of tunneling experiments: i) Garcia et al. [17] and ii) Fein et al.[18]. The studies of Y-Ba-Cu-O surface by Garcia et al. by means of STM have demonstrated, that the grains formed in the synthesis of material do not show bulk superconductivity but even within the grain there occurs coexistence of semiconducting and metallic regions. The metallic, and then superconducting regions are formed at twinned domain boundaries. The conductive percolating network of grain and domain boundaries is responsible for the superconductivity. Therefore, it is clear, that the point contact tunneling can give quite different results when probing only the local properties of sample.

According to Fein et al. [18] even if the nonconducting surface layer is penetrated in a tunneling experiment a wide variety of tunneling characteristics is observed. But if care is taken to select the data that looks most "ideal", that is, most like what would be predicted from standard calculations using the BCS-like DOS, fairly reproducible gap values can be attained.

The paper of Fein et al. [18] is a good illustration of the tunneling data analysis performed on $La_{2-x}Sr_xCu-O_{4-y}$. The tunneling measurements were made using a low temperature STM. Authors have reduced the problems associated with surface contaminations by breaking samples at low temperature "in situ". Even with this procedure authors were unable to convincingly tunnel into the bulk material at low voltages and they never saw topographic images on an atomic scale. The samples broke at low temperatures in a clean environment along the grain boundaries, which were not metallic, but once the tip broke through this nonconducting surface region, the tunneling into proper superconducting material was observed. A great deal of different dI/dV characteristics was collected and carefully selected using the following criteria: small conductance at zero bias, overshoots in conductance at the gap edge and symmetry in the positions of the conductance overshoots with respect to voltage bias.

After such selection the results were consistent: the background conductance rises linearly with voltage, the gap in conductance although smeared relative to what would be expected from the BCS-like

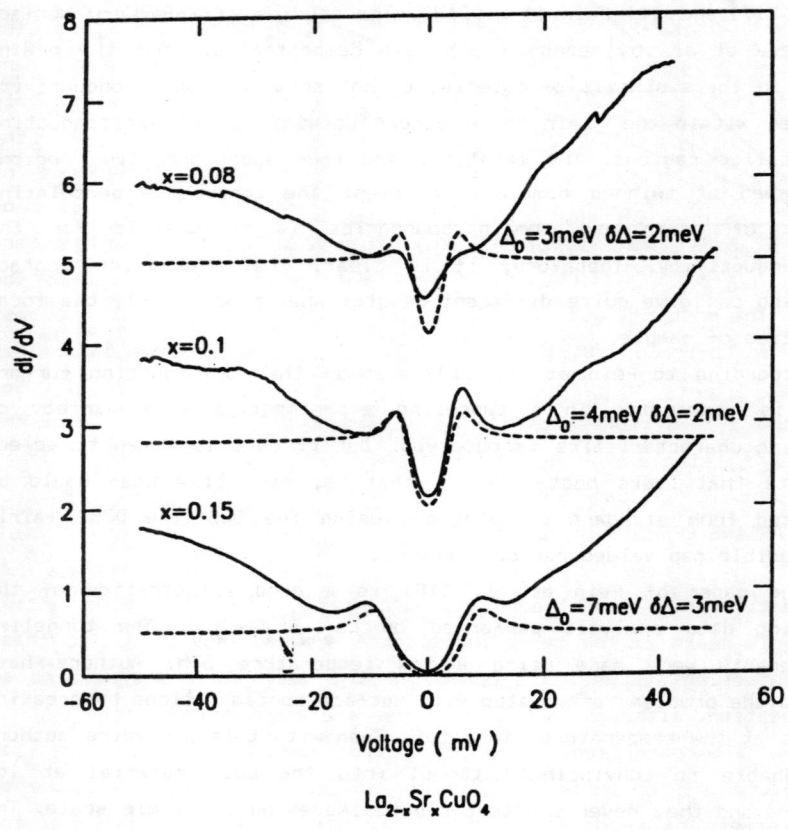

Fig. 2. Fitting of Δ and $\delta\Delta$ to experimental data [17]

The dashed curves in Fig.2 are predictions of standard tunneling theory using distribution of the BCS gap Δ_o [21] described by a weighting function $f(\Delta) = C \exp[-(\Delta - \Delta_o)^2/\delta\Delta^2]$. The values for Δ_o and $\delta\Delta$ were chosen to fit the voltage positions and widths of the observed conductance overshoots. Varying the Sr concentration x in the investigated $La_{2-x}Sr_xCu O_{4-y}$ varies the hole concentration and through the hole concentration the critical temperature of material. The results of $\Delta(T_c)$ measurements are presented in Fig. 3. The dashed line is the best fit to the data. This gives $\gamma = 4.95 \pm 0.96$. The slope $\gamma = 8$ represents the IR [19] and NMR [20] results on the $YBa_2Cu_3O_{7-\delta}$ single crystals.

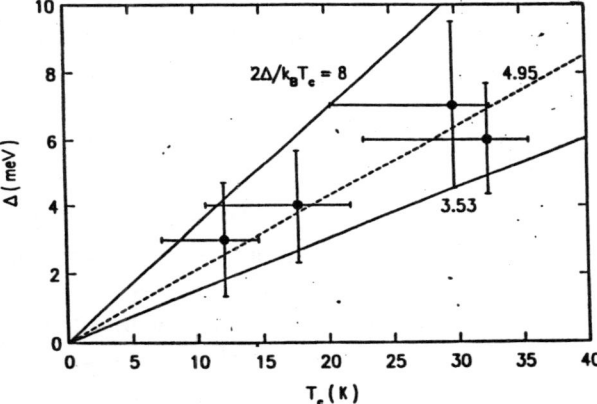

Fig. 3. The most reliable values of $\Delta(T_c)$ in La-Sr-Cu-O [18]

6. CHARGING EFFECTS IN TUNNELING

Particularly troublesome to interpret are the observations of multiple peaks in the tunneling spectra - phenomenon unique to these superconducting systems. The nature of the peak structure remains an open question, although some suggestions are worth discussing.

The ETS experiments are conducted mostly employing the low temperature STM to perform the point contact tunneling. In spite of large variety of reported values of γ, one mechanism underlying the observed features has to be taken into account - charging effects of small particles. This model is applicable in the case where tunneling occurs to a number of small particles with a finite-size distribution. This is the model of Zeller-Giaever [22], applied recently by many other authors [12,15]. It predicts a linear dependence of tunneling conductance dI/dV *vs* bias voltage V up to a voltage e/C, where C is the mean grain capacitance. The usual BCS behaviour - zero conductance up to, and a peak of conductance overshoot at, $V = \Delta/e$ - is modified so that, although a gap in conductance still appears, there is no associated peak in DOS at $V = \Delta/e$. This behaviour should operate particularly in HTS tunneling systems, whose granular structure and granular superconductivity are generally accepted.

The Zeller-Giaever model did not explain the presence of multiple peak structure in tunneling data. Recently the theory of Mullen et al. [23] (and confirming experiments of Ruggiero et al. [24] and van Bentum et al. [25]) predicts that under appropriate circumstances the multiple peaks should appear in dI/dV (or in dV/dI) tunneling characteristics of a small particle system with a bias-voltage spacing of e/C (Fig.4). The overall prediction is that there should be a series of peaks at voltages of e/2C, 3e/2C, 5e/2C, 7e/2C, etc. The superconducting tunneling conductance does not reflect a BCS-like DOS, but still exhibits a gap at voltage Δ/e. It may then be that the first peak, representative for the gap, is modified by charging effects and the subsequent peaks are due to the charging.

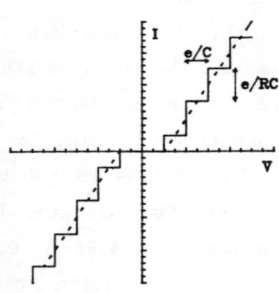

Fig. 4. Coulomb staircase with e/2C ofset

If this model of charging effects is taken into account in examination of experimental data, it turns out that in many point contact junctions both on La-Sr-Cu-O and Y-Ba-Cu-O ([12] for example) and in planar junctions experiments [26] a series of peaks appears close to the 1,3,5, sequence expected for charging. However, the energy of the first peaks appears to scale with T_c in going from the La-Sr-Cu-O to Y-Ba-Cu-O materials. Superconductivity sets the energy scale, at least for the first gap.

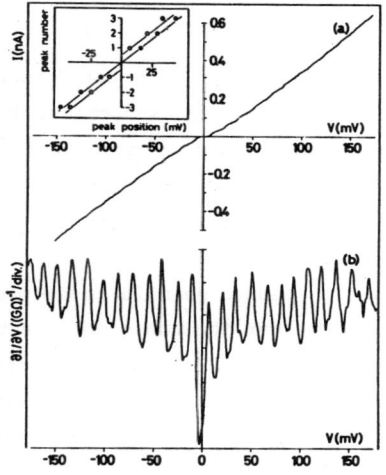
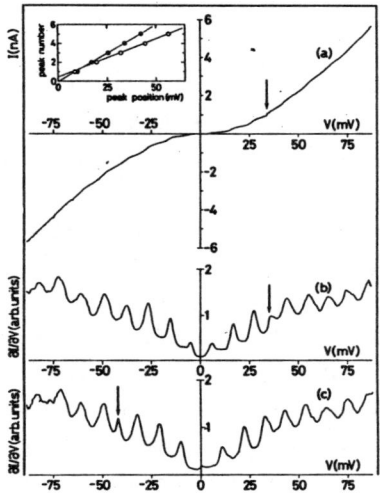

Fig. 5. I-V and dI/dV of point contact on granular Al [25]

Fig. 6. I-V and dI/dV of point contact on single crystal Y-Ba-Cu-O [25]

Assuming that the superconducting ceramic consists of particles spherical in shape embedded in dielectric medium of constant $\varepsilon = 8$ (arbitrarily taken like for Al_2O_3), the peak spacing $\Delta V = 3.6/d$, d being the mean particle diameter (ΔV and d in units of volts and **Å**, respectively). Taking into account peaks positions from [12] as 36, 108, 180 mV, or from [26] as 44, 138, 219 mV, the average particle size is in the range of 100 **Å**. This simple relation between peak spacing ΔV and particle diameter d could be applied to an estimation of effective mean particle size in HTS.

7. CONCLUSIONS

In conclusion I would like to emphasize that ETS can really encounter many difficulties when applied to high-T_c superconductors. But frankly speaking, all experimental methods were in similar situations - even the simple measurements of electrical resistivity of ceramic pellets or thin films have been meeting some problems with contact resistance. All this these traps are due to peculiar properties of HTS.

It is clear that in ETS special attention should be paid to surface degradation phenomena. It turned out very soon that the charging effects of small particles are also of great importance. However, as our understanding of properties of HTS increases, the investigation methods will be better applied and bring more reliable and properly interpreted results. The same concerns the ETS, of course.

ACKNOWLEDGEMENTS

Thanks are due to Profs. H. Szymczak and A. Pajączkowska for their stimulating interest and to Drs Przysłupski, R. Czajka and A. Reich for cooperation and valuable discussions. This work was financially supported by the Governmental Program RPBP 01.09.

REFERENCES

[1] Giaever, I., Phys.Rev. Lett. **5**, 147 (1960); Rev.Mod.Phys. **46**, 245 (1984).
[2] Bednorz, J.G. and Müller, K.A., Z.Physik **64**, 189 (1986).
[3] Cava, R.J., van Dover, R.B., Batlog, B. and Rietman, E.A., Phys. Rev.Lett. **58**, 408 (1987).
[4] Wu, M.K., Ashburn, J.R., Torn, C.J., Hor, P.H., Meng, R.L., Gao, L., Huang, Z.J., Zang, Y.Q. and Chen C.Z., Phys.Rev.Lett. **58**, 908 (1987).
[5] Maeda, M., Tanaka, S., Fukutomi, S. and Asano, T., Jpn J.Appl. Phys.Lett. **27**, L209 (1988).

[6] Koch, R.H., Umbach, C.P., Clark, G.J., Chaudhari, P. and Leibowitz Appl.Phys.Lett.**51**, 200 (1987).
[7] Koch, R.H., Cantor, R., Marek, J.F., Eikenbush, H. and Schöllhorn, R., Phys.Rev. **B36**, 722 (1987).
[8] Moreland, J., Clark, A.F., Ku, H.C. and Shelton, R.N., Cryogenics **27**, 227 (1987).
[9] Barone, A., Physica C **153-155**, 1712 (1988).
[10] Tozer, S.W., Kleinsasser, A.W., Penney, T., Kaiser, D. and Holtzberg, F., Phys.Rev.Lett. **59**, 1768 (1987).
[11] Chaudhari, P., Collins, R.T., Freitas, P., Gambino, R.J., Kirtley, J.R., Koch, R.H., Leibowitz, R.B., Le Goues, F.K., McGuire, T.R., Penney, T., Schlessinger, Z., Segemüller, A.P., Foner,S. and McNiff, E.J., Phys.Rev. **B36**, 8903 (1987).
[12] Kirtley, R., Collins, R.T., Schlessinger, Z., Gallagher W.J., Sandstrom, R.L., Dinger T.R. and Chance, D.A., Phys.Rev. **B35**, 8846 (1987).
[13] Ekino, T. and Akimitsu, J., Jpn J.Appl.Phys. **26**, L452 (1987).
[14] Kirk, M.D., Smith, D.P.E., Mitzi, D.B., Sun, J.Z., Webb, D.J., Char, K., Hahn, M.R., Naito, M., Oh, B., Beasley M.R., Geballe T.H., Hammond R.H., Kapitulnik A. and Quate C.F., Phys.Rev. **B35**, 8850 (1987).
[15] Moreland, J., Ekin, J.W., Goodrich, L.F., Capobianco, T.E., Clark, A.F., Kwo, J., Hong, M. and Liu, S.H., Phys.Rev. **B35**, 8856 (1987).
[16] Schlessinger, Z., Green, R.L., Bednorz,J.G. and Müller. K.A., Phys.Rev. **B35**, 5334 (9187).
[17] Garcia, N., Vieira, S., Baro, A.M., Tornero, J., Pazos, M., Vazquez, L., Gomez J., Aguiló, A., Bourgeal, S., Buendia, A., Hortal, M., López de la Torre, M.A., Ramos, M.A., Villar, R., Rao, K.V., Chen, D,-X., Nogues, J. and Karpe, N., Z.Physik **70**, 9 (1988)
[18] Fein, A.P., Kirtley, J.R. and Schafer, M.W., Phys.Rev. **B37**, 9738 (1988).
[19] Schlessinger, Z., Collins, R.T., Kaiser, D.L. and Holtzberg, F., Phys.Rev.Lett. **59**, 1958 (1987).
[20] Warren, W.W., Walsted, R.E., Brennert, G.F., Espinoza, G.F. and Remeika, J.P., Phys.Rev.Lett. **59**, 1860 (1987).

[21] Kirtley, J.R., Feenstra, R.M., Fein, A.P., Raider S.I., Gallagher, W.J., Sandstrom, R., Dinger, T., Shafer, M.W., Koch, R., Leibowitz, R. and Bumble, B., J.Vac.Sci.Technol. **A6**, 259 (1988).
[22] Giaever, I. and Zeller, H.R., Phys.Rev.Lett. **20**, 1504 (1968); Zeller, H.R. and Giaever, I., Phys.Rev. **181**, 789 (1969).
[23] Mullen, K., Ben-Jacob, E., Jaklevic, R.C. and Schuss,Z., Phys.Rev. **B37**, 98 (1988).
[24] Ruggiero, S.T. and Barner, J.B., Phys.Rev. **B36**, 8870 (1987).
[25] van Bentum, P.J.M., Smokers, R.T.M. and van Kempen, H., Phys.Rev. Lett. **60**, 2543 (1988).
[26] Raułuszkiewicz, J. and Reich, A., this Conference, paper A-17.

#

HEAT CAPACITY AND TRANSPORT PROPERTIES OF COPPER OXIDE SUPERCONDUCTORS

S. von Molnár, J. M. D. Coey* and P. Strobel**

IBM Research Division
Thomas J. Watson Research Center
Yorktown Heights, NY 10598

ABSTRACT

Results on the electrical conductivity, Hall effect and specific heat of ceramic and single crystal samples of the main families of copper oxide superconductors are reviewed, with a view to establishing which intrinsic properties are characteristic of all superconductors with $(CuO_2)_\infty$ planes. These include an in-plane resistivity linear in temperature and p-type in-plane Hall effect. The linear low-temperature specific heat found for many ceramic and single crystal samples is, in part, an intrinsic effect of electronic origin that does not appear to be directly related to superconductivity. A possible origin is localized states associated with disordered oxygen vacancies in these compounds.

---

*Permanent address: Physics Department, Trinity College, Dublin, Ireland
**Permanent address: Laboratoire de Cristallographie C.N.R.S., Grenoble, France.

INTRODUCTION

Oxide superconductors have provided an unprecedented stimulus for solid state physicist over the past two years. The search for new materials, the efforts to characterize their intrinsic physical properties and the attempts to explain them theoretically have led to an avalanche of results - seven hundred papers in the proceedings of one recent conference alone.[1] Here we attempt to review results in two related areas, transport properties and specific heat, which should offer some insight into the electronic structure of this new and still puzzling class of materials. No attempt is made at a full literature survey: we have simply selected some representative data that are of interest.

So far, it appears that all materials with a superconducting transition temperature T_c above the boiling point of liquid nitrogen possess a common motif in their crystal structures, namely infinite two dimensional $(CuO_2)_\infty$ sheets where the copper is at the center of corner-sharing oxygen squares. Several of these may be intercalated by sheets of calcium cations; n sheets of (CuO_2) and (n-1) alkaline earth sheets form a stack of composition $A_{n-1}Cu_nO_{2n}$. The superconducting transition temperature is found to increase with increasing n, such that all compounds with $T_c > 77$ K have $n \geq 2$. Multiple sheets of bismuth or thallium oxide with Ba or Sr can be inserted between the stacks leading to an extensive family of layered compounds, all with structures related to that of perovskite $CaTiO_3$.

Superconducting oxides with a single $(CuO_2)_\infty$ sheet having transition temperatures in the range 5-40 K include the original Zurich oxide $(La_{1-x}Ba_x)_2CuO_4$,[2] and the 1-layer bismuth compound $Bi_2Sr_2CuO_x$.[3,4] Superconducting perovskite structure oxides containing no $[CuO_2]_\infty$ sheets also exist (n = 0), but superconductivity in these barium bismuth oxides appears in a narrow composition range. Table I provides a classification of the various oxide superconductors, and some structures are illustrated in Fig. 1.

Table I: Classification of Oxide Superconductors

n	Compound	T_c (K)	References
0	$Ba(Bi_xPb_{1-x})O_3$	11 (x = 0.25)	5
	$(Ba_xK_{1-x})BiO_3$	30 (x = 0.6)	6
1	$(La_xBa_{1-x})_2CuO_4$	35 (x≈0.05)	2
	$Bi_2Sr_2CuO_{6+y}$	10	3, 4
2	$YBa_2Cu_3O_7$	92	7
	$Tl_2Ba_2CaCu_2O_7$	91	8
	$Bi_2Sr_2CaCu_2O_{8+y}$	85	3, 9
3	$Tl_2Ba_2Ca_2Cu_3O_{10+y}$	120	8, 10
4	$Tl_2Ba_2Ca_2Cu_4O_{11}$	122	11

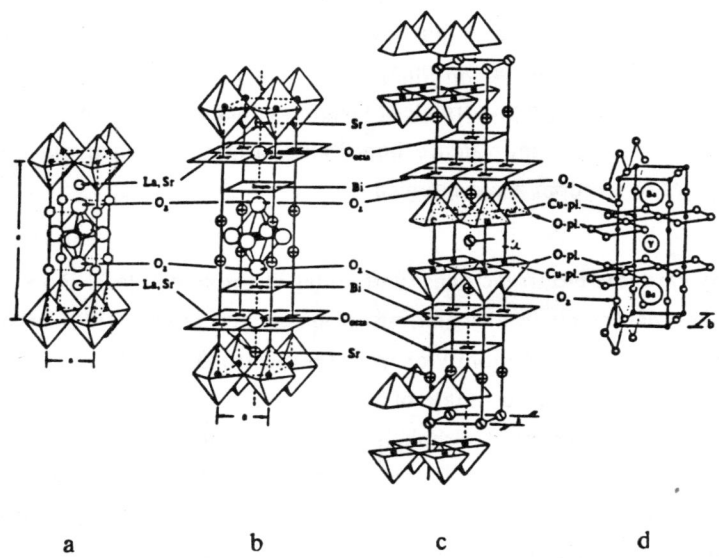

Figure 1. Crystal structures of superconducting oxides with different numbers of $(CuO_2)_\infty$ planes: a) $(La,Sr)_2CuO_4$ and b) $Bi_2Sr_2CuO_{6+y}$ (n = 1); c) $Bi_2Sr_2CaCu_2O_{8+y}$ and d) $YBa_2Cu_3O_7$ (n = 2).

To make progress towards an understanding of the superconductivity, it is desirable to have a clear picture of the electronic properties of the normal state. Except for n = 0 compounds, the layer structures are anisotropic and transport measurements on single crystals are a good starting point. The materials are poor metals, with conductivities that are 2-4 orders of magnitude less than those of copper. When there is only one partly-filled band to consider, the sign and the number of carriers can be obtained directly from the Hall coefficient. The Seebeck coefficient, however, may be of either sign since it depends on the detailed curvature of the density of states.[12] If more than one band contributes to the transport, or if the mobility is anisotropic, additional information is needed to analyze the data.

The nature of the electronic states of the Fermi energy is still a matter of intense controversy.[13] In the original work of Bednorz and Müller, they were thought to be polaronic (or bipolaronic) 3d(Cu) states.[2] Now it seems more likely that the electrons are highly correlated, yielding either a two-dimensional Fermi liquid, or a novel 'resonant valence bond' (RVB) state[14,15] where the spin and charge degrees of freedom are decoupled.

In any case, the electronic density of states at the Fermi level in the normal state is a key parameter in any analysis of superconductivity. It is usually obtained from the Pauli susceptibility or from the electronic specific heat coefficient γ_N in the normal state. An estimate of γ_N can also be made from the magnitude of the specific heat jump at T_c although the constant of proportionality depends somewhat on theory. Far below T_c the electronic specific heat is expected to disappear exponentially with T as a superconducting energy gap opens up at the Fermi level. In principle the normal state can be restored by applying a magnetic field in excess of the upper critical field H_{c2}. All of these classic techniques pose problems when applied to the present materials, however. Susceptibilities are difficult to interpret because the large number of Cu ions present in these compounds are close to having localized magnetic moments and often order antiferromagnetically following small changes in composition or oxygen stoichiometry. Furthermore, H_{c2} is so large in high T_c copper oxide superconductors that quenching is impossible in normal laboratory fields. Finally, a finite linear specific heat term is observed in the superconducting state of most high T_c materials. If intrinsic, this result suggests unusual superconducting properties or at the very least (localized) electronic states at E_F. It also means that the specific heat jump does not define γ_N uniquely. Nevertheless the magnitude of the jump in heat capacity at T_c provides virtually the only reliable method of determining, albeit qualitatively, whether the superconductivity is a bulk effect in a material in question, rather than a filamentary or surface phenomenon. These matters are discussed in detail in the following paragraphs.

TRANSPORT PROPERTIES

Conductivity

A feature of copper oxide superconductors is their poor and highly anisotropic normal state conductivity.[16] Dense ceramic samples typically have resistivities in the $\mu\Omega$-cm range at room temperature. However, the Yoffe-Regel criterion for diffusive transport, $k_F \ell > 1$, is obeyed below room temperature at least in the Zürich oxide and $YBa_2Cu_3O_7$, where enough information exists to evaluate k_F, the Fermi wavevector and ℓ, the mean free path. It may, therefore, be more useful to think of these materials as degenerate semiconductors rather than exceptionally poor metals. Furthermore, ℓ, the mean free path derived from a simple parabolic band approximation is in the range of 5 - 10 Å for both compounds and in either two or three dimensions. This value is smaller than the c-axis lattice constant which raises the question whether the conductivity is two or three dimensional.

The best crystals and epitaxial films show in-plane resistivity ρ_{ab} of order 200 $\mu\Omega$-cm; resistivity in the c-direction is much higher. The resistivity has a large positive temperature coefficient, $(1/\rho_{ab})(d\rho_{ab}/dT) \simeq 3 \times 10^{-3}$, and the linear variation extends down to T_c and up to at least 600K.[17] The extrapolated residual resistivity is close to zero.[18] Typical data on a $YBa_2Cu_3O_7$ crystal[19] are shown in Fig. 2. Similar dependence of the in-plane resistivity is reported for epitaxial $YBa_2Cu_3O_7$ films,[20,21] $(La,Sr)_2CuO_4$ films[22] and $Bi_2Sr_2CaCu_2O_x$ crystals.[9,23]

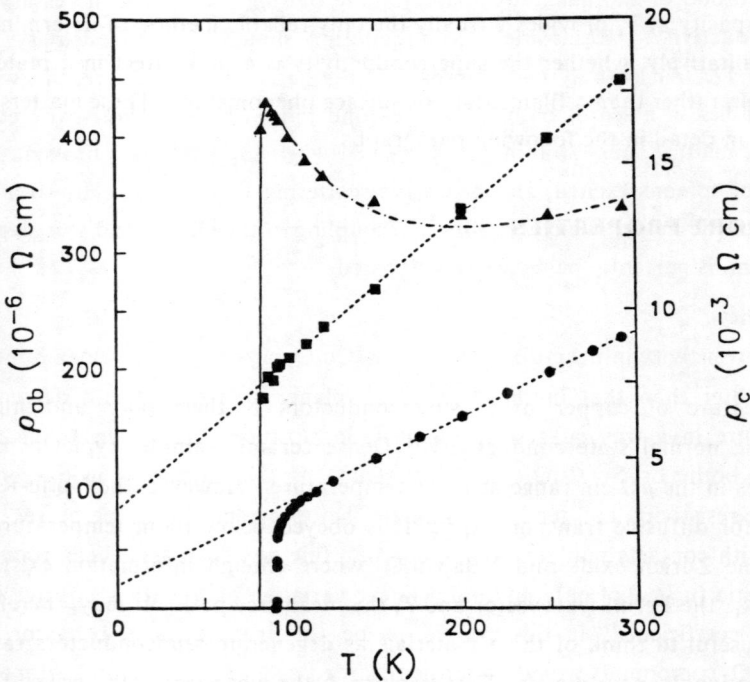

Figure 2. Resistivity of crystals of $YBa_2Cu_3O_7$ in twinned ab plane (■ and ●), and along the c-direction (▲) (from ref. 19).

The usual theory for a metal, based on electron-phonon scattering, gives a resistivity effectively linear in T for $T \gtrsim 0.2\Theta_D$, and a T^n term at low temperatures (n = 3-5, Θ_D is the Debye temperature).[24] Electron-phonon scattering with a two-dimensional electronic structure permits the linear term to be extended to yet lower temperatures.[25] Thus, even the lower T_c materials, such as $(La, Sr)_2CuO_4$ might be explained in this manner. However, because the linear term extends to such high temperatures without evidence of saturation, it has been argued[17] that electron-phonon coupling must be weak and therefore cannot account either for the magnitude of the linear term nor for the pairing mechanism necessary for superconductivity. Electron-electron scattering should give a resistivity that varies as T^2. An entirely new approach is the gapless RVB picture,[14] in which spinless holes in a sea of mobile valence-bond pairs ('holons') are scattered by 'spinons', neutral spin 1/2 fermions produced when the bonds are broken. The number of spinons excited, and hence the scattering varies as T. However this analysis of the model, which requires a decoupling of the charge and spin degrees of freedom, is currently being actively debated.

The resistivity in the c-direction of $YBa_2Cu_3O_7$ crystals is typically a hundred times greater than that in the twinned ab plane; it increases with decreasing temperature in a way suggesting incipient localization. However it has been found in both $(La, Sr)_2CuO_4$[22] and $YBa_2Cu_3O_7$ [26] that annealing in oxygen reduces the resistivity, and tends to eliminate the upturn at low temperature. One crystal with contacts made by spark bonding fine gold wires has been found to have a ratio (ρ_c/ρ_{ab}) of only 40, and a linear variation of ρ_c with a <u>positive</u> temperature coefficient.[27] If this proves to be the true intrinsic behavior, the $RBa_2Cu_3O_7$ compounds would better be regarded as highly anisotropic 3d metals, rather than 2d metals. In any case, measurements along c (perpendicular to the layers) must be considered with caution, since thick crystals of layered compounds such as $RBa_2Cu_3O_7$ or Bi/Tl superconductors tend to exhibit various imperfections (stacking faults, delamination) along c.

Anisotropy of the resistivity in the a-b plane is difficult to measure because of twinning. There is a report of in-plane anisotropy[28] in an untwinned crystal of $Y(Ba,Sr)_2Cu_3O_7$; a sample with $T_c = 82$ K shows $\rho_b << \rho_a \lesssim \rho_c$, indicating that conduction is easiest along the direction of the Cu_1 chains ($\rho_b = 2\mu\Omega cm$) There is also evidence from studies of optical dichroism that the optical conductivity is much greater along $\vec{b}$ than along $\vec{a}$ or $\vec{c}$.[29] If the in-plane anisotropy is confirmed, one may have to seek an explanation of ρ_{ab} in terms of a temperature-dependent twin boundary contribution to the resistivity. At present it appears more likely that the large resistivity, linear in T, is an intrinsic property of the $(CuO_2)_\infty$ sheets, since it is found in the layered bismuth and thallium copper oxides and doped $LaCuO_4$, as well as $YBa_2Cu_3O_7$.

The states at the Fermi level, responsible for the conductivity, have recently been identified in an angular-resolved resonant photoemission experiment on a $Bi_2CaSr_2Cu_2O_8$ crystal.[30] They are holes with oxygen 2p character, which occupy a narrow band with a Fermi surface that occupies approximately half the Brillouin zone, suggesting a strongly correlated Fermi liquid state. The copper 3d electrons are thought by the authors to be localized, as in a Mott insulator. Additional photoelectron and polarization dependent near edge absorption spectroscopies have confirmed that the oxygen holes have $p_{x,y}$ character, where x, y defines the CuO planes.[31]

Hall Effect

In many studies of ceramic samples, the sign of the Hall coefficient R_H has been found to be positive (p-type) for all the families of copper oxide superconductors, indicating that holes make the dominant contribution.[16] The Hall coefficient R_H in a one-band model gives the carrier concentration n_H directly; $R_H = 1/n_H e$.

Interpretation of the Hall data is rather straightforward in the $(La_{2-x}Sr_x)CuO_{4-\delta}$ system, which retains the K_2NiF_4 structure over a wide range of x, $0 \leq x \leq 1.33$. Substitution of Sr for La introduces holes. At values of x greater than 0.15, however, compensating oxygen vacancies begin to appear.[32] Formally the number of holes/Cu may be written as $p = x - 2\delta$. The accessible range of hole concentration may be extended by preparing samples under oxygen pressure so that a wide range, $0 \leq p \leq 0.4$ may be explored.[33] The excess charge on the Cu-O units is best determined by iodometric titration.[32,34] Charge measured in this way is due to any holes relative to ($Cu^{2+} - O^{2-}$), whether these are predominantly of Cu^{3+} character, or, as is now more widely accepted, of O^{1-} (peroxide) character.[35] The chemically-determined hole concentration p, when plotted against x, follows a straight line with a slope of unity up to the point where oxygen vacancies form.[33] The variation of T_c in $(La_{2-x}Sr_x)CuO_4$ with p is plotted in Fig. 3. There is a broad maximum around $p \approx 0.2$ and the material is not superconducting when the hole concentration is either too small $p < 0.05$, or too large $p > 0.30$.

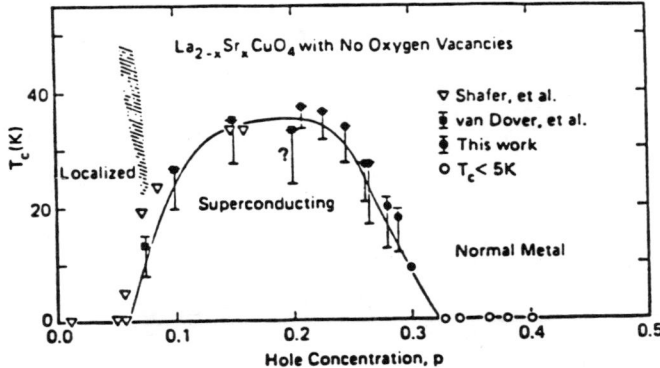

Figure 3. The variation of T_c with hole concentration in $(La, Sr)_2CuO_4$ (from ref. 33).

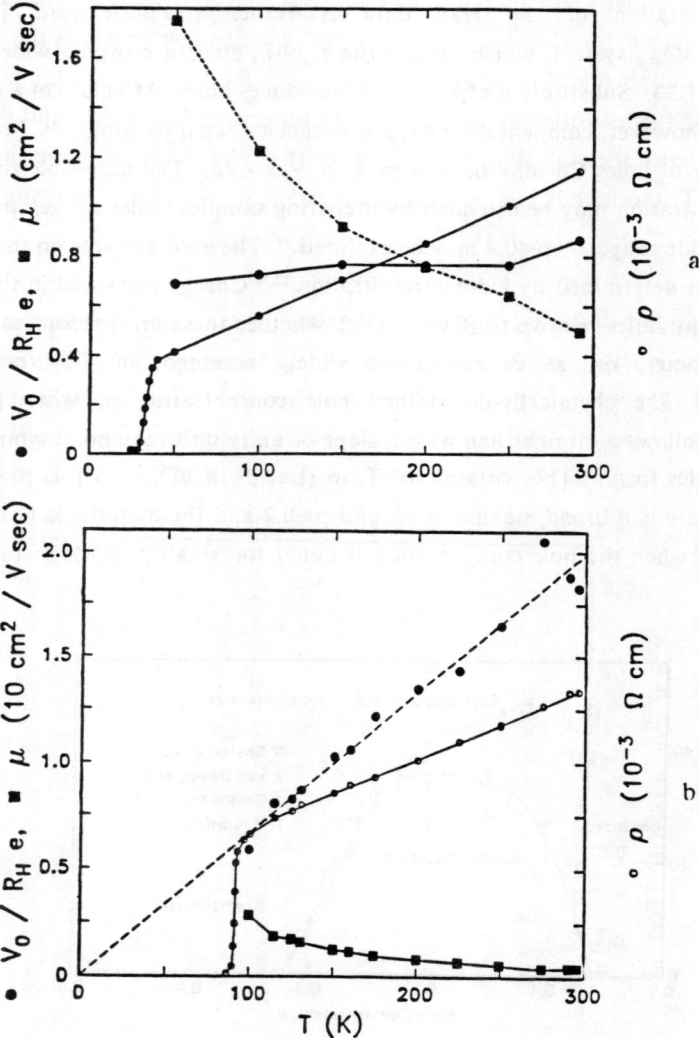

Figure 4. Resistivity (○), Hall number (•) and Hall mobility (■) for a) $(La, Sr)_2CuO_4$ and b) $YBa_2Cu_3O_7$ (from ref. 36).

The Hall coefficient of polycrystalline $(La,Sr)_2CuO_4$ shows little variation with temperature; the Hall number n_H, defined as V_0/R_He where V_0 is the volume of the unit cell, is found to be practically equal to the hole concentration p indicating that Sr substitution creates holes in an otherwise-filled band, and that one band dominates the transport.[32] Since the resistivity of the oxide varies approximately as T, the mobility $\mu_H = (\sigma/n_H e)$ derived from the Hall effect, must vary as $\sim 1/T$.

The Hall effect in ceramic samples of $YBa_2Cu_3O_7$ is also p-type, but differs from $(La,Sr)_2CuO_4$ in that the Hall number n_H is not constant, but varies proportionally to T ($R_H \propto 1/T$). Since the resistivity also varies as T, a one-band model yields the result[36] that $\mu_H \sim 1/T^2$. Data on ceramic samples of $(La,Sr)_2CuO_4$ and $YBa_2Cu_3O_7$ may be compared in Fig. 4.

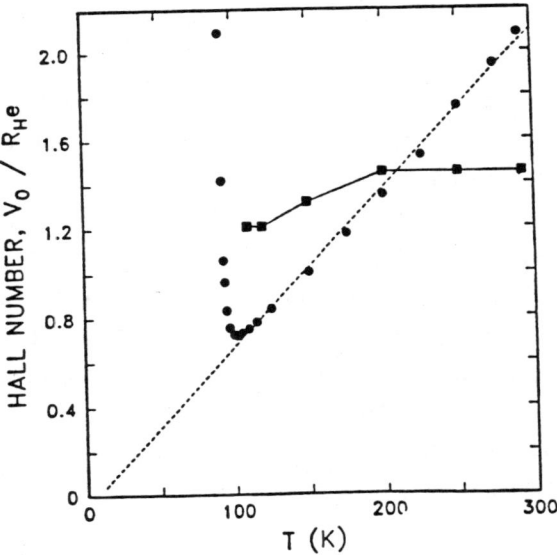

Figure 5. Hall number for a crystal of $YBa_2Cu_3O_7$, with the field applied parallel to $\vec{c}$ (•) or perpendicular to $\vec{c}$ (■) (from ref. 19).

Hall effect measurements on single-crystal $YBa_2Cu_3O_{7-\delta}$ (Fig. 5) show quite different results according to whether the magnetic field is applied parallel or perpendicular to $\vec{c}$.[19,37] The Hall coefficient when the magnetic field is parallel to $\vec{c}$ is p-type in crystals and epitaxial films[20,21] and varies as $1/T$, like that of the ceramic samples. However, when the field is applied perpendicular to $\vec{c}$, R_H is n − type[37] and its magnitude decreases with decreasing temperature.[38] Two-band models[39] with a very narrow peak in the density of states at E_F[21] or with additional localized states[40] have been suggested to account for the observed Hall effect. The signs of R_H are consistent with RVB theory.[14]

Quantitative analysis of the Hall effect in the $YBa_2Cu_3O_7$ system is complicated by the temperature dependent Hall number. The Hall effect does however, give the trend in the number of mobile carriers.[41,42] The situation in $RBa_2Cu_3O_y$ is further complicated by the possibility of localized holes, probably situated in the planes occupied by Cu_1 (chain site), in addition to the mobile holes in the Cu_2 planes. In a recent study, Tokura et al.[43] have established a line separating metals from insulators by examining numerous samples in the $(Y_{1-x}Ca_x)Ba_2Cu_3O_y$ and $Y(Ba_{2-x}La_x)Cu_3O_y$ systems (Fig. 6). (Note that oxygen in excess of 7 can be accommodated in the Cu_1 planes).
The number of holes, p, per copper is determined by wet chemical analysis, as for the $(LaSr)_2CuO_4$ series. $YBa_2Cu_3O_7$ is indicated by the star. The behavior of $YBa_2Cu_3O_y$ indicates that T_c increases on moving to the right away from the metal-insulator line. Other solid solutions (e.g., Zn or Ga for Cu; Nd for Ba) suggest that there is a ridge of maximum T_c (dotted) that runs parallel to the metal-insulator line at a distance $\Delta p \simeq 0.25$. The optimum number of mobile holes per copper in the $(CuO_2)_\infty$ sheets for maximum T_c is therefore quite similar to that found in $(La,Sr)_2CuO_4$.[42,44]

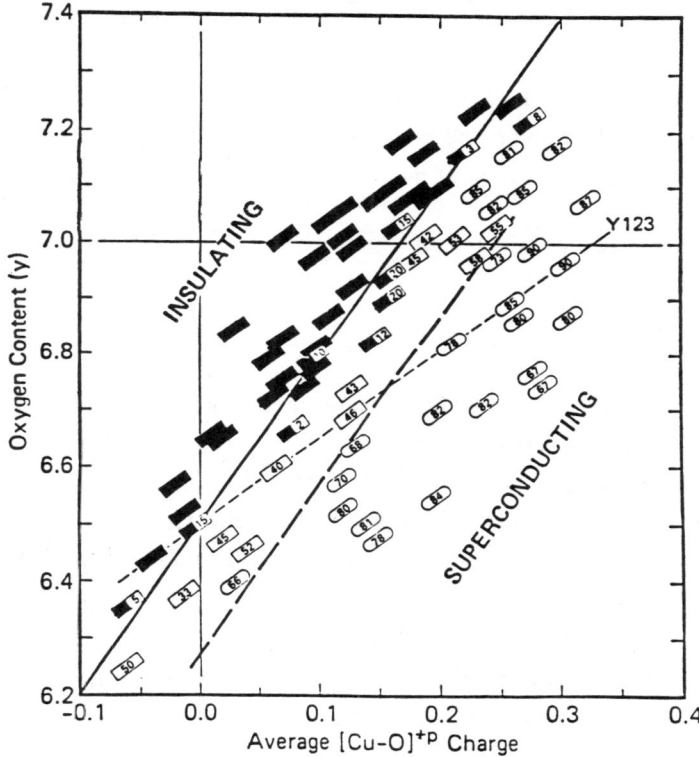

Figure 6. The oxygen content (±0.03) measured for a series of isostructural samples plotted versus the average $[Cu - O]^{+p}$ charge (±0.02). The short dashed line labeled Y123 is for $YBa_2Cu_3O_y$. The solid line is drawn to separate insulating and superconducting phases, while the long dashed line is drawn to separate high T_c phase from that with $20K < T_c < 65K$, where T_c is the temperature of zero resistance (from ref. 43).

In the bismuth oxides, R_H has been found to vary as $1/T$ in $Bi_2Ba_2CaCu_2O_x$ ceramics,[45] although a less-marked temperature dependence is found in a Bi crystal[46] or in a $Tl_2Ca_2Ba_2Cu_3O_x$ ceramic.[45] Both systems give a value of n_p per copper near T_c of 0.1, distinctly less than in the old established families. This raises the question of whether it might be possible to increase T_c still further in these systems by doping and controlling the oxygen content in such a way as to augment the concentration of mobile holes in the $(CuO_2)_\infty$ sheets.

SPECIFIC HEAT

Specific Heat at T_c

The specific heat anomaly where the energy gap falls continuously to zero in a conventional superconductor is a simple discontinuity, of the type expected from the mean field theory of a second-order phase transition. In this respect the superconducting transition is unlike a magnetic Curie or Néel point, where short-range order persists some way above T_c, and fluctuations of the short range order occur in a critical region of temperature extending on either side of T_c.

The jump in the specific heat at T_c, ΔC, is related to the electronic density of states in the normal state, γ_N, through the relation

$$\Delta C = K\gamma_N T_c \tag{1}$$

where K is a constant depending on the coupling regime. A classical argument gives $K = 2$; the BCS theory in the weak-coupling limit gives $K = 1.43$, but strong coupling corrections increase this value somewhat.[47]

Specific heat anomalies at T_c reported on polycrystalline oxide superconductors are usually rounded. Typical data by Fisher et al.[48] on polycrystalline $YBa_2Cu_3O_7$ in the range 0.5-100 K are shown in Fig. 7. Junod et al. have studied many ceramic samples of $YBa_2Cu_3O_7$ prepared in different ways.[49] The largest measured jump $\Delta C/T_c$ is about 40 mJ/mole K^2, but when the data are extrapolated via a constant entropy construction to obtain the idealized jump, the discontinuity becomes 57 mJ/mole K². Hence, from Eq. 1, $\gamma_N \lesssim 40$ mJ/mole K². The exact number depends on how the large lattice specific heat is extrapolated in the vicinity of T_c, but the value is plausible, and demonstrates that the superconductivity is essentially a bulk phenomenon. However, it is difficult to estimate the density of states in the normal state independently. The critical field $H_{c2}(0)$ is too high for low temperature specific heat measurements in the normal state to be feasible; estimates from the paramagnetic susceptibility require a knowledge of the contribution of the copper ions; estimates from dH_{c2}/dT near T_c are subject to uncertainty due to the choice of transition point on the curves, and depend on the BCS model.

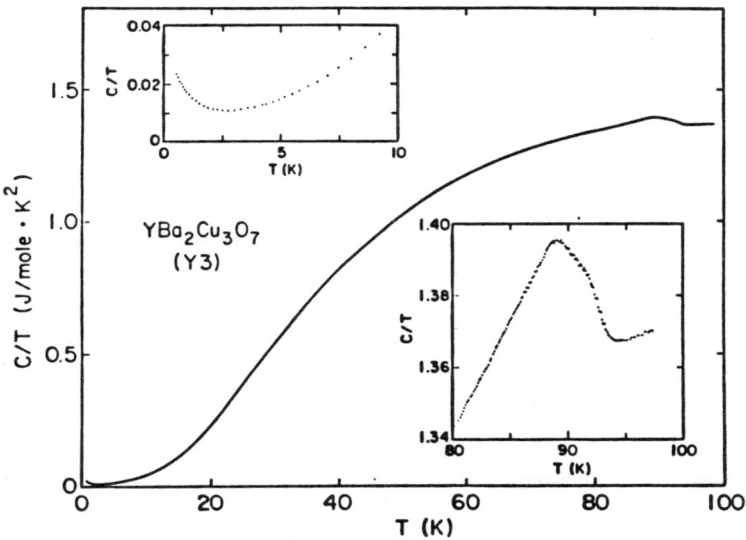

Figure 7. Heat capacity for a ceramic sample of $YBa_2Cu_3O_7$ in the range from 0 - 100 K. Inserts show the anomaly near T_c, and the low temperature variation (from ref. 48).

Measurements by Inderhees et al.[50] on small crystals of $YBa_2Cu_3O_7$ in the range 87-93 K were the first to suggest that critical fluctuation may be observable in high T_c superconductors. Some of their data are shown in Fig. 8a.
Critical fluctuations are normally absent in conventional superconductors because of the long zero-temperature coherence length, ξ_0, of order 1000 Å. Ginzburg gives a criterion for the width of the critical region $\delta T/T_c$ for an isotropic three-dimensional superconductor,

$$\delta T/T_c \sim [k/\Delta C \rho \xi_0^3]^2 \tag{2}$$

where ρ is the density and ΔC is the jump in specific heat at T_c. For the $YBa_2Cu_3O_7$ under consideration in ref. 50, $\rho = 6.4$ g/cc, $T_c = 89.7$ K and $\Delta C = 6.2$ mJ/g K. The correlation length ξ is of order 10 Å[51] hence the critical region given by Eq. 2 is some hundredths of a degree in width. The heat capacity on either side of T_c after subtraction of the BCS term and the background is shown in Fig. 8b fitted to

$$\Delta C = C^{\pm}(\pm\varepsilon)^{-1/2} \tag{3}$$

where $\varepsilon = (T/T_c - 1)$. From the ratio C^+/C^-, it is concluded that the order parameter of $YBa_2Cu_3O_7$ is not the order parameter of classical Ginzburg Landau theory. The number of components is a matter of dispute.[52]

There is little data available on specific heat anomalies of compounds with $n<2$. No anomaly whatever was observed at T_c for $Ba(PbBi)O_3$,[53] or in some crystals of $Bi_2Sr_2CuO_6$ we have examined. The absence of the anomaly is not understood although perhaps due to sample inhomogeneity.

Low-temperature specific heat

Although the electronic specific heat in the superconducting state, $C_{es} \sim e^{-b/T}$, is expected to disappear as $T \to 0$, almost all the La and Y based

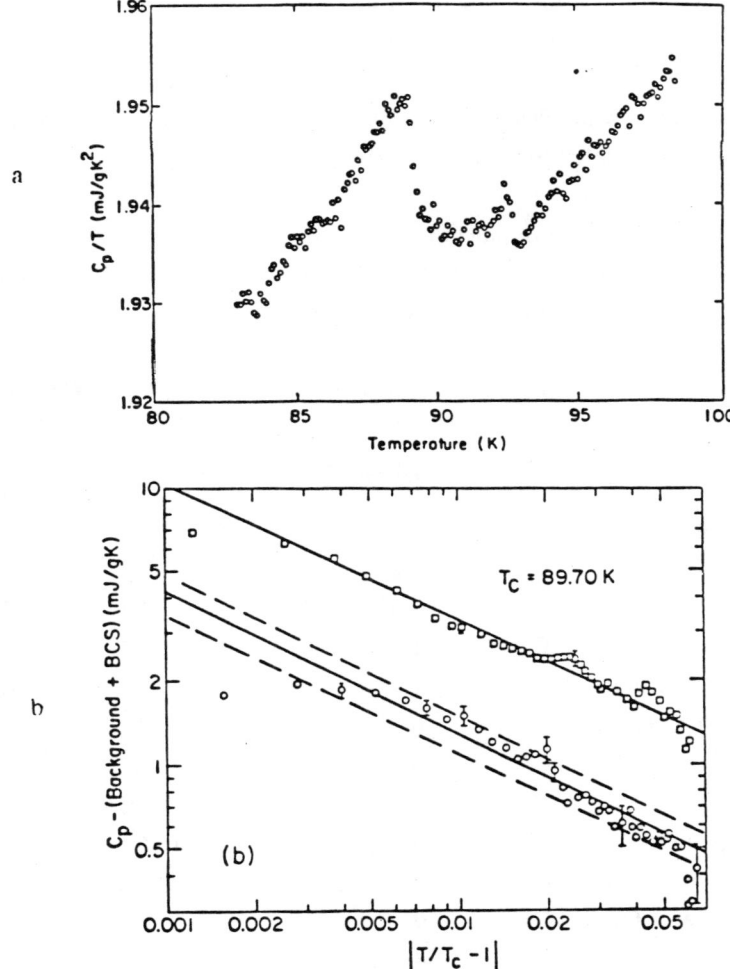

Figure 8. a) Heat capacity, c_p, of a small crystals of $YBa_2Cu_3O_7$ in the vicinity of T_c and b) analysis of the data above and below T_c for a similar crystal to show the contribution of critical fluctuations (from ref. 50).

high-T_c oxides are found to exhibit a linear term in the specific heat, $\gamma(0)T$, at low temperature.[44] This might argue against conventional theories for superconductivity in these materials, and, in our view, still remains one of the interesting unsolved problems. In the following paragraphs we consider some of the evidence which leads us to believe that this effect is, in part, intrinsic, but probably unrelated to superconductivity.

A compendium of most of the data on ceramics may be found in ref. 44; selected values of $\gamma(0)$, γ_N and Θ_D derived from specific heat data on a variety of copper oxide superconductors are given in Table II.

Table II: Specific Heat Parameter for Various High T_c Oxide Compounds.

Material	Type	$\gamma(0)$	γ_N	$\Theta(K)$	$T_c(K)$	Reference
La_2CuO_4	C*	1.1		460	-	55
La_2CuO_4	C	0.2		280	-	56
$La_{1.85}Ca_{0.15}CuO_4$	C	3.05	4.4	450	22	55
$La_{1.85}Sr_{0.15}CuO_4$	C	1.54	8.6	430	37	55
$La_{1.85}Sr_{0.15}CuO_4$		1.8	6±2	285	37	56
$La_{1-x}Sr_xCuO_4$	C	0-5				57
$La_{1-x}Ba_xCuO_4$	C	0-9		345†		57
$YBa_2Cu_3O_7$	C	5.6	30	410	90-95	55, 58
$YBa_2Cu_3O_{7-\delta}$	MC*	5-9		300-410	0-89	59
$TlCaBaCu_2O_{5.5}$	C	16		270	114	60
$Bi_2Sr_2CaCu_2O_8$	C	0		250	84	60
$Bi_2Sr_2CaCu_2O_8$	C	4±2		260	86	61
$Bi_2Sr_2CaCu_2O_8$	MC	10±2		240	85	62
$Bi_2Sr_2CuO_6$	MC	~0	10±2	195	11	62

* C = ceramic; MC = multi-crystal
† Quoted value corrected for the number of atoms/formula

As an illustration of the data analysis, we refer again in Fig. 7 to the work of Fisher et al.[48] for $YBa_2Cu_3O_7$. The low-temperature data are fitted to a function of the form

$$C = A_{-3}T^{-3} + A_{-2}T^{-2} + A_1T + A_3T^3 + A_5T^5 + \quad (4)$$

where the first two terms represent the upturn at low temperature, $A_1 = \gamma(0)$, and A_3 and A_5 are lattice terms, with $A_3 = 1944r/\Theta_D^3$ where r is the number of atoms performula and Θ_D the familiar Debye temperature. If all these terms are present in the specific heat the extraction of $\gamma(0)$ and Θ_D becomes difficult.

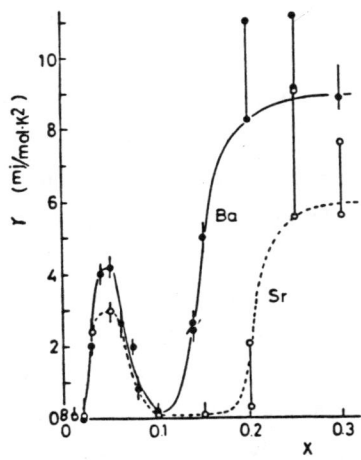

Figure 9. Concentration dependence of γ in $(La_{2-x}Ba_x)CuO_4$ and $(La_{2-x}Sr_x)CuO_4$ (from ref. 57).

Eq. 4 presupposes that the data are taken at sufficient low temperature for the exponential term C_{es} to be neglected. A common procedure is to deduce $\gamma(0)$ and Θ_D from a plot of C/T vs T^2. Such plots show a kink near 22K in

RBa$_2$Cu$_3$O$_7$,[63] which is associated with a dimensional crossover from three to two dimensions. The kink occurs at $\Theta_1/3.9$, where Θ_1 is an effective Debye temperature for the transverse modes propagating parallel to $\vec{c}$. In the n = 1 bismuth copper oxide Θ_1 is even smaller ($\simeq$30K) which means that a T^3 lattice term is observed only in a very limited temperature range.[62]

The question concerning the $\gamma(0)$ term is twofold; a) Is the effect intrinsic to the superconducting phase and b) if so, does it have anything to do with superconductivity? Let us first examine the most obvious extrinsic cause; that the samples are multiphase. Ramirez et al.[64] have shown that only a few percent of BaCuO$_{2+y}$ impurity can account for the observed $\gamma(0)$. However, $\gamma(0)$T terms are found both in Ca[55] and Sr[56,57] doped La$_2$CuO$_4$. Thus, only if unknown Ca or Sr copper oxides also have anomalous low temperature contributions, can this be the principal low-temperature effect. A particularly curious result has been obtained be Kumagai et al.,[57] Fig. 9, who show that $\gamma(0)$ oscillates in value depending on the Sr or Ba concentration, x, in La$_{2-x}$(Sr, Ba)$_x$CuO$_4$. They attribute the vanishing of $\gamma(0)$ for x$\leq$0.02 to antiferromagnetic order, and conclude that non-zero values are intrinsic. Their discussion however, does not address the disappearance of $\gamma(0)$, as in a classic superconductor, for some compositions with x$\gtrsim$0.1. Consequently the finite $\gamma(0)$ values observed for 0.02$\leq$x$\leq$0.1 might indicate a mixture of both normal and superconducting phases. Fig. 10 exhibits data for ceramic YBa$_2$Cu$_3$O$_y$ samples. The correlation between antiferromagnetism, which occurs at about y = 6.3,[65] and $\gamma(0)$ is absent, as is any decrease in $\gamma(0)$ in the superconducting regime. However, the sharp decrease near y = 6 argues against a BaCuO$_{2+y}$ impurity phase being solely responsible for the effect, since all ceramic samples used for this study came from the same preparation. Different oxygen contents were achieved by annealing at controlled oxygen pressure (between 1 and 10^{-4} bar) and by heating in a closed tube containing a buffer sample. The oxygen content was determined gravimetrically.

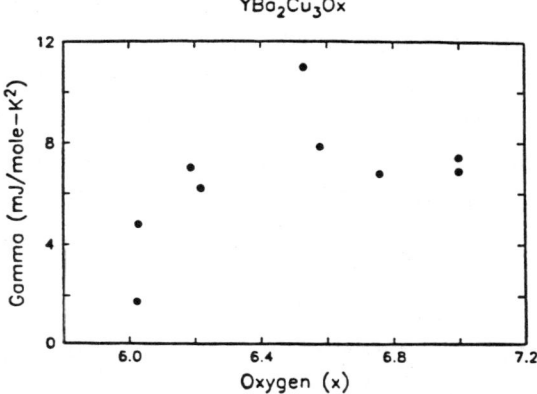

Figure 10. Variation of γ with oxygen content of a ceramic sample of $YBa_2Cu_3O_x$ subjected to different heat treatments.

Perhaps the most compelling evidence for an intrinsic effect comes from Fisher et al.[54] who were able to correlate the low temperature impurity upturn with $\gamma(0)$ for many $RBa_2Cu_3O_{7-y}$ samples. For no impurities they extrapolate an intrinsic $\gamma(0)$ of ~ 7 mJ/mole K² for the 123 type structure.

Single crystal data on $YBa_2Cu_3O_7$ are shown in Fig. 11. The motivation for these experiments rested on the fact that magnetic as well as resistive transitions in individual small crystals were sharper than in comparable ceramics and on the assumption that single crystals would be cleaner (no grain boundaries), thereby reducing impurity effects. Furthermore, the 21 single crystals used in the measurement were annealed (at 600° C for 120 hrs) in vacuum, destroying superconductivity and thereby addressing the effects due to oxygen concentration only. The $\gamma(0)$ term, which in disordered materials is often associated with low energy tunneling,[66] is very large in $YBa_2Cu_3O_7$; $\gamma(0) \simeq 9$ mJ/mole K² $= (87 \times 10^{-6}$ J/ccK²) compared to various insulating and metallic glasses[67] for which $\gamma(0)$ varies from 1-10×10^{-6} J/ccK². Furthermore Golding et al.[68] have recently estimated the tunneling density of states using acoustic techniques and find $\sim 10 \times 10^{-6}$ J/ccK², again almost an order of magnitude smaller than expected from the linear specific heat term. We conclude, therefore, that tunneling states cannot quantitatively explain the low temperature linear specific heat. Furthermore, antiferromagnetic spin fluctuations, even if primarily two-dimensional are unlikely to lead to a linear T term (a T^2 dependence is expected).

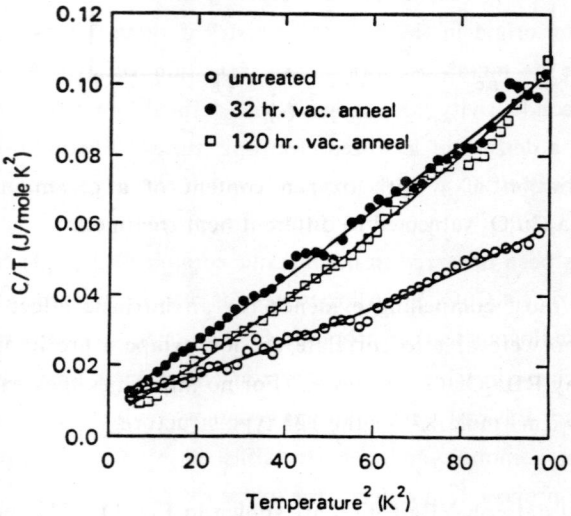

Figure 11. Low temperature specific heat of a sample composed of 21 small crystals of $YBa_2Cu_3O_x$ and its variation with vacuum annealing time as oxygen is removed (from ref. 59).

The foregoing discussion implies that the linear term is at least in part intrinsic and of electronic origin in the La and Y superconducting compounds. It is, therefore, significant that in some recent reports on ceramic samples of the high T_c n = 2 Bi compound[60,69,70] no linear term was observed. We have recently examined crystals of both n = 1 ($Bi_2Sr_2CuO_{6+y}$; $T_c \sim 11K$) and n = 2 ($Bi_2Sr_{1.5}CaCu_2O_x$; $T_c \sim 85K$) compounds.[62] In the n = 1 compound it is possible to measure γ_N directly, since superconductivity can be suppressed above 2K in a 5T applied magnetic field. There $\gamma(0) \sim 0$, and $\gamma_N \simeq 10$ mJ/mole K². However, in the $n=2$ compound, we find $\gamma(0) \simeq 10$ mJ/mole K². Together with the ceramic data these results lead to the conclusion that the linear term is electronic in origin but not evidently connected with superconductivity.

The fact that many single crystal and ceramic copper oxide superconductors, including the $n=0$ $BaPb_{1-x}Bi_xO_3$ superconductors, show comparable $\gamma(0)$ values, suggests a common origin in the defect chemistry if these oxides. The specific heat of Si:P near the metal-insulator phase transition varies only very weakly compared to the conductivity, and shows a well-defined linear term on the insulating side due to a density of localized electronic states. Furthermore, large γT terms in nonstoichiometric vanadium bronzes, which are hopping conductors, are well established.[71] Finally, variable range hopping transport of the form $\exp[(T_0/T)^{1/4}]$ has been observed in nonmetallic ceramic and single crystal forms of nonstoichiometric La_2CuO_4.[72,73] All these results may be explained by a density of states picture analogous to that proposed for $Gd_{3-x}v_xS_4$, a Th_3P_4-type structure containing vacancies v in which the carrier concentration is increased by filling of v by Gd atoms.[74] Basically, the material becomes conducting when the Fermi energy E_F crosses a mobility edge E_μ, established by the random potentials creased by v. We propose that defects, the oxygen vacancies endemic in all these structures, play a similar role.

Finally, we note that insulating strontium copper oxide $Sr_{14}Cu_{24}O_{41}$, which also has planes of copper in square planar coordination in its structure shows a linear term in the specific heat with $\gamma = 100$ mJ/mole K².[75] Because of the highly two dimensional character of the conductivity in the (CuO_2) planes of the high T_c superconductors, we propose that localized defect states can coexist with the extended states in these materials, giving rise to the observed $\gamma(0)$.

Conclusions

(i) The characteristic feature of the electrical resistivity of the copper-oxide superconductors in the normal state is an in-plane resistivity which is large and linear in T. The c-axis resistivity is much higher, and is usually activated at low temperature. However, present evidence precludes characterizing the compounds as either two or three dimensional metals.

(ii) From Hall effect and chemical analysis it is concluded that superconductivity appears over a limited range of concentration of mobile holes in the Cu planes. The maximum T_c is achieved when the number of mobile holes per copper in the plane lies in the range (0.1-0.3).

(iii) The body of evidence on the low temperature specific heat suggests that the linear term is at least in part intrinsic and of electronic origin, but is related localized defect states rather than to superconductivity.

Acknowledgements: We are very grateful to A. Torressen for technical assistance in all of the specific heat measurements, and T. Penney for his active involvement in this work. We also thank G. Burns, D. Newns and R. Greene for helpful discussions.

REFERENCES

1. *"High Temperature Superconductors and Materials and Mechanisms of Superconductivity"*, J. Müller and J. L. Olsen, eds, North Holland, Amesterdam, 1988.
2. J. G. Bednorz and K. A. Müller, Z. Phys. B **64**, 189 (1986);
   J. G. Bednorz and K. A. Müller, Europhys. Lett. **3**, 379 (1987).

3. C. Michel, M. Hervieu, M. M. Borel, A. Grandin, F. Deslandes, J. Provost and B. Raveau, Z. Phys. B **68**, 421 (1987).
4. P. Strobel, K. Kelleher, F. Holtzberg and T. Worthington, Physica C, in press.
5. B. Batlogg, Physica **126B**, 275 (1984).
6. R. J. Cava, B. Batlogg, J. J. Krajewski, R. C. Farrow, L. W. Rupp, Jr., A. E. White, K. T. Short, W. F. Peck, Jr. and T. Y. Kometani, Nature **332**, 814 (1988).
7. M. K. Wu, J. R. Ashburn, C. J. Torng, P. H. Hor, R. L. Meng, L. Gao, Z. J. Huang, Y. Q. Wang and O. W. Chu, Phys. Rev. Lett. **58**, 908 (1987); J. Caponi, C. Chaillout, A. W. Hewat, P. Lejay, M. Marezio, N. Ngyuyen, B. Raveau, J. L. Soubeyroux, J. L. Tholence and R. Tournier, Europhs. Lett. **12**, 1301 (1987).
8. R. M. Hazen, L. W. Finger, R. J. Angel, C. T. Prewitt, N. L. Ross, C. G. Hadidiacos, P. J. Heaney, D. R. Veblen, Z. Z. Sheng, A. El Ali and A. M. Hermann, Phys. Rev. Lett. **60**, 1657 (1988).
9. B. Batlogg, T. T. M. Palstra, L. F. Schneemeyer, R. B. Van Dover and R. J. Cava, Physica C **153-155**, 1063 (1988).
10. S. S. P. Parkin, V. Y. Lee, E. M. Engler, A. I. Nazzal, T. C. Juang, G. Gorman, R. Savoy and R. Beyers, Phys. Rev. Lett. **60**, 2639 (1988).
11. H. Ihara, R. Sugise, M. Hirabayashi, N. Terada, M. Jo, K. Hayashi, A. Negishi, M. Tokumoto, Y. Kimura and T. Shimomura, Nature **334**, 510 (1988).
12. U. Gottwick, R. Held, G. Sparn, F. Steglich, H. Rietschel, D. Ewert, B. Renker, W. Bauhofer, S. von Molnár, M. Wilhelm and H. E. Hoenig, Europhys. Lett. **4**, 1183 (1987);
C. Uher and A. B. Kaiser, Phys. Rev. B **36**, 5680 (1987).
13. T. M. Rice, Z. Phys. B **67**, 141 (1987).
14. P. W. Anderson and Z. , Phys. Rev. Lett. **60**, 132 (1988);

P. W. Anderson, G. Baskaran, Z. Zou and T. Hsu, Phys. Rev. Lett. **58**, 2790 (1987).

15. S. A. Kivelson, D. S. Rokhsar and J. P. Sethna, Phys. Rev. B **35**, 8865 (1987); S. A. Kivelson, ibid. **36**, 7237 (1987).

16. For an excellent review, see Yasuhino Iye, Technical Report of ISSP, A2022, Sept. 1988, to be published in Intrnl. J. Mod. Phys. **B**.

17. M. Gurvitch and A. T. Fiory, Phys. Rev. Lett. **59**, 1337 (1988).

18. J. Halbritter, M. R. Dietrich, M. Küpfer, B. Runtsch and H. Wuhl, Z. Physik B **71**, 411 (1988).

19. T. Penney, S. von Molnár, D. Kaiser, F. Holtzberg and A. W. Kleinsasser, Phys. Rev. B **38**, 2918 (1988).

20. P. Chaudhari, R. T. Collins, P. Freitas, R. J. Gambino, J. R. Kirtley, R. H. Koch, R. B. Laibowitz, F. K. LeGoues, T. R. McGuire, T. Penney, Z. Schlesinger, Armand, P. Segmuller, S. Foner, and E. J. McNiff, Jr., Phys. Rev. B **36**, 8903 (1987).

21. H. L. Stormer, A. F. J. Levi, K. W. Baldwin, M. Arzlowar and G. S. Boebinger, Phys. Rev. B **38**, 2472 (1988).

22. T. Murakami, M. Suzuki and Y. Enomoto, Jap. J. Appl. Phys. Series 1, Super. Mater., 65 (1988).

23. X. Z. Wang, K. Donnelly, T. Bakas, J. M. D. Cocy, I. Rosenmen and C. S. Simon, preprint.

24. Frank J. Blatt, *"Physics of Electronic Conduction in Solids"*, McGraw-Hill Book Company, New York, 1968.

25. R. Micnas, J. Ramiger and S. Robaszkiewicz, Phys. **Rev.** B **36**, 4051 (1987).

26. L. I. Buravov, L. Ya. Vinnikov, G. A. Emeltchenko, P. A. Kononovitch, V. N. Laukhin, Y. A. Ossipyan and I. F. Schegolev, preprint.

27. Y. Iye, T. Tamegai, H. Torkeya and H. Takei, Jap. J. Appl. Phys. Series 1, Super. Mater., 46 (1988).

28. V. K. Vlasko-Vlasov, M. V. Indenbom and Y. A. Ossipyan, preprint.
29. Y. A. Ossipyan, V. B. Timofeev and I. F. Schegolev, preprint.
30. T. Takahashi, H. Matsuyama, H. Katayama - Yoshida, Y. Okabe, S. Hosoya, K. Seki, H. Fujimoto, M. Sato and H. Inokuchi, Nature **334**, 691 (1988).
31. F. J. Himpsel, G. V. Chandrashekar, A. Taleb-Ibrahimi, A. B. McLean and M. W. Shafer, preprint.
32. M. W. Shafer, T. Penney and B. Olson, Phys. Rev. B **36**, 4047 (1987).
33. J. B. Torrance, Y. Tokura, A. L. Nazzal, A. Bezinge, T. C. Huang, and S. S. P. Parkin, Phys. Rev. Lett. **61**, 1127 (1988).
34. N. Nguyen, J. Choiset, M. Hervieu and B. Raveau, J. Solid State Chem. **39**, 120 (1981);
N. Nguyen, F. Studer and B. Raveau, J. Phys. Chem. Solids **44**, 389 (1983).
35. M. W. Shafer, R. A. deGroot, M. M. Plechaty and G. J. Scilla, Physica C **153-155**, 836 (1988).
36. T. Penney, M. W. Shafer, B. L. Olson and T. S. Plaskett, Adv. Cheramic Mater. **2**, 577 (1987).
37. S. W. Tozer, A. W. Kleinsasser, T. Penney, D. Kaiser and F. Holtzberg, Phys. Rev. Lett. **59**, 1768 (1988).
38. L. Forro, M. Raki, C. Ayache, P. C. E. Stamp, J. Y. Henry and J. Rossat-Mignod, Physica C **153-155**, 1357 (1988).
39. A. Davidson, A. Palevski, M. J. Brady and P. Santhanam, Phys. Rev. B **38**, 2828 (1988).
40. D. Y. Xing and C. S. Ting, preprint.
41. Z. Z. Wang, J. Clayhold, N. P. Ong, J. M. Tarascon, L. H. Greene, W. R. McKinnon and G. W. Hull, Phys. Rev. B **36**, 7222 (1988).
42. M. W. Shafer, T. Penney, B. L. Olson, R. L. Greene and R. H. Koch, preprint.

43. Y. Tokura, J. B. Torrance, T. C. Huang and A. I. Nazzal, preprint.
44. T. Penney, M. W. Shafer and B. L. Olson, preprint.
45. J. Clayhold, N. P. Ong, P. H. Hor and C. W. Chu, preprint.
46. H. Takagi, H. Eisaki, S. Uchida, A. Maeda, S. Tajuna, K. Uchinokura and S. Tanaka, Nature **332 (6161)**, 236 (1988).
47. See e.g. E. S. R. Gopal, *"Specific Heats at Low Temperatures"* (Plenum Press, New York 1966).
48. R. A. Fisher, J. E. Gordon, S. Kim, N. E. Phillips and A. M. Stacey, Physica C **153-155**, 1092 (1988).
49. A. Junod, A. Bezinge and J. Muller, Physica C **152**, 50 (1988).
50. S. E. Inderhees, M. B. Salamon, Nigel Goldenfeld, J. P. Rice, B. G. Pazol, D. M. Ginsberg, J. Z. Liu and G. W. Crabtree, Phys. Rev. Lett. **60**, 1178 (1988); ibid., Phys. Rev. B **60**, 2445 (1988);
S. E. Inderhees, M. B. Salamon, T. A. Friedmann and D. M. Ginsberg, Phys. Rev. B **36**, 2401 (1988).
51. R. J. Cava, B. Batlogg, R. B. Van Dover, D. W. Murphy, S. Sunshine, T. Siegrist, J. P. Remeika, E. A. Rietman, S. Zakurak and G. P. Espinosa, Phys. Rev. Lett. **58**, 1676 (1987).
52. Comments, Phys. Rev. Lett. **61**, 480 (1988).
53. C. E. Methfessel, G. R. Stewart, B. T. Matthias and C. K. N. Patel, Proc. Natl. Acad. Sci. USA **77**, 6307 (1980).
54. For a detailed review of specific heat of high T_c oxide superconductors see R. A. Fisher, J. E. Gordon, and N. E. Phillips, to appear in J. Superconductivity.
55. N. E. Phillips, R. A. Fisher, S. E. Lacy, C. Marcenat, J. A. Olsen, W. K. Ham, A. M. Stacy, J. E. Gordon and M. L. Tan, Jap. J. Appl. Phys. **26**, Suppl. 26-3, 115 (1987);

N. E. Phillips R. A. Fisher, S. E. Lacy, C. Marcenat, J. A. Olsen, W. K. Ham and A. M. Stacy, "Novel Superconductivity," eds. S. A. Wolf and V. Z. Kresin (Plenum, NY 1987), p. 739.

56. Y. P. Feng, A. Jin, D. Finotello, K. A. Gillis, M. H. W. Chan and J. E. Greedan, preprint.

57. K. Kumangai, I. Watanabe, H. Aoki, Y. Nakamura, T. Kimura, Y. Nakamichi and H. Nakajima, Jap. J. Appl. Phys, Series 1, "Superconducting Materials," 37 (1988).

58. A. Junod, A. Bezinge, D. Cattani, M. Decroux, D. Eckert, M. Francois, A. Hewat, J. Muller and K. Yoon, Helvetica Physica Acta **61**, 460 (1988).

59. S. von Molnár, A. Torressen, D. Kaiser, F. Holtzberg and T. Penney, Phys. Rev. **B37**, 3762 (1988).

60. R. A. Fisher, S. Kim, S. E. Lacy, N. E. Phillips, D. E. Morris, A. G. Markelz, J. Y. T. Wei and D. S. Ginley, preprint.

61. S. J. Colcott, R. Driver, C. Andrikidis and F. Pavese, preprint.

62. J. M. D. Coey, S. von Molnár and P. Strobel, unpublished.

63. N. V. Anshukova, Y. V. Bugoslavskiy, V. G. Veselago, A. I. Golovashkin, O. V. Ershov, I. A. Zaytzev, O. M. Ivanenko, A. A. Minakov and K. V. Mitzen, preprint.

64. A. P. Rameriz, R. J. Cava, G. P. Espinosa, J. P. Remaka, B. Batlogg, S. Zahurak and E. A. Rietman, Mat. Res. Soc. Symp. Proc. **99**, 459 (1987).

65. J. M. Tranquada, D. E. Cox, W. Kunnmann, H. Moudden, G. Shirane, M. Suenaga and P. Zolliker, Phys. Rev. Lett. **60**, 156 (1988).

66. e.g., W. A. Phillips, J. Low Temp. Physics **7**, 351 (1972).

67. R. O. Pohl, Phase Transitions **5**, 239 (1985);
    A. C. Anderson; ibid pg. 302;
    John E. Graebner, Brage Golding, R. J. Schutz, F. S. L. Hsu and H. S. Chen, Phys. Rev. Lett. **39**, 1480 (1977).

68. B. Golding, N. O. Birge, W. H. Haemmerle, R. J. Cava and E. Rietman, Phys. Rev. B **36**, 5606 (1987).

69. M. Sera, S. Kondoh, K. Fukuda and M. Sato, Sol. State. Commun. **66**, 1101 (1988).

70. K. Kumagai and Y. Nakamura, Physica C **152**, 286 (1988).

71. B. K. Chakraverty, M. J. Sienko and J. Bonnerot, Phys, Rev. B **17**, 3781 (1988).

72. M. A. Kastner, R. J. Birgeneau, C. Y. Chen, Y. M. Chiang, D. R. Gabbe, H. P. Jensen, T. Junk, C. J. Peters, P. J. Picone, Tineke Thio, T. R. Thurston and H. L. Tuller, Phys. Rev. B **37**, 111 (1988).

73. C. Uher and A. B. Kaiser, Phys. Rev. B **37**, 127 (1988).

74. S. von Molnár and T. Penney, in *"Localization and Metal Insulator Tranistions"*, eds. Hellmut Frtizsthe and David Adler (Plenum, New York, 1985) p. 183 and references therein.

75. M. W. McElfresh, J. M. D. Coey, P. Strobel and S. von Molnár, unpublished.

LIST OF CONTRIBUTED PAPERS

presented at the 4th International Conference
on Physics of Magnetic Materials (4ICPMM)

The asterisk $^{(*)}$ denotes the papers preliminary accepted for publication in Acta Physica Polonica (July and September 1989).

Session A: Superconducting Oxides

1. N.E. Alekseevskii, A.V. Mitin, E.P. Khlybov,
    V.V. Evdokimova, and G.M. Kuzmicheva
A correlation between T_c and lattice parameters. Study of high temperature superconducting metal oxides

2. N.E. Alekseevskii, A.V. Mitin, V.I. Nizhankovskii,
    E.P. Khlybov, A.I. Kharkovskii and V.V. Evdokimova
Magnetic properties of superconducting metal oxides

3. V. Alexandrov, V. Veselago, L. Vinokurova, V. Ivanov,
    L. Klimova, V. Osiko and V. Udovenchik
Magnetoresistance of high T_c superconductors

4.* N.V. Anshukova, Y.V. Bugoslavskiy, V.G. Veselago,
    I. Golovashkin, O.V. Ershov, I.A. Zaytzev, O.M. Ivanenko, A.A. Minakov and K.V. Mitzen
The effect of oxygen content variation and rare earth ion substitution on the magnetic properties and specific heat of the granular high-T_c superconductors

5.* M. Baran, M. Heinonen, J. Pełka and A. Wiśniewski
XPS and magnetic study of $YBa_2Cu_3O_{7-x}$ irradiated by fast neutrons

6. M. Baran, P. Przysłupski, A. Wiśniewski, J. Górecka, K. Godwod, W. Dobrowolski and A. Pajączkowska
Magnetic, structural and transport properties of superconducting $Bi_{1.1}Sr_1Ca_{0.4}Cu_1O_x$

7. J. Baszyński
Crystal structure of the high-T_c superconductor $(Y_{1.2}Ba_{0.8})_4Cu_4O_{16-\delta}$

8. Z. Bąk, M. Piasecki and J. Kasperczyk
Phenomenological description of magnetic system with periodically modulated composition

9. A.V. Belushkin, E.A. Goremychkin, I. Natkaniec, I.L. Sashin and W. Zajac
Lattice dynamics and crystalline electric field effects in high temperature superconducting ceramics Tm-La-Sr-Cu-O studied by inelastic neutron scattering

10. H. Börner, M. Wurlitzer, B. Lippold and H.-Ch. Semmelhack
The influence of some preparation conditions on the behaviour of polycrystalline $YBa_2Cu_3O_x$

11. A. Chełkowski, A. Winiarska and A. Winiarski
Structural, electric and magnetic properties of $YBa_2Cu_3O_{6+x}$

12. B.P. Gorshunov, G.V. Kozlov, S.I. Krasnosvobodtsev, E.V. Pechen, A.M. Prokhorov, A.S. Prokhorov, O.I. Syrotynsky and A.A. Volkov
Submillimeter properties of high-T_c superconductors

13. V. Nekvasil
Magnetism of rare earths in REBaCuO

14.* I. Onyszkiewicz, M. Koralewski, P. Czarnecki
Comparative study of Y-Ba-Cu-O and Y-Ba-Cu-O-S superconducting ceramics

15.   M. Pękała, K. Pękała and A. Pajączkowska
Thermoelectric power of high-T_c superconductors Bi-Sr-Ca-Cu-O

16.* M. Piasecki, Z. Bąk and J. Kasperczyk
Magnetoelastic hole pairing in the copper oxides

17.   P. Przysłupski and P. Nowicki
Superconducting Bi-Sr-Ca-Cu-O thin films

18.* J. Raułuszkiewicz and A. Reich
Tunneling studies of high-T_c superconductors

19.   Than Duc Hien, Than Hoai Anh, Nguyen Hun Duc and Phan Phuong Mai
Studies of the high-temperature superconductors in Y-Ba-Cu-O and Y-Ba-Cu-Cl-O systems

20.   Z. Tomkowicz, M. Bałanda, A. Bajorek, Kim Chol Sik, A. Szytuła and M. Turowska
The influence of intercalate atoms and of fluoride on superconducting properties in the $YBa_2Cu_3O_{7-y}$ type structure

21.* A. Witek and J. Raułuszkiewicz
Surface imaging of Bi-Sr-Ca-Cu-O superconducting ceramics by means of scanning tunneling microscope

Session B: **Magnetic Oxides**

22. J.J. Bara, B.F. Bogacz, T. Jaworska, J. Leciejewicz, M. Styczyńska and A. Szytuła
Crystal structure and magnetic properties of β-NaFe$_{1-x}$Al$_x$O$_2$ solid solution

23.* S.N. Barilo, A.P. Ges, S.A. Guretskii, N.K. Danshin, G.G. Kramarchuk, A.M. Louguinets, M.A. Sdvizhkov and V.V. Fedotova
Magnetic resonance in the substituted orthoferrites

24. I.V. Baryakhtar
Nonlinear waves in two sublattice magnets

25. V.G. Baryakhtar, A.N. Bogdanov, D.A. Yablonskii
Thermodynamic theory of domain structures in magnetically ordered systems

26.* V.V. Eremenko, K.L. Dudko, N.V. Gapon, V.N. Savitskii, V.V. Soloviev and V.L. Fingold
On the influence of the rare-earth subsystem on the orientation transition in samarium orthoferrite

27.* A.P. Ges, V.V. Fedotova, V.N. Derkachenko and S.A. Buzhynsky
Effect of scandium ions on magnetic properties of holmium and terbium orthoferrites

28. S.L. Gnatchenko, A.B. Chizhik and N.F. Kharchenko
Kinetics of phase transition antiferromagnet – weak ferromagnet in DyFeO$_3$

29. V.N. Golubev, M.D. Kaplan, B.G. Vekhter
Magnetic properties of crystals with Jahn-Teller structural transition stimulated by magnetic field

30.* J. Hankiewicz, Z. Pająk, J. Radomski
NMR study of effective fields at ^{57}Fe nuclei in cadmium substituted copper ferrite

31. S. Juszczyk
Exchange in noncolinear $Zn_{1-x}Cu_xCr_2Se_4$

32. A.M. Kadomtseva, A.A. Mukhin, I.A. Zorin, A.K. Zvezdin
Magnetism of orthochromites with the rare-earth Ising ions

33. M.D. Kaplan
Magnetic reorientational phase transitions and cooperative Jahn-Teller effect

34. A.N. Kocharian
X-ray absorption and resonance emission theory in variable valency compounds

35.* G.V. Kozlov, S.P. Lebedev, A.A. Mukhin, A.Yu. Pronin, A.S. Prokhorov
Submillimeter dynamics and spin-reorientation transitions in the rare-earth orthoferrites

36.* J. Lech, K. Dziliński, A. Ślęzak, B. Wysłocki and H. Krukiewicz
Grain motions induced in Fe_3O_4/SiO_2 mixtures by combined steady and alternating magnetic fields

37.* V.S. Lutovinov
Magnon-magnon and magnon-phonon interaction induced by paramagnetic impurities in magnetics

38.* A. Maryanowska, J. Pietrzak, H. Sitarz and Z. Kałuski
Structural and FMR studies of $Sr_2FeVO_{5.6}$ single crystals

39. V.P. Miroshkin, V.V. Passynkov, K. Perzyńska
The influence of Mn ions on the dielectric properties of Mn-Zn ferrites

40.* V.P. Miroshkin, K. Perzyńska, I.I. Zyatkov
The polycrystalline ferrites donor centers concentration estimation according to the DC conductivity

41. A.I. Mitsek, K.Yu. Guslienko
Magnon relaxation by paramagnetic rare-earth impurities

42.* E. Mosiniewicz-Szablewska, S. Piechota, G.I. Vinogradova, V.G. Veselago
Photoferromagnetic effect in $CdCr_2Se_4$ ferromagnetic semiconductor

43. J.M. Mucha
Magnetic anisotropy and rotational hysteresis in barium ferrites

44. I. Onyszkiewicz, J. Janicki, J. Pietrzak, J. Suwalski
Mossbauer study of new magnetic materials with hollandite and ilmenite type structures

45. K. Piotrowski, R. Szymczak, H. Szymczak, V.V. Eremenko, S.L. Gnatchenko, N.F. Kharchenko, P.P. Lebedev
Spin reorientations induced by skew magnetic field in $DyFeO_3$

46. A.V. Andrienko and L.V. Poddyakov
Parametric excitation of magnetoelastic waves in the antiferromagnet $FeBO_3$

47. V.L. Safonov
Phase transition in a spin system with indirect Suhl-Nakamura interaction

48. J. Suwalski, M. Łukasiak, Z. Kucharski, C. Michalk, A. Purew, Sh. Chadrabaal
Multi-layered and magnetic structure of biotite

49. A. Wiśniewska, E. Okoniewska-Pszczółkowska, T. Postupolski
Magnetization behaviour in a polycrystalline material with zero crossing magnetocrystalline anisotropy energy

Session C: <u>Rare-Earth Based Hard Magnetic Materials</u>

50. H. Bala, S. Szymura, J.J. Wysłocki
Corrosion behaviour of Fe-Nd-B permanent magnets

51. J.J. Bara, B.F. Bogacz and A. Szytuła
Mössbauer effect investigations of the $Y_2Fe_{14}B$-based intermetallic compounds

52. H. Broda, J. Warchulska
Magnetic properties of the low iron content $Lu(Fe_{1-x}Al_x)_2$ compounds

53. N.H. Chau, N.H. Duc, T.D. Hien, K. Krop and J. Żukrowski
The compensation point in the $Er_{0.6}Y_{0.4}(Fe_xNi_{1-x})_2$ compounds

54.* K. Chróst, J. Kłodaś
Deformation-aged Fe-Cr-Co type hard magnetic materials

55. Z. Drzazga
Magnetic anisotropy measurements with rotating-sample magnetometer

56. Z. Drzazga, A. Winiarska
Magnetic properties of Fe substituted $DyCo_4B$ compound

57. M. Duczmal, S. Pokrzywnicki and L. Pawlak
The magnetization of Er_3Te_4

58.[*] J. Dudáš, A. Feher
The temperature hysteresis of the electrical resistance of dysprosium near magnetic phase transitions

59. T.T. Dung, N.P. Thuy, N.M. Hong, T.D. Hien and J.J.M. Franse
Influence of the Fe substitution on the magnetic anisotropy of the YCo_4B compound

60.[*] M. Duraj, A. Szytuła
Magnetic properties of $Sm_{1-x}R_xMn_2Ge_2$ (R = Nd, Gd, Tb)

61. T. Dymkowski, K. Kłodaś
Ways of the structural interpretation of the hard magnetic material with the bcc lattice

62.[*] A. Handstein, K.-H. Müller, J. Schneider, R. Szymczak and J. Zawadzki
Domain structure and magnetization behaviour of sintered Nd-Fe-B permanent magnets

63.[*] U. Heinecke, A. Handstein, J. Schneider and P. Nothnagel
Magnetization processes in NdFeB-based magnets

64. B. Idzikowski, A. Wrzeciono and A. Szlaferek
Crystallization behaviour of amorphous magnetic $Sm_xFe_{80-x}B_{20}$ alloys

65. M. Jurczyk and T. Mydlarz
Magnetic studies of Sm-Co-B system

66. Cz. Kapusta, H. Figiel
The NMR analysis of local magnetic properties of Co in $(Y_{1-x}Nd_x)_2Fe_{14}B$

67. Cz. Kapusta, H. Figiel, Z. Kąkol
Analysis of neodymium single ion anisotropy in $(Y_{1-x}Nd_x)_2Fe_{14}B$

68. J. Kłodaś, K. Chróst
Changes of the magnetic properties of $Fe_{22}Cr_{15}Co_1Nb_1Ti$ alloy during low temperature ageing

69. J. Kosiorowska, A. Bajorek, A.W. Pacyna and A.J. van Duyneveldt
Magnetization and susceptibility studies of La-Cu-Mn

70. A. Kowalczyk, P. Stefański and A. Wrzeciono
Crystal structure and magnetic properties of $R_2Fe_{14-x}Cr_xB$ compounds (R = Y, Nd, Gd)

71. E. Kulatov, V. Veselago, L. Vinokurova
Electronic structure of ternary silicides with rare-earth and manganese

72. M. Leonowicz, S. Heisz and G. Hilscher
Effect of Al-rich precipitate on magnetic properties and coercivity mechanisms of sintered Nd-Fe-Al-B magnets

73. Luu Tuan Tai, Nguyen Hoang Luong, Than Duc Hien
Ordering phenomena in $R_2Fe_{14-x}Al_xB$ compounds (R = Nd, Er)

74. M. Łukasiak, J. Pszczoła, J. Suwalski, K. Krop
^{161}Dy Mössbauer effect studies of $Dy_{1-x}Y_xCo_3$ intermetallic compounds

75.* E. Machowska and S. Nadolski
Spin echo defocusing spectroscopy

76. J.M. Mucha, A. Szytuła, S. Bednarski
Magnetic properties of $Fe_{3-x}Mn_xSi$ alloys

77.* K.H. Müller
Coercivity of sintered Nd-Fe-B

78. J. Olszewski, J. Wójcik, B. Wysłocki and S. Szymura
High coercivity state in low-cobalt FeCrCo alloys

79.* D. Płusa, R. Pfranger
Domain structure in sintered $Nd_{15}Fe_{77}B_8$ permanent magnet at different points of magnetization reversal curve

80. O. Popov and M. Mikhov
Critical single domain size for Nd-Fe-B type magnetic materials

81.* J. Pszczoła and K. Krop
Ordering temperatures of R-3d intermetallics

82. R.J. Radwański, J.J.M. Franse and R. Verhoef
Magnetic cone structure in Ho_2Co_{17} and $Ho_2Fe_{14}B$

83. N. Scheludko, M. Mikhov, A. Apostolov
Investigation of the magnetocrystalline anisotropy of $Mm_2Co_{17-x}Zr_x$ compounds

84.* P. Stefański, A. Szlaferek and A. Wrzeciono
Magnetocrystalline anisotropy in $RFe_{10}V_2$ compounds

85. W. Suski and A. Baran
Magnetic properties of the $UFe_{10}Si_2$ and $UFe_{10}Mo_2$ compounds

86. A. Szewczyk, Z. Henkie
Different forms of the domain structure in U_3P_4

87. A. Szewczyk, R. Szymczak, H. Szymczak, J. Zawadzki, D. Gignoux, B. Gorges, R. Lemaire
Ising domain walls in $SmNi_5$ compound

88. R. Szymczak, J. Zawadzki and Z. Drzazga
A study of the rapid change of the domain wall energy in $Ho(Co_{1-x}Ni_x)$ alloys

89.* A. Szytuła, W. Bażela, J. Leciejewicz
Magnetic structures of orthorhombic rare earth compounds

90. M. Ślepowroński, A. Modrzejewski and S. Warchoł
Growth of sizable single crystals of $(Y,Nd)_2Fe_{14}B$ and $Y_2(Fe,M)_{14}B$ compounds

91. M. Wójcik, E. Jędryka and M. Ślepowroński
Hyperfine field distribution in cubic $LaCo_{13}$

92. J.J. Wysłocki, D. Płusa, B. Wysłocki, R. Pfranger, B. Szafrańska-Miller
Angular dependence of the magnetic rotational hysteresis energy in $Nd_{15}Fe_{77}B_8$ magnet

Session D: Amorphous Magnetics

93. Y.G. Abov, M.I. Bulgukov, S.P. Boroblev, A.D. Gul'ko, B.M. Garochkin, F.S. Dzheparov, S.V. Stepanov, S.S. Trosbin and V.E. Shestopal
Polarization transfer in disordered systems and β-NMR spectroscopy

94. B. Alessandro and Z. Kaczkowski
Power losses of the $Fe_{78}Si_8B_{14}$ amorphous alloy strips in the audio frequency range from 50 Hz to 20 kHz

95. P. Allia, C. Beatrice, Z. Kaczkowski and F. Vinai
Magnetic disaccommodation in $Fe_{79}Si_{12}B_9$ amorphous alloy

96. Z. Arnold and J. Kamarád
The high pressure behaviour of Curie temperature of Fe-based metallic glasses

97.[*] K. Brzózka, M. Gawroński, K. Jezuita and J. Szlanta
The shape of the hyperfine field distribution in amorphous ferromagnets near T_c

98. W. Ciurzyńska, J.W. Moroń, J. Zbroszczyk, B. Wysłocki and Y. Yamashiro
The influence of silicon content on magnetic after-effect at 230-400 K in microcrystalline Fe-Si ribbons

99.[*] W. Ciurzyńska, J. Zbroszczyk, J.W. Moroń, B. Wysłocki and Y. Yamashiro
Some magnetic properties of microcrystalline 6.5% Si-Fe ribbons

100. P. Czarnecki, B. Idzikowski and A. Wrzeciono
Transport properties of Re_xFe_{80-x} and $Ho_xCo_{70-x}B_{30}$ amorphous alloys

101. Duong hai Trieu and A. Maksymowicz
Magnetostriction of dilute amorphous alloys with Ce or Yb impurities

102. M. Duś-Sitek
Effect of changes of iron admixture at the cost of metaloids on the activation energies of relaxation processes and structural transformation in magnetic materials of Co-Fe system

103.* M. Duś-Sitek, B. Wysłocki, Z. Olszowski and B. Sujak
Thermostimulated exoelectron emission (TSEE) kinetics from (Co-Fe)-(Si-B)-type magnetic metallic glass

104.* J. Fink-Finowicki, G. Konczos, J. Krzywiński, A. Siemko, T. Tarnóczi and Z. Vértesy
Stress dependent permeability after-effect in metallic glasses with induced anisotropy

105.* J. Fink-Finowicki, B. Lisowski, S.M. Zakharov and L. Załuski
Field induced magnetic anisotropy in Co-based amorphous alloy

106. W. Gawior, R. Kolano and N. Wójcik
Magnetic properties and thermal stability of $Fe_{66}Co_{19-x}B_{15}Si_x$ amorphous alloys

107.* M. Ghafari, M. Matsuura, W. Keune, P.K. Schletz and K. Fukamichi
Magnetic studies of amorphous Fe_2ErH_3 and Fe_2CeH_4 alloys

108. Cz. Górecki, T. Górecki and Z. Michno
Thermal stability of the $Fe_{80-x}V_xB_{20}$ metallic glasses as studied by the EEE and DTA methods

109. Cz. Górecki and Z. Michno
Exoelectron emission accompanying the structural transformations in metallic glasses $Fe_{80-x}Cr_xB_{20}$

110. E. Jarocki
X-Ray K_β emission spectra of iron in some of amorphous iron alloys

111. A. Jezierski
The electronic density of states in transition metal-metalloid alloys

112. Y. Jirásková and T. Zemčik
Activation enthalpy of the relaxation in amorphous $Fe_{40}Ni_{40}B_{20}$ by magnetic and Mössbauer methods

113.* Z. Kaczkowski and P. Duhaj
Some piezomagnetic properties of the $Fe_{73}Co_{12}B_{15}$ metallic glasses

114. Z. Kaczkowski and Ho Su Nam
Influence of the annealing temperature on the magnetomechanical coupling in $Fe_{78}B_{12}Si_{10}$ metallic glasses

115.* Z. Kaczkowski, L. Potocky and É. Kisdi-Koszó
Moduli of elasticity of the $Fe_{83.2}Cr_{2.5}B_{14.3}$ metallic glass ribbons produced by applying magnetic field during quenching

116. V. Kislov, Yu. Levin, A. Serebryakov, M. Tejedor and B. Hernando
Magnetic anisotropy of as-quenched ribbons of amorphous near-zero-magnetostrictive alloys

117.* S. Kiss, L. Małkiński, G. Posgay and L. Pogany
Influence of tensile stresses on magnetic field dependences of internal friction and of shear modulus of the Fe-Si-B metallic glass

118. L. Kraus
A new method for measurement of saturation magnetostriction of amorphous ribbons

119.* L. Kraus
Effect of external stress and magnetic field on the saturation magnetostriction of nearly zero magnetostriction metallic glasses

120. J. Krzywiński, A. Siemko, M. Baran, B. Lisowski and A. Waniewska
Influence of stress annealing on magnetic properties of low magnetostrictive metallic glasses

121. T. Kulik, J. Lisiecki and H. Matyja
Stability of magnetic permeability of zeromagnetostrictive Co-Fe-Mn-Mo-Si-B glass

122.* K. Kułakowski and A. Węgrzyn
Linear magnetostriction in random field model

123. V.S. Lutovinov
The long-wavelength spin wave damping in disordered ferromagnetic metals

124.* Z. Michno and A. Baranowski
Positron lifetime study of amorphous transition metal-metaloid alloys

125. M. Mihalik, A. Zentko, I. Škorvánek and A. Lovas
Effect of pressure on the Curie temperature of Fe-Cr-B and Fe-W-B amorphous alloys

126. E. Pilipczuk
The dilatometric investigation of the high magnetostrictive amorphous alloys

127. A. Siemko and H.K. Lachowicz
Resistometric method of estimation of the Young's modulus in metallic glass ribbons

128. A. Ślawska-Waniewska
Crystallization of amorphous $Fe_{76}B_{15}Si_9$ under high pressure

129. J. Świerczek, S. Szymura, J. Zbroszczyk, B. Wysłocki and H. Lampa
The effect of internal and bending stresses on magnetic properties in amorphous Co-Fe-Si-B alloys

130. Z. Vértesy and É. Kisdi-Koszó
Effect of mechanical stress on domain structure of metallic glasses

131. G. Vlasak and Z. Kaczkowski
Linear and volume magnetostriction characteristics of the iron rich metallic glasses

132.* K. Witański and T.J. Panek
Influence of Co content on magnetic hyperfine field in Fe-Co-B and Fe-Co-Si-B metallic glasses

133.* J. Wolny, T. Freltoft and B. Lebech
SANS experiments on amorphous Fe-Ni and Co-Ni based materials

134. N. Wójcik, R. Kolano and W. Gawior
Influence of the heat treatment parameters on magnetic properties of the $(CoFeMnMo)_{77}(SiB)_{23}$ amorphous ribbons

135. Y. Yamashiro and K.I. Arai
Magnetic properties and grain texture in rapidly quenched 4.5% Si-Fe ribbons

136.* L. Załuski, A. Załuska and J. Krzywiński
Initial stages of crystallization in current heated Fe-Ni-Si-B metallic glass

137. K. Zaveta, K. Jurek, L. Kraus, H. Lachowicz, A. Siemko and P. Kral
Domain structures of zero-magnetostrictive metallic glasses after creep annealing

138. J. Zbroszczyk, W. Ciurzyńska, Y. Yamashiro and B. Wysłocki
Influence of surface conditions on some magnetic properties of microcrystalline 6.5% Si-Fe ribbons

139.* J. Zbroszczyk, J. Świerczek, W. Ciurzyńska, B. Wysłocki, S. Szymura and H. Lampa
Improvement of magnetic properties in the amorphous $Co_{78}Si_{11}B_{11}$ alloy after surface layer etching

140. A.P. Zhukov and B.K. Ponomaryov
Start fields distribution in a bistable amorphous ribbons

Session E: **Magnetic Films, Multi-Layered Structures and Surface Phenomena**

141.* L.T. Baczewski, M. Piecuch, J. Durand, G. Marchal and P. Delcroix
Magnetization in Fe-Nd compositionally modulated thin films

142.* T. Balcerzak, G. Wiatrowski and J. Mielnicki
The magnetic correlations in thin diluted films

143. J. Baszyński, F. Stobiecki and Z. Frait
Magnetic surface anisotropy of Co in multilayers Co/Ti

144. J. Ben Youssef, H. Le Gall and J.M. Desvignes
Growth conditions investigation for very high uniaxial anisotropy in $(BiY)_3Fe_5O_{12}$ LPE films

145. A.V. Boltushkin, V.G. Shadrov and V.M. Fedosyuk
Hard-magnetic Co-W, Co-Ni-W films with perpendicular magnetic anisotropy

146.* L.M. Dedukh, V.S. Gornakov, Yu.P. Kabanov and V.I. Nikitenko
Effect of external magnetic field on behavior of Bloch lines

147.* L.M. Dedukh, V.I. Nikitenko and V.T. Synogach
Experimental study of spectrum of elementary excitations of Bloch wall in yttrium iron garnet

148. E.B. Dokukin, D.A. Korneev, W. Lobner, V.V. Pasjuk, A.V. Petrenko and H. Rzany
Neutron depolarization study of static magnetization fluctuations in ferromagnets

149.* J. Dubowik, Yu.V. Kudryavtsev and R. Gontarz
Magnetic and optical properties of amorphous and crystalline Co-W films

150.* J. Dubowik, Yu.V. Kudryavtsev and F. Stobiecki
Ellipsometric evidence of the Zr/Co amorphization by a solid state reaction

151. G.A. Gehring, M. Gajdek and W. Wasilewski
Coupled domain and modulated structures in thin ferromagnetic films

152. R. Gemperle, J. Šimšová, J. Káczer, J.C. Lodder, L. Murtinová, S. Saic and I. Tomáš
Domain period determination in CoCr films

153. R. Gieniusz
Magnetostatic waves in garnet films characterized by mixed anisotropies

154. I. Gościańska and H. Ratajczak
Uniaxial anisotropy in amorphous $Nd_{12}Co_{82}B_6$ thin films

155. E.I. Il'yashenko and I.N. Kondratev
Magnetization processes in thin permaloy narrow strips

156. E. Jędryka, P. Panissod, P. Guilmin and G. Marchal
NMR study of solid state reaction in CoSn multilayers

157. M. Jirsa
The temperature dependence of domain FMR in uniaxial garnet films

158.* W.A. Kaczmarek, J. Pietrzak and H. Ratajczak
Ferromagnetic and spin waves resonances in amorphous $Fe_{80-x}Sm_xB_{20}$

159.* M. Kisielewski, A. Maziewski and P. Görnert
AC susceptibility analysis of YIG+Co films

160.* R.A. Kosiński and A. Sukiennicki
Chaotic solutions in domain wall dynamics with vertical Bloch lines

161.* H. Le Gall, F. Chevrier, A. Rakii, J. Gieraltowski and
J. Loaec
Sputtering and annealing parameters dependence of magnetic properties of amorphous CoZr thin films

162. N.A. Lesnik and S.Ya. Kharitonski
Effect of layer interaction on resonance characteristics of two-layer ferromagnetic films

163. M. Lubecka, W. Powroźnik, L.J. Maksymowicz and
R. Żuberek
FMR studies of thin $CdCr_2Se_4$ films doped with indium

164. T. Luciński
Electric resistivity of amorphous Fe-Ti films

165.* W. Maciejewski
Spin wave resonance spectra of ferromagnetic film with coverlayer

166. W. Maj, J. Meiresonne and J. Aarts
Transport measurements on amorphous GdAl thin films

167. L.J. Maksymowicz, T. Stobiecki, F. Stobiecki and
     H. Hoffmann
Surface anisotropy energy of thin amorphous CoZr films.
SWR experiment

168. A. Maziewski, K. Brzosko, M. Tekielak and P. Görnert
Anisotropy of Co-doped YIG films

169. A. Maziewski, A. Stankiewicz and E. Kubajewska
Domain structure analysis in garnets by digital image processing

170. A. Milewski, J. Samuła, L.J. Maksymowicz, J. Wenda,
     H. Jankowski and A. Kułak
Multilayer structure of metallic glass films in variable SAW delay line

171. A.I. Mitsek and K.Yu. Guslienko
Magnetic phase diagrams and high-frequency properties of double-layer films

172. A.I. Mitsek and V.N. Pushkar
Nonlinear model of Bloch line

173.* Z. Onyszkiewicz and A. Wierzbicki
A simple spin model of spontaneous metamagnetism in thin film

174. M. Pardavi-Horvath and P.E. Wigen
Ferrimagnetism of diluted uncompensated yttrium iron garnet

175. K. Pátek and I. Tomáš
Magnetization process of a single Bloch wall with a single Bloch line

176. H. Ratajczak and I. Gościańska
Angular dependence of coercive field in amorphous $Nd_{12}Co_{82}B_6$ thin films

177. H. Ratajczak, I. Gościańska, T. Luciński and A. Kowalczyk
Saturation magnetization of amorphous Nd-Co-B thin films

178.* W. Schmidt
The susceptibility of magnetic multilayers

179. L. Smardz, J. Baszyński and A.N. Bazhan
Magnetic properties of compositionally modulated Cu/Co films

180. L. Smardz, J. Baszyński and B. Rauschenbach
Ion beam mixing effects in metallic magnetic superlattices

181.* L. Smardz, J. Baszyński, B. Szymański and H. Reuther
Structure and temperature dependence of magnetization for compositionally modulated Cu/Fe films

182.* F. Stobiecki, T. Stobiecki and H. Hoffmann
Diffusion in Co/Zr multilayers studied by Auger electron depth profiling and magnetic measurements

183. T. Stobiecki and H. Hoffmann
Magnetic contribution to the electrical resistivity in amorphous FeZr and CoZr films

184.* B. Szymański and F. Stobiecki
Magnetization and coercive force in amorphous Fe-B and Fe-B-Si films of different thickness

185.* K. Szymański, L. Dobrzyński, J. Waliszewski and A. Wiśniewski
Mössbauer investigation of $Fe_{3-x}Si_{1-x}$ and $Fe_{3-x}Cu_xSi$ series

186.* Z. Šimša, J. Šimšová, P. Görnert and A. Maziewski
Absorption and Faraday rotation of YIG:Co(Ti,Ge,Ca) garnet films

187. I. Tomáš, Z. Vértesy and L. Pust
Domain wall coercivity field and domain wall parameters

188. P. Tomczak and A.R. Ferchmin
Critical temperature and critical concentration in a diluted Ising superlattice. Interface magnetic ordering

189. T. Tymosz, K. Brańska and Z. Petykiewicz
Lorentz microscope study of evaporated Co-Cr thin films

190. S. Uba, L. Uba and P. Gerard
A comparative study of Kerr effect in garnet films implanted with H_2^+, He^+ and Ne^+ ions

191.* N. Vukadinovic and H. Le Gall
Domain wall resonance in garnet films with orthorhombic anisotropy

192.* J. Wenda and K. Kułakowski
Magnetic anisotropy in thin films of metallic glasses of FeBSi

193. J. Wenda, L.J. Maksymowicz, H. Jankowski and A. Kułak
Magnetic anisotropy in single films and multilayer structure of FeBSi alloys

194. S.E. Yurchenko and E.E. Chepurova
Magnetic relief in garnet films formed by laser annealing

195. S.E. Yurchenko, G.Yu. Zharkov and O.N. Melnikov
Study on bending instability of stripe domains in garnet films

196. R. Żuberek, H. Szymczak, G. Suran and K. Ounadjela
Magnetostriction of amorphous $Co_{1-x}Ti_x$ soft ferromagnetic thin films

Session F: <u>Spin Glasses</u>

197. B. Fechner and M. Rashed
Specific heat of the long-range Ising spin glass. Exact results for small samples.

198. S. Kobe and A. Schütte
Ground states of neural network models

199.* Mai Suan Li
Spin glass in a two-sublattice model

200.* A.A. Minakov, I.V. Shvets and V.G. Veselago
The antiferromagnetic domains dynamics in helical antiferromagnets

201.* J. Richter
Magnetic order-disorder transition induced by magnetic field in antiferromagnets with random anisotropy

202.* G.A. Takzei, I.I. Sych and Yu.P. Grebenyuk
Paramagnetic-spin glass phase transition in amorphous and crystalline alloys of $3d$-transition metals

203.* G.A. Takzei, I.I. Sych and Yu.P. Grebenyuk
Reentrant ferromagnetic-spin glass and antiferromagnetic-spin glass transition in disordered f.c.c. Fe based alloys

204.* L. Vinokurova, V. Ivanov, E. Kulatov, M. Pardavi-Horvath and E. Svab
Magnetic ordering of Pt-Fe alloys

205. W. Zapart and M.B. Zapart
Exchange interactions in $KMn_xZn_{1-x}F_3$ perovskites by EPR of Mn^{2+} ions

206.* W. Zapart and M.B. Zapart
EPR investigation of magnetic phase transition in $KMn_xZn_{1-x}F_3$

LIST OF PARTICIPANTS

BULGARIA

M.T. Mikhov
Faculty of Physics
Sofia University
5 Anton Ivanov Blvd.
BG - 1126 SOFIA

O.A. Popov see Mikhov
N.A. Scheludko see Mikhov

CZECHOSLOVAKIA

Z. Arnold
Institute of Physics
Czech. Academy of Sciences
Na Slovance 2
180 40 PRAHA 8

J. Dudáš
Technical University
Dept. of Theoretical Electrotechnics
and Circuit Techniques
park Komenského 3
043 89 KOŠICE

M. Hrdina
Zbrojovka Brno, k.p.
VVVT
Lazaretni 7
656 17 BRNO

Y. Jirásková
Czechoslovak Academy of Sciences
Institute of Physical Metallurgy
Žižkova 22
616 62 BRNO

M. Jirsa see Arnold
K. Jurek see Arnold
J. Kadlecová see Arnold
L. Kraus see Arnold
S. Krupička see Arnold
V. Nekvasil see Arnold
K. Pátek see Arnold
H. Polášková see Hrdina
L. Pust see Arnold

Z. Šimša   see Arnold

I. Škorvánek
Institute of Experimental Physics
Slovak Academy of Sciences
Solovjevova 47
040 01 KOŠICE

P. Švec
Institute of Physics
of Electro-Physical Research Centre
Slovak Academy of Sciences
Dúbravska cesta 9
842 28 BRATISLAVA

I. Tomáš   see Arnold
K. Závěta   see Arnold

------ FRANCE

I.A. Campbell
Université Paris-sud
Laboratoire de Physique des Solides
Associé au C.N.R.S.
Bâtiment 510
91405 ORSAY-CEDEX

M. Guyot
Laboratoire de Magnetisme
C.N.R.S.
1, Place Aristide-Briand
92195 MEUDON Principal Cedex

R. Krishnan   see Guyot

H. Le Gall
7, Place de Bretagne
91430 IGNY

R. Lemaire
CNRS-USTMG
Laboratoire Louis Néel
25, avenue des Martyrs
166 X, 38042 GRENOBLE Cedex

I.B. Puchalska   see Guyot

------ **FRG (West Germany)**

M. Fähnle
Institut für Physik
Max Planck Inst. für Metallforschung
Heisenbergstr. 1
Postfach 80 06 65
D-7000 STUTTGART 80 (BUSNAU)

M. Ghafari
Universität Duisburg
Fachbereich 10, Angewandte Physik
Postfach 101629
Lotharstr. 65
D-4100 DUISBURG 1

M. Ohkoshi   see Fähnle

------ **GDR**

L. Fritzsch
Akademie der Wissenschaften der DDR
Physikalisch-Technisches Institut
Postfach 129
Helmholtzweg 4
6900 JENA

O. Gebhardt   see Fritzsch
P. Görnert   see Fritzsch

U. Heinecke
Hochschule für Verkehrswesen
"Friedrich List"
Wissenschaftsbereich Physik
PSF 103
8072 DRESDEN

R. Höhne
Sektion Physik der
Karl-Marx-Universität
Linnèstr. 5
7010 LEIPZIG

S. Kobe
Technische Universität Dresden
Sektion Physik
WB Teoretische Physik
Mommsenstr. 13
DDR-8027 DRESDEN

T. Moch
Technische Universität Magdeburg
"Otto von Guericke"
Sektion Physik
Postfach 124
3010 MAGDEBURG

K.H. Müller
Akademie der Wissenschaften der DDR
Zentralinstitut für Festkörperphysik
und Werkstofforschung
Helmholtzstr. 20
DDR-8027 DRESDEN

H. Nörenberg
Department of Physics
WPU Rostock
Universitätsplatz 3
2500 ROSTOCK

J. Richter   see Moch
J. Schneider   see Müller

H. Siegel
Magnetbandfabrik
Kochstedterstr. 1
4500 DESSAU

M. Wurlitzer   see Höhne

——————— HUNGARY

E. Kisdi-Koszó
Hungarian Academy of Sciences
Central Res. Institute for Physics
P.O. Box 49
H-1525 BUDAPEST 114

G. Konczos   see Kisdi-Koszó
M. Pardavi-Horváth   see Kisdi-Koszó
T. Tarnóczi   see Kisdi-Koszó
Z. Vértesy   see Kisdi-Koszó

------ ISRAEL

I. Felner
The Hebrew University of Jerusalem
Racah Institute of Physics
Givat Ram
91904 JERUSALEM

------ ITALY

Ying-Shan Yang
TIB-CHIA
Processi Preparativi Materiali Amorfi
C.R.E. CASACCIA - S.P. Anguillarese, 301
Casella Postale N. 2400- 00100 ROMA A.D.

------ JAPAN

K. Fukamichi
The Research Institute for Iron, Steel
and other Metals
Tohoku University
SENDAI 980

T. Morishita
NHK Science and Technical
Research Laboratories
1-10-11 Kinuta, Setagaya-ku
TOKYO 157

M. Takahashi
Tohoku University
Department of Electronic Engn.
Faculty of Engineering
Aoba Aramaki
SENDAI 980

Y. Yamashiro
Ryukyu University
Dept. Electrical Engn.
Fac. Engn.
1 Senbaru Nishihara
OKINAWA 903-01

POLAND

L.T. Baczewski
Institute of Physics
Polish Academy of Sciences
Al. Lotników 32/46
02-668 WARSAW

A. Bajorek
Institute of Nuclear Physics
ul. Radzikowskiego 152
31-342 KRAKÓW 23

T. Balcerzak
Łódź University
Institute of Physics
al. Nowotki 149/153
90-236 ŁÓDŹ

M. Bałanda   see Bajorek

J.J. Bara
Jagiellonian University
Institute of Physics
al. Reymonta 4
30-059 KRAKÓW 16

M. Baran   see Baczewski

J. Baszyński
Institute of Molecular Physics
Polish Academy of Sciences
ul. Smoluchowskiego 17/19
60-179 POZNAŃ

W. Bażela-Wróbel
Politechnika Krakowska
Instytut Fizyki
ul. Podchorążych 1
30-084 KRAKÓW

K. Blinowski
Institute of Atomic Energy
Dept. Solid State Physics
05-400 OTWOCK - Świerk

B.F. Bogacz   see Bara

K. Brańska
Warsaw Polytechnics
Institute of Physics
ul. Koszykowa 75
00-662 WARSZAWA

H. Broda
Silesian University
Institute of Physics
ul. Uniwersytecka 4
40-007 KATOWICE

N.H. Chau
Institute of Mining and Metallurgy
Dept. Solid State Physics
Al. Mickiewicza 30
30-059 KRAKOW

A. Chełkowski   see Broda

K. Chróst
Warsaw Polytechnics
Institute of Material Engineering
ul. Narbutta 85
02-524 WARSZAWA

W. Cieśliński
Plant of Magnetic Materials "POLFER"
ul. Dzielna 60
01-029 WARSZAWA

W. Ciurzyńska
Częstochowa Polytechnics
Department of Physics
ul. Zawadzkiego 19
42-200 CZĘSTOCHOWA

P.M. Czarnecki
Adam Mickiewicz University
Institute of Physics
ul. Grunwaldzka 6
60-780 POZNAN

Z. Drzazga   see Broda
J. Dubowik   see Baszyński
M. Duraj   see Bażela-Wróbel
M. Duś-Sitek   see Ciurzyńska
T. Dymkowski   see Chróst

B. Fechner
Adam Mickiewicz University
Institute of Physics
ul. Matejki 48/49
60-769 POZNAN

A.R. Ferchmin see Baszyński

J.E. Frąckowiak
Silesian University
Institute of Physics and Chemistry of Metals
ul. Bankowa 12
40-007 KATOWICE

M. Gajdek
Politechnika Świętokrzyska
Department of Physics
Al. 1000-lecia Państwa Polskiego 7
25-314 KIELCE

W. Gawior
Silesian Polytechnics
Institute of Non-ferrous Metals
ul. Sowińskiego 5
44-100 GLIWICE

R. Gieniusz
Branch of Warsaw University
Department of Physics
ul. Lipowa 41
15-424 BIAŁYSTOK

R. Gontarz see Baszyński
W. Gorzkowski see Baczewski
M.W. Gutowski see Baczewski
J. Hankiewicz see Czarnecki
P. Hutny see Cieśliński
B. Idzikowski see Baszyński

A. Ingram
Engineering School
Department of Physics
ul. Ozimska 75
45-370 OPOLE

Z. Jacyna-Onyszkiewicz see Czarnecki
E. Jarocki see Chau

K. Jezuita
Engineering School
Department of Physics
ul. Malczewskiego 29
26-600 RADOM

E. Jędryka   see Baczewski
M. Jurczyk   see Baszyński
S.R. Juszczyk   see Broda
Z.I. Kaczkowski   see Baczewski
C. Kapusta   see Chau

J. Kasperczyk
Pedagogical University
Institute of Physics
al. Zawadzkiego 13/15
42-200 CZĘSTOCHOWA

A. Kasprzyk   see Broda
J. Kłodaś   see Chróst
A. Kołodziejczyk   see Chau
M. Konwicki   see Cieśliński
R.A. Kosiński   see Brańska
J. Kosiorowska   see Bajorek
L. Kowalewski   see Fechner
Z. Kozioł   see Baczewski
K. Krop   see Chau
T. Kulik   see Chróst
J. Kulikowski   see Cieśliński
K. Kułakowski   see Chau
H.K. Lachowicz   see Baczewski
J. Lech   see Ciurzyńska
M. Leonowicz   see Chróst
A. Leśniewski   see Cieśliński
T. Luciński   see Baszyński
K. Łatka   see Bara
M. Łukasiak   see Blinowski
K. Łukawski   see Cieśliński
E. Machowska   see Baczewski
W. Maciejewski   see Fechner
W. Maj   see Baczewski
A. Maksymowicz   see Chau
L.J. Maksymowicz   see Chau
L. Małkiński   see Baczewski
A. Maryanowska   see Czarnecki
H. Matyja   see Chróst
A. Maziewski   see Gieniusz
J. Mielnicki   see Baczewski
A. Modrzejewski   see Blinowski
E. Mosiniewicz-Szablewska   see Baczewski
R. Moskalewicz   see Cieśliński

H. Mrowiec see Cieśliński
J.M. Mucha see Bara

T. Mydlarz
International Laboratory
of Strong Magnetic Fields & Low Temperatures
ul. Próchnika 95
53-529 WROCŁAW

S.C. Nadolski see Baczewski

J. Nowak
Polish Academy of Sciences
Department of Solid State Physics
ul. Kawalca 3
41-800 ZABRZE

E. Okoniewska-Pszczółkowska see Cieśliński
A. Oleś see Chau
Z. Olszowski see Ciurzyńska
I. Onyszkiewicz see Czarnecki

R. Pac
Tele- & Radiotechnical Institute
ul. Ratuszowa 11
03-950 WARSZAWA

A. Pajączkowska see Baczewski

K. Perzyńska
UNITRA-BIAZET
Sz. Płn. Obwodowa 38
15-113 BIAŁYSTOK

M. Pękała
Warsaw University
Department of Chemistry
al. Żwirki i Wigury 101
02-089 WARSZAWA

R. Pfranger see Ciurzyńska
M. Piasecki see Kasperczyk
J. Pietrzak see Czarnecki
E. Pilipczuk see Chróst
K. Piotrowski see Baczewski
D. Płusa see Ciurzyńska

S. Pokrzywnicki
Engineering School
Institute of Math., Physics & Chemistry
ul. ZSP 5
45-233 OPOLE

T. Postupolski   see Cieśliński
P. Przysłupski   see Baczewski
J. Pszczoła   see Chau
J. Radomski   see Czarnecki
R.J. Radwański   see Chau
H. Ratajczak   see Baszyński
J. Raułuszkiewicz   see Baczewski
A. Reich   see Baczewski
W. Schmidt   see Baszyński
A. Siemko   see Baczewski

Z. Siermiński
Magnets & Magnetic Core Plant "POLFER"
ul. Zwierzyniecka 1
SKIERNIEWICE

A. Stankiewicz   see Gieniusz
P. Stefański   see Baszyński
F. Stobiecki   see Baszyński
T. Stobiecki   see Chau
A. Sukiennicki   see Brańska

W. Suski
Institute of Low Temperatures
& Structural Research
Polish Academy of Sciences
P.O. Box 937
pl. Katedralny 1
50-950 WROCŁAW 2

J. Suwalski   see Blinowski
A. Szewczyk   see Baczewski
B.S. Szymański   see Baszyński
K. Szymański   see Gieniusz

M. Szymański
Warsaw University
Institute of Experimental Physics
ul. Hoża 69
WARSZAWA

R. Szymczak   see Baczewski
A. Szytuła   see Bara
M. Ślepowroński   see Blinowski
U. Świderska   see Cieśliński

J. Świerczek   see Ciurzyńska
M. Tekielak   see Gieniusz
Than Duc Hien   see Chau
K. Turek   see Chau
T. Tymosz   see Brańska
L. Uba   see Gieniusz
S. Uba   see Gieniusz
A. Urbaniak-Kucharczyk   see Balcerzak
W. Wasilewski   see Jezuita
G. Wiatrowski   see Balcerzak
A. Winiarska   see Broda
A. Wiśniewska   see Cieśliński
A. Wiśniewski   see Baczewski
K.J. Witański   see Frąckowiak
A. Witek   see Baczewski
J. Wolny   see Chau
J. Wójcik   see Ciurzyńska
M. Wójcik   see Baczewski
A. Wrzeciono   see Baszyński
J.J. Wysłocki   see Ciurzyńska
A. Załuska   see Chróst
L. Załuski   see Baczewski
M.B. Zapart   see Ciurzyńska
W. Zapart   see Ciurzyńska
W. Zarek   see Broda
J. Zawadzki   see Baczewski
J. Zbroszczyk   see Ciurzyńska

J. Żebrowski
Institute of Biocybernetics
& Biomedical Engineering
Polish Academy of Sciences
ul. KRN 55
00-818 WARSZAWA

J.J. Żebrowski   see Brańska
R. Żuberek   see Baczewski

------- SWEDEN

K.V. Rao
The Royal Institute of Technology
Dept. of Solid State Physics
Teknikringen 14
100 44 STOCKHOLM 70

K. Tsushima   see Rao

------- THE NETHERLANDS

J.J.M. Franse
Natuurkundig Laboratorium
der Universiteit van Amsterdam
Postbus 20215 1000 HE Amsterdam
Valckenierstraat 65
1018 XE AMSTERDAM

------- UNITED KINGDOM

R. Carey
Coventry Polytechnic
Faculty of Applied Science
Dept. of Appl. Physical Sciences
Priory Street
COVENTRY CV1 5FB

W.W. Clegg
Electrical Engineering Laboratories
The University
MANCHESTER M13 9PL

P.J. Grundy
Dept. of Pure & Appl. Phys.
University of Salford
Applied Magnetic Materials Group
SALFORD M5 4WT

------- USA

A.I. Braginski
Westinghouse Electric Corporation
Research and Development Center
1310 Beulah Road
PITTSBURGH, PA 15235

E.M. Gyorgy
AT&T Bell Laboratories
Room 1B-303
600 Mountain Avenue
MURRAY HILL, NJ 07974-2070

S. von Molnár
IBM Research Division
T.J. Watson Research Center
P.O. Box 218
YORKTOWN HEIGHTS, NY 10598

M. Pomerantz   see von Molnàr

H.T. Savage
Naval Surface Weapons Center
code R45
SILVER SPRING, MD 20903-5000

J.C. Slonczewski   see von Molnàr

P.E. Wigen
The Ohio State University
Department of Physics
174W 18th Avenue
COLUMBUS, OH 43210-1106

------ USSR

A.V. Andrienko
The I.V. Kurchatov Institute
of Atomic Energy
Kurchatov sq.
123182 MOSCOW, D-182

I.V. Baryakhtar
Kharkov Institute of Physics
and Technology
ul. Akademicheskaya 1
KHARKOV 310108

K.P. Belov
Department of Physics
Moscow State University (MGU)
Leninskiye Gory
117 234 MOSCOW

A.B. Chizhik
Institute for Low Temperature
Physics & Engineering
Ukr. SSR Academy of Scences
47 Lenin Ave.
310164 KHARKOV

L.M. Dedukh
Institute of Solid State Physics
of the Academy of Sciences USSR
CHERNOGOLOVKA, MOSCOW distr. 142432

E.B. Dokukin
Joint Institute for Nuclear Research
Laboratory of Neutron Physics
P.O. Box 79 jinr
101000 MOSCOW

F.S. Dzheparov
Institute of Theoretical
and Experimental Physics
117259 MOSCOW

V.V. Fedotova
Institute of Physics
of Solids and Semiconductors
The BSSR Acad. Sci.
P. Brovki 17
220 726 MINSK

A.P. Ges   see Fedotova
S.L. Gnatchenko   see Chizhik

A.G. Gurevich
A.F. Ioffe Physico-Technical Institute
Academy of Sciences of the USSR
ul. Politekhnicheskaya 26
194021 LENINGRAD, K-21

E.I. Il'yashenko
Institute of Control Sciences
USSR Academy of Sciences
Profsoyuznaya 65
117806 MOSCOW, GSP-312

O.M. Ivanenko
USSR Academy of Sciences
Lebedev Physical Institute
Leninskiy prosp. 53
117 333 MOSCOW

M.D. Kaplan
Institute of Chemistry
of Academy of Sciences
Moldavian SSR
ul. Grosul 3
277028 KISHINEV

E.P. Khlybov
The Institute for High Pressure Physics
USSR Academy of Sciences
142092 TROITSK, Moscow region

A.N. Kocharian
Yerevan Physics Institute
Markarian str. 2
Yerevan 375036, Arm. SSR

A.P. Kovaleva
General Physics Institute
Academy of Sciences of the USSR
Vavilova st. 38
117 942 MOSCOW

E.T. Kulatov see Kovaleva

V.S. Lutovinov
Moscow Institute of Radio Engineering,
Electronics and Automation (MIREA)
pr. Vernadskogo, 78
117454 MOSCOW

A.A. Minakov see Kovaleva

A.V. Mitin
Academy of Sciences of the USSR
Institute for Physical Problems
ul. Kosygina, 2
117973 GSP-1, MOSCOW, V-334

K.V. Mitsen see Ivanenko

A.N. Pogorelyi
Institute of Metal Physics
Academy of Sciences Ukrainian SSR
Dept. Solid State Physics
pr. Vernadskogo 36
252680 KIEV

A.S. Prokhorov see Kovaleva
I.L. Sashin see Dokukin
A. Serebryakov see Dedukh

Y. Shaldin
Institute of Crystallography
USSR Academy of Sciences
Leninskii prosp. 59
MOSCOW 117333

V.I. Sokolov
Department of Physics
Moscow State University (MGU)
Leninskiye Gory
119899 MOSCOW

G.A. Takzei   see Pogorelyi

E.A. Turov
Institute of Metal Physics
Academy of Sciences of the USSR
Ural Branch
ul. S. Kovalevskoy 18
620219 SVERDLOVSK GSP-170

L.I. Vinokurova   see Kovaleva

------ **VIETNAM**

Mai Suan Li
Thai Nguyen Technical Institute
77, Mai Hăc Dê
HANOI